How well can we reconstruct the appearance, movements, and behavior of extinct vertebrates from studies of their bones and other, more rarely preserved parts? Where is the boundary between the scientific evidence for reconstruction and the need to resort to imagination? In this book, sixteen paleontologists and biologists discuss these questions, review the current status of functional studies of extinct vertebrates in the context of similar work on living animals, and present a broad philosophical view of the subject's development within the framework of phylogenetic analysis. The authors describe and debate methods for making robust inferences of function in fossil vertebrates and present examples where we may be confident that our reconstructions are both detailed and accurate. The area of greatest success is in reconstructing masticatory mechanics in mammals and their cynodont ancestors; several chapters address aspects of this work. Further chapters consider the cranial and postcranial skeletons of a range of vertebrates.

The detailed studies are placed in the context of their contribution to the understanding of evolutionary processes and will be valuable reading for vertebrate paleontologists, comparative anatomists, and evolutionary biologists.

Functional morphology
in vertebrate paleontology

Functional morphology in vertebrate paleontology

Edited by

Jeff Thomason
University of Guelph

Published by the Press Syndicate of the University of Cambridge
The Pitt Building, Trumpington Street, Cambridge CB2 1RP
40 West 20th Street, New York, NY 10011–4211, USA
10 Stamford Road, Oakleigh, Melbourne 3166, Australia

First published 1995

Printed in the United States of America

Library of Congress Cataloging-in-Publication Data
Functional morphology in vertebrate paleontology / edited by J. J.
Thomason.

  p. cm.
ISBN 0–521–44095–5
1. Vertebrates, Fossil. I. Thomason, J. J. (Jeffrey J.)
QE841.F86 1994   93–47981
566–dc20   CIP

A catalog record for this book is available from the British Library.

ISBN 0–521–44095–5 hardback

Contents

Contributors

Harold N. Bryant
Vertebrate Morphology Research Group
Department of Biological Sciences
The University of Calgary
Calgary, Alberta T2N 1N4
Canada
Current address:
Provincial Museum of Alberta
Mammalogy Program
12845–102 Avenue
Edmonton, Alberta T5N 0M6

Arthur B. Busbey
Department of Geology, Box 30798
Texas Christian University
Fort Worth, TX 76129

Alfred W. Crompton
Department of Organismic and Evolutionary
   Biology
Harvard University
Cambridge, MA 02138

Stephen M. Gatesy
Museum of Comparative Zoology
Harvard University
Cambridge, MA 02138

Emily B. Giffin
Department of Biological Sciences
Wellesley College
Wellesley, MA 02181

Walter S. Greaves
Department of Oral Biology, Anatomy, and Cell
   Biology and Department of Biology
University of Illinois at Chicago
801 S. Paulina St.
Chicago, IL 60612

Christine M. Janis
Program in Ecology and Evolutionary Biology
Division of Biology and Medicine

Brown University
Providence, RI 02912

Rolf E. Johnson
Milwaukee Public Museum
800 W. Wells St.
Milwaukee, WI 53233

George V. Lauder
School of Biological Sciences
University of California
Irvine, CA 92717

Virginia L. Naples
Department of Biological Science
Northern Illinois University
DeKalb, IL 60115

John H. Ostrom
Peabody Museum
Yale University
170 Whitney Avenue
New Haven, CT 06511

Kevin Padian
Department of Integrative Biology
University of California
Berkeley, CA 94720

John M. Rensberger
Department of Geological Sciences and Burke
   Museum
University of Washington, DB-10
Seattle, WA 98195

Anthony P. Russell
Vertebrate Morphology Research Group
Department of Biological Sciences
The University of Calgary
Calgary, Alberta T2N 1N4
Canada

Jeffrey J. Thomason
Department of Biological Sciences and College of
  Osteopathic Medicine
Ohio University
Athens, OH 45701
Current address:
Department of Biomedical Sciences
Ontario Veterinary College
University of Guelph
Guelph, Ontario N1G 2W1
Canada

Keith S. Thomson
Academy of Natural Sciences
1900 Benjamin Franklin Parkway
Philadelphia, PA 19103–1195

David B. Weishampel
Department of Cell Biology and Anatomy
The Johns Hopkins University School of Medicine
Baltimore, MD 21205

Lawrence M. Witmer
Department of Cell Biology and Anatomy
The Johns Hopkins University School of Medicine
Baltimore, MD 21205
Current address:
Department of Anatomy
New York College of Osteopathic Medicine
New York Institute of Technology
Old Westbury, NY 11568

Preface

The primary intent of this book is to provide a snapshot of the current status of functional studies of fossil vertebrates in North America. Half of the contributions are based on presentations at a symposium of the same title as this book that was held in conjunction with the Annual Meeting of the Society of Vertebrate Paleontology in Toronto, October 1992. The other contributions were solicited. To give coherence to both the symposium and this volume I sent some questions for each author to consider while writing:

1. On what philosophical premises are functional interpretations of fossil vertebrates based, and what contributions do they make to our understanding of evolutionary processes? For example, what do functional studies draw from phylogenetic reconstruction, and what do they add back to phylogenetic methods?
2. What are the conceptual and methodological links between functional studies of fossil vertebrates and corresponding neontological work? Is it a two-way interchange, or unidirectional from extant to extinct?
3. How does functional morphology relate to the other subdisciplines of paleontology? The interaction with phylogenetic methods has been mentioned; what interactions may exist with paleoecology, stratigraphy, taphonomy, etc.?
4. What methodologies are appropriate for assessing function in an extinct beast? Describe any relevant assumptions and caveats in addition to areas where functional inference may be robust.

As a result, the book does more than illustrate the kinds of functional work currently under way in vertebrate paleontology. It is a forum of discussion on how to study functional morphology of fossils, the levels of resolution we might expect in reconstructing function from structure, and the position of functional studies in the broader context of paleontology and evolutionary biology. The strongest theme to emerge is that of methodology, with several authors proposing or discussing general protocols for robustly inferring function from structure in fossils.

Traditionally, the functional reconstruction of fossils has been open to the criticism of excessive subjectivity. Even though individual workers have usually had good backgrounds in the osteology of extinct vertebrates and the comparative anatomy of living ones, reconstructions often still came down to individual interpretation and opinion. The same was true of systematic methods, but the past 20 years have seen a revolution in systematics, resulting in the quantifiable and repeatable methods of phylogenetic analysis. Procedures have been developed for linking the inference of function in fossils to phylogenetic analyses, providing paleontologists with some confidence in the robustness of their reconstructions. Witmer and Weishampel describe one such procedure; Bryant and Russell and Gatesy present variants. (Lauder and Padian dissent, for entirely different reasons.)

In marked contrast to the emerging confidence among "functional paleontologists" is the situation among neontologists (here represented by George Lauder, though Gatesy, Janis, and Crompton work from a strong base of experiments on living animals when inferring function in fossils). Current technology has allowed neontologists to study some of the functions of living organisms in precise detail, so function and form (morphology) can be assessed independently and compared (rather than the circular procedure of some studies in which function was inferred from morphology, then compared with it). The more animals that are investigated, the looser the coupling between form and function appears to be. Lauder documents examples from his own work that demonstrate that details of function are controlled to a greater extent by motor innervation than by morphology. The nervous system is never pre-served, and is only marginally represented in the preserved parts of vertebrates (see Giffin).

Lauder concludes that paleontologists are largely confined to determining function in only the most general terms for fossils. This pessimistic statement opens the book. The subsequent chapters are by paleontologists presenting, discussing, and defend-ing their methods for assessing function in specific cases. There is, therefore, a tension in the book that I have made no attempt to resolve.

Witmer and Weishampel follow Lauder, present-ing a method for inferring function that is linked to phylogenetic assessment: a fossil is bracketed between living animals on a cladogram and their function is used to infer that in the extinct form (cf. Bryant & Russell 1992 and herein). One of the areas remaining problematical to this procedure is when a fossil appears to have functioned differently from the bracketing living forms. An example cited by Weishampel is the evolution of the mammalian jaw joint, which is described in detail in the chapter by Crompton. His chapter provides a transition to a series of chapters on functional studies of the jaws, skull, and dentition. Functional analysis of mamma-lian dentitions, in particular, is one area in which the interpretation of extinct material has been ac-complished successfully and at quite a detailed level of resolution.

Janis reviews the past few decades of work on inferring dental function from increasingly finer de-tails of structure: jaw mechanics, occlusal morphol-ogy of the teeth, general wear patterns, microwear patterns, and crystalline orientation in enamel. She also shows how studies of fossils can enhance the interpretation of population structure and ecology of living forms. Specific examples of the methods she reviews follow in the chapters by Greaves on jaw mechanics, Bryant and Russell on carnassial wear in sabertoothed cats, Naples on producing dental wear artificially, and Rensberger on comparing stress patterns and crystal alignment in enamel. Greaves's piece is a prime example of an alternate methodol-ogy to those based on phylogeny: the paradigm method described by Rudwick (1964). Greaves in-fers the constraints on "natural design" of (prima-rily) mammalian jaws from geometrical models illustrating the principles of their mechanical behavior. Busbey also uses mechanical modeling to interpret the evolutionary changes in the crocodilian skull, whereas Thomson devises an entirely novel method for analyzing sutural orientation in the der-mal skull of lower tetrapods, interpreting his results in terms of cranial function.

We then move back from the skull to a lesser number of papers on the postcranial skeleton. I be-lieve the relative weighting of cranial to postcranial

chapters reflects the preponderance of cranial re-search, as does the general emphasis on mammals. Among nonmammalian taxa, dinosaurs are certainly underrepresented here compared with the amount of scientific and public interest in them, but they are well covered elsewhere. Johnson and Ostrom and Gatesy provide the only two pieces focused on dino-saurs and their avian descendants. Johnson and Ostrom remind us that the structural data set for fossil vertebrates is largely limited to the skeleton, and we had better make the most of the functional inferences that can be made directly from osteology. Gatesy returns us to the themes of the earlier chap-ters, describing his own methods for combining ana-tomical, phylogenetic, and experimental methods for inferring the change in function of the tail from theropods to birds. Giffin then tackles the problem of the nervous system. Although nerves are not preserved, the cranial cavity and neural canal do contain evidence of the gross anatomy of the central nervous system. Giffin shows how estimates of the change in cross-sectional area of the spinal cord along its length can be used to generate or support infer-ences of general locomotor patterns in fossil mam-mals and reptiles. My own chapter argues that, given that bones are our primary data base, we could be missing a whole category of functionally relevant structural information: that encoded in the internal structure of bones.

Kevin Padian rounds out the book with a histori-cal essay that traces the influence of contemporane-ous philosophy on functional interpretations of form, from Aristotle to the present day. He ends by presenting a prospectus for the future of functional studies in paleontology that is more optimistic than the opening statements of George Lauder and in-cludes a role for functional work in systematics.

Several authors build their discussions around the phrase "form and function." For functional studies to be viable, the two have to be related in ways that are amenable to interpretation. Older studies were largely based on the assumption that the relation-ship was tight, regular, and easy to reconstruct for both fossils and living animals. The situation has changed. There is no longer any excuse for pale-ontologists to reconstruct fossils based on informed opinion. Rigorous tests of the robusticity of func-tional inferences are now available. At the same time, "functional paleontologists" need to keep a wary eye over their shoulder to see what the neontologists are reporting. Their work defines the extent to which function can be inferred from structure, and that question seems to be undergoing considerable flux at the moment. If this book achieves its intended purpose of assessing the status quo of functional morphology in vertebrate paleontology in North America, I hope it will also provide a guide and stimulus to workers in the field. The field is changing and will continue to change; if this volume stimulates ideas and work that contribute to the change it will have achieved a measure of success.

The idea for this work came from a similar venture: the Symposium *Biomechanics in Evolution* organized by Jeremy Rayner for a joint meeting of the Society of Experimental Biology and the Palaeontological Association in Manchester, England, 1987 (now published under the same title with R.J. Wootton as coeditor).

Thanks to all the symposium participants, other contributors to this volume, and the editing and production staff for their enthusiastic participation and cooperation during all stages of the organization, presentation, and preparation for publication. It has been a pleasure working with you.

Thanks to R.M. Sullivan (Program Director) and C.S. Churcher (President) of the Society of Vertebrate Paleontology for their cooperation and assistance in scheduling the symposium in Toronto, and for partial support of the costs. The two institutions I have worked at during this project provided secretarial assistance, stationery, and mailing costs.

In addition to those authors who doubled as reviewers of other chapters, I would like to thank the following referees for their time and expertise: P. Ulinski, W.P. Coombs, Jr., R. MacDougall, C. McGowan, M. Ruse, A. Biknevicius, B. Van Valkenburgh, A.P. Russell, M. Fortelius, N. Solounias, and J. Damuth.

And finally, a personal vote of thanks to K.E. Joysey and A.E. Friday, who sparked my interest in this subject as an undergraduate, and to Chris McGowan, who fanned the spark to flame.

Jeff Thomason
Guelph, Ontario

REFERENCES

Bryant, H.N., & Russell, A.P. 1992. The role of phylogenetic analysis in the inference of unpreserved attributes of extinct taxa. *Philosophical Transactions of the Royal Society of London* B337, 405–418.

Rudwick, M.J.S. 1964. The inference of function from structure in fossils. *British Journal of the Philosophy of Science* 15, 27–40.

Functional morphology
in vertebrate paleontology

1

On the inference of function from structure

GEORGE V. LAUDER

ABSTRACT

Inferring the function of structures in extinct organisms from analyses of morphology has been a goal of many paleontological studies, and several methods have been proposed to guide the inference of function from structure. Success has been greatest where general predictions of organismal behavior or ecology are desired. However, predicting precise aspects of organismal function from morphology depends on the assumption of a close match between structure and function. The central theme of this chapter is that this assumption is unwarranted in many cases. I present three case studies from research on the feeding mechanisms of ray-finned fishes and salamanders that illustrate the difficulties of inferring function from structure. These examples all involve the experimental determination of function in living taxa, and the results in each case may be compared directly to functional predictions based on morphology. The case studies range from broadly comparative phylogenetic analyses in salamanders to intraspecific comparisons of structure and function in sunfishes.

All three case studies illustrate that even when relatively simple predictions are made about function and behavior from morphology, these predictions are often entirely erroneous. Function and structure are often not tightly linked, and we may have been overly optimistic in the past regarding the possibility of inferring function from the morphology of extinct taxa, even where phylogenetically appropriate extant "model taxa" exist. A key reason for the lack of correlation between structure and function in the musculoskeletal system, even in living species, is the lack of information on motor programs in the central nervous system. Identical morphology in two species may be involved in very different functions and behaviors if changes have occurred in the nervous system. Our ability to infer changes in the motor programs of fossil taxa is limited, and for this reason alone we are unlikely to be very successful in making detailed predictions of function and performance from morphological data alone.

INTRODUCTION

The alliterative appeal of the phrase "form and function" is such that virtually every field has used it to describe its subject matter. Thus, we may read about architecture in *Form and Function: A Source Book for the History of Architecture and Design 1890–1939* (Benton 1975); about linguistics in the book *Language Development: Form and Function in Emerging Grammar* (Bloom 1970); about literature in *The Phenomenon of the Grotesque in Modern Southern Fiction: Some Aspects of its Form and Function* (Haar 1983), and about digestion in *Guts: The Form and Function of the Digestive System* (Morton 1979).

But the widespread use of the phrase "form and function" has not been accompanied by much understanding of the relationship between form and function. Are form and function inseparably linked? Does form determine function? Does function determine form? This debate has raged for centuries, and is one of the most enduring in biology (Russell 1916). And yet there is little agreement on just what the relationship is between form and function.

The association between form and function is of special importance to paleobiologists and comparative anatomists for whom the primary available data on organisms are structural. If one has access only to structural data, then any attempt to understand how animals work and how they move must involve the inference of function from form. We must then seek methods to infer function from form and a better understanding of how form and function are interrelated.

In much of the literature, form and function are used as closely related concepts that, when considered together, promise to provide a reasonably complete characterization of any heterogeneous system. Many authors believe that these two properties have an intimate relationship and that it is not fruitful to consider form and function as distinct entities (e.g., Arber 1950; Dullemeijer 1974; Thomas 1978).

This view of form and function has a clear implication for the direction of research in functional and evolutionary morphology. If form and function are tied closely together and map onto each other in a relatively simple way, then it should be possible to make accurate predictions of biological function from a study of form or structure. Similarly, a study of organismal function (or the generation of theoretical models of function) should enable us to predict the design of biological structures associated with those functions. The tightness of the linkage between biological form and function will determine the ease and accuracy with which these types of predictions can be made.

The central theme of this paper is that we have placed unwarranted faith in our understanding of the relationship between form and function, and are thus more confident of our ability to predict function from form than is legitimate, given current data. This is especially true at the organismal level where prediction of organismal function, performance, and behavior from structure is desired. I will argue that biological structures and their associated functions often have a very complex relationship to each other, and, at certain levels of analysis, very little relationship at all. Hence, prediction of one from the other may be extremely difficult. We tend to *assume* that form and function have a mutually predictable relationship even though few specific tests of the fit of structures to *experimentally measured* functions have been conducted. Of particular concern in this chapter are the nature and accuracy of predictions of function from morphology in extinct taxa. In fossil animals the skeletal system makes up the majority of the available preserved material, and thus predictions of function are of necessity derived from skeletal remains. In many cases it is necessary to infer from these skeletal remains the arrangement of muscles and ligaments in order to generate predictions of muscle function, skeletal movement, and thus behavior. The concerns I express here apply equally well to the inference of function from anatomical analyses of the musculoskeletal system in living taxa.

The relationship between form and function in the musculoskeletal system will be the focus of this paper because of the dominant role that bones and muscles play in paleontological and comparative anatomical investigations. First, I will review two general approaches that have been used to study the relationship between form and function (especially in fossil taxa) and I will evaluate the assumptions on which these methods are based. I will also discuss definitions of the terms form and function, as well as the related concepts of performance and behavior. Second, I will briefly present the results of three case studies in different clades to illustrate the difficulties of inferring function from structure even in extant taxa for which structure can be studied comprehensively. These case studies demonstrate the value of experimentally measuring function to test directly predictions from structure. Finally, I will discuss the implications of these conclusions for the

Table 1.1. *Definitions of terms used in Chapter 1*

Term	Definition
Biological role	Role of phenotypic features in a specific environmental or ecological setting (e.g., escape from predators)
Behavior	Actions and/or responses of whole organisms
Performance	A measure of the execution of an ecologically relevant activity (e.g., feeding rate, maximal burst speed)
Function	Use or action of phenotypic features; the mechanical role of phenotypic features
Structure	Topological relationships among phenotypic features; internal organization of phenotypic features

general enterprise of form–function studies, and the extent to which the results from these case studies are likely to be of wide applicability.

STRUCTURE AND FUNCTION

Before we can evaluate the success with which we might predict function from structure, it is critical to arrive at a definition of function. The definitions of the terms form and function (and the related words, performance, behavior, and biological role) as used in this chapter are given in Table 1.1. Form is defined in terms of phenotypic features and their component anatomical parts, and structural analysis describes the topological relationships among features. Structures might be bones, muscles, or amino acids in a protein.

Function is more troublesome to define. As discussed previously (Lauder 1986, 1990), many investigators (especially in ethology) use the term function as synonymous with "selective advantage." The function of a structure under this view is the effect that the structure has on fitness: How did selection act to produce the structure? If selection on the hindlimb, for example, acted to increase the length of the limb due to the fact that longer-limbed individuals were better able to escape from predators, then the function of the hindlimb would be "predator escape." There are two main difficulties with this definition. First, we must know both that selection acted specifically on the structure in the first place, *and* the direction of that selection before we can speak of the function of a structure. This is exceedingly unlikely information to possess, especially in any retrospective comparative study of extinct species

(Lauder et al. in preparation). Second, this definition of function leaves us without a word to describe what morphologists often mean by function: the use or action of a structure (Bock & von Wahlert 1965; Hickman 1991).

In this paper, the function of a structure will be defined as the *mechanical role* or *physical role* that it performs in the organism: that is, how a phenotypic feature is used (Lauder 1990). A bone might have the function or mechanical role of stiffening the limb, or of providing a rigid element on which muscular forces can act. In describing mechanical roles, we might measure kinematic and kinetic variables – force, displacement, velocity, and acceleration – to quantify how a structure is used. Or we might analyze when a structure is used in a particular behavior in comparison to other structures that contribute to the behavior. Experimentally modifying structures is also useful for understanding mechanical roles. Under this definition of function, which is derived from that of Bock and von Wahlert (1965), our ability to measure the occurrence and direction of selection on structures is irrelevant to determining their function. Functional data, then, are measurements of mechanical role and use (Lauder 1990). This definition of function thus does not include structural data that may be given a functional *interpretation* or may assist in defining possible limits to mechanical role: that is, lever arms, muscle masses, allometric relationships, bone shapes, or interconnections of central nervous system nuclei.

While the analysis of structure and function is often at the level of individual features of organisms, the study of animal performance, behavior, or biological role may occur at more integrative levels (Table 1.1). For example, performance (Arnold 1983) is the measured score of an individual on an ecologically relevant task (such as maximal feeding rate or maximum sprint speed). Individuals (or species) that differ in mean performance on a specific task might do so because of changes at many other levels. For example, feeding rates on the same prey in two species might differ because of changes in muscle physiology, muscle lever arms, or nervous system structure in feeding and/or locomotor morphology; many structural and functional features may be involved in causing a difference in performance.

A similar situation holds for behavioral differences among individuals or taxa. If we define behavior as the actions and/or responses of whole organisms (Table 1.1), then it is apparent that behavioral differences among species may also be the result of differences at both structural and functional levels. Different patterns of muscle activity, for example, will generate different locomotor behaviors in two species, even given similar limb morphology (Lauder 1991).

The term *biological role* (Table 1.1) was advanced by Bock & von Wahlert (1965) to indicate the role that a structure plays during behavior in a specific environmental and/or ecological setting. The biological role of a particular hindlimb morphology in a species, for example, might be given as "enhancing escape from predators." Biological role can also be viewed as synonymous with selective advantage. Given the difficulty of measuring selection in comparative studies (especially in extinct taxa), the study of biological roles is likely to be quite difficult in practice.

THE INFERENCE OF FUNCTION FROM STRUCTURE IN PALEONTOLOGY: METHODS AND THEIR DIFFICULTIES

Two main methods have been used to infer function, behavior, or performance from structure in extinct taxa: the phylogenetic method and the paradigm method.

The phylogenetic method

This method relies on comparative studies of related living taxa to estimate the function of a structure. The observed relationship between structure and function in extant species is then applied to extinct taxa of interest. Structures that are homologous among living and extinct taxa are presumed to share homologous functions, allowing the inference of function in a fossil taxon (Stanley 1970). As summarized by Raup and Stanley (1971; p. 166) in their text on principles of paleontology, "Often we can observe the mechanical function of a structure in Recent taxa and, by homology, apply our observations to closely related fossil taxa."

The phylogenetic method as it is often practiced is illustrated in Figure 1.1a. The extinct taxon E can be shown by a study of its morphology to share structure S3 with taxa A and B. Structure S3 is thus a shared derived feature of taxa E, A, and B and provides evidence that these three taxa form a monophyletic group. S3 is thus considered to be homologous in these three taxa (Lauder in press; Patterson 1982). The function of structure S3 may then be studied experimentally in taxa A and B. If similar functions (F3) are found in these two taxa, then one may by inference ascribe function F3 to taxon E. Figure 1.1a also illustrates that taxon E shares as primitive traits (plesiomorphies) structures S2 and S1 with taxa A and B. We could thus also investigate the functional correlates of S2 and S1 in A and B as well as in taxon C to assist in the inference of the functional attributes of taxon E.

The phylogenetic method as illustrated in Figure

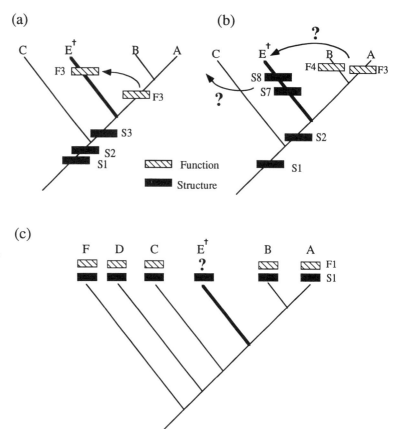

Figure 1.1. (a) One application of a phylogenetic approach to inferring function from form. Taxon E is extinct and together with taxa A and B possesses structure S3. Inference of the function of this structure in extant taxa A and B provides the basis for inferring the function of structure S3 in taxon E. (b) Potential difficulties with a comparative approach to inferring function from structure. Uniquely derived structures may exist in the taxon of interest, E, such as S7 and S8, and the inference of the function of these structures must then be based on a search for "similar" (analogous) structures in other clades (*left arrow*). Also, if taxa A and B are shown by the experimental measurement of function to possess different functions for S2 (F3 and F4), then the inference of function in taxon E is greatly complicated (*top arrow*). (c) An appropriate use of phylogenetic methods (parsimony) to infer that extinct taxon E possesses function F1.

1.1a for inferring function from structure relies on a crucial assumption: a tight link between structure and function. If homologous structures have similar functions, then the phylogenetic method may work well for characters that are not unique (apomorphic) to extinct taxa. However, two difficulties exist that may severely cripple our ability to infer function using this methodology. First, the extinct taxon E might possess many uniquely derived characters that have no homolog in extant species. Figure 1.1b shows extinct taxon E with two unique structures S7 and S8. If we are interested in understanding the functional significance of these structures, then we must reach outside this clade to find similar (analogous) structures elsewhere. When we do this, however, we have moved beyond comparing homologous structural features to infer function: We are now comparing analogous structural features in the hopes of identifying analogous functions (Lauder in press). This is surely a risky procedure as we no longer have even the assurance of an homologous structural base on which to base our inferences. Even with this caveat, many structures that draw the attention of paleontologists are unique to a clade: It is their very novelty that attracts research in the first place. But novel features pose special problems if

similar structures cannot be found in other clades. Inferring function in this case becomes more a matter of speculation than inference with a firm empirical phylogenetic foundation.

The second problem with the phylogenetic method is the assumption of a close link between structure and function. Figure 1.1b illustrates a situation in which the experimental measurement of the function of structure S2 in taxa A and B results in the discovery of different functions for S2: F3 in clade A, and F4 in clade B. What then can be concluded about the function of S2 in extinct clade E? Furthermore, a similar divergence in function for structure S1 among the terminal extant taxa would leave us in the same position of being unable to infer the functional consequences of S1 in clade E. If the mapping between structure and function, or among structure, function, performance, and behavior is loose, then our ability to predict the appropriate functional or behavioral correlates of structures will be weak. A major aim of this paper is to indicate that the situation illustrated in Figure 1.1b, a phylogenetic divergence or discordance between structure and function, is actually a common situation. This pattern has not been widely recognized because most analyses that follow the method illustrated in Figure

1.1a use structure to infer functions, such as F3 and F4, in extant species. Thus, morphology is used to infer functions in related taxa, and then these inferences are reapplied to the clade and structure of interest: No independent test of function is available.

The difficulties with phylogenetic inferences of function from structure are not inherent in phylogenetic methodology. Functional characters, just like structures and behaviors, may be subjected to phylogenetic analysis and the inference, by parsimony, of states in extinct taxa (Lauder 1990, in press). The methodology for making such inferences is now widely known and has been recently discussed in Brooks and McLennan (1991), Felsenstein (1985), Harvey and Pagel (1991), Maddison and Maddison (1992), and Swofford and Maddison (1992) (also see Witmer, this volume). Figure 1.1c illustrates the reconstruction of function in extinct taxon E using functional data from adjacent extant taxa. Outgroup taxa C, D, and F all possess structure S1 and function F1 as do taxa A and B (Figure 1.1c). Taxon E may be shown by studying its morphology to possess structure S1, and in this case it is simple (by parsimony) to infer that extinct taxon E possessed function F1. But actual cases such as this in which functional characters have been measured for relevant extant outgroup and ingroup taxa in the clade of interest are few in number. Most often, functional characters are inferred, even in extant taxa, and function in outgroup taxa is even more rarely measured. Functions inferred from structures are not independent of those structures, and basing inferences of function in an extinct clade of interest on inferences of function in extant taxa is not likely to lead to robust conclusions.

The paradigm method

This is a second method developed to allow the inference of function from structure. Rudwick (1964b) first outlined the paradigm procedure, and indicated that his method was especially useful for structures for which there are no analogs or homologs in extant taxa. This method has been widely discussed in the paleontological literature (Gould 1970; Raup & Stanley 1971; Grant 1972; Thomas 1978; Cowen 1979; McGhee 1980; Hickman 1988; Levinton 1988). According to Rudwick's (1964a,b) procedure, the first step in inferring the function of a novel structure is to develop several mechanical abstractions or models (usually from physical and mechanical first principles) that can be used to predict what a structure should look like, given the constraints and functional goals of the model. This predicted structure is called a paradigm structure. Thus, if a novel bone were discovered in the jaw of a fossil taxon, one might make a mechanical model

of the jaw by assuming a specific pattern of movement, and this model could be used to predict the specific morphology of a bone in the location of the new structure. Rudwick advocated making several alternative models so that the several predicted structures could be compared with the real one. The function of the new structure would then be inferred to be that function whose paradigm most closely matches the real structure.

Overall, the paradigm method may be considered as a variant of the general approach of making mechanical or mathematical models to "deduce" the shapes and types of biological structures that might perform the functions included in the model (Dullemeijer 1974; Gutmann 1981). A priori model-building approaches have not been very successful in functional morphology, due in part to the enormous difficulty in building a model that reflects biological reality from first principles, and to the fact that a given model often does not generate a sufficiently limited set of predicted structures.

The paradigm method makes several important assumptions, one of which is that there is a strong correlation between function and structure at the level of the traits under study: Any model developed from first principles should only allow one paradigm structure. If multiple functions can be associated with a given structure and if a given function can be performed by many structures, then model-building is unlikely to yield successful predictions. A further difficulty is that the structure is known before the model(s) are built, making a truly independent prediction of structure from function impossible. In addition, the "functions" assumed by the paradigm method are often general behaviors or biological roles (Table 1.1), and usually do not lead to specific structural specifications.

A variant on the paradigm method is the use of basic physical principles to predict general functional or behavioral properties of extinct taxa (Alexander 1989; McGowan 1991; also see the papers in Rayner & Wootton 1991). This approach has been more successful (although the predictions can usually only be indirectly tested) because a considerable body of data on extant species may be used to generate the predictions. An example is Alexander's (1989) analysis of the locomotor speeds of dinosaurs: an attempt to predict a functional attribute from structural data. Alexander used fossil trackway data, data on limb length and stride length from living animals, and estimates of leg length in dinosaurs to predict their locomotor speeds. Similarly, Thomason (1991) used the extensive data base on the biophysics of extant horse locomotion to infer functional properties of the limb in two extinct horse taxa (*Mesohippus* and *Merychippus*). Even using biophysical principles to predict function may be difficult, however, as there may be many possible predictions and interpretations even when these are based on a single morphological system. One example of such a situation is the controversy over the aerodynamic properties of morphological features in pterosaurs and *Archaeopteryx* (Padian 1991; Rayner 1991).

LEVELS OF RESOLUTION IN PREDICTIONS FROM STRUCTURE

Given the above caveats regarding our ability to infer function from structure, it is reasonable to ask what conditions are most likely to lead to successful predictions of organismal function, behavior, or performance, and which past investigations have yielded the most success. It is my view that predictions appear to have been most successful at both the most general (ecological/behavioral) and at the finest (within tissue) levels of resolution.

At a general level, success has been obtained in predicting basic ecological categories (such as herbivore vs. carnivore; climbing vs. cursorial) from gross and microscopic features of fossil taxa and from data such as limb proportions (e.g., Van Valkenburgh 1988, 1991; Damuth & MacFadden 1990). Tooth wear patterns as well as general tooth morphology and other aspects of skull structure may be reliable indicators of ecological situation, and are often based on comparative analyses of many extant taxa. Similarly, analyses of performance may be based on measurements from extant taxa and may thus allow estimates of performance in fossils (or extant taxa for which performance cannot be measured). For example, Garland and Janis (1993) tested the ability of measurements of limb proportions in 49 species of cursorial mammals to predict maximal running speed. They did have some success in making predictions of speed from hind limb length, but end on a cautionary note: "Prediction of locomotor performance of extinct forms, based solely on their limb proportions, should be undertaken with caution" (p. 133).

At a specific level, prediction of function from structure within biological tissues also appears to have been relatively successful. Basic principles of mechanical design do allow precise statements about function to be made, or at least the definition of functional limits within which structures must work (Wainwright et al. 1976). One example is the highly predictable relationship between the physiological cross-sectional area of vertebrate skeletal muscle and the maximum tetanic tension developed by that muscle (Powell et al. 1984; also see Gans & Bock 1965 and Alexander 1968). Another example would be the general predictability of muscle contractile performance from a knowledge of myosin ATPase

types, and mitochondrial, glycogen, and capillary density. Unfortunately, this type of prediction is one that is not often of use to paleontologists. Of more use to paleontologists are analyses of bone mechanical properties. At least some of the functional properties of bone tissue appear to be generally predictable from an understanding of structure. For example, Bertram & Biewener (1988) analyzed bone curvature as a response to loading pattern, and Biewener (1982) examined the regularity of the interspecific relationship between bone safety factors and body size (see Thomason, this volume).

However, it is our attempt to understand biological function at an *intermediate* level of generality that causes the greatest difficulties, and unfortunately, this is precisely the area where many workers wish to make inferences. An example of prediction at such an intermediate level might be the inference of specific movement patterns of jaws or limbs from morphology alone (in order to reconstruct feeding or locomotor biomechanics and behavior). While the case studies considered below do not give much cause for optimism in predicting functional characteristics at this level, there are approaches that have proven useful for defining the physical limits to function given a specific morphology. Much progress, for example, has been made by modeling locomotion using well-established biomechanical and hydrodynamic principles, as has been discussed by Briggs et al. (1991) for fossil arthropods. Interspecific measurements of drag and lift forces on models of extinct taxa permit general comparative statements about locomotor habit. At the least, such modeling does assist in delimiting the boundary conditions for organismal function. In addition, tooth wear patterns and striations allow inferences to be made about the movement patterns of jaws in fossil vertebrates, providing some indirect indications of the results of muscle activity (see chapters by Janis, Naples, and Rensberger, this volume).

Much of the difficulty in predicting function from structure by any method stems from the existence of multiple confounding factors that increase the chance that many functions could be performed by a given structure. In the case of muscles and bones, the nervous system serves as an important locus of these difficulties. In fossil taxa we have no access to the motor programs that the nervous system is capable of producing, and yet changes at the level of central nervous structure and function may have a dramatic impact on the function of morphological features (Lauder 1990, 1991). For example, the evolution of a novel pattern of motor output could easily alter stride length by changing patterns of limb-muscle activity (thus compromising predictions of speed), or change jaw mechanics with few visible structural effects (leading us to incorrectly infer jaw function based on living taxa).

THREE CASE STUDIES

How closely linked are structure and function? This is the key issue that must be addressed if we are to make progress in developing methods for inferring functional characteristics from morphological data. One way to address this question is to undertake direct experimental studies of function in living species to measure the functional properties of structures. In this way, we can determine empirically just how predictable functional features are from morphology, and we can do so by the measurement of function. It is vital that functional inferences be tested by experimental measurement to obtain an independent assessment of function, and that functional hypotheses themselves not be tested by inference from other morphological features (a circular procedure). The three case studies that follow all address the issue of predicting function from structure, given particular configurations of muscles and bones.

Case Study 1: Prey manipulation in osteoglossomorph fishes

The Osteoglossomorpha is a monophyletic clade of fishes that, as presently understood, represents the most primitive living clade of teleost fishes (Figure 1.2a; Lauder & Liem 1983). Outgroup taxa to the Osteoglossomorpha include the bowfin (*Amia*) and the clade including gar (Ginglymodi: *Lepisosteus*). There are approximately 210 species of osteoglossomorphs (Nelson 1984) including such taxa as the knifefishes (Notopteridae), arawana (Osteoglossidae), and elephantfishes (Mormyridae).

The Osteoglossomorpha are of interest for the analysis of structure–function relationships because most members of this clade possess a rather dramatic suite of apomorphic morphological modifications to the feeding apparatus: a bite between the basihyal (BH), or "tongue," and the base of the skull (Figure 1.2b). (In addition, shearing action may take place between the basihyal and lateral teeth on the palate.) This tongue-bite is a significant evolutionary novelty that, if present in an extinct taxon, would provoke functional speculation and hypotheses as to its role in the feeding mechanism. Given that many members of the clade Osteoglossomorpha possess this structural novelty, we may hypothesize that, if structure and function are closely linked (as indicated in Figure 1.1a), these taxa will share a common functional novelty associated with the tongue bite.

The teeth on the tongue of osteoglossomorphs may be quite large (Figure 1.2b), and experimental research has shown that these teeth are used very effectively to puncture and shred prey once they have been trapped within the mouth cavity (Sanford & Lauder 1989). A number of muscles attach to bones surrounding the tongue bite (Figure 1.2b), including the sternohyoideus (SH) which retracts the hyoid (and attached teeth), the posterior intermandibularis (PIM) which protracts the hyoid, and the hypaxial muscles (HY) which retract the pectoral girdle. In addition, the epaxial muscles (EP) insert on the skull and elevate the cranium and the attached teeth on the base of the skull.

By recording the electrical activity pattern of these (and other) muscles in the knifefish *Notopterus* using electromyography, and by analyzing high-speed films of prey manipulation, Sanford and Lauder (1989) documented the use (function) of the tongue-bite apparatus. The pattern of motion of the teeth on the hyoid is shown by black arrows inside the mouth labeled "Tongue-bite" (Figure 1.2b). A key finding of these experiments was that the major movement of the hyoid teeth was anteroposterior (not dorsoventral): The teeth were dragged (raked) posteriorly through the body of the prey while it was clamped in a fixed position by adduction of the mandible and upper jaw. The basihyal teeth were moved posteriorly by extensive posterior rotation of the whole pectoral girdle pulling the hyoid with it. These kinematic results were surprising; such extensive rotation is not seen in outgroup taxa. *Notopterus,* therefore, possesses an extensive suite of functional novelties (including pectoral rotation) correlated with the obvious structural novelty of the teeth on the hyoid and skull (the tongue-bite), which is shared with other osteoglossomorphs. This result begs the question: Do other osteoglossomorph species with a tongue-bite morphology also share the functional novelties identified in *Notopterus?*

Sanford and Lauder (1990) extended this study to a comparative kinematic analysis of two other osteoglossomorph taxa: *Pantodon* (the freshwater "butterfly-fish"), and *Osteoglossum* (the arawana). From high-speed video records of prey manipulation behavior using the tongue-bite, we derived eight kinematic variables that measured various aspects of tongue-bite function (such as the extent of hyoid movement, the relative timing of mouth opening and head lifting, etc.). Statistical analyses (analysis of variance and principal components) were then used to compare the function (pattern of bone movement) of the tongue-bite apparatus among the three taxa. The result of the principal-components analysis is summarized in Figure 1.3a: The taxa differ significantly in the function of the tongue-bite morphology. Each of the three genera cluster in a significantly different portion of multivariate kinematic space, and

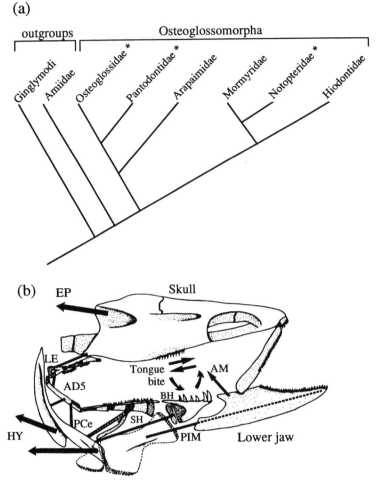

(a)

outgroups   Osteoglossomorpha

Ginglymodi Amiidae Osteoglossidae * Pantodontidae * Arapaimidae Mormyridae Notopteridae * Hiodontidae

(b) EP   Skull

LE Tongue AM

AD5 bite

BH

HY PCe SH PIM Lower jaw

Figure 1.2. (a) Phylogenetic relationships of the major groups of osteoglossomorph fishes. Asterisks indicate the three osteoglossomorph taxa that have been studied experimentally; outgroup taxa (Ginglymodi and Amiidae) have been investigated also (Lauder & Wainwright 1992). (b) Diagram of the skull of the knifefish, *Notopterus*, to show the teeth on the basihyal (BH) or "tongue" and on the base of the skull. During intraoral prey manipulation the prey is held in a fixed position by the front jaws, while the basihyal teeth move dorsally into the prey and then posteriorly as indicated by the arrows. Muscles that power the tongue-bite include the epaxial muscles (EP), the sternohyoideus (SH), the posterior intermandibularis (PIM), the adductor mandibulae (AM), and the hypaxial muscles (HY). The pharyngeal jaws are located posteriorly to the tongue-bite and are controlled by several muscles of which three are shown: the pharyngocleithralis externus (PCe), the fifth branchial adductor (AD5), and the levatores externi muscles (LE).

the analysis of variance on the individual variables confirmed that there were extensive differences among the genera.

The conclusion from these experimental results is that, while the three osteoglossomorph taxa all share the common structural novelty of a tongue-bite, they use this morphological adaptation in very different ways. The phylogenetic interpretation (Figure 1.3b) is that, despite the presence of a shared plesiomorphic morphological character, the taxa possess divergently derived functional features. This must be due, at least in part, to novelties in the nervous system that result in changed motor output to the jaw muscles in the three taxa. Differences among taxa in the relative timing and amplitude of activity in homologous muscles (not yet quantified) would test this hypothesis.

The ability to analyze directly the use (function) of the tongue-bite in osteoglossomorph fishes has thus resulted in a pattern similar to that illustrated in Figure 1.1b, where taxa with a common morphology do *not* share common functions. Given these results, I would feel unable to predict functional patterns in other osteoglossomorph taxa, and would need to conduct an experimental measurement of function in order to empirically assess how the tongue-bite morphology was being used.

While the specific movement pattern used by osteoglossomorph fishes might be difficult to predict from structure alone, we might wish to infer a general behavioral pattern of intraoral prey maceration using the teeth on the skull and hyoid in all osteoglossomorph taxa. We could be reasonably confident given these results, that a fossil taxon possessing a tongue-bite morphology used these teeth to manipulate and puncture prey. Such a general behavioral prediction would be reasonable, even if we could not venture a specific functional reconstruction.

Case Study 2: Aquatic prey capture in salamanders

Aquatic prey capture in salamanders is of interest for the analysis of structure–function relationships because divergent head morphology across families permits an analysis of the extent of functional

(a)

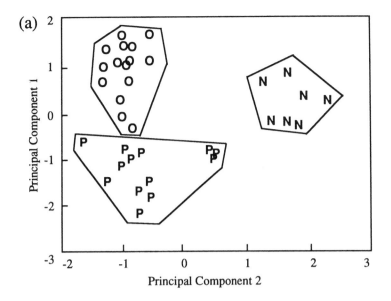

(b)

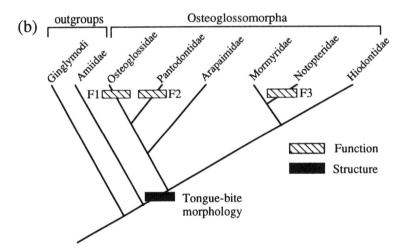

Figure 1.3. (a) Results of a principal components analysis of kinematic variables measured from high-speed films and video recordings of intraoral prey manipulation (raking) behavior using the tongue-bite in three taxa of osteoglossomorph fishes: *Notopterus* (N), *Pantodon* (P), and *Osteoglossum* (O). (b) Phylogenetic interpretation of the results of the principal components analysis. Each of the three taxa studied possesses a different functional pattern for the common morphology, rendering the inference of function of the tongue-bite in other members of this clade risky.

congruence with morphological features. Thus, this case study differs from the previous one in that the salamander families being compared are divergent in morphology, and we wish to examine the extent of functional divergence (if any).

Reilly and Lauder (1992) conducted a comparative analysis of head morphology and aquatic prey capture kinematics in six families of salamanders (Cryptobranchidae, Dicamptodontidae, Ambystomatidae, Amphiumidae, Proteidae, and Sirenidae). The goal of the study was to examine the relationship between morphology and kinematic patterns during prey capture, and the theoretical framework outlined below was used to assess the relationship between morphology and kinematics. Varying numbers of individuals were studied from each of the six families. Morphological variables (e.g., head width and depth, gill slit number) were measured from the

head to capture the size and shape of the feeding apparatus. Seven kinematic variables were derived from a frame-by-frame analysis of high-speed video recordings (200 fields per second) of prey capture. The function of individual head bones was defined as movement contributing to the overall feeding behavior, and was, therefore, quantified directly. All individuals from each family fed in the water on the same type and size of prey (earthworm pieces) dropped in front of the head. Both analysis of variance (ANOVA and MANOVA) and principal components analysis were conducted to examine differences among the taxa.

Before presenting the results of this study, it is useful to consider some of the possible theoretical relationships between structure and function among taxa. Figure 1.4 illustrates one way of visualizing these data. This format has the advantage of allowing

Kinematic (functional) level

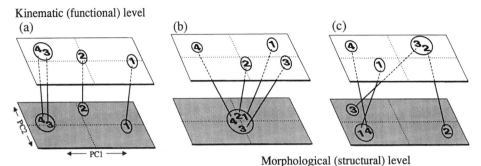

Morphological (structural) level

Figure 1.4. Using principal component analysis to compare the congruence of variation in morphological space (*lower*, shaded planes) with that in kinematic or functional space (*upper*, clear planes) for four species (numbered 1 to 4). The mean position of a taxon in each plane is indicated by a number, and ellipses around each number indicate the 95 percent confidence interval about the taxon mean. Thus, taxa included in the same ellipse are not significantly different from each other in that principal component plane. Each of the three pairs of planes illustrates a different possible result from a study of morphology and function in a clade of four species. (a) A situation in which function can at least broadly be inferred from structure: the relationships *among* taxa are the same across planes. (b) Function cannot be predicted from morphology as the four taxa are all similar morphologically but possess divergent functions. (c) A second between-level pattern in which function cannot be predicted from structure. Taxa 1 and 4 share similar morphologies but divergent functions, whereas taxa 2 and 3 are similar functionally but divergent morphologically. Abbreviations: PC1 and PC2, principal component axes 1 and 2.

us simultaneously to see patterns of variation among numerous morphological and kinematic variables, and also to map the degree of congruence between both classes of data. If we combine the individual morphological variables measured from each taxon into a reduced number of variables using a principal components analysis, we can plot the position of each taxon (as the mean of all individuals studied) in a plane representing principal components 1 and 2. This plane represents a multivariate morphological space. Similarly, we may use the kinematic variables for all the taxa in a principal components analysis and plot the mean position of each taxon in multivariate kinematic (functional) space. The mapping of taxon position across the two planes provides an indication of the extent of congruence between structure and function.

In this theoretical example, the position of the number in the plane indicates the mean value for that taxon, and the circle around the number represents the 95 percent confidence interval about the mean. Taxa whose circles do not overlap are considered significantly different from each other in multivariate space. Panel (a) in Figure 1.4 illustrates a situation in which structure and function are closely matched. Taxa 1 and 2, for example, are divergent both in kinematic and morphological space from the other two taxa, while taxa 3 and 4 share similar positions in both morphological and kinematic space. In no case does a taxon share space in the kinematic plane with a morphologically divergent taxon.

Panel (b) of Figure 1.4 illustrates one possible pattern of incongruence between structure and function where taxa that share a common morphology

possess divergent kinematic patterns (similar to the result of case study 1 above). In such a case, knowledge of the position of a taxon in the morphological plane tells us nothing about its position in the kinematic plane. Panel (c) shows a pattern of incongruence in which there is no clear relationship between morphological variation and functional variation among taxa. Taxa 1 and 4 share a common morphology but are divergent kinematically, while taxa 2 and 3 show the reverse pattern.

The results of the salamander study are presented in Figure 1.5 (Reilly & Lauder 1992). One pair of taxa (A, *Ambystoma* and D, *Dicamptodon*) are grouped together in both morphological and kinematic space, indicating that shared structure of the feeding mechanism is associated with a common kinematic pattern. *Cryptobranchus,* which is divergent morphologically, also possesses a divergent kinematic pattern. However, *Siren* (Figure 1.5; S), which groups with *Ambystoma* and *Dicamptodon* in the morphological plane, possesses a divergent pattern of feeding kinematics. Thus, the kinematic pattern of *Siren* feeding is not predictable from its morphology. The genera *Necturus* and *Amphiuma* (Figure 1.5; N, P) do share common morphological and kinematic patterns. However, together they are divergent from the other salamander taxa morphologically and yet similar in kinematics to *Ambystoma* and *Dicamptodon*. The morphological data do not allow prediction of this similarity in feeding behavior.

The overall conclusion from this experimental study of structure and function in the salamander feeding mechanism is that, for some taxa, there is evidence of congruence between the structural and

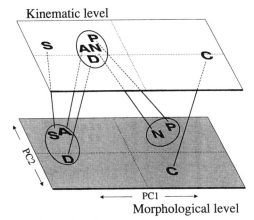

Kinematic level

Morphological level

Figure 1.5. Results of an analysis of head morphology (shaded plane) and functional or kinematic variation in aquatic feeding (top plane) in six families of aquatic salamanders: A, Ambystomatidae, *Ambystoma*; C, Cryptobranchidae, *Cryptobranchus*; D, Dicamptodontidae, *Dicamptodon*; N, Proteidae, *Necturus*; P, Amphiumidae, *Amphiuma*; and S, Sirenidae, *Siren*. Note that taxa such as *Ambystoma* and *Necturus* are divergent morphologically but share a common region of kinematic space, while *Siren* and *Dicamptodon* are similar morphologically but divergent kinematically. A pattern such as this makes prediction of function from structure difficult. Other conventions and abbreviations match those of Figure 1.4.

functional levels, but for others there is not. Perhaps most disturbing is the lack of any clear a priori indication of the taxa in which one might find congruence, and the taxa in which one might not find congruence. If these six salamander families are in any way representative of other animal taxa, then empirical investigations of both structure *and* function will be the only way to determine their relationship; this prevents the formulation of a broad, general framework of congruence to use in making functional predictions from structure.

Case Study 3: Snail crushing in pumpkinseed sunfish

The North American sunfish family Centrarchidae contains about 35 species, including a well-known piscivore, the bass *Micropterus*. Also included in this clade are two sister species that eat snails: the pumpkinseed sunfish (*Lepomis gibbosus*) and the redear sunfish (*Lepomis microlophus*). In previous work I have studied interspecific morphological and functional differentiation among sunfish species in their feeding mechanisms (Lauder 1983a,b), but (as is the case with most research in functional morphology) little work has been done at the popula-

tion level within a species. The pumpkinseed sunfish populations in lakes of southern Michigan are of special interest in this regard because there is considerable interlake variation in prey abundance. For example, Wintergreen Lake contains few snails, while a nearby lake, Three Lakes II, possesses a very large snail population (Wainwright et al. 1991a,b). Both lakes contain pumpkinseed sunfishes, and both populations of sunfishes will readily feed on snails in the laboratory. Wainwright et al. (1991b) discovered that the populations of pumpkinseed sunfishes in these two lakes possessed a distinctive trophic polymorphism: Individuals from Three Lakes II (with abundant snails) possess hypertrophied pharyngeal muscles relative to individuals in Wintergreen Lake. (The morphological differences between sunfish populations appear to be an ecophenotypic response to prey differences.) In this section I describe the results of the morphological analyses of Wainwright et al. (1991b) and our subsequent functional analyses of the patterns of muscle activity during snail crushing (Wainwright et al. 1991a).

The mechanism of snail crushing in pumpkinseed sunfish is of interest for the analysis of structure–function relationships because intraspecific comparisons (i.e., among populations) should provide us with one of the strongest possible tests of the correlation between structure and function. There is little possibility of confounding phylogenetic factors as there has not been much time for differentiation among taxa. Thus, it would be difficult to argue that demonstrating a lack of congruence between structure and function is really due to uncontrolled phylogenetic differences among the taxa. In this case, the difference among pumpkinseed populations appears to be very recent, the result of a winter snail die-off after 1977 (Wainwright et al. 1991a). The key question in this case study is: Do interpopulational differences in morphology of the trophic musculature correlate with functional differences in those muscles?

Figure 1.6a illustrates the morphology of the part of the feeding mechanism involved in eating snails. Pumpkinseed sunfishes capture snails by suction feeding and transport them back to the pharyngeal jaw apparatus just anterior to the esophagus by creating a current of water through the oral cavity (Lauder 1983a). Once the snail reaches the pharyngeal region it is positioned in between the upper and lower pharyngeal jaws (Fig 1.6a; UPJ, LPJ). Activation of many of the pharyngeal jaw muscles in a distinctive motor pattern (Lauder 1983a,b) results in forces which crush the snail's shell. Shell pieces are separated from the body of the snail by the pharyngeal jaws and by water currents created by head-bone movements. The pharyngeal jaws then move the snail body posteriorly into the esophagus.

Figure 1.6. (a) Schematic lateral
view of head of a percomorph
teleost fish such as a
pumpkinseed sunfish
(Centrarchidae) illustrating
position of the pharyngeal jaws
(stippled), some of the muscles
that control the pharyngeal jaw
apparatus, and position of a snail
prey item between upper
pharyngeal jaw (UPJ) and lower
jaw (LPJ). Pharyngeal jaws are
used to crack the snail's shell
and to separate it from the
snail's body which is then
swallowed. (b) Procedure
followed to analyze the
morphology of pharyngeal jaw
apparatus in two populations of
sunfishes (*Lepomis gibbosus*)
that live in two lakes differing
greatly in the number of snail
prey available. (c) Procedure
followed to analyze the function
of PJA muscles during feeding
on snail prey. (d) The result of
both procedures was a ranking
of muscles by significance of the
difference in either structure or
function between lake
populations. Note that there is
little correlation between
structure and function even in
this intraspecific case study.
Muscle abbreviations:
GH, geniohyoideus; LE,
levatores externi; LP, levator
posterior; PCe,
pharyngocleithralis externus; PCi,
pharyngocleithralis internus; RD,
retractor dorsalis; SH,
sternohyoideus.

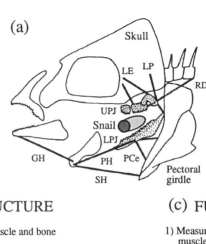

(a)

(b) STRUCTURE

1) Measure muscle and bone
   morphology

2) Analyze muscle morphology
   statistically (ANCOVA)

3) Rank muscles by differences
   among populations

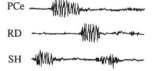

(c) FUNCTION

1) Measure activity of pharyngeal
   muscles during prey processing

2) Analyze muscle activity pattern
   statistically (nested ANOVA)

3) Rank muscle activity by
   differences among populations

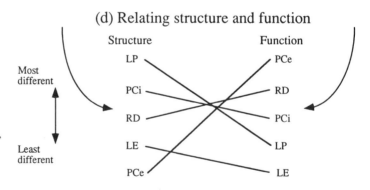

(d) Relating structure and function

ANALYZING THE STRUCTURE OF THE PHARYNGEAL
JAWS (FIGURE 1.6B). Pharyngeal morphology was
measured in a series of individuals from each lake
that ranged in size from 40 to 132 mm standard
length: ten muscles and five bones from the pharyngeal jaw apparatus were dissected out and weighed
(Wainwright et al. 1991a). Previous interspecific work
(Lauder 1983a) had shown that the mass of several
pharyngeal jaw muscles is correlated with the proportion of snails in the diet, but no comprehensive
analysis of both the bones and muscles between
populations had previously been undertaken. Second, an analysis of covariance (ANCOVA) was used
to assess significant interpopulational differences in
muscle and bone mass. Three of the five bone masses
differed between lakes, with the masses of three

pharyngeal bones mechanically involved in snail
crushing being on average 1.5 times greater in the
Three Lakes fish. In addition, six of the ten pharyngeal muscle masses were significantly greater in the
Three Lakes fish, with one muscle (the levator posterior) possessing a mass 2.3 times greater than that
of a fish of similar size from Wintergreen Lake.
Third, the muscles were ranked using the degree of
statistical difference between lakes as the criterion.
Thus, the levator posterior muscle is at the top of
the list of muscles (Figure 1.6b; LP) because it exhibited the most significant differences in mass between the lake populations.

ANALYZING THE FUNCTION OF THE PHARYNGEAL JAWS
(FIGURE 1.6C). Electrical activity patterns were

recorded (using the electromyographic procedures described in Lauder [1983a,b]) from five pharyngeal jaw muscles during snail crushing. This allowed the timing of muscle activity to be quantified. We measured 16 statistical variables from each feeding on each of six individuals from each lake to capture variation in relative timing of muscle activity, duration of activity, and intensity of activity. These data were then analyzed using a nested ANOVA (individuals were nested within each lake), and the significance of the "lake effect" tested. Four of the 16 variables showed a significant lake effect, and the muscle that showed the greatest differentiation between lakes in function was the pharyngocleithralis externus (PCe). The five muscles were then ranked by the extent of functional differentiation between lake populations using the statistical difference between populations as the criterion. The pharyngocleithralis externus muscle is at the top of the list (Figure 1.6c; PCe) because it showed the most significant difference in function between lake populations.

ASSESSING THE STRUCTURE–FUNCTION RELATIONSHIP IN THE PHARYNGEAL JAW APPARATUS. By connecting homologous muscles in the two lists (Figure 1.6d) we can see that there is little relationship between structure and function in the pharyngeal jaw apparatus of pumpkinseed sunfishes. The levator posterior muscle (LP) shows the greatest differences between lakes in mass, but almost the least difference in function. Conversely, the pharyngocleithralis externus (PCe) is functionally the most differentiated between lake populations but is the least differentiated structurally.

This case study of sunfishes is of special concern because the relative sizes of muscles and bones often are used as functional or behavioral indicators. A massive muscle (by implication, one of large physiological cross-sectional area) will exert greater maximal force than a similarly designed muscle of smaller cross-sectional area. But such comparative physiological facts, while true, say little about how a muscle is *used* during behaviors such as locomotion and feeding. The use of muscles is a critical determinant of organismal behavior, and different morphological substrates may produce very similar behaviors, given an appropriate pattern of muscle activity (Lauder & Shaffer 1986; Lauder 1991; Reilly & Lauder 1992). The finding of considerable discordance in patterns of structure and function even at an intraspecific level indicates that interspecific predictions of function based on structure alone should be treated with caution.

It is nonetheless true, that if prediction at a general level is desired, one could accurately infer the difference among populations in diet and ecological situation based on morphological differences measured in the pharyngeal jaws. Snail-eating populations do possess hypertrophied bones and muscles relative to outgroup populations. Such general inferences are valuable, especially if we wish to understand the life habits of extinct taxa, even if we cannot accurately infer function.

DISCUSSION

On the importance of functional inference

The desire to infer function, performance, or behavior from morphological data is common to many fields. Ecologists, ethologists, systematists, and population biologists, in addition to morphologists and paleontologists, have used structural data on organisms as a basis for inferences about the movements and actions of living animals (Hopson & Radinsky 1980; Fisher 1985; Grant 1986; Boucot 1990; Wainwright & Reilly in press). While the urge to make such inferences is a natural one (after all, the goal of gathering morphological data in the first place is usually to contribute to our understanding of what animals do and how they live), it is also reasonable to ask just how accurate such inferences are likely to be.

The primary means of evaluating the accuracy of inferences of function from structure should be a comparison of functional predictions based on form to measured functions in living taxa. Only in a situation where we can measure actual movements, actions, and experimentally evaluate alternative functional hypotheses can we test predictions from structure. The conceptual methods proposed to date as aids to the inference of function from structure contain no means of verifying or testing the functional predictions independent of morphological data. I have argued elsewhere (Lauder 1990) that using only morphological data to test hypotheses of function that have themselves been predicted from morphological data is circular. Functional data are of a fundamentally different character than morphological data (Lauder 1990, 1991), and a further examination of morphological characteristics (or a manipulation of morphological characters alone) is not likely to provide a strong test of a hypothesized function. Functional data need to be generated by direct observation, measurement, and experimentation. Methods such as the paradigm approach serve only to provide a mechanical model based on initial morphological inputs: they do not provide data on actual performance or function.

This is not to say that the extensive studies to date on the biomechanics and functional morphology of fossil taxa have no value. Quite the contrary. Research (much of it experimental in character) on how structures in fossils *might* have been used and the ecological roles that might have resulted from use of these structures (e.g., Stanley 1970; Fisher 1977; LaBarbera 1981; contributions in Rayner & Wootton 1991) is an important component in the analysis of organismal design. These studies may allow rejection of certain classes of functional hypotheses, or may provide an indication that a given structure may have had other functions than those currently under consideration. Biomechanical research on fossils helps to define the realm of the possible. But, given the results from experimental work on extant taxa, we must be extremely cautious in drawing the conclusion that these types of studies definitively show how structures were used.

One reason that the inference of function is so widespread may be that organismal function and performance are difficult things to measure. Even most extant taxa do not lend themselves to experimental work and may be located in inaccessible habitats. Individual species in a clade of interest may not adjust well to laboratory conditions or to experimental protocols. Finally, conducting experimental work such as a description of limb movements, an analysis of muscle function, an analysis of ground reaction forces, and the related biophysical calculations on a clade of even a few species is a time-consuming task. It is often easier to generate a model or make general inferences of function from structure.

In addition to satisfying our natural urge to understand the actions and movements of living organisms, functional inferences are beginning to play an increasingly important role in many aspects of comparative biology (e.g., Brooks & McLennan 1991; Harvey & Purvis 1991). For instance, Greene (1986), Coddington (1988), and Baum and Larson (1991), all advocate a method for recognizing adaptive characters in organisms that depends in part on our ability to measure organismal function and/or performance. Coddington (1988; p. 3) even defines the term *adaptation* in terms of organismal function: an adaptation is "apomorphic function promoted by natural selection." Baum and Larson (1991; p. 1) note that in order for a character to be an adaptation it must "provide current utility to the organism" and that "the criterion of current utility is applied by comparing the performance of a derived trait to that of its phylogenetically antecedent state." Measurement of function and performance is one aspect of identifying adaptations in organisms.

But how are such analyses of function or performance carried out in practice? Baum and Larson (1991; pp. 12–13) recommend use of the paradigm method so that "performance is assessed by comparing a character state with an abstract model, or 'para-

digm,'" and "the closer the observed character comes to the paradigm, the higher is its inferred performance." However, as indicated above, inferences of function or performance from morphology via the paradigm method do not provide a reliable means of estimating how organisms function. In fact, in working out the example of adaptation in the limbs of plethodontid salamanders, Baum and Larson (1991; p. 13) simply assert that the derived state is "mechanically superior"; no model is presented that allows such predictions for primitive and derived limb morphologies, nor do functional data to support such a statement exist in the literature. Demonstration of mechanical "superiority" in limb function by experimental functional analysis would be a difficult task indeed, and identifying the precise morphological basis for relative performance differences among species would be even more laborious. In fact, a signal difficulty in comparative performance analyses of organismal function is identifying the underlying morphological bases of those differences (Arnold 1986; Jayne & Bennett 1989). Unfortunately, without such experimental data, historical or retrospective methods for the analysis of adaptation cannot be applied.

The case studies

The three case studies presented earlier were chosen to indicate the difficulty of inferring function from structure even in extant species on which we can perform experimental functional analyses. These examples show that functional characteristics may not be easily correlated with (apparently) associated morphological features, and that structure and function may be relatively uncorrelated phylogenetically. These case studies are not an isolated body of data. Numerous examples, especially in the musculoskeletal system, exist to corroborate the main conclusion of each case study: Function is often not predictable from structure.

One excellent example from the mammalian literature further illustrates the difficulty in predicting function even in the best of circumstances. O'Donovan et al. (1982) conducted a comprehensive experimental functional study of two muscles in cats: the flexor digitorum longus (FDL) and an anatomical synergist, the flexor hallucis longus (FHL). Based on morphological evidence, one would predict that the two muscles should show very similar activity patterns. Both muscles extend in parallel around the ankle joint, and their tendons join together on the plantar surface of the foot to insert together onto the distal phalanges. From morphology alone and from simple manipulative experiments (such as muscle stimulation), it was predicted that the two muscles would show similar activity patterns: "For a variety of reasons the FDL and FHL muscles in the cat have been considered as more-or-less interchangeable 'extensor' muscles" (O'Donovan et al. 1982; p. 1140).

The function of these muscles was assessed experimentally by recording electromyographic activity of each muscle during a variety of locomotor behaviors as well as by directly measuring muscle force (using implanted strain transducers on the muscle tendon) and taking simultaneous video images of each behavior. O'Donovan et al. (1982; p. 1127) described their results as "surprising in that the functional activity of the FDL muscle during locomotion proved to be complex, including stereotyped flexor behavior and facultative activity that appeared to respond to perturbations in the step cycle. Neither aspect was present in the activity of the FHL, which instead behaved as a stereotyped antigravity extensor." The authors conclude that these two muscles, defined by anatomists as synergists, have fundamentally different activity patterns (O'Donovan et al. 1982).

The feeding mechanism in terrestrial salamanders illustrates another point. The subarcualis rectus 1 (SAR) muscle in salamanders is a muscle that has been predicted to cause projection of the hyobranchial apparatus since turn of the century morphologists began to study salamander structure (Drüner 1902, 1904). This muscle, based on anatomical analyses (Severtsov 1971; Lombard & Wake 1976) should protract the hyobranchial apparatus thus extending the tongue toward the prey during feeding. Reilly and Lauder (1989, 1990; 1991a) conducted morphological, functional, and experimental investigations of the role of this muscle during initial prey capture in the tiger salamander, *Ambystoma tigrinum*, and confirmed that this was indeed the case, although the electrical activity pattern for the SAR showed an interesting pattern with two peaks: one during mouth opening, and the other during the mouth closing phase of the gape cycle.

Reilly and Lauder (1991b) then investigated the function of the SAR muscle during prey transport. One would have predicted that, because the tongue is not projected from the mouth when previously captured prey are transported from within the buccal cavity posteriorly to the esophagus, the SAR muscle would not be active during prey transport. Quite unexpectedly, we found that the SAR muscle is indeed active during prey transport behavior, with a single sharp peak in activity during mouth opening, a time when the hyoid is moving posteroventrally. The mechanical role of this muscle during prey transport is still unknown. The SAR muscle thus functions during two different behaviors with different

kinematics, although activity in just one behavior was expected from a priori anatomical analyses.

This last example illustrates yet another problem with inferences of function from structure, and one that has long been recognized (Darwin 1859; Gans 1974; Levinton 1988; Raup & Stanley 1971): Structures may have many functions. Although this idea is well known, the serious consequences of this fact for specific inferences of function from structure have not hindered functional speculation based on morphology. In practice, even in extant taxa on which functional studies can be performed, organisms exhibit constant surprises: Structures are used in novel and unexpected ways. There is hardly a paper in the literature that contains measurements of the function or functions of a structure or complex of structures, in which unexpected and surprising results about the use of those structures are not found. Given this track record, it is perhaps unwise to place much faith in detailed predictions of function from structure, especially those predictions involving taxa and structures in which no experimental test can be conducted to test function independently.

Why is function not predictable from structure?

A key reason for the discordance described above between structure and function in the musculoskeletal system is the nervous system. The nervous system is the "wildcard" in comparative and historical analyses of musculoskeletal systems. Changes in central nervous structure and function alone have the capability of radically altering the accuracy of predictions based only on musculoskeletal morphology (Lauder 1990, 1991). Alterations in the output of central neuronal circuits can (and do) thoroughly alter the function of structures, the effect of structures on organismal performance, and the behavioral pattern of the whole organism. In addition, changes in sensory feedback pathways from peripheral morphological structures may also have a profound influence on movement patterns, even in the absence of changes in central motor circuits. Where nervous system structure and function can be investigated along with musculoskeletal function, analysis has often shown that unexpected functional patterns may have a basis in neuronal reorganization (O'Donovan et al. 1982; Vilensky 1989; Vilensky & Larson 1989; Arbas, Meinertzhagen & Shaw 1991; Katz 1991; Paul 1991). Unfortunately, methods of assessing changes in neural circuitry for fossils are extremely circumscribed (Giffin, this volume) and permit only broad inferences of behavior, not of specific functional patterns (Table 1.1).

PROSPECTUS: LIMITS TO COMPARATIVE METHODS

Our desire to understand the biological significance of newly discovered structures (in both fossils and extant organisms), and the lack of a clear understanding of the difficulties that even experimental functional morphologists have in elucidating the function of structures in extant taxa, have provided much of the basis for optimism regarding the prediction of function from structure. Functions of structures, the mechanical roles or properties of structures, and the effect of structure on performance can best be studied by direct measurement in extant taxa. Even in extant taxa, functional morphologists have been constantly surprised by unexpected functions and mechanical roles for structures.

Perhaps the best case scenario for functional inference would be one in which a clade contains mostly extant species that can be subjected to experimental studies of function. If we can verify by experimental analysis that a certain structure has a specific function in extant taxa, and if a newly discovered taxon in this clade (either a fossil or living species that is not accessible for functional study) shares homologous morphology, then it is most parsimonious to infer that the new taxon shares the function for that structure (Figure 1.1c). Unfortunately, this will not be a case of much interest to most workers, as it is the truly novel structures, or at least structures that have no obvious known counterparts, that evoke the greatest interest.

But it is exactly this latter situation in which the limits to comparative and retrospective investigation are the greatest and in which our inferences are weakest. The results of the case studies presented above provide arguments indicating that the inferences may be so weak that it might be best to limit the extent of functional inference and speculation.

Perhaps now is a good time, as we gather increasing amounts of functional data from living organisms and as functional morphologists make progress in understanding the functional consequences of organismal design, to step back from extensive functional inference based on morphology alone, and recognize the limits that exist on the inference of both functional and evolutionary mechanisms from comparative and historical data. Comparative data on morphology have numerous valuable uses which are subject to fewer limitations than functional inference, including the study of ontogenetic and evolutionary trends to size and shape, and the testing of models of morphological change through time (Gould 1985). Similarly, neontological data on organismal structure contribute in many ways to some of the most important questions in evolutionary biology, contributions that do not depend on our ability to infer function from structure.

If there are limits on the ability of comparative and retrospective investigations to infer mechanistic processes, then perhaps more overt recognition of these limits in studies of structure–function relationships will act as a positive force, and stimulate the discovery of new directions and methods for understanding how organisms work.

ACKNOWLEDGMENTS

I thank Gary Gillis and Peter Wainwright for comments on the manuscript, and Kevin Padian for his thoughtful and constructive review. The writing of this chapter was supported by NSF grants IBN 91-19502 and DCB 87-10210. Lance Grande provided many very helpful comments following the symposium on which the volume is based.

REFERENCES

Alexander, R.M. 1968. *Animal Mechanics.* London: Sidgwick and Jackson.

Alexander, R.M. 1989. *Dynamics of Dinosaurs and Other Extinct Giants.* New York: Columbia University Press.

Arbas, E.A., Meinertzhagen, I.A., & Shaw, S.R. 1991. Evolution in nervous systems. *Annual Reviews of Neuroscience* 14, 9–38.

Arber, A. 1950. *The Natural Philosophy of Plant Form.* Cambridge University Press.

Arnold, S.J. 1983. Morphology, performance, and fitness. *American Zoologist* 23, 347–361.

Arnold, S.J. 1986. Laboratory and field approaches to the study of adaptation. In *Predator-Prey Relationships: Perspectives and Approaches from the Study of Lower Vertebrates,* ed. M.E. Feder & G.V. Lauder, pp. 157–179. Chicago: University of Chicago Press.

Baum, D.A., & Larson, A. 1991. Adaptation reviewed: a phylogenetic methodology for studying character macroevolution. *Systematic Zoology* 40, 1–18.

Benton, T. 1975. *Form and Function: A Source Book for the History of Architecture and Design 1890–1939.* London: Crosby Lockwood Staples.

Bertram, J., & Biewener, A.A. 1988. Bone curvature: sacrificing strength for load predictability? *Journal of Theoretical Biology* 131, 75–92.

Biewener, A.A. 1982. Bone strength in small mammals and bipedal birds: do safety factors change with body size? *Journal of Experimental Biology* 98, 289–301.

Bloom, L. 1970. *Language Development: Form and Function in Emerging Grammars.* Cambridge, Mass.: MIT Press.

Bock, W., & von Wahlert, G. 1965. Adaptation and the form-function complex. *Evolution* 19, 269–299.

Boucot, A.J., ed. 1990. *Evolutionary Paleobiology of Behavior and Coevolution.* Amsterdam: Elsevier.

Briggs, D., Dalingwater, J.E., & Selden, P.A. 1991. Biomechanics of locomotion in fossil arthropods. In *Biomechanics in Evolution,* ed. J.M.V. Rayner & R.J. Wootton, pp. 37–56. Cambridge University Press.

Brooks, D.R., & McLennan, D.A. 1991. *Phylogeny, Ecology and Behaviour*. Chicago: University of Chicago Press.

Coddington, J.A. 1988. Cladistic tests of adaptational hypotheses. *Cladistics* 4, 3–22.

Cowen, R. 1979. Functional morphology. In *The Encyclopedia of Paleontology*, ed. R.W. Fairbridge & D. Jablonski, pp. 487–492. Stroudsburg, PA: Dowden, Hutchinson, and Ross.

Damuth, J., & MacFadden, B.J., ed. 1990. *Body Size in Mammalian Paleobiology*. Cambridge University Press.

Darwin, C. 1859. *On the Origin of Species by Means of Natural Selection, or, the Preservation of Favored Races in the Struggle for Life*. London: John Murray.

Drüner, L. 1902. Studien zur Anatomie der Zungenbein, Kiemengbogen, und Kehlopfmuskelen der Urodelen. I. Theil. *Zoologische Jahrbücher Abteilung für Anatomie* 15, 435–622.

Drüner, L. 1904. Studien zur Anatomie der Zungenbein, Kiemengbogen, und Kehlopfmuskelen der Urodelen. II. Theil. *Zoologische Jahrbücher Abteilung für Anatomie* 19, 361–690.

Dullemeijer, P. 1974. *Concepts and Approaches in Animal Morphology*. The Netherlands: Van Gorcum.

Felsenstein, J. 1985. Phylogenies and the comparative method. *American Naturalist* 125, 1–15.

Fisher, D. 1977. Functional significance of spines in the Pennsylvanian horseshoe crab *Euproops danae*. *Paleobiology* 3, 175–195.

Fisher, D. 1985. Evolutionary morphology: beyond the analogous, the anecdotal, and the *ad hoc*. *Paleobiology* 11, 120–138.

Gans, C. 1974. *Biomechanics: an Approach to Vertebrate Biology*. Philadelphia: J.B. Lippincott.

Gans, C., & Bock, W. 1965. The functional significance of muscle architecture: a theoretical analysis. *Ergebnisse der Anatomie und Entwicklungsgeschichte* 38, 116–142.

Garland, T., & Janis, C.M. 1993. Does metatarsal/femur ratio predict maximal running speed in cursorial mammals? *Journal of Zoology, London* 229, 133–151.

Gould, S.J. 1970. Evolutionary paleontology and the science of form. *Earth-Science Reviews* 6, 77–119.

Gould, S.J. 1985. The paradox of the first tier: an agenda for paleobiology. *Paleobiology* 11, 2–12.

Grant, P. 1986. *Ecology and Evolution of Darwin's Finches*. Princeton: Princeton University Press.

Grant, R.E. 1972. Methods and conclusions in functional analysis: a reply. *Lethaia* 8, 31–33.

Greene, H.W. 1986. Diet and arboreality in the Emerald Monitor, *Varanus prarisinus*, with comments on the study of adaptation. *Fieldiana: Zoology n.s.* 31, 1-12.

Gutmann, W.F. 1981. Relationships between invertebrate phyla based on functional-mechanical analysis of the hydrostatic skeleton. *American Zoologist* 21, 63–81.

Haar, M. 1983. *The Phenomenon of the Grotesque in Modern Southern Fiction: Some Aspects of its Form and Function*. Amsterdam: North-Holland.

Harvey, P.H., & Pagel, M.D. 1991. *The Comparative Method in Evolutionary Biology*. Oxford: Oxford University Press.

Harvey, P.H., & Purvis, A. 1991. Comparative methods for explaining adaptations. *Nature* 351, 619–624.

Hickman, C.S. 1988. Analysis of form and function in fossils. *American Zoologist* 28, 775–793.

Hickman, C.S. 1991. Functional analysis and the power of the fourth dimension in comparative evolutionary studies. In *The Unity of Evolutionary Biology*, ed. E.C. Dudley, pp. 548–554. Portland: Dioscorides Press.

Hopson, J.A., & Radinsky, L.B. 1980. Vertebrate paleontology: new approaches and new insights. *Paleobiology* 6, 250–270.

Jayne, B.C., & Bennett, A.F. 1989. The effect of tail morphology on locomotor performance of snakes: a comparison of experimental and correlative methods. *Journal of Experimental Zoology* 252, 126–133.

Katz, P.S. 1991. Neuromodulation and the evolution of a simple motor system. *Seminars in the Neurosciences* 3, 379–389.

LaBarbera, M. 1981. The ecology of Mesozoic *Gryphaea*, *Exogyra*, and *Ilymatogyra* (Bivalvia: Mollusca) in a modern ocean. *Paleobiology* 7, 510–526.

Lauder, G.V. 1983a. Functional and morphological bases of trophic specialization in sunfishes. *Journal of Morphology* 178, 1–21.

Lauder, G.V. 1983b. Neuromuscular patterns and the origin of trophic specialization in fishes. *Science* 219, 1235–1237.

Lauder, G.V. 1986. Homology, analogy, and the evolution of behavior. In *The Evolution of Behavior*, ed. M. Nitecki & J. Kitchell, pp. 9–40. Oxford: Oxford University Press.

Lauder, G.V. 1990. Functional morphology and systematics: studying functional patterns in an historical context. *Annual Review of Ecology and Systematics* 21, 317–340.

Lauder, G.V. 1991. Biomechanics and evolution: integrating physical and historical biology in the study of complex systems. In *Biomechanics in Evolution*, ed. J.M.V. Rayner & R.J. Wootton, pp. 1–19. Cambridge University Press.

Lauder, G.V. In press. Homology, structure, and function. In *Homology*, ed. B.K. Hall. San Diego: Academic Press.

Lauder, G.V., Leroi, A.M., & Rose, M.R. In preparation. Adaptations and history.

Lauder, G.V., & Liem, K.F. 1983. The evolution and interrelationships of the actinopterygian fishes. *Bulletin of the Museum of Comparative Zoology* 150, 95–197.

Lauder, G.V., & Shaffer, H.B. 1986. Functional design of the feeding mechanism in lower vertebrates: unidirectional and bidirectional flow systems in the tiger salamander. *Zoological Journal of the Linnean Society of London* 88, 277–290.

Lauder, G.V., & Wainwright, P.C. 1992. Function and history: the pharyngeal jaw apparatus in primitive ray-finned fishes. In *Systematics, Historical Ecology, and North American Freshwater Fishes*, ed. R. Mayden, pp. 455–471. Stanford: Stanford University Press.

Levinton, J. 1988. *Genetics, Paleontology, and Macroevolution*. Cambridge University Press.

Lombard, R.E., & Wake, D.B. 1976. Tongue evolution in the lungless salamanders, family Plethodontidae. I. Introduction, theory, and a general model of dynamics. *Journal of Morphology* 148, 265–286.

Maddison, W.P., & Maddison, D.R. 1992. *MacClade: analysis of phylogeny and character evolution*. Sunderland, Mass.: Sinauer.

McGhee, G.R. 1980. Shell form in the biconvex articulate Brachiopoda: a geometric analysis. *Paleobiology* 6, 57–76.

McGowan, C. 1991. *Dinosaurs, Spitfires, and Sea Dragons.* Cambridge: Harvard University Press.

Morton, J.E., ed. 1979. *Guts: The Form and Function of the Digestive System.* Baltimore: University Park Press.

Nelson, J.S. 1984. *Fishes of the World*, 2nd ed. New York: John Wiley & Sons.

O'Donovan, M.J., Pinter, M.J., Dum, R.P., & Burke, R.E. 1982. Actions of FDL and FHL muscles in intact cats: functional dissociation between anatomical synergists. *Journal of Neurophysiology* 47, 1126–1143.

Padian, K. 1991. Pterosaurs: were they functional birds or functional bats? In *Biomechanics in Evolution*, ed. J.M.V. Rayner & R.J. Wootton, pp. 145–160. Cambridge University Press.

Patterson, C. 1982. Morphological characters and homology. In *Problems of Phylogenetic Reconstruction*, ed. K.A. Joysey & A.E. Friday, pp. 21–74. London: Academic Press.

Paul, D.H. 1991. Pedigrees of neurobehavioral circuits: tracing the evolution of novel behaviors by comparing motor patterns, muscles and neurons in members of related taxa. *Brain, Behavior and Evolution* 1991, 226–239.

Powell, P.L., Roy, R.R., Kanim, P., Bello, M.A., & Edgerton, V.R. 1984. Predictability of skeletal muscle tension from architectural determinations in guinea pig hindlimbs. *Journal of Applied Physiology* 57, 1715–1721.

Raup, D.M., & Stanley, S.M. 1971. *Principles of Paleontology.* San Francisco: W.H. Freeman and Co.

Rayner, J.M.V. 1991. Avian flight evolution and the problem of *Archaeopteryx*. In *Biomechanics in Evolution*, ed. J.M.V. Rayner & R.J. Wootton, pp. 183–212. Cambridge University Press.

Rayner, J.M.V., & Wootton, R.J., ed. 1991. *Biomechanics in Evolution.* Cambridge University Press.

Reilly, S.M., & Lauder, G.V. 1989. Kinetics of tongue projection in *Ambystoma tigrinum*: quantitative kinematics, muscle function, and evolutionary hypotheses. *Journal of Morphology* 199, 223–243.

Reilly, S.M., & Lauder, G.V. 1990. The strike of the tiger salamander: quantitative electromyography and muscle function during prey capture. *Journal of Comparative Physiology* 167, 827–839.

Reilly, S.M., & Lauder, G.V. 1991a. Experimental morphology of the feeding mechanism in salamanders. *Journal of Morphology* 210, 33–44.

Reilly, S.M., & Lauder, G.V. 1991b. Prey transport in the tiger salamander: quantitative electromyography and muscle function in tetrapods. *Journal of Experimental Zoology* 260, 1–17.

Reilly, S.M., & Lauder, G.V. 1992. Morphology, behavior, and evolution: comparative kinematics of aquatic feeding in salamanders. *Brain, Behavior, and Evolution* 40, 182–196.

Rudwick, M.J.S. 1964a. The function of zigzag deflexions in the commissures of fossil brachiopods. *Paleontology* 7, 135–171.

Rudwick, M.J.S. 1964b. The inference of function from structure in fossils. *British Journal of the Philosophy of Science* 15, 27–40.

Russell, E.S. 1916. *Form and Function: A Contribution to the History of Animal Morphology.* Reprinted in 1982 with a new Introduction by G.V. Lauder. London: John Murray.

Sanford, C.P.J., & Lauder, G.V. 1989. Functional morphology of the "tongue-bite" in the osteoglossomorph fish *Notopterus*. *Journal of Morphology* 202, 379–408.

Sanford, C.P.J., & Lauder, G.V. 1990. Kinematics of the tongue-bite apparatus in osteoglossomorph fishes. *Journal of Experimental Biology* 154, 137–162.

Severtsov, A.S. 1971. The mechanism of food capture in tailed amphibians. *Proceedings of the Academy of Science, USSR* 197, 185–187.

Stanley, S.M. 1970. Relation of shell form to life habits in the Bivalvia. *Memoirs of the Geological Society of America* 125, 1–296.

Swofford, D.L., & Maddison, W.P. 1992. Parsimony, character-state reconstructions, and evolutionary inferences. In *Systematics, Historical Ecology, and North American Freshwater Fishes*, ed. R.L. Mayden, pp. 186–223. Stanford: Stanford University Press.

Thomas, R.D.K. 1978. Shell form and the ecological range of living and extinct Arcoidea. *Paleobiology* 4, 181–194.

Thomason, J.J. 1991. Functional interpretation of locomotory adaptations during equid evolution. In *Biomechanics in Evolution*, ed. J.M.V. Rayner & R.J. Wootton, pp. 213–227. Cambridge University Press.

Van Valkenburgh, B. 1988. Trophic diversity in past and present guilds of large predatory mammals. *Paleobiology* 14, 155–173.

Van Valkenburgh, B. 1991. Iterative evolution of hypercarnivory in canids (Mammalia: Carnivora): evolutionary interactions among sympatric predators. *Paleobiology* 17, 340–362.

Vilensky, J. 1989. Primate quadrupedalism: how and why does it differ from that of typical quadrupeds? *Brain, Behavior, and Evolution* 34, 357–364.

Vilensky, J.A., & Larson, S.G. 1989. Primate locomotion: utilization and control of symmetrical gaits. *Annual Reviews of Anthropology* 18, 17–35.

Wainwright, P.C., Lauder, G.V., Osenberg, C.W., & Mittelbach, G.G. 1991a. The functional basis of intraspecific trophic diversification in sunfishes. In *The Unity of Evolutionary Biology*, ed. E.C. Dudley, pp. 515–529. Portland: Dioscorides Press.

Wainwright, P.C., Osenberg, C.W., & Mittelbach, G.G. 1991b. Trophic polymorphism in the pumpkinseed sunfish (*Lepomis gibbosus*): effects of environment on ontogeny. *Functional Ecology* 5, 40–55.

Wainwright, P.C., & Reilly, S.M., ed. In press. *Ecological Morphology: Integrative Organismal Biology.* Chicago: University of Chicago Press.

Wainwright, S.A., Biggs, W.D., Currey, J.D., & Gosline, J.M. 1976. *Mechanical Design in Organisms.* London: Edward Arnold.

2

The Extant Phylogenetic Bracket and the importance of reconstructing soft tissues in fossils

LAWRENCE M. WITMER

ABSTRACT

Fossils usually provide paleontologists with little more than bones and teeth as primary data. Because the broad aim of functional morphological analyses of extinct organisms is to breathe life into fossils, paleontologists are challenged to reconstruct those portions of the animal not normally preserved in the fossil record – in particular, soft tissues. Soft-tissue considerations are important for two major reasons: (1) Soft tissues often have morphogenetic primacy over skeletal tissues, and (2) such considerations often lie at the base of numerous paleobiological inferences. Furthermore, soft-tissue analysis can provide causal hypotheses of phylogenetic character correlation. A methodology for reconstructing soft anatomy in fossils – the Extant Phylogenetic Bracket approach – is proposed here. This method makes explicit reference to at least the first two extant outgroups of the fossil taxon of interest. It is based firmly on the relation of biological homology, provides a means of rigorously establishing the limits of our inferences, and allows construction of a hierarchy of inference. The approach can be applied to virtually any unpreserved trait of a fossil taxon with little or no modification.

INTRODUCTION

Recently, extinct organisms, in a sense, have come alive. That is, there has been renewed interest in viewing fossils not just as collections of bones but rather as living, functioning animals. This resurgence has its foundation, at least in part, in the explosion of dinosaur paleobiology in both the scientific literature (Weishampel, Dodson, & Osmólska 1990; Carpenter & Currie 1990; Halstead 1991) and the popular press (Norman 1985; Bakker 1986; Paul 1988; Benton 1989). The success of dinosaur research has led to similar popular treatments for other groups of extinct animals (Savage & Long 1986; Dixon et al. 1988; Benton 1991; Wellnhofer 1991). In fact, the proliferation of fossils fleshed out with soft tissues has even elicited a primer for illustrators for

reconstructing soft tissues (Paul & Chase 1989). It probably is of little scientific consequence whether or not robotic sauropods have trunks, and it is relatively harmless if they do. However, the question remains that, given this justifiable interest in fossils as once living organisms, what relevance do considerations of soft tissues have for the analysis and interpretation of organisms known only from fossils?

This paper seeks an answer to this question by addressing two major issues in turn. First, what importance does the study of soft anatomy hold for disciplines such as vertebrate paleontology that traditionally have bones and teeth as their primary (if not only) data? In other words, why worry about soft tissues at all? Obviously, living organisms are more than collections of bones and teeth. As a result, the evolutionary interpretation of the paleobiology of extinct organisms often *requires* explicit reference to anatomical systems other than the skeleton, that is, to those portions of the organism not normally preserved in the fossil record.

Second, since the necessary soft tissues are rarely preserved as fossils, how is this requisite information obtained? Probably because many older soft-tissue reconstructions had no explicit methodological basis, a common sentiment among paleontologists has developed that any reference to soft tissues in fossils involves little more than unsubstantiated guesswork. However, if approached correctly, more useful information can be extracted from fossils than previously has been appreciated (see Nicholls & Russell 1985; Bryant & Seymour 1990; Bryant & Russell 1992, this volume). A phylogenetically rigorous means of inferring these attributes – the Extant Phylogenetic Bracket (EPB) approach – is proposed here, based in large measure on the anatomy of the extant relatives of a fossil taxon. While speculation cannot always be eliminated, the method outlined here identifies those soft-tissue inferences that are speculative and those that are well founded. In fact, the EPB approach provides a clear way to characterize the different sorts of speculation that must be invoked to reconstruct an extinct organism with a particular soft tissue.

The extant phylogenetic bracket approach was developed specifically for the inference of soft anatomical properties (Witmer 1992), but is generalized in the last section for the inference of any aspect of paleobiology. A very similar general approach was devised independently by Bryant and Russell (1992), and the reader is referred to their paper for an excellent and complementary discussion.

THE IMPORTANCE OF SOFT-TISSUE INFORMATION

Knowledge about the general form of soft anatomical components and their relationships to the skeleton in extinct (as well as extant) organisms is important for at least three general reasons: (1) Soft tissues largely are responsible for the existence, maintenance, and form of bones; (2) judgments about the form and actions of soft tissues are (implicitly if not explicitly) the basis for a host of paleobiological inferences; and (3), with regard to systematics, soft-tissue relationships may provide testable hypotheses on independence or nonindependence of phylogenetic characters.

The morphogenetic primacy of soft tissues

At perhaps the most fundamental level, soft tissues, not bones, may well be the proper focus of analysis. In recent years it has been recognized that skeletal tissues are largely responsive to their epigenetic systems and are correctly viewed as products of epigenetic cascades involving inductive interactions among nonosseous tissues (Hall 1983, 1988, 1990). In other words, bones do not self-differentiate (Smith & Hall 1990). Carlson (1981) noted that the inductive influence provided by the developing brain is necessary for initiation of ossification of the dermal skull roof, and that pressure from the brain and cerebrospinal fluid subsequently determines the final form of the bones. This simple, well-understood example illustrates the hierarchical nature of soft-tissue influences on bony morphology: first, the specification of the existence and spatial patterning of skeletal elements (Wolpert 1983; Wedden et al. 1988; Hall 1988), and second, the maintenance and form of these elements once present (Moss 1968, 1971; Hall 1990). The two major tiers of this hierarchy are examined in more detail below.

THE FIRST TIER: EXISTENCE AND PATTERN OF BONES. Since bones are not self-differentiating entities,

understanding the presence (or absence) and positions of bones requires some knowledge of the epigenetic system specifying their existence, for example, the nature of inductive tissue interactions, the rate and timing of cellular migrations or programmed cell death, and the pattern and size of preskeletogenic mesenchymal condensations (Oster et al. 1988; Hall 1991; Atchley & Hall 1991). Bones tend to form relatively late in ontogeny, after the differentiation of other tissues such as, in the head, the cartilaginous chondrocranium, nasal epithelium, brain, and eye. As a result, bones tend to originate in close association with particular soft tissues. Although the causal mechanisms of these associations are being worked out for a few model systems (e.g., chick scleral ossicles; Pinto & Hall 1991), in most cases, these associations remain largely phenomenological. In embryos of *Alligator mississippiensis*, for example, many of the dermal skull bones ossify next to very specific portions of the chondrocranium: the frontal bone along the *planum supraseptale* and sphenethmoidal commissure, the vomer next to the primary nasal concha, the prefrontal at the juncture of the sphenethmoidal commissure and nasal capsule, and the postorbital bone at the tip of the rostrodorsal process of the quadrate cartilage (L.M. Witmer, unpublished data). The chondrocranium itself is specified even earlier in ontogeny by aspects of epithelial folding (see Born 1883, Kamal & Abdeen 1972, and Noden 1983 for the importance of nasal epithelium morphogenesis) and the rate and timing of neural crest migration (Thorogood 1988).

Thus, whether a particular bony element is present or absent probably depends more on those tissues that precede bone formation in morphogenesis than on factors intrinsic to the particular bone. For example, most of the sites of bone formation observed in alligator embryos are conserved across the major clades of vertebrates, probably because chondrocrania are so conservative (Thorogood 1988). In crocodilians, the ascending ramus of the maxilla develops after and in proximity to the *lamina transversalis rostralis* (LTR) of the chondrocranium; other sauropsid amniotes generally resemble crocodilians in this feature (de Beer 1937; Bellairs & Kamal 1981). Ornithurine birds, however, represent an exception in that they have both lost the ascending ramus of the maxilla (Cracraft 1986; Witmer 1990) and suppressed the LTR (Witmer 1992); given the situation in other sauropsids, loss of the ascending ramus in birds is probably causally associated with suppression of the LTR.

THE SECOND TIER: THE MAINTENANCE AND FORM OF BONES. Once formed, the subsequent maintenance and detailed form of bony structures are largely controlled by associated soft tissues. Hall (1990) referred to this level of control as second-order epigenetic control. This notion has been captured effectively (although in a somewhat extreme form) in the Functional Matrix Hypothesis championed by Moss and colleagues (Moss 1968, 1971, 1972a,b; Moss & Salentijn 1969). Under the Functional Matrix Hypothesis, each specific function is performed by a particular *functional component*, which is itself composed of a *functional matrix* (which actually carries out the function and incorporates all nonskeletal tissues and "functioning spaces" that directly constrain bony morphology) and its *skeletal unit* (which responds to its particular functional matrix, playing a role in protection and/or support).

According to Moss (1981, p. 370), "the origin, growth, and maintenance of all skeletal tissue and organs are always secondary, compensatory, and obligatory responses to temporally and operationally prior events or processes that occur in specifically related nonskeletal tissues, organs, or functioning spaces (functional matrices)." Atchley and Hall (1991, p. 138) viewed Moss's functional matrices as "the sum of the epigenetic [heritable] and environmental [nonheritable] influences that impinge on skeletal development," and although they disagreed with Moss's extreme stance on genomic involvement, they agreed that functional matrices are "an important component(s) of the interactive, integrative, hierarchical network that governs morphogenesis."

No one has questioned the existence of functional matrices (even if we have difficulty characterizing them), and they have been demonstrated empirically (Moss 1972b; Ranly 1988). The now-classic example is the dramatic regression and ultimate loss of a skeletal unit (the coronoid process of the mandible) when the functional matrix (the temporalis muscle) has been eliminated via denervation or surgical resection (Schumacher & Dokládal 1968; Moss & Salentijn 1969). The profound influence of numerous soft anatomical systems on the growth and form of skeletal tissues is well documented, particularly with respect to the central nervous system (Ross 1941; Noetzel 1949; Young 1959), the jaw musculature (Scott 1957; Humphrey 1971; Spyropoulus 1977; Hall & Herring 1990), and the eye (Taylor 1939; Coulombre & Crelin 1958; Hanken 1983).

Although there may be formal problems with correctly delimiting skeletal units and functional matrices (since each is based in large measure on reference to the other), the *perspective* of the Functional Matrix Hypothesis is not only conceptually appropriate but perhaps even required for generating the correct explanations in evolutionary osteology and functional morphology. Although Moss's (1968, p. 198) slogan – "functional matrices evolve; skeletal units respond" – is probably overstated, it properly emphasizes the primacy of soft-tissue

considerations in osteology. Thus, I see the Functional Matrix Hypothesis less as an analytical tool and more as a conceptual framework for interpreting bony morphology and evolution. Many of the well-known morphological trends documented in the fossil record clearly have their foundations in particular soft anatomical systems. An obvious example is the expansion of the bones forming the neurocranial vault of hominids associated with increasing brain size (Le Gros Clark 1964). It should be emphasized, however, that the causality vector does not point only from soft to skeletal tissues, but rather the two interact. Particular biomechanical loading regimes, for instance, often require bony buttressing to function properly. Thus, whereas the functional matrix (i.e., soft tissues) may often or even usually be a "target" of natural selection (as declared in Moss's slogan), this need not always be the case.

Soft tissues and paleobiological inference

When paleontologists ask questions about the functional morphology of an extinct organism, they often are asking more than just "how does it work?" Often more general paleobiological information is sought, which requires a dependent sequence of questions and answers. These often start with, How does an observed anatomical structure work? and are followed by, What does this functional assessment tell us about the behavior and mode of life of the organism? What kind of paleoecological interactions did the organism have with other members of its community? and, How did this community evolve? Thus, there is a chain of inference, with each successive inference assuming accuracy of the previous one. Figure 2.1 presents such a chain of inference as an inverted pyramid, with each level (except the lowest) representing, more or less, a tier of the ecological hierarchy (sensu Eldredge & Salthe 1984; Eldredge 1986; Liem 1989). The lowest level of the pyramid (i.e., the one closest to observable data) is soft-tissue inference because it is argued here that accurate soft-tissue reconstructions often are the foundation of such paleobiological inferences as behavior, ecology, and community structure.

This point is not typically appreciated, and errors of soft-tissue reconstruction cascade up the ecological hierarchy. Thus, the pyramid in Figure 2.1 is inverted to emphasize this amplification of initial mistakes. For example, whether the jaw musculature and craniofacial biomechanics of the gigantic bird *Diatryma*·point to its being an herbivore or a carnivore (Witmer & Rose 1991) bears importance beyond the functional morphology of a single taxon. These two behavioral alternatives have drastically different implications for the ecological associations

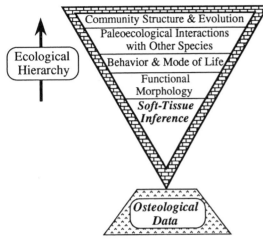

Figure 2.1. Inverted pyramid of inference. Inferences about the soft-tissue attributes of fossil organisms are often the basis for a host of paleobiological inferences. Inferences made at lower levels of the pyramid form the justification for larger-scale inferences. Pyramid is inverted to emphasize how mistakes in soft-tissue inference or at lower levels of the pyramid cascade upwards, amplifying the error. Levels of the pyramid correspond roughly to the ecological hierarchy (*arrow at left*).

of *Diatryma* with contemporary mammals: If it was a herbivore, *Diatryma* browsed peacefully or perhaps even competed for resources with *Hyracotherium* and *Phenacodus*, whereas if it was a carnivore, it ate them. Furthermore, larger issues are at stake, such as the guild structure of the community (sensu Van Valkenburgh 1988) to which *Diatryma* and its relatives belonged and what effect extinction of these birds had on the evolving predator–prey interactions during the Tertiary (Bakker 1983). Our interpretation of these issues would vary depending on whether we regard *Diatryma* as a herbivore or a carnivore, a determination that is at least partly founded on assessment of soft-tissue relations. Similarly, Weishampel's (1984a) treatment of the evolution of craniofacial kinesis in ornithischian dinosaurs addressed not only the musculature but also the soft anatomy of joints. Again, the deployment of different feeding strategies among ornithischians (Weishampel 1984a, 1985) and the resulting inferences regarding evolving plant–herbivore interactions throughout the Mesozoic (Weishampel 1984b; Norman & Weishampel 1985; Weishampel & Norman 1987, 1989) have soft-tissue considerations in large measure at their base.

Certainly, attention to aspects of soft anatomy and probing the nature of their relations to the skeleton rightly have held a prominent place – implicitly if not always explicitly – in several research programs, such as in biomechanics and paleoneurology. How-

ever, the structure and even the existence of a pyramid of inferences may not always be appreciated, and ignoring soft-tissue considerations amounts to making unrealized (and hence untested) assumptions of perhaps questionable validity.

Soft tissues and systematics

With the growing popularity of numerical cladistics, identifying the precise basis of comparison – "the character" – becomes a central issue. In particular, it is axiomatic that any parsimony-based system carries the fundamental assumption of statistical independence of its data points, characters in this case. Obviously, if many correlated features are counted separately, the composite of which they are a part is weighted more than independent characters, overestimating the branch length of the taxon supported by the correlated character complex. This problem is virtually identical to that emphasized recently by Harvey and colleagues (Clutton-Brock & Harvey 1977; Pagel & Harvey 1988; Harvey & Pagel 1991) with regard to the use of species as independent data points in quantitative comparisons. Character splitting may boost a consistency index to comforting levels, but it increases the likelihood of invalidating the entire analysis by violating the assumption of independence. The relevance here is that knowledge of soft tissues in some cases will help tie together correlated features into a single phylogenetic character.

Taking soft tissues into account, therefore, allows formulation of causal hypotheses of character correlation. Such a priori hypotheses are falsifiable by further cladistic analysis showing that the features are not phylogenetically correlated at all in that they do not specify a single node on a cladogram but instead occur sequentially in phylogeny (as Cracraft [1990] found for the avian "flight apparatus"). Furthermore, additional anatomical research could falsify the correlation hypothesis by demonstrating the inaccuracy of the proposed causal network in that the observed osteological features are actually associated with other aspects of soft anatomy. Alternatively, assessments of character correlation can be made a posteriori based on the pattern of character distribution, after which the causal (i.e., process-oriented) explanation of correlation may be postulated.

Summary of the importance of soft-tissue information

Evolutionary morphological analyses of bones alone are in some sense incomplete. Reconstructing (or at least considering) soft anatomy in fossils is important because skeletal tissues are largely responsive to the influence of their soft-tissue functional matrices and thus may be only subject indirectly to natural selection. Furthermore, researchers are making assumptions (explicitly or not) about nonskeletal tissues in probing the nature of functioning anatomical systems, and indeed these soft-tissue considerations often are the foundation on which a whole suite of paleobiological inferences are based. Soft tissues even can play a role in determining the phylogenetic relationships of taxa. Thus, if features not typically preserved in fossils are so important, then considerable care must be taken in their reconstruction. It is thus surprising that there has been little attention paid to this issue in the literature until very recently (Bryant & Russell 1992).

METHODOLOGICAL CONSIDERATIONS IN SOFT-TISSUE RECONSTRUCTION

For the reasons outlined above, many questions in evolutionary biology require knowledge of the soft-tissue attributes of organisms known only as fossils. The question then becomes, what is the best way of obtaining this knowledge? Sometimes the taxon of interest may happen to be found in *Lagerstätten*, that is, in unusual preservational environments that preserve soft tissues. In these cases we may need only to observe the fossil material for the required information. Although *Lagerstätten* can provide critical clues (as described below), the vast majority of fossils are known only from hard parts, such as bones and teeth. Thus, the only soft tissues that can be reconstructed are those that may be inferred from the fossils themselves, that is, those that have osteological correlates.

The following section presents first an overview of a method for inferring soft tissues in fossils, examining its basis in phylogenetic principles, the homology relation and the importance of extant taxa. Then the method is discussed in more detail, beginning with an analysis of the causal associations of soft tissues and osteological correlates as determined by study of the extant taxa, proceeding through hypothesis formulation and testing, and finally discussing special cases where the hypothesis does not survive testing and where data from *Lagerstätten* are available.

Overview

The methods proposed here involve an application of basic cladistic principles such as outgroup comparison and parsimony (Wiley 1981; Maddison, Donoghue, & Maddison 1984; Wiley et al. 1991) and the traditional techniques of comparative anatomy. Although the question being asked seemingly is focused solely on fossil taxa, the proposed method

requires and makes explicit use of extant taxa as these are the only organisms for which we can obtain precise information about the soft tissues and their relations to the bones. The method also presupposes an existing, independently corroborated hypothesis of the phylogenetic relationships of the fossil and extant taxa.

Briefly, the approach seeks to reconstruct the soft-tissue attributes of extinct vertebrate organisms (1) by determining the causal associations between soft tissues and their osteological correlates in the extant relatives of the fossil taxon, (2) by formulating a hypothesis that similarities among the extant taxa in these associations are due to inheritance from a common ancestor (i.e., the associations are homologous), and (3) by testing this hypothesis by surveying the fossil taxa for the osteological correlates. If the hypothesis survives the test and the associations are indeed causally based, then the soft tissue can be inferred in the fossil taxon with confidence.

The appropriate extant taxa are, minimally, the first two outgroups of the fossil taxon of interest having living representatives (Figure 2.2a). The character assessment of the most recent common ancestor of the fossil taxon and its first extant outgroup (i.e., its extant sister group) is critical for determining the ancestral condition of the fossil clade (see Maddison et al. 1984, who designated this hypothetical common ancestor as the outgroup node). The second, more distantly related, extant taxon serves as the outgroup to the first two together, allowing estimation of the ancestral features at the outgroup node. As always, additional extant outgroups may be necessary to resolve conflicts in situations where there is character variation between the first two outgroups (Maddison et al. 1984).

It is of some heuristic use to take advantage of the free rotation around cladogram nodes and rearrange Figure 2.2a to bring the extant outgroups to the periphery (Figure 2.2b). Thus, we may refer to the two extant outgroups as the extant phylogenetic bracket (EPB) of the fossil taxon of interest. The extant taxa in a very real sense form a "bracket" in that they constrain any inferences about the fossil taxon. For the sake of discussion, the most recent common ancestor of the extant phylogenetic bracket may be termed the bracket ancestor, which is located at the bracket node (Figure 2.2b).

Often the extant taxa are highly diverse and may vary internally with respect to the features being compared. In theory, each extant taxon used in the analysis is not a single or even several species but rather the whole monophyletic supraspecific taxon. The characters of each extant taxon are thus abstractions of its hypothetical common ancestor deduced from all subtended taxa (see Maddison et al. 1984).

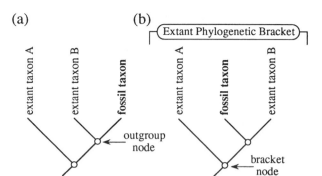

Figure 2.2. (a) Phylogenetic relationships of a fossil taxon and its first two extant sister-groups. (b) Rotation around the outgroup node in (a) brings the extant taxa to the periphery, forming the Extant Phylogenetic Bracket.

The basic approach is little more than a particular application of homology determination: In order to reconstruct a particular soft-tissue attribute in a fossil taxon, the resemblances in osteological correlates between the fossil and its EPB must not only pass the test of similarity (Patterson 1982), or better, of "1:1 correspondence" (Stevens 1984), but must also pass the test of congruence (Patterson 1982) by characterizing a monophyletic group. But in addition to homology among the various osteological features, the *causal associations* of the soft and hard tissues must be homologous. Put simply, homologous soft tissues must produce homologous osteological correlates. As soft-tissue reconstructions of extinct taxa are inferential in nature, the aim of the following method is to put these inferences in the form of hypotheses that can be subjected to testing, in particular the homology tests proposed by Patterson (1982) and others.

Determination of the causal associations of soft tissues and osteological correlates in extant taxa

Routine comparative anatomical research of the extant taxa proceeds with determination of the topographical relationships and 1:1 correspondences of all of those soft tissues that are both relevant to the osteological system of interest and have clear osteological correlates that may be assessed in fossils. Osteological evidence for particular soft tissues include such typical features as tuberosities, crests, grooves, fossae, foramina, fenestrae, and septa. These features often have unambiguous relationships to the soft tissues producing them, such that the causal association is clear and can be demonstrated experimentally. For example, as discussed earlier, the mammalian temporalis muscles certainly are causally associated with coronoid processes. As another example, in *A. mississippiensis* the postvestibular recess and its pneumatic foramen are caused directly by pneumatization by an epithelial paranasal air sac; in some individuals, the air sac fails to appear during ontogeny, and in these cases the bony recess and

foramen are hence also absent (Witmer 1992). A potential hazard, however, is that there sometimes may be more than one soft-tissue system that may produce a particular feature. In these cases, it may be difficult or impossible to choose without directly studying the actual soft tissues of the extant taxa (e.g., via dissection, histology, ontogenetic analysis). For example, grooving on bones may be caused by neurovascular bundles, muscle tendons, attachment of connective-tissue sheets, pneumatic diverticula, ducts of glands, glands, or other organs and structures.

A more insidious pitfall, perhaps, is the "hole-or-the-doughnut" problem, in which it may be difficult to determine from the bones alone whether the soft tissues are directly responsible for a particular bony feature (the doughnut) or if the feature is actually an epiphenomenon (the hole) produced by the effect of other soft tissues on the area. In other words, the causal relationships are uncertain. As a general example, an elongate bony elevation could result either directly (as from muscle insertion) or may be merely the region "left over" after excavation of adjacent surfaces. Furthermore, unexpected, alternative soft-tissue systems potentially may be responsible for observed osteological features. This is more of a psychological than a methodological issue and involves over-reliance on a search image tied to the system of interest. For example, an uncritical worker may interpret all bony scars as muscle attachments, all craniofacial cavities as air sinuses, and all braincase foramina as exits for cranial nerves.

As has been emphasized throughout, the approach advocated here requires that the osteological correlates be causally related to a particular soft tissue. Ideally, the goal is to identify those soft-anatomical elements that are both necessary and sufficient to explain a particular osteological feature. That some causal relationship exists is predicted by Moss's Functional Matrix Hypothesis and has been demonstrated by evidence that bones are products of their epigenetic systems. But as the above examples illustrate, the causal association may not be particularly apparent if only dried skeletal or fossil material is studied. The proposed causal basis of a particular association is itself a hypothesis, perhaps requiring experimental approaches such as extirpation studies (e.g., Coulombre & Crelin 1958; Schumacher & Dokládal 1968) to be tested adequately. All of the above hazards are significantly lessened by direct comparative studies (via dissection, etc.) of the extant taxa. Comparative ontogenetic analysis also may be helpful in determining how the soft and skeletal tissues interact during morphogenesis, providing highly detailed information on the changing topographical relationships of structures.

All of the structures that are used as topographical landmarks obviously must be homologous among the extant taxa in the bracket if they are to have any bearing on the fossil taxon. As with any such hypothesis of homology, these must pass the tests of similarity (i.e., 1:1 correspondence of topographical relationships), congruence, and conjunction (i.e., putative homologs may not co-occur in a single organism) as outlined by Patterson (1982, 1988), Stevens (1984), and Rieppel (1988), among others. Additional extant outgroups will be required for this procedure, although in many cases the homologies of the soft-anatomical systems are so well established as to require little further justification.

Formulating and testing hypotheses about the bracket ancestor

Having assessed the soft tissues and their relationships to their osteological correlates in the extant taxa, it is possible to formulate hypotheses about the soft-tissue attributes of the common ancestor of the extant phylogenetic bracket. All similarities in soft tissues (and correlated osteological features) between the two extant taxa can be hypothesized to have been present (minimally at least) in their common ancestor and all its descendants, including the fossil taxon of interest (Figure 2.3: broken arrows). The assumption of this hypothesis is that the bracket ancestor had these attributes and passed them down to all its descendants, two clades of which still have living representatives. Since the hypothesis predicts that other descendants of this common ancestor also should have inherited these attributes, we may test this hypothesis by surveying these other descendants – the fossil taxa – for the osteological correlates of the soft-tissue attributes (Figure 2.3: solid arrows). If all of the fossil taxa (including the taxon of interest) exhibit the osteological correlates of the soft tissues (or clear apomorphic transformations), then we have a sound basis for inferring and reconstructing these soft tissues in the fossil taxa (Figure 2.4). In situations where one or more clades of fossil taxa lack the osteological correlates, we may appeal to parsimony to determine the fate of the initial hypothesis.

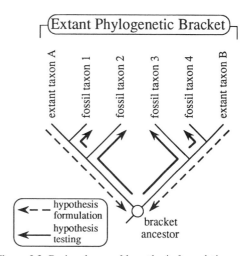

Figure 2.3. Basic scheme of hypothesis formulation and testing in the Extant Phylogenetic Bracket approach. Similarities between the components of the EPB are hypothesized as being present in the bracket ancestor (*broken arrows*). This hypothesis is tested for its congruence with the phylogenetic pattern by surveying the fossil taxa (*solid arrows*). See text.

Since osteological correlates and their presumably causally associated soft tissues are being mapped onto an existing cladogram, the EPB procedure involves the well-understood principles of a posteriori character optimization (see Swofford & Maddison 1987). The above method clearly resembles two-pass systems (Wiley et al. 1991) such as Farris optimization (Farris 1970): The downward pass involves formulating hypotheses about the bracket ancestor based on study of the extant taxa, and the upward pass involves hypothesis testing by surveying the fossil taxa for the specified osteological correlates. Optimization procedures allow determination of the character assessment at internal (ancestral) nodes, and in this case, the relevant node is the outgroup node (Figure 2.2a) because the assessment here decides whether or not the soft tissue can be inferred in the fossil taxon of interest. Three types of soft-tissue assessments at the outgroup node are possible (Figure 2.5). In the first case, both of the extant members of the bracket exhibit the soft tissue (and its causally associated bony features) that is suspected to occur in the fossil taxon; thus, the assessment at the outgroup node is decisive and positive (Figure 2.5a). In the second case, only one of the extant taxa has the suspected soft tissue and other extant outgroups lack it; here the assessment at the outgroup node is equivocal (Figure 2.5b). In the third case, neither extant taxon has the soft-tissue attribute suspected to occur in the fossil; thus, here the assessment is decisive and negative (Figure 2.5c). We will return to these three situations shortly

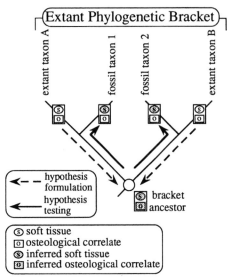

Figure 2.4. Cladogram showing the inference of soft-tissue attributes in fossil taxa using the EPB approach. Similarities in soft tissues and osteological correlates between the extant taxa are hypothesized as having been present in the bracket ancestor (*broken arrows*), as in Figure 2.3. Fossil taxa possess the osteological correlates, and thus the hypothesis of homology of the osteological correlates survives the congruence test. If soft tissues and osteological correlates indeed are causally associated, then soft tissue can be inferred in the fossil taxa with confidence.

because they provide a means of ordinating soft-tissue inferences.

As indicated earlier, the proposed method (and indeed optimization procedures) are based simply on the homology relation. The 1:1 correspondences within the EPB in causal associations between hard and soft tissues comprise the similarity test, allowing the formulation of a hypothesis of homology. Surveying the fossil taxa for the osteological correlates basically follows the congruence test of homology (Patterson 1982): For the hypothesis to be accepted, the 1:1 correspondences in soft and hard tissues hypothesized to be present in the bracket ancestor must be congruent with the phylogenetic structure of the EPB itself (i.e., found in the other descendants of the common ancestor). In other words, the 1:1 correspondences must characterize the monophyletic group comprised by the bracket. Even if a character passes the similarity test, if it fails to characterize a monophyletic group, then it is judged a homoplasy, not a homology. For example, the mammalian paranasal sinuses resemble somewhat those of archosaurs and so might pass the similarity test. However, paranasal pneumaticity is absent in nonarchosaurian sauropsids and basal synapsids, and thus on the basis of parsimony is judged as independent acquisitions in the clades leading to

mammals and to archosaurs because it does not characterize the monophyletic group Amniota (Witmer 1992).

For the sake of simplicity, the EPB approach will be briefly illustrated using an example in which the soft-tissue inference seems more or less obvious; the logic for more complex situations is the same (see Witmer 1992 for several examples relating to the soft-tissue relations of the antorbital cavity of archosaurs). What is the basis for inferring an eyeball in a fossil vertebrate, such as, say, *Tyrannosaurus rex*? The EPB of *T. rex* comprises birds and crocodilians. We know from numerous ontogenetic, experimental, and teratological studies that the eyeballs of extant birds and crocodilians are causally associated with specific osteological features (e.g., a complex cavity laterally within the skull communicating with the endocranial cavity via a foramen that conveys the optic nerve). The homology of these causal associations between birds and crocodilians (indeed among all vertebrates) is well established. Thus, our hypothesis is that the common (bracket) ancestor of birds and crocodilians possessed an eyeball with certain, specified osteological correlates. This hypothesis is tested by surveying other descendants of the bracket ancestor (i.e., fossil archosaurs, including *T. rex*) for these correlates. In this case, the osteological correlates are found in all fossil archosaurs, the hypothesis thus survives testing, and an eyeball may be inferred with confidence in *T. rex*. The simplicity of this example highlights the close links of the EPB approach to the homology relation and phylogenetic parsimony. The conclusion is that eyeballs and their osteological correlates are homologous in birds, crocodilians, and *T. rex*, and furthermore, since the assessment at the outgroup node is decisive and positive, it would be less parsimonious to argue differently. The stricture of causal association assures that the osteological features are indeed both necessary and sufficient to infer the soft-anatomical component in the extinct taxon.

Situations in which the initial hypothesis fails or is equivocal

The proposed method for reconstructing the soft-tissue attributes of extinct taxa may seem so rigorous that few hypotheses more complex than eyeballs in tyrannosaurs would survive. Is the method too stringent? Is there any way to save a hypothesis in which the assessment at the outgroup node is equivocal or even decisive and negative? The answer to the latter question lies in the researcher's willingness to speculate. It should be noted that the inference of soft tissues requires speculation even when the hypothesis survives all of the above tests. In many

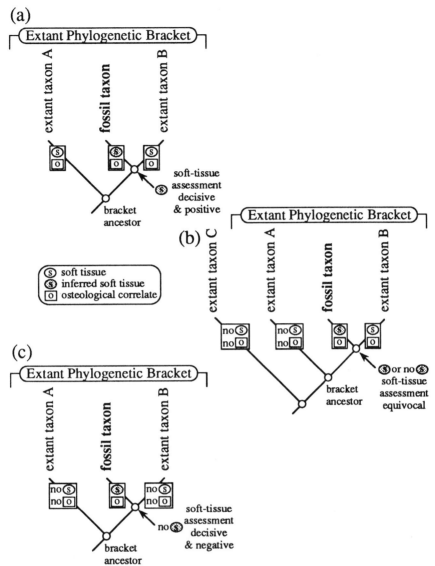

Figure 2.5. Assessment at the outgroup node and levels of inference. (a) Both extant components of the EPB possess the soft-tissue attribute (and causally associated osteological correlates) suspected to occur in a fossil taxon, leading to a decisive positive assessment at the outgroup node. (b) Only one of the extant components of the EPB has the soft-tissue attribute, leading to an equivocal assessment at the outgroup node. (c) Neither component of the EPB has the suspected soft tissue, leading to a decisive negative assessment. In (b) and (c), inference of the suspected soft tissue in the fossil may be justified if there is compelling morphological evidence. Reconstruction of the soft tissue in (a) requires a Level I inference, whereas reconstruction in (b) requires a Level II inference and in (c) a Level III inference.

respects, *any* inference of something that cannot be observed directly involves a measure of speculation (Rudwick 1964), that is, the inference of a brain within the cranial cavity of, say, *Homo erectus* is a speculation. Levels or degrees of speculation should be recognized based on the phylogenetic assessment at the outgroup node. Thus, the term speculation is not used here in its more common, pejorative sense, and implies no de facto absence of testability. One

important point of this chapter is that we need greater methodological rigor in order to determine the limits of our objective inferences – that is, to constrain, not completely eliminate, speculation. The EPB method makes the best, most economical use of the data at hand, and correctly identifies those soft-tissue inferences that are equivocal.

Furthermore, all speculation need not be idle. There are cases where the morphological clues from hard parts are so compelling that they would seem to point to the presence of specific soft tissues regardless of the condition in outgroups; I call this an argument of compelling morphological evidence (Witmer 1992). Bryant and Russell (1992, p. 406) regard this type of approach as extrapolatory analysis because it "depends on established biological generalizations" that allow the inference of attributes of extinct organisms by extrapolating from extant taxa. Bryant and Russell's (1992) extrapolatory analysis was explicitly intended to be separate from (but employed in parallel with) phylogenetic analysis. In the EPB approach, phylogenetics and compelling morphological evidence work together to give us a sense of the relative strength of our conclusions, that is, to ordinate our inferences.

Thus, a hierarchy of inference can be envisioned corresponding to the three possible assessments at the outgroup node that were noted earlier: (I) almost no speculation in situations where the EPB approach yields a decisive positive assessment at the outgroup node (Figure 2.5a, for example, eyeballs in tyrannosaurs), (II) more speculation when a compelling morphological evidence argument is advanced and the assessment is equivocal (Figure 2.5b), and (III) even more speculation when there is compelling morphological evidence and a decisive negative assessment (Figure 2.5c). Examples of soft-tissue inferences requiring speculations at levels II and III are presented below. In general, as long as the speculation is explicitly noted, situations involving levels II and III would seem to be admissible, and in many cases may be the only alternative short of ignoring the morphology altogether (Dodson et al. 1990). The notion of a hierarchy of inference is generalized and amplified in the Discussion.

A LEVEL II INFERENCE. This case occurs in situations where the soft tissue suspected to occur in a fossil taxon is found in its extant sister group but not in any other outgroups, leading to an equivocal assessment at the outgroup node (Figure 2.5b). In other words, the soft tissue feature indeed may characterize a monophyletic group including a living taxon but not the entire EPB of the fossil taxon. However, specific morphological attributes of the fossil may resemble those in the extant taxon so closely that, given the causal association of the attributes and soft tissues, the inference of the soft tissue in the fossil may be justified. For example, it has always been assumed that the Cretaceous bird *Ichthyornis* was feathered and capable of flight, suggesting that feathers characterize a more inclusive group than neornithine birds alone. However, applying the methodology yields an equivocal assessment with regard to feathers on *Ichthyornis* because one of the components of its extant phylogenetic bracket, crocodilians, lacks feathers. Nevertheless, *Ichthyornis* manifests so many of the morphological attributes causally associated with feathers and flight in extant birds that we may confidently clothe *Ichthyornis* with feathers, although this inference still involves a measure of speculation. (The obvious significance of *Archaeopteryx* in this example is discussed below.)

A LEVEL III INFERENCE. In other cases, the soft tissue may have appeared independently of any extant taxon but still can be inferred in a fossil taxon with a high probability using a similar argument of compelling morphological evidence (Figure 2.5c). For example, the presence of cheeks in ornithischian dinosaurs has been postulated on the basis of, first, distinctive morphology also found in mammals (e.g., medial emargination of the maxillary and dentary dentitions), and, second, the presumed functional necessity of such structures in animals that apparently masticated plant material (Galton 1973). However, application of the above methodology results in a decisive *negative* assessment because the EPB of ornithischians (birds and crocodilians) lacks cheeks, as do all other sauropsids. Nevertheless, many workers regard the morphological evidence as sufficiently compelling to reconstruct cheeks in ornithischians, probably because it is difficult to imagine how ornithischians could chew effectively without the assistance of cheeks. In fact, Galton (1973) suggested that the development of cheeks was instrumental (i.e., a key innovation) in not only the radiation of these herbivorous dinosaurs but also their competitive replacement of prosauropods. Such weighty paleobiological inferences, however, carry the *caveat* that they are founded on speculation – no matter how compelling the morphological evidence.

The significance of *Lagerstätten*

Fossil organisms known from *Lagerstätten* were mentioned earlier as often being easy cases for reconstructing soft tissues. However, they bear much greater significance for the methodology proposed here in that they can stand in some cases as "extant" taxa when composing the extant phylogenetic bracket. The method is basically the same, although the amount of morphological detail that can be extracted from fossils from *Lagerstätten* almost always will be less than for extant taxa. Nevertheless, these rare and unusual fossils can be critical in turning an equivocal assessment at the outgroup node into a decisive one. For example, the Cretaceous toothed bird *Hesperornis* has virtually no wings and was certainly a flightless diver. Despite its being part of the avian clade, what is our basis for reconstructing

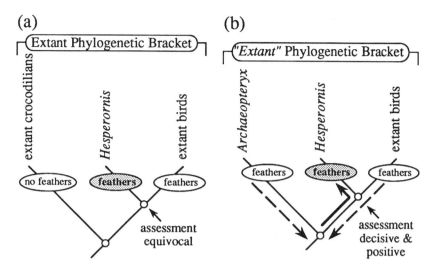

Figure 2.6. Cladograms showing the relevance of fossils known from *Lagerstätten* to the EPB approach. (a) Cretaceous flightless diving bird *Hesperornis* lacks compelling morphological evidence for the presence of feathers, and the EPB approach yields an equivocal assessment at the outgroup node. Feathers in *Hesperornis* are a Level II inference. (b) Adding *Archaeopteryx*, which is from a *Lagerstätten* and is known to have had feathers, the "E"PB approach yields a decisive positive assessment. Feathers in *Hesperornis* thus becomes a Level I inference.

Hesperornis with feathers? Given its EPB of neornithine birds and crocodilians, we have no strict basis for inferring feathers in *Hesperornis*: the presence of feathers is equivocal at the outgroup node (Figure 2.6a). If it had a good wing skeleton like its contemporary *Ichthyornis*, we might be tempted to clothe it in feathers on the basis of compelling morphological evidence (a Level II inference). Fortunately, we have *Archaeopteryx* that we know from multiple specimens from one of the most spectacular *Lagerstätten* (the Solnhofen lithographic limestone) had feathers. Thus, *Archaeopteryx* may stand as an "extant" taxon in the analysis such that the "extant" phylogenetic bracket of *Hesperornis* is *Archaeopteryx* and neornithine birds (Figure 2.6b). Proceeding through hypothesis formulation and testing, we now conclude with confidence that *Hesperornis* had feathers (i.e., a decisive positive assessment of feathers at the outgroup node – a Level I inference).

Summary of methodological considerations

An extant phylogenetic bracket allows formulation of hypotheses about the soft-tissue relations of extinct taxa that may be tested by reference to the known osteological correlates of the soft tissues in fossil taxa enclosed by the bracket. In cases where an assessment at the outgroup node is decisive and positive speculation is minimized – for example, when the two extant outgroups both exhibit the soft-

tissue attribute and form a doublet in the sense of Maddison et al. (1984). Speculation increases when the assessment is equivocal or when it is decisive and negative; in the latter case, inferring the soft tissue in the fossil may be regarded as unfounded. However, speculation may be fruitful (and even correct) if there is compelling morphological evidence for a particular soft-anatomical feature. Again, the point is to constrain speculation, not necessarily eliminate it.

DISCUSSION

The previous sections outline why paleontologists often need to look beyond fossil bones, and toward the soft-tissue functional matrices that influence bony morphology. Reconstructing the soft-anatomical properties of extinct organisms is inferential and by necessity involves a certain amount of speculation – such is the nature of historical science – and requires no apology. The EPB approach advocated here for obtaining this soft-tissue information draws heavily on the extant outgroups of the fossil taxon of interest. The next section addresses "a central worry" raised by Pagel (1991, p. 432) in a slightly different context; namely, the possibility that using extant taxa to bracket fossil taxa "guarantees that the animals that get constructed are a kind of average animal – an 'everyanimal' of sorts." The final section generalizes the EPB approach for any traits not normally preserved in fossil taxa.

Pagel's "everyanimal"

For many questions in evolutionary biology, the dichotomy between paleontology and neontology is not only artificial but actually may impede an investigator's arrival at appropriate inferences. Paleontology is traditionally the domain of only hard parts, whereas neontology theoretically can examine virtually any aspect of an organism. This dichotomy presents a problem for paleontologists because critical information on soft-tissue attributes is seemingly out of reach. Thus, to interpret extinct organisms correctly, we are challenged to recover whatever relevant soft-anatomical information we can from fossils. Gauthier, Kluge, and Rowe (1988) effectively demonstrated the importance of fossils in phylogeny reconstruction. The approach advocated here is in some respects the other side of the same coin in that it emphasizes the importance of extant taxa in paleobiology.

According to the EPB approach, confidently reconstructing an extinct organism with a particular soft tissue requires (in the simplest case) that both of its extant outgroups possess both the suspected soft-tissue element and its causally associated osteological correlates. Higher levels of inference are required in cases where one or both extant outgroups lack the specified soft tissue. Thus, Pagel's (1991, p. 532) concern seems justified in this case: "use of the present to reconstruct the past condemns the past to be like the present. Worse, perhaps, the past that we get from looking backwards is a very ordinary past, an average past."

What, then, are our alternatives, since Pagel offers none? As emphasized throughout, the EPB approach seeks not only to constrain our inferences and characterize levels of inference, but also to ground these inferences in organisms for which the data are richest – those living today. The EPB approach does not bar the recognition of the novelties of extinct taxa; rather, it allows us to assess critically the empirical basis of the inference. Level I inferences such as eyeballs in tyrannosaurs are relatively safe, whereas those requiring higher levels require more caution. We *should* feel better about inferring eyeballs in tyrannosaurs than cheeks in ornithischians. No one should (or would) deny that the orbit of *T. rex* housed an eyeball, but a researcher would be justified in questioning the existence of ornithischian cheeks and certainly Galton's (1973) claims about cheeks as key innovations.

To cite a more concrete example that was in fact the impetus for developing the EPB approach, debate about the function of the antorbital cavity (an opening in the snout) of fossil archosaurs has revolved around three major hypotheses: The cavity housed (1) a gland, (2) a jaw adductor muscle, or (3) a paranasal air sac. This is clearly a soft-tissue problem, and the EPB approach was applied to each of these hypotheses (Witmer 1992). The results showed that the bracket ancestor indeed possessed a nasal gland, a large jaw muscle, and an air sac, but only the paranasal air sinus produced osteological correlates that involved the antorbital cavity; thus, it was concluded that the function of the cavity in extinct taxa was to house an air sac (Witmer 1992). The point here is that although fossil archosaurs indeed were reconstructed to be similar to their living relatives, the EPB approach discriminated the hypothesis that most economically accounted for the data and yielded a robust result (i.e., a Level I inference). One is still free to argue for glands, muscles, or anything else, but to do so requires higher levels of inference and, since the approach is grounded in phylogenetics, more homoplasy (Witmer 1992).

The generality of the Extant Phylogenetic Bracket approach

Although the EPB approach was developed for the reconstruction of soft-tissue features of fossil taxa and soft tissues have been emphasized throughout, the approach can be generalized with little modification to any other attribute not normally preserved in the fossil record (e.g., function, behavior, some ecological parameters) but that has causally associated features (e.g., osteological correlates) that can be checked in extinct and extant animals alike. As mentioned, Bryant and Russell (1992) also presented a very similar general approach (see their paper for examples), and Weishampel (this volume) uses the EPB approach to probe functional inferences.

A special, yet important, case needs to be mentioned here. Although, the soft-tissue focus of the previous discussion emphasized the role of osteological correlates, not all attributes of organisms have reliable bony indicators. In the absence of such data, one may formulate a hypothesis about unpreserved traits on strictly phylogenetic grounds (i.e., presence of the trait in the bracket ancestor), but this hypothesis can be tested only by additional phylogenetic analysis and not by direct reference to fossil material. According to the protocol outlined earlier, inferring the trait in the bracket ancestor amounts to formulating a hypothesis of homology. Only when the trait has osteological correlates can this hypothesis of homology be tested directly for its congruence with the phylogeny of extinct and extant taxa alike. Furthermore, arguments of compelling morphological evidence obviously cannot be advanced without osteological correlates.

These limitations do not necessarily invalidate the

inference of those unpreserved traits that lack osteological correlates. However, such inferences must require additional speculation. The inference of a four-chambered heart (which lacks osteological correlates) in, say, *Hyracotherium* seems relatively safe, but strictly requires more speculation than the inference of a brain (which has such correlates). Thus, to be consistent, a hierarchy of inference parallel to the one outlined earlier can be constructed, again based on whether the assessment at the outgroup node is (I') decisive and positive, (II') equivocal, or (III') decisive and negative, but this time in the absence of osteological correlates. The levels of this hierarchy are given prime (') designations because it is not easy to ordinate all six levels. On the one hand, a Level I inference clearly is preferred to a Level I' because the former draws additional support from osteological evidence, and similarly a Level III' inference is probably untenable because it involves a decisive negative assessment and lacks osteological corroboration. On the other hand, it is unclear a priori if Level I' inferences are more robust than those at Level II, or if Level II' inferences are more robust than those at Level III. What remains clear, however, is that there are relatively rigorous means now available for inferring soft tissues and other normally unpreserved traits in extinct taxa, providing an explicit characterization of the level or amount of speculation that a researcher must invoke.

ACKNOWLEDGMENTS

Thanks are due to H.N. Bryant and A.P. Russell for friendly and productive discussion and correspondence and for supplying a preprint of their paper. Thanks are also due to D. Krause and S. Leigh for useful suggestions. The manuscript was read by D.B. Weishampel, N. Solounias, and an anonymous referee. The research presented in this chapter was part of my doctoral dissertation, undertaken at the Johns Hopkins University School of Medicine under the direction of D.B. Weishampel. I thank D.B. Weishampel, A. Walker, P. Dodson, N.C. Fraser, and R.L. Zusi for their discussions and for their critical comments on those portions of this chapter that are derived from my dissertation. Funding was provided by NSF Dissertation Improvement Grant BSR-9112070, the Alexander Wetmore Fund of the American Ornithologists' Union, Sigma Xi, and a fellowship from the Lucille P. Markey Charitable Trust.

REFERENCES

Atchley, W.R., & Hall, B.K. 1991. A model for development and evolution of complex morphological structures. *Biological Review* 66, 101–157.

Bakker, R.T. 1983. The deer flies, the wolf pursues: incongruencies in predator-prey coevolution. In *Coevolution*, ed. D.J. Futuyma & M. Slatkin, pp. 350–382. Sunderland: Sinauer Associates, Inc.

Bakker, R.T. 1986. *The Dinosaur Heresies*. New York: William Morrow.

Bellairs, A.d'A., & Kamal, A. 1981. The chondrocranium and development of the skull in Recent reptiles. In *Biology of the Reptilia*, Vol. 11, ed. C. Gans & T.S. Parsons, pp. 1–264. New York: Academic Press.

Benton, M.J. 1989. *On the Trail of the Dinosaurs*. New York: Crescent Books.

Benton, M.J. 1991. *The Rise of the Mammals*. London: New Burlington Books.

Born, G. 1883. Die Nasenhöhlen und der Thränennasengang der amnioten Wirbelthiere. III. *Gegenbaurs Morphologisches Jahrbuch* 8, 188–232.

Bryant, H.N., & Russell, A.P. 1992. The role of phylogenetic analysis in the inference of unpreserved attributes of extinct taxa. *Philosophical Transactions of the Royal Society of London B* 337, 405–418.

Bryant, H.N., & Seymour, K.L. 1990. Observations and comments on the reliability of muscle reconstruction in fossil vertebrates. *Journal of Morphology* 206, 109–117.

Carlson, B.M. 1981. Summary. In *Morphogenesis and Pattern Formation*, ed. T.G. Connelly, L.L. Brinkley, & B.M. Carlson, pp. 289–293. New York: Raven Press.

Carpenter, K., & Currie, P.J. 1990. *Dinosaur Systematics: Approaches and Perspectives*. Cambridge University Press.

Clutton-Brock, T.H., & Harvey, P.H. 1977 Primate ecology and social organisation. *Journal of Zoology* 183, 1–39.

Coulombre, A.J., & Crelin, E.S. 1958. The role of the developing eye in the morphogenesis of the avian skull. *American Journal of Physical Anthropology* 16, 25–37.

Cracraft, J. 1986. The origin and early diversification of birds. *Paleobiology* 12, 383–399.

Cracraft, J. 1990. The origin of evolutionary novelties: pattern and process at different hierarchical levels. In *Evolutionary Innovations*, ed. M.H. Nitecki, pp. 21–44. Chicago: University of Chicago Press.

de Beer, G.R. 1937. *The Development of the Vertebrate Skull*. London: Oxford University Press.

Dixon, D., Cox, B., Savage, R.J.G., & Gardiner, B. 1988. *The Macmillan Illustrated Encyclopedia of Dinosaurs and Prehistoric Animals*. New York: Macmillan Publishing.

Dodson, P., Coombs, W.P., Jr., Farlow, J.O., & Tatarinov, L.P. 1990. Dinosaur paleobiology. In *The Dinosauria*, ed. D.B. Weishampel, P. Dodson, & H. Osmólska, pp. 31–62. Berkeley: University of California Press.

Eldredge, N. 1986. Information, economics, and evolution. *Annual Review of Ecology and Systematics* 17, 351–369.

Eldredge, N., & Salthe, S.N. 1984. Hierarchy and evolution. In *Oxford Surveys in Evolutionary Biology*, Vol. 1, ed. R. Dawkins & M. Ridley, pp. 184–208. Oxford: Oxford University Press.

Farris, J.S. 1970. Methods of computing Wagner trees. *Systematic Zoology* 19, 83–92.

Galton, P.M. 1973. The cheeks of ornithischian dinosaurs. *Lethaia* 6, 67–89.

Gauthier, J.A., Kluge, A.G., & Rowe, T. 1988. Amniote phylogeny and the importance of fossils. *Cladistics* 4, 105–209.

Hall, B.K. 1983. Epigenetic control in development and evolution. In *Development and Evolution,* ed. B.C. Goodwin, N. Holder, & C.C. Wylie, pp. 353–379. Cambridge University Press.

Hall, B.K. 1988. The embryonic development of bone. *American Scientist* 76, 174–181.

Hall, B.K. 1990. Genetic and epigenetic control of vertebrate embryonic development. *Netherlands Journal of Zoology* 40, 352–361.

Hall, B.K. 1991. What is bone growth? In *Fundamentals of Bone Growth: Methodology and Applications,* ed. A.D. Dixon, B.G. Sarnat, & D.A.N. Hoyte, pp. 605–612. Boston: CRC Press.

Hall, B.K., & Herring, S.W. 1990. Paralysis and growth of the musculoskeletal system in the embryonic chick. *Journal of Morphology* 206, 45–56.

Halstead, L.B. 1991. Dinosaur Studies, Commemorating the 150th Anniversary of Richard Owen's Dinosauria. *Modern Geology,* Volume 16.

Hanken, J. 1983. Miniaturization and its effects on cranial morphology in plethodontid salamanders, genus *Thorius* (Amphibia, Plethodontidae): II. The fate of the brain and sense organs and their role in skull morphogenesis and evolution. *Journal of Morphology* 177, 255–268.

Harvey, P.H., & Pagel, M.D. 1991. *The Comparative Method in Evolutionary Biology.* Oxford: Oxford University Press.

Humphrey, T. 1971. Development of oral and facial motor mechanisms in human fetuses and their relation to craniofacial growth. *Journal of Dental Research* 50, 1428–1441.

Kamal, A.M., & Abdeen, A.M. 1972. The development of the chondrocranium of the lacertid lizard, *Acanthodactylus boskiana. Journal of Morphology* 137, 289–334.

Le Gros Clark, W.E. 1964. *The Fossil Evidence for Human Evolution.* Chicago: University of Chicago Press.

Liem, K.F. 1989. Functional morphology and phylogenetic testing within the framework of symecomorphosis. *Acta Morphologica Neerlandoscandinavica* 27, 119–131.

Maddison, W.P., Donoghue, M.J., & Maddison, D.R. 1984. Outgroup analysis and parsimony. *Systematic Zoology* 33, 83–103.

Moss, M.L. 1968. A theoretical analysis of the functional matrix. *Acta Biotheoretica* 18, 195–202.

Moss, M.L. 1971. Functional cranial analysis and the functional matrix. *American Speech and Hearing Association Reports* 6, 5–18.

Moss, M.L. 1972a. The regulation of skeletal growth. In *Regulation of Organ and Tissue Growth,* ed. R.J. Goss, pp. 127–142. New York: Academic Press.

Moss, M.L. 1972b. Twenty years of functional cranial analysis. *American Journal of Orthodontics* 62, 479–485.

Moss, M.L. 1981. Genetics, epigenetics, and causation. *American Journal of Orthodontics* 80, 366–375.

Moss, M.L., & Salentijn, L. 1969. The primary role of functional matrices in facial growth. *American Journal of Orthodontics* 55, 566–577.

Nicholls, E.L., & Russell, A.P. 1985. Structure and function of the pectoral girdle and forelimb of *Struthiomimus altus* (Theropoda: Ornithomimidae). *Palaeontology* 28, 643–677.

Noden, D.M. 1983. The role of the neural crest in patterning of avian cranial skeletal, connective, and muscle tissues. *Developmental Biology* 96, 144–165.

Noetzel, H. 1949. Ueber den Einfluss des Gehirns auf die Form der benachbarten Nebenhöhlen des Schädels. *Deutsche Zeitschrift für Nervenheilkunde* 160, 126–136.

Norman, D.B. 1985. *The Illustrated Encyclopedia of Dinosaurs.* New York: Crescent Books.

Norman, D.B., & Weishampel, D.B. 1985. Ornithopod feeding mechanisms: their bearing on the evolution of herbivory. *American Naturalist* 126, 151–164.

Oster, G.F., Shubin, N.H., Murray, J.D., & Alberch, P. 1988. Evolution and morphogenetic rules: the shape of the vertebrate limb in ontogeny and phylogeny. *Evolution* 42, 862–884.

Pagel, M.D. 1991. Constructing "everyanimal." Review of *Body Size in Mammalian Paleobiology,* edited by J. Damuth & B. MacFadden, Cambridge University Press, 1990. *Nature* 351, 532–533.

Pagel, M.D., & Harvey, P.H. 1988. Recent developments in the analysis of comparative data. *Quarterly Review of Biology* 63, 413–440.

Patterson, C. 1982. Morphological characters and homology. In *Problems of Phylogenetic Reconstruction,* ed. K.A. Joysey & A.E. Friday, pp. 21–74. New York: Academic Press.

Patterson, C. 1988. Homology in classical and molecular biology. *Molecular Biology and Evolution* 5, 603–625.

Paul, G.S. 1988. *Predatory Dinosaurs of the World.* New York: Simon & Schuster.

Paul, G.S., & Chase, T.L. 1989. Reconstructing extinct vertebrates. In *The Guild Handbook of Scientific Illustration,* ed. E.R.S. Hodges, pp. 239–256. New York: Van Nostrand.

Pinto, C.B., & Hall, B.K. 1991. Toward an understanding of the epithelial requirement for osteogenesis in scleral mesenchyme of the embryonic chick. *Journal of Experimental Zoology* 259, 92–108.

Ranly, D.M. 1988. *A Synopsis of Craniofacial Growth,* 2nd Edition. Norwalk: Appleton & Lange.

Rieppel, O. 1988. *Fundamentals of Comparative Biology.* Boston: Birkhäuser Verlag.

Ross, A.T. 1941. Cerebral hemiatrophy with compensatory homolateral hypertrophy of the skull and sinuses, and diminution of cranial volume. *American Journal of Roentgenology* 45, 332–341.

Rudwick, M.J.S. 1964. The inference of function from structure in fossils. *British Journal for the Philosphy of Science* 15, 27–40.

Savage, R.J.G., & Long, M.R. 1986. *Mammal Evolution: An Illustrated Guide.* New York: Facts on File.

Schumacher, G.-H., & Dokládal, M. 1968. Ueber unterschiedliche Sekundärveränderungen am Schädel als Folge von Kaumuskelresektionen. *Acta Anatomica* 69, 378–392.

Scott, J.H. 1957. Muscle growth and function in relation to skeletal morphology. *American Journal of Physical Anthropology* 15, 197–234.

Smith, M.M., & Hall, B.K. 1990. Development and evolutionary origins of vertebrate skeletogenic and odontogenic tissues. *Biological Review* 65, 277–373.

Spyropoulus, M.N. 1977. The morphogenetic relationship of the temporal muscle to the coronoid process in human embryos and fetuses. *American Journal of Anatomy* 150, 395–410.

Stevens, P.F. 1984. Homology and phylogeny: morphology and systematics. *Systematic Botany* 9, 395–409.

Swofford, D.L., & Maddison, W.P. 1987. Reconstructing ancestral states under Wagner parsimony. *Mathematical Bioscience* 87, 199–229.

Taylor, W.O.G. 1939. Effect of enucleation of one eye in childhood upon the subsequent development of the face. *Transactions of the Ophthalmology Society of the United Kingdom* 59, 361–371.

Thorogood, P.V. 1988. The developmental specification of the vertebrate skull. *Development* 103 Supplement, 141–153.

Van Valkenburgh, B. 1988. Trophic diversity in past and present guilds of large predatory mammals. *Paleobiology* 14, 155–173.

Wedden, S.E., Ralphs, J.R., & Tickle, C. 1988. Pattern formation in the facial primordia. *Development* 103 Supplement, 31–40.

Weishampel, D.B. 1984a. Evolution of jaw mechanisms in ornithopod dinosaurs. *Advances in Anatomy, Embryology and Cell Biology* 87, 1–110.

Weishampel, D.B. 1984b. Interactions between Mesozoic plants and vertebrates: fructifications and seed predation. *Neues Jahrbuch für Geologie und Paläontologie, Abhandlungen* 167, 224–250.

Weishampel, D.B. 1985. An approach to jaw mechanics and diversity: the case of ornithopod dinosaurs. In *Vertebrate Morphology,* ed. H.-R. Duncker and G. Fleischer, pp. 261–263. New York: Gustav Fischer Verlag.

Weishampel, D.B., & Norman, D.B. 1987. Dinosaur-plant interactions in the Mesozoic. In *Fourth Symposium on Mesozoic Terrestrial Ecosystems, Short Papers*, ed. P.J. Currie & E.H. Koster, pp. 167–172. Drumheller: Occasional Paper of the Tyrrell Museum of Palaeontology 3.

Weishampel, D.B., & Norman, D.B. 1989. Vertebrate herbivory in the Mesozoic: Jaws, plants, and evolutionary metrics. *Geological Society of America Special Paper* 238, 87–100.

Weishampel, D.B., Dodson, P., & Osmólska, H. 1990. *The Dinosauria.* Berkeley: University of California Press.

Wellnhofer, P. 1991. *The Illustrated Encyclopedia of Pterosaurs.* New York: Crescent Books.

Wiley, E.O. 1981. *Phylogenetics: The Theory and Practice of Phylogenetic Systematics.* New York: John Wiley and Sons.

Wiley, E.O., Siegel-Causey, D., Brooks, D.R., & Funk, V.A. 1991. *The Compleat Cladist: A Primer of Phylogenetic Procedures.* Lawrence: University of Kansas Museum of Natural History Special Publication 19.

Witmer, L.M. 1990. The craniofacial air sac system of Mesozoic birds (Aves). *Zoological Journal of the Linnean Society* 100, 327–378.

Witmer, L.M. 1992. *Ontogeny, Phylogeny, and Air Sacs: The Importance of Soft-Tissue Inferences in the Interpretation of Facial Evolution in Archosauria.* Unpublished Ph.D. Dissertation. Baltimore: Johns Hopkins University School of Medicine.

Witmer, L.M., & Rose, K.D. 1991. Biomechanics of the jaw apparatus of the gigantic Eocene bird *Diatryma*: implications for diet and mode of life. *Paleobiology* 17, 95–120.

Wolpert, L. 1983. Constancy and change in the development and evolution of pattern. In *Development and Evolution*, ed. B.C. Goodwin, N. Holder, & C.C. Wylie, pp. 47–57. Cambridge University Press.

Young, R.W. 1959. The influence of cranial contents on postnatal growth of the skull in the rat. *American Journal of Anatomy* 105, 383–415.

3
Fossils, function, and phylogeny

DAVID B. WEISHAMPEL

ABSTRACT

Understanding the function(s) of organ systems of extinct vertebrates in most cases has been based on biomechanical modeling. Because modeling is, at its root, analogy and therefore inherently ahistorical, this kind of approach to paleontological functional morphology can be seen as transcending phylogenetic boundaries among taxa. Studies using modeling (as beams, machines, and animal models) are the most common among functional morphologic research on extinct organisms. Two examples – jaw mechanics and locomotor mechanics in ornithopod dinosaurs – are discussed with respect to such modeling.

Function also has a historical context, which can be assessed by phylogenetic analysis. When function in extant organisms is treated as characters that have a phylogenetic disposition (i.e., plesiomorphic versus apomorphic distribution among particular extant taxa), extinct taxa can be bracketed within this phylogenetic hierarchy, and function can be assessed in these fossil forms on the basis of cladistic parsimony. Constriction in a fossil snake and food prehension in a fossil squamate are discussed in terms of this kind of phylogenetic approach.

While phylogenetic approaches based solely on extant organisms ignore any former functional attributes of fossils, these methods do include characteristics of presently living, fully functional animals. Modeling, on the other hand, gives extinct forms some say about their functional qualities, even though the reality of such function can never be directly tested. Thus, it is desirable that these two approaches be brought into conflict or combination whenever possible. When modeling and phylogeny have no intersection, modeling may be used to extend plesiomorphic function to deeper phylogenetic levels, while in turn phylogeny can provide boundary conditions on modeling parameters. When modeling and plesiomorphic function intersect, they may provide concordant or discordant results. When concordant, phylogeny and modeling can mutually reinforce each other, thus doubling support for particular functions in extinct organisms and for their evolutionary/ functional transformations. Discordance between modeling and phylogeny forces the investigator to make decisions

about the relative significance of modeled function and its phylogenetic implications for particular extinct taxa.

INTRODUCTION

Two boxing sika deer are obviously fully functional animals, while a bloated dead elephant no longer has any functional qualities whatsoever. And although not obviously so, an articulated skeleton of the ceratosaurian dinosaur *Coelophysis bauri* is a nonfunctional nonanimal. In this context, the problem for the paleontological functional morphologist is to bridge these two gaps, ultimately turning inanimate petrifactions into not just formerly living but also functioning organisms. This business of paleontological functional morphology is at the very least a partial fulfillment of Nicolaus Steno's recognition of the biological reality of fossils (Edwards 1967). Functional reconstitution of extinct organisms also places us in the position of expanding our understanding of the paleobiology of particular organisms and, if we are lucky, of discovering some of the subtleties of their evolutionary context.

This chapter will explore two somewhat disparate approaches to the problem of the functional morphology of extinct organisms and the means by which they combine or act as tests of each other. In each case, paleontological functional morphology will be examined first from an ahistorical, Newtonian mechanical perspective, and second from a historical, phylogenetic viewpoint. In this chapter, I will treat function in extinct organisms in a very general sense, one that also involves not only the function sensu stricto (following Lauder, this volume: the biomechanics of phenotypic features), but also behavior (i.e., the sum of the function of parts of an organism in a particular context; Lauder, this volume). For the case studies under consideration, it should be clear from the context of each whether I am referring to strict function or behavior.

The ahistorical approach

There are many variants on this kind of approach, which treats structure–function relationships as if they were free of historical legacy or phylogenetic constraint. Epitomized by Rudwick's (1964) paradigm method, this approach emphasizes analogies (with beams, machines, and animals as models/modern analogs, etc.) as universals from which to attribute functional qualities directly to extinct organisms. Because nearly all studies using analogy are implicitly ahistorical at their roots, such an approach has been applied without regard for phylogenetic boundaries, in much the same way as architectural statics and dynamics can cut across the stylistic traditions behind building construction. In paleontological functional morphology, studies that use analogy have a long and rich history, dating at least to Cuvier's conceptualization of correlation of parts, and they certainly have remained strong to the present day (see, e.g., Van Valkenburgh, in press).

A good deal of paleontological functional morphology is based on ahistorical models (often generated on principles of Newtonian mechanics) which are used in an attempt to "revitalize" extinct organisms or parts of them. For any particular model to be reasonable, it must be based upon explicit, well-understood parameters that come from engineering, biology, or other relevant sources. Predictions about the real world (both present and past) that come out of these exercises provide an indication of the robustness of the model. Such predictions can be used to test the accuracy of the model and to make further statements about the topic at hand. Some of these subsequent statements may be empirical in nature or may be heuristic in that they may lead to interpretations that would have at first appeared counter-intuitive.

Most often, functional morphologic modeling is based on machine analogies that are harnessed to understand the operation of a particular anatomical system (Weishampel 1993). As such, it is not surprising that many of these studies use actual physical models (e.g., Bock & Kummer 1968; Nobiling 1977; Pauwels 1980; Demes 1984; Demes, Preuschoft, & Wolff 1984). Other approaches include thought-experiments (Gingerich 1971; Hylander 1975), graphical representation and mathematical computation (Kripp 1933a; DeMar & Barghusen 1973; Anker 1974; Lombard & Wake 1976, 1977; Thomason & Russell 1986), and computer-based simulations (Elshoud 1980; Otten 1983, 1991; Weishampel 1984a). Many of these perspectives are discussed in later chapters in this volume.

The phylogenetic approach

Much less common than these modeling efforts, the phylogenetic approach to function in extinct organisms emphasizes extant organisms and what they do with their structures. Only a few such studies are yet available. Greene and Burghardt (1978) used such an approach to constriction in boid snakes and the extinct taxon *Dinilysia,* and I will outline another example of how feeding in living squamates might be used to infer food prehension in an extinct iguanian *Pristiguana.* By treating function as homology and projecting the functional features of extant organisms into the past, this approach provides a direct assessment of past functions from truly functioning, extant organisms (Greene & Burghardt 1978; Bryant & Russell 1992; Witmer 1992, this volume).

This use of phylogeny to assess paleontological functional morphology falls within an emerging research arena where unpreserved aspects of extinct organisms – including inferred functions – are interpreted along genealogical lines (Bryant & Russell 1992; Witmer 1992, this volume). As these authors argue, a phylogenetic approach requires first an explicit hypothesis of relationship of both extant and extinct taxa. From this phylogeny, functions in extinct organisms are inferred for extinct taxa by mapping and optimizing functions of extant taxa in a way that includes the extinct taxa.

In this chapter, I first focus on function(s) as consequences of ahistorical Newtonian mechanics. This discussion then turns to phylogenetic approaches to paleontological functional morphology. At the end of the chapter, I turn to the nature of both approaches, as well as to how they might be used to reciprocally illuminate function in fossils.

AHISTORICAL FUNCTIONAL MORPHOLOGY

Assessing the function in extinct organisms represents a difficult, yet important problem in the study of macroevolution (Hickman 1988). The extraction of functional information from fossils suffers from lack of critical information. What were the details of muscular anatomy and muscle-firing sequences? What were the loading regimes? In what behavioral context were structures utilized? For these and other questions, only plausible suggestions can begin to remedy the situation, but how are they to be tested? Biomechanical modeling of function may offer a solution to some of these otherwise intractable problems. Two examples include computer modeling of the skull in ornithopod dinosaurs to understand the evolution of mastication among these animals, and biomechanical modeling of femoral cross-sectional

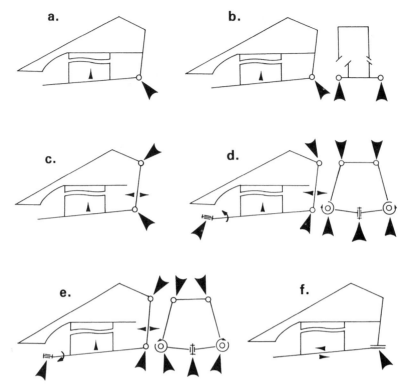

Figure 3.1. Kinematic abstractions (in left lateral and caudal views) of jaw mechanics models proposed for members of Ornithopoda. (a) Rotational mobility solely at the jaw joint, with isognathy (sensu Lambe 1920). (b) Rotational mobility solely at the jaw joint, with anisognathy (sensu Galton 1974). (c) Rotational mobility at the quadrate–squamosal and jaw joints, with isognathy (sensu Nopcsa 1900). (d) Rotational mobility at the quadrate–squamosal, mandibular symphyseal (note direction of arrow), and jaw joints (sensu Versluys 1923). (e) Rotational mobility at the quadrate–squamosal, mandibular symphyseal (note direction of arrow), and jaw joints (sensu Kripp 1933b). (f) Translational mobility of the jaw joint (sensu Ostrom 1961). Large arrows indicate sites of intracranial mobility; small arrows show possible directions of jaw movement. (After Weishampel 1984a.)

geometry to understand the ontogeny of locomotion in the dinosaur *Dryosaurus lettowvorbecki*.

Jaw mechanics in ornithopod dinosaurs

Ornithopoda comprises a monophyletic clade of dinosaurs consisting of more than 75 species dating from the Early Jurassic through the end of the Cretaceous and distributed on all continental masses (Weishampel 1990). There are three major subclades within Ornithopoda: heterodontosaurids, hypsilophodontids, and iguanodontians (Weishampel & Witmer 1990; Sues & Norman 1990; Norman & Weishampel 1990; Weishampel & Horner 1990; Weishampel, Norman, & Grigorescu in press). These animals represent an important group of herbivores during the Mesozoic (Weishampel 1984b; Norman & Weishampel 1985, 1991; Weishampel & Norman 1989) and studies of their biology have had a significant impact on our understanding of parental care, development, and social behavior among dinosaurs. Ornithopods have been the focus of a number of

studies of their food processing abilities (summarized in Norman 1984; Weishampel 1984a). Proposed jaw mechanisms range from simple jaw closure (Lambe 1920) to complex movement of the quadrate and partially independent longitudinal rotation of the lower jaws (Kripp 1933b), with many mechanisms intercalated in between (Figure 3.1).

In order to separate implausible from plausible mechanisms, ornithopod skulls were modeled as three-dimensional, multilink, chewing machines whose task was to break down plant food (Weishampel 1984a, 1993). This work, using computer-based kinematic analyses (Integrated Mechanisms Program; Uicker, Denavit, & Hartenberg 1964), was carried out on as wide an array of ornithopod taxa as available in order to assess the phylogenetic disposition of different jaw systems.

In these kinematic analyses, each ornithopod skull was broken down into sites of potential mobility among skull elements specified by a given mechanism. These sites of mobility were located in three-dimensional space and the kind of mobility (hinge joints, spheroidal joints, planar joints, etc.) was assessed prior to kinematic manipulation. Because each ornithopod skull could be abstracted kinematically and subjected to different kinds of biomechanical manipulation, each of these abstractions represents a possible model of ornithopod feeding. Figure 3.1 shows a few of the possible models that were subjected to testing. Obviously independent testing of these models is needed to stem the ever-increasing complexity of kinematic models. For these studies, tooth wear is that critical data base.

For all tetrapods that occlude their teeth, that part of the feeding cycle during which food items are broken apart begins with upper and lower teeth being brought into contact with the food by jaw closure. In the case of ornithopods, this occlusion appears to have been bilateral, with the upper and lower teeth meeting on both sides of the jaws at the same time. During the power stroke, these occluded teeth moved past one another, during which food items were subdivided prior to swallowing. The jaws were then opened again and the cycle continued. In all chewing vertebrates, the teeth are subjected to wear as food is fragmented. Some aspects of tooth wear – specifically wear striations, the configuration of the enamel–dentine interface, and facet angulation, among other features – provide evidence of relative tooth-to-tooth movement. This power-stroke direction in turn is ultimately controlled by the configuration of the skeletal framework of the jaw system. Thus, tooth wear can provide information that can be used to test a given model of jaw mechanics.

Turning back to the kinematic analyses, each modeled jaw system was analyzed for tooth-to-tooth movement that might occur during the power stroke. The kinds of tooth wear that might develop from this occlusal movement constituted specific predictions about how wear might be inscribed on the dentition during chewing for a modeled mechanism. These predictions were then tested against tooth wear on actual ornithopod dentitions.

Using this approach, two jaw mechanics models (Figure 3.2) were used to predict the kinds of tooth wear exhibited by actual ornithopod dentitions, which consisted of nearly transversely directed striations and enamel-dentine interfaces that indicated a "lower-dentition in/upper-dentition out" power stroke. Kinematically, euornithopods (that is, hypsilophodontids and iguanodontians) appear to have had jaw mechanisms that included hingelike rotation of the maxillae and associated elements against the premaxillae and skull roof (Figure 3.2a). This kind of intracranial mobility, termed pleurokinesis (Norman 1984), ultimately allows the production of a transverse power stroke during chewing (as indicated by tooth wear) through the lateral

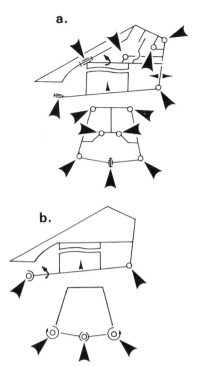

Figure 3.2. (a) Kinematic abstraction of pleurokinesis, the jaw mechanics model proposed for Euornithopoda. (b) Kinematic abstraction of the jaw mechanics model proposed for Heterodontosauridae. Large arrows indicate sites of intracranial mobility; small arrows show possible directions of jaw movement. (After Weishampel 1984a.)

rotation of the maxillae in concert with continued adduction of the lower jaws.

Pleurokinesis is not found in Heterodontosauridae, the remaining ornithopod clade. In this group, transverse chewing was achieved by a slight rotational mobility of the lower jaws that was facilitated by a modified symphyseal joint (Figure 3.2b; see also Crompton & Attridge 1986; Norman & Weishampel 1991). This system seems to be restricted solely to these basal ornithopods.

Both kinds of jaw systems, pleurokinesis and heterodontosaurid, were revealed through ahistorical modeling based on Newtonian mechanics. Similarly, locomotor mechanics in extinct organisms can also be assessed using similar approaches.

Locomotion in an iguanodontian ornithopod dinosaur

Dryosaurus lettowvorbecki is an iguanodontian ornithopod dinosaur known from abundant skeletal material from a monospecific bonebed in rocks of Late Jurassic age near Tendaguru, Tanzania. This material comprises a reasonably dense growth series, which in turn provides the basis for a study of

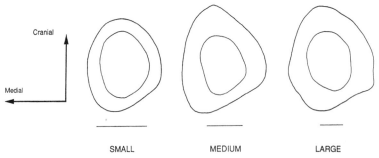

Figure 3.3. Cross-sectional shape of representative femora of the three size classes that come from the *Dryosaurus lettowvorbecki* quarry of Tendaguru, Tanzania. Scale = 1 cm. (From Heinrich et al. in press, with permission.)

the ontogeny of locomotor biomechanics in this species (Heinrich, Ruff, & Weishampel in press). More specifically, the study attempts to relate ontogenetic changes in the cross-sectional geometry of a suite of femora to the three-dimensional bending stresses imposed on the element during normal locomotion. Consequently, we rely heavily on the ahistorical precepts of beam theory (see Thomason, this volume). The ahistorical nature of this study – and many others like it – stems from the way bone appears to respond to changes in mechanical strain in a precise, systematic, and taxon-independent manner (Woo et al. 1981; Alexander et al. 1984; Lanyon 1984; Lanyon & Rubin 1985). It is well known that the basic geometry of bony elements is altered when subjected to both normal (i.e., growth-related) and abnormal loadings (Goodship, Lanyon, & McFie 1979; Lanyon et al. 1982; Rubin & Lanyon 1985; Biewener, Swartz, & Bertram 1986). The geometric distribution, amount, and material properties of bone within a load-bearing element are thought to reflect the magnitude and orientation of the mechanical loadings acting on it (Lanyon & Rubin 1985). The resultant cross-sectional morphology of a load-bearing element is thought to capture this unique interaction (see Thomason, this volume, for further discussion). Beam theory is used to relate cross-sectional geometry to load; that is, long bone diaphyses are modeled as hollow beams (Timoshenko & Gere 1972; Ruff & Hayes 1983 and references cited therein).

Cross-sectional geometries, then, provide information on the mechanical loadings that may have acted on long bones in vivo (Ruff & Hayes 1983), and an ontogenetic series of cross sections provides a means of reconstructing the pattern of loadings that accompanied growth and development. If young animals maintain similar weight distributions and utilize the same locomotor gaits as mature animals, then we might expect cross-sectional properties of bone to change in some linear proportion to increases in body weight, all other things being equal. Conversely, if the pattern of change in cross-sectional morphol-

ogy deviates significantly from such unidirectional trends, factors other than increasing size are likely to affect femoral loadings and modifications to bone architecture. These then are the relationships we attempted to assess in the Tendaguru sample of *D. lettowvorbecki*.

Geometric data, obtained from 27 femora with naturally broken cross sections, were obtained by digitizing endosteal and periosteal perimeters. Several biomechanical statistics were calculated from these perimeter data (using a modified version of the computer program SLICE; Nagurka & Hayes 1980): cross-sectional area, cortical area, second moments of area (estimates of resistance to bending loads) about x- and y-axes (I_x, I_y), maximum and minimum second moments (I_{max}, I_{min}), and the angle expressing the disposition of greatest bending strength relative to the mediolateral plane. All provide information on the general orientation of bone distribution and its resistance to bending forces.

Based on the values calculated using SLICE, we found that ontogenetic growth of the femur in *D. lettowvorbecki* is accompanied by adjustments in both the relative amount and distribution of bone (Figure 3.3). Going first from small to medium-sized classes, the cross-sectional properties of medium-sized femora differ most noticeably from those of small femora in two ways. First, there is a 33 percent increase in area of cortical bone, which corresponds to femoral lengthening from 150 to 180 mm and a doubling of body weight from an estimated 9 to 17 kg. Second, the cross-sectional shape of the femur is modified, as indicated by an increase in the ratios I_{max}/I_{min} (from 1.269 to 1.444) and I_x/I_y (from 1.208 to 1.307).

Additional differences in the cross-sectional properties are associated with the transition between femora of medium and large size. Although percentage cortical area remains constant over almost a doubling in femoral length and an estimated 50 kg increase in body weight, I_{max}/I_{min} ratios decrease, from 1.444 to 1.241, as did I_x/I_y ratios, from 1.307 to 1.016. Taken as reflecting in vivo, biomechanical

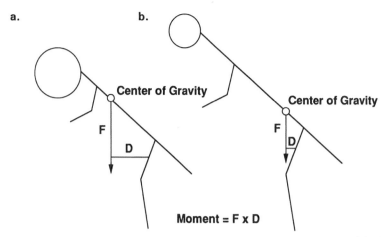

Figure 3.4. Schematic drawing of a hatchling (a) and a fully adult *Dryosaurus lettowvorbecki* (b), showing the biomechanical effect of the position of the center of gravity on the bending stresses incurred by the femur. (After Heinrich et al. in press.)

modifications during ontogeny, both changes suggest a more mediolateral orientation in greatest bending strength.

Increases in both relative amount of bone and maximum bending strength across the small-to-medium-size transition suggests that loadings on the femur had increased significantly and rapidly, but decreased from medium to large size. We attempted to relate these changes to shifting limb posture and locomotor grade or style. If young, subadult, and adult *D. lettowvorbecki* all used similar stance and locomotor styles, then we also expect our measures of femoral bending strength to change unidirectionally as earlier indicated. This is clearly not the case in the comparisons we made. Percentage cortical area increases dramatically in relatively small animals, whereas subsequent increases of over 300 percent in body mass (20–70 kg) produce no increases in the relative amount of bone (i.e., percent cortical area). Instead, as we have noted, cross-sectional geometry as measured by I_{max}/I_{min} and I_x/I_y ratios exhibit not a linear but parabolic pattern of change (low in small and large individuals, high in medium-sized individuals), which suggests that relative bending loads do not continue to increase with increasing size. Increasing body size alone, therefore, seems an unlikely explanation for the observed architectural changes in cross-sectional morphology.

In the place of ontogenetic size increase, we looked to the effects of an ontogenetic shifting of the center of gravity. Our clue to look for effects from changes in the center of gravity comes from the relatively large heads and small tails of embryonic and hatchling *Orodromeus makelai* and *Maiasaura peeblesorum*, two other euornithopod species (no embryonic or hatchling material is known for *D. lettowvorbecki*). These body proportions, unlike

those of fully adult individuals, dictate that the center of gravity was far forward of the hip and that with growth (and relative reduction of head size) this position would shift caudally during development. This pattern is illustrated schematically in Figure 3.4. For a bipedal dinosaur, the farther the center of gravity is from the hip joint, the larger the moment acting about the femur and the greater the bending stress on the element. We first considered the case of a bipedal hatchling. Since the center of gravity was farthest cranial in these smallest individuals, bending loads should have been relatively greater than at any other stage of ontogeny. This proposition was easily rejected, since none of our cross-sectional trends support it. We next turned to hatchlings with a quadrupedal stance, under the supposition that a center of gravity positioned well in front of the hip may have made habitual bipedality difficult, if not impossible, for hatchlings.

Shifting from quadrupedal locomotion to bipedality would essentially double the forces acting on the hindlimb, unless there were some sort of special mechanism that might compensate for these forces. Because we could find none, it is likely that bending and compressional stresses would increase significantly in the bones of the limb, requiring more normal kinds of structural compensation to maintain acceptable safety factors. We believe that what we are identifying over the small to medium-size transition is increased bending stresses requiring compensation in the form of an increase in relative bone mass and an increase in the relative magnitude of greatest bending strength. Since neither increases with increasing length among small femora, the abruptness of these changes in cross-sectional properties fits a hypothesis of novel mechanical loadings acting on the femur at a specific body size.

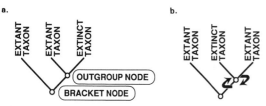

Figure 3.5. (a) Phylogenetic relationships of an extinct taxon and its first two extant outgroups. (b) Free rotation around the outgroup node to produce the extant phylogenetic bracket of the extinct taxon. (After Witmer 1992, this volume.)

In fact, we regard this relatively rapid transition as involving a threshold effect of combined body and tail size increases which effectively shifted the center of gravity to a position closer to the hip, thus making bipedal posture possible.

Heinrich et al. (in press) go on to discuss later modifications in femoral architecture after the small-medium size transition, as well as boundary conditions to growth rates and several aspects of life history strategies in *D. lettowvorbecki,* but the point to be emphasized here is the same as with ornithopod jaw mechanics: The Newtonian world of biophysics, particularly at the level of modeling the workings of the musculoskeletal system, can be (and is regularly) applied to estimate function in extinct organisms. Yet there is another approach to function and fossils.

HISTORICAL FUNCTIONAL MORPHOLOGY

Phylogeny constitutes the only direct link between extant, functioning organisms and their extinct, nonfunctioning forebears. It therefore should be possible to use phylogenetic relationships to assess function in extinct organisms. Bryant and Russell (1992) and Witmer (1992, this volume) all show that we must begin with an explicit hypothesis of phylogenetic relationship of both extant and extinct taxa in order to see how phylogenetically distant extant taxa can inform about unpreserved attributes – including function – in extinct taxa. Bryant and Russell (1992) called this relationship encompassing the extinct and extant taxa a closed community of descent, and Witmer (1992, this volume) termed it the Extant Phylogenetic Bracket (EPB). Thus, if we wanted to assess aquatic locomotion in *Pinnarctidion bishopi,* an extinct pinniped from the Oligocene–Miocene transition of California (Barnes 1979; Berta 1991), we need to position this species within its EPB (using Witmer's terminology) whose extant members can inform about the evolutionary distri-

bution of locomotor function (i.e., limb kinematics and swimming within Pinnipedia).

As noted by Witmer and Bryant and Russell, phylogenetic analysis of functions in extinct taxa first requires the identification of those living taxa that comprise the first two extant outgroups of the fossil taxon of interest (following Maddison, Donoghue, & Maddison 1984). The first outgroup – the immediate living outgroup of the extinct taxon – and the extinct taxon itself are linked together at what has been called the outgroup node. The second outgroup to the extinct taxon serves to interpret polarity at this outgroup node, forming what Witmer (1992, this volume) calls the bracket node (Figure 3.5). As he further also points out, from a heuristic perspective, the free rotation of the outgroup node positions the extinct taxon within the "confines" of the two outgroup taxa. It is this rearrangement at the outgroup node that produces the EPB of the extinct taxon.

The next job is to assess how function can be treated as homology and to determine its distribution among the taxa under consideration. "Function as homology" may sound like unlikely bedfellows but, as is becoming clear, individual function can have plesiomorphic and apomorphic distributions (viz. Lauder 1990 and references therein). That is, we can speak of preferred body temperature, muscle activity patterns, and pattern of application of coils in constriction as having uniquely evolved (i.e., are apomorphic) during phylogeny as surely as we can say that powered flight has evolved at least three times among vertebrates. No matter how we identify functional characters as homologies, they must pass the test of phylogenetic congruence by characterizing monophyletic taxa (Patterson 1982). If they do so, then the functional characters can be hypothesized to have been present minimally in the common ancestor of the extant organisms and in all remaining descendants including the fossil forms.

The pattern of acquisition of functional characters can be assessed using what is known as character optimization, which is overlaid on a cladogram for the organisms under consideration (Farris 1970; Swofford & Maddison 1987; Wiley et al. 1991). Character optimization is a two-step technique in which morphological, functional, ecological, behavioral, and other kinds of characters are first mapped onto a cladogram on the basis of their link to terminal taxa and then generalized downward toward the base of the tree. Then, moving up from the basal node, ambiguities in successive node values are resolved by reference to the value of the node directly below it. For equally shortest phylogenetic trees, it then becomes possible to identify areas of ambiguities that come from interpretations of character reversals and convergences. In either case, optimization will yield the best supported sequences of evolutionary transformations for the characters at hand.

Such a phylogenetic approach necessarily treats extinct taxa as "slaves" to extant taxa. As such, optimization of functional characters to include extinct taxa is not influenced by what the fossils themselves might have to reveal about their own functional condition. Instead, it is the strict flow of functional information from organisms that themselves are capable of functioning. The separation of function and morphology obviously eschews osteological correlates of function (sensu Witmer, this volume) for the following reasons. First, I hope to make pure cases of how modeling and phylogeny each can be used to ascribe function to fossils and that can be done only through isolated treatments. Second, I want to insure, at least from a methodological perspective, that functional (and behavioral) characters are free to vary with respect to the pattern of acquisition of morphological features (viz. Lauder 1990, 1991, this volume). By their very nature, osteological correlates do not automatically have such a relationship with function. Instead, form and function must closely coevolve for osteological correlates of function to exist. Yet this pattern of coevolution means that the pattern of acquisition of osteological characters (based on both extant and extinct taxa) be congruent with the pattern of acquisition of functional characters (knowable only in extant taxa). We are still stuck with extinct organisms which, because they do not *do* anything, are unable to directly contribute information about their functional qualities. Nevertheless, it would be foolish to suggest that such correlates do not exist (however they might be discovered). If known, they would certainly be a great aid in inferring function to extinct organisms (see Witmer, this volume).

Examples of how a purely phylogenetic approach to the problem of inferring function from extant to extinct organisms are detailed in the next two sections, leaving aside until the final section of this chapter the issue of how modeling and phylogenetic approaches to interpreting function in extinct organisms might be brought into intersection.

Constriction in the extinct snake *Dinilysia*

Although there are few works that use phylogeny to approach function in extinct species, the study on the phylogeny of constriction by Greene and Burghardt (1978) provides an good example. This work poses a number of questions, among them: How did *Dinilysia patagonica,* a Late Cretaceous snake from Argentina (Woodward 1901; Estes, Frazzetta, & Williams 1970), dispatch its prey? Using a phylogenetic approach, the study begins with a survey of patterns of constriction among extant snake

taxa thought to bracket *Dinilysia.* These taxa (Acrochordidae, Aniliidae, Boidae, Colubridae, and Xenopeltidae; Alethinophidia sensu Rieppel 1988) amount to nearly 150 species. Remaining snakes (Scolecophidia) do not constrict. Greene and Burghardt subdivided constriction into 27 combinations of individual modal action patterns. These are based on (1) application movement (winding or wrapping), (2) presence or absence of initial twisting, (3) anterior or posterior coil composition, and (4) horizontal, vertical, or inclined long axis of the coiling. From these, the authors searched for similarities in within-species coiling patterns.

All extant alethinophidians use a single constriction pattern to subdue their prey (Figure 3.6): winding, which uses the anterior part of the snake's body; the coils are horizontal (i.e., parallel to the ground); and the cranial part of the snake's body has an initial twist. Colubrids also use this pattern of constriction, as well as several other combinations. Treated as homologous functional complexes, these constriction patterns were then mapped against the phylogenetic relationships then available for snakes. Perhaps not surprisingly, the winding, anterior, horizontal coil, initial twist pattern was identified as primitive for the Alethinophidia clade.

We now turn to the impact of the phylogenetic disposition of constriction patterns to the fossil snake *Dinilysia.* As indicated by Greene and Burghardt (1978), Acrochordidae and a taxon formed of Aniliidae + Boidae + Xenopeltidae comprise the EPB of *Dinilysia* (Figure 3.7a). This diagram can be converted to a cladogram (Figure 3.7b) based solely on the proximity of branching in Figure 3.7a, and onto which constricting patterns can be qualitatively optimized. Plesiomorphic for this clade (therefore present at the bracket node), winding from anterior, horizontal coils, and an initial twist must also bracket *Dinilysia.* Thus, from an EPB perspective, *Dinilysia* probably dispatched its prey using such a killing tactic. Said another way, retention of this plesiomorphic condition in the extinct species represents the most cladistically parsimonious interpretation of the evolutionary pattern of constriction.

The same cannot be said for identifying constriction patterns in *Dinilysia* using Rieppel's (1988) cladogram (Figure 3.8; see also Kluge 1991). Serpentes in this case becomes the bracket node for *Dinilysia,* with Alethinophidia as its sister taxon and Scolecophidia as its next outgroup. Members of Scolecophidia do not constrict at all (Greene, personal communication) and neither do either of the two likely candidates for the next outgroup (and thus the sister taxon to Serpentes), Scincomorpha or Anguimorpha. In all, *Dinilysia* falls outside the EPB comprising constricting snakes and therefore – based on this phylogeny alone – cannot be unambiguously

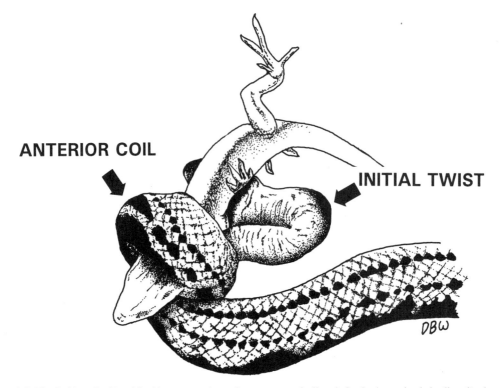

Figure 3.6. The boid snake *Tropidophis greenwayi* preying on an anole lizard. Snake has seized the lizard's thorax, applied its coils by winding and using the anterior part of its body. Snake's coils are horizontal (i.e., parallel to the ground) and cranial part of its body has an initial twist. This constriction pattern is found in all alethinophidian snakes. (Drawing made from a photograph courtesy of H. Greene.)

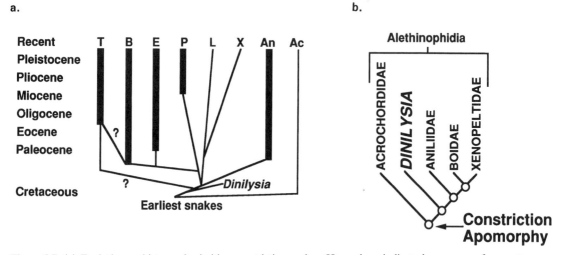

Figure 3.7. (a) Evolutionary history of primitive constricting snakes. Heavy bars indicate known age of separate lineages. Taxa: T, Tropidophiinae; B, Boinae; E, Erycinae; P, Pythoninae; L, Loxoceminae; X, Xenopeltinae; An, Aniliidae; Ac, Acrochordidae. (From Greene & Burghardt 1978; with permission.) (b) Conversion of (a) into a cladogram, with constriction mapped and optimized onto it.

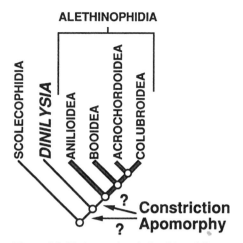

Figure 3.8. Phylogenetic relationships of Serpentes, including *Dinilysia*, according to Rieppel (1988), with constriction mapped and optimized onto it.

identified as having a constricting prey-dispatching behavior.

Lingual food prehension in *Pristiguana*, an extinct squamate

Like the *Dinilysia* example, a case can be made for the use of phylogenetic approaches to aspects of tongue movement and feeding in extinct squamates. The following represents a combination of systematic studies by Estes, de Queiroz, and Gauthier (1988), and Schwenk's (1988) work on lingual kinematics and phylogenetics in living members of Squamata (see also Schwenk & Throckmorton 1989). I point out that neither study area has been used to explore function in fossils. For this, I will discuss feeding behavior in a reasonably well-known extinct squamate *Pristiguana brasiliensis*. This species, from the Late Cretaceous of Brazil, was described by Estes and Price (1973) as an iguanid more primitive than the most primitive living members of the clade. Although not strictly true, I treat this species as in fact the most basal member of Iguania for the sake of the following arguments.

In order to use phylogeny to assess feeding in *Pristiguana*, we need to know the distribution of functional homologues among its extant relatives. Among lepidosaurs (Figure 3.9a; Estes et al. 1988; Schwenk 1988), the closest extant outgroup taxon is the clade consisting of remaining iguanians. Next is Scleroglossa (Gekkota, Anguimorpha, Scincomorpha, and probably also snakes), and more distantly is *Sphenodon*. Most extant lepidosaurs feed using lingual prehension and have a fleshy tongue (Gorniak, Rosenberg, & Gans 1982; Schwenk 1988; Schwenk & Throckmorton 1989). Only in scleroglossans are the jaws alone used in food ingestion.

Looked at in detail, lingual ingestion is begun when the animal first locates a food item and the jaws begin to be opened (Gorniak et al. 1982; Schwenk & Throckmorton 1989). Already in its characteristic ventrally curled and dorsally arched configuration, the tongue moves forward as the jaws begin to part. As gape reaches approximately a third of its maximum, the anterodorsal part of the tongue clears the margin of the jaws. It continues to protrude beyond the mandible at a fairly constant rate while the jaws increase their gape. Midway toward maximum gape, jaw movement pauses, and tongue movement slows as it reaches maximum protrusion and contact with the food item (contact occurs on the anterior third of the dorsal surface of the tongue).

Gape increases rapidly again immediately after contact. During this time, there is a simultaneous retraction of the tongue. The jaws are finally closed as soon as the tongue and food have been drawn into the oral cavity. Oral transport involving additional tongue movement and unilateral, alternating biting over a variable number of cycles follows thereafter.

This interplay between jaw and tongue movement can be illustrated using kinematic profiles (Figure 3.10). Lingual prehension and a single transport cycle involves a common sequence of jaw movement: slow opening, fast opening, fast closing, and a slow close-power stroke phase. It is during slow opening, particularly the pronounced slow open-II phase that the tongue protrudes to its maximal extent and food is stuck to its tip. Tongue retraction occurs during the fast-open phase, while transport and any oral preparation occurs during fast and slow close phases.

This pattern of lingual prehension is common to all living members of Iguania, the first extant outgroup to *Pristiguana*. However, it is not found in the next available outgroup, Scleroglossa, which ingests solely by jaw prehension. Here then we have identified an ambiguous plesiomorphic condition for Squamata. Since it might involve either tongue or jaw prehension, we have no clear sense of the functional state of the bracket node for *Pristiguana*. Therefore it is appropriate to turn to the next available extant outgroup, Sphenodontida or more specifically *Sphenodon* itself. Here we again encounter lingual prehension of the kind seen in living iguanians. Optimizing this information up the cladogram, it is possible to resolve the condition at the Squamata node, with plesiomorphic retention of lingual ingestion characterizing Iguania and jaw prehension apomorphically characterizing Scleroglossa.

On the basis of the EPB resolution of ingestion characters among extant lepidosaur taxa alone, we might expect that *Pristiguana* had a fleshy tongue and that it ingested food items using lingual prehension (Figure 3.9b). Further, this lingual prehension is likely to have exhibited the same slow open, fast

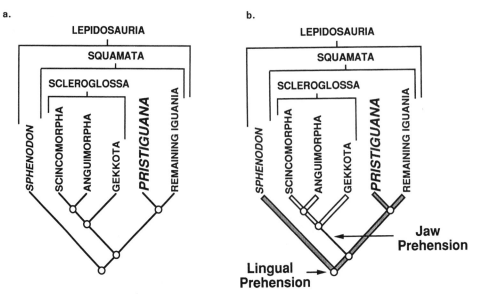

Figure 3.9. (a) Phylogenetic relationships of Lepidosauria, including *Pristiguana* (After Estes et al. 1988; Schwenk 1988.) (b) Lingual and jaw prehension mapped and optimized onto the Lepidosauria cladogram.

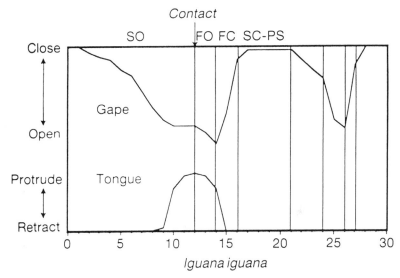

Figure 3.10. Kinematic profile of interplay between jaw and tongue movement in *Iguana iguana*. Notations: contact, tongue–prey contact; FC, fast close; FO, fast open; SC-PS, slow close-power stroke; SO, slow open. Cine-film frame numbers are shown along horizontal axis. Films were taken at 48 frames/second. (From Schwenk & Throckmorton 1989)

open, fast close, and slow close-power stroke phases, with tongue protrusion accompanying a pronounced slow open-II phase at large gape distance.

DISCUSSION

In looking back over these two approaches – phylogeny and modeling – there are some obvious advantages and limitations to what each can and cannot

reveal about function in fossils. Because phylogenetic approaches fall under the more general issue of historical analysis of function sensu Lauder (1981, 1990, 1991), extant phylogenetic bracketing brings about direct contact between extant, functioning organisms and those known solely from the fossil record. By identifying patterns of evolutionary transformation of function, phylogenetic analyses also identify how deeply nested a given function is within a clade. At the same time, extinct organisms nested within this hierarchy can be claimed to also have exhibited the

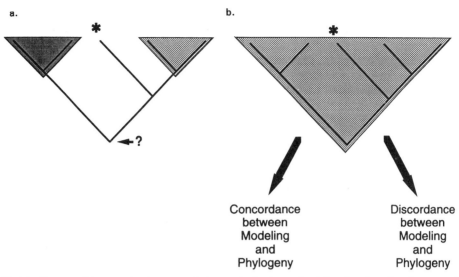

Figure 3.11. (a) Absence of intersection between phylogenetically determined function (i.e., function not sufficiently plesiomorphic to bracket extinct taxon) and that available from modeling for an extinct organism (indicated by asterisks). (b) Complete intersection between function as determined through phylogeny and modeling. Concordance between phylogeny and modeling yields double support for identification of function in the extinct taxon (indicated by asterisk). Discordance between phylogeny and modeling yields conflict between modeled function and that identified from plesiomorphic distribution of extant terminal taxa.

plesiomorphic function as we have seen in the *Dinilysia* and *Pristiguana* examples.

Such a view, it might be objected, makes the past little more than a lock-step version of the present, with fossils deprived of evidential claims about their own function. Instead, ascribing function to them on the basis of extant organisms is done in order to maintain the most parsimonious character distribution on the cladogram that contains them. A strictly phylogenetic perspective of the functional qualities of extinct taxa also places stringent limits on which taxa may be analyzed: They must be members of a relevant crown group whose bracketing also resolves functional homologs. When the EPB of a fossil taxon yields an unresolved ancestral functional condition (i.e., extant functional homologies are not deeply enough nested), the possibility of a phylogenetic approach based solely on extant taxa is ruled out (Figure 3.11a). For the latter case, where phylogenetic assessment of function at the bracket node is equivocal, many important structure–function transitions may well be out of reach from a phylogenetic perspective. Lack of sufficiently deep phylogenetic nesting of functional homologies requires what Witmer (this volume) terms Level II inference (also referred to as Category 2 comparisons by Bryant and Russell [1992]). For example, using the Gauthier, Kluge, and Rowe (1988) cladogram for Tetrapoda, the functional details of the origin of terrestrial locomotion might be deemed equivocal and hence lost from purely phylogenetic considerations (although see Fricke et al. [1987] for the possibility that limb cadence may be homologous at least at the level of Sarcopterygii).

When extant function does not bracket a fossil taxon, as in this example as well as for pleurokinesis in ornithopod dinosaurs, locomotion in *Dryosaurus,* powered flight in pterosaurs, or bipedal locomotion in australopithecines (viz. McHenry 1991; Bryant & Russell 1992; Witmer this volume), we must turn to modeling as the only reasonable means to "draw down" function to the level of the extinct taxon. These modeling approaches, when also treated in isolation, represent the paleontological consequences of Lauder's (1981, 1990, 1991) equilibrium or biophysical analyses. By calling upon environmental equilibrium and ecological correlates to explain biological form, the functional context of features represent the solution end of questions about ecological problems and organismal solutions. With this in mind, modeling can be viewed as a double-prong tool for understanding function in both extant and extinct organisms. For the former, it comprises a means to understand biomechanical construction and its parameters. Because living organisms provide the only recourse for testing a given model, these exercises have tended to draw their strength from analyses of extant taxa (with animal models constituting the epitome of models as synonymous with extant organisms). In a strict sense, models need not rely on such support, because as analogy these approaches are ahistorical and this is how modeling of jaw mechanics in ornithopods and locomotion in *Dryosaurus* have been employed here. However, in practice, testing against extant taxa is usually seen as prerequisite for assessing function in extinct organisms (cf. the quadric crank model of "lizard" intracranial mobility; Frazzetta 1962; Smith & Hylander 1985; Condon 1987). A model that passes the test of the recent is then used to evaluate function in the past. Although there are no "guarantees" to modeling function in extinct organisms (which is itself untestable), it does provide the forum for fossils to "speak for themselves" on the basis of their construction (albeit with some degree of taphonomic information loss). In many cases, however, the fossils themselves may not be amenable to modeling particular kinds of function (e.g., respiration, digestion, hormonal and enzymatic function, muscle firing/recruitment patterns). The only likely recourse to such instances is to rely on plesiomorphic function determined on the basis of the phylogeny of extant relatives to identify the presence of a given function in the extinct taxon.

Although very little pursued, it is obvious that when thrown together in a way that phylogeny and modeling illuminate the subject, full intersection of these approaches presents a much more interesting perspective on the problem of ascribing function to extinct organisms than either do in isolation. Two kinds of intersections are possible (Figure 3.11b). On the one hand, modeling and phylogeny may be fully concordant in their assessment of function. Here we have decisive characterization of function at the bracket node via phylogenetic analysis and its positive association with similar inferred function from modeling (Category 1 comparisons sensu Bryant & Russell 1992; Level I speculation sensu Witmer, this volume). On the other hand, we may discover that these two perspectives clash with each other: Function decisively suggested at the particular bracket function contrasts with that suggested by modeling. Such a decisive and negative relationship between phylogeny and modeling requires what Witmer (this volume) calls Level III inference (see also Category 3 comparisons, Bryant & Russell [1992]). Are there examples in which either possibility – concordance or discordance – can be recognized?

Turning first to concordance between modeling and phylogeny, the few examples that are available, as far as they go, include gliding flight in Cenozoic teratornithid birds from the late Cenozoic (Campbell & Tonni 1983) and digging in proscalopid moles from the Miocene (Barnosky 1981). In both cases, there is a reasonable connection between biomechanical models (scaling of wing dimensions in teratornids, forelimb kinematics for proscalopids) and the phylogenetic relationships of the extinct taxa within extant clades (discussed at least implicitly in the context of the evolution of function). In other examples (e.g., Greene & Burghardt 1978), one or the

other approach was not carried out to its fullest: Either modeling or plesiomorphic function was viewed as not contradicting, therefore allowing, the assessment of function provided by the other perspective. Why further examples of engagement appear not to be undertaken may stem from the fact that concordance automatically lends twice the support for extinct function. Lack of practical engagement between modeling and phylogeny may seem redundant, particularly for uncontroversial assessments of function. Though in great need of further study, concordance may come about in a variety of ways. Common identification of function by phylogenetic and modeling analyses may be associated with relatively proximate relationships among the concerned taxa, coupled with little extinction intervening between the fossil organism and extant taxa. However, it may also result from long-term stability in functional characters, for example, the alternating synchronous locomotor cycle of tetrapods (Hildebrand 1976) and the continuous axial undulatory locomotion of gnathostomes (Braun & Reif 1985).

Discordance between modeling and phylogeny is the more interesting of the two possible combinations, if for no other reason than it forces the investigator to come to terms with the meaning, implications, and resolution of conflicting evaluations. If modeling is held to somehow have more significance than phylogenetic nesting of function, then the modeled function would overthrow that provided by phylogeny, thus falsifying at least some of the functional transformations and therefore the functional homologs identified for terminal taxa. When phylogeny is thought to be powerful enough to overthrow modeling, then modeling must be deemed incorrect. In contrast to concordance between modeling and phylogeny of function, discordance, more times than not, is likely to be found in clades with deep relationships, where extinction is high and widely distributed in the clade.

Real examples obviously would be beneficial to illustrate this kind of discordant interplay between modeling and phylogeny. However, I have found none that exemplify the points about discordance noted here. In its place, a hypothetical example that would help exemplify such conflict would be, say, an extinct snake with a rigid skull whose phylogenetic position was bracketed by extant taxa with high levels of cranial kinesis. Foregoing a real example, I resort to how these approaches can be brought to bear on a well-known evolutionary sequence in which fossils and function have played a large part: jaw mechanics in Cynodontia through the origin of Theria (viz. DeMar & Barghusen 1973; Bramble 1978; Crompton & Hylander 1986). This important, reasonably dense, and phylogenetically well-investigated clade can be

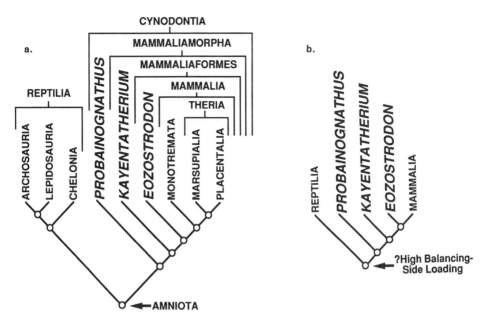

Figure 3.12. (a) Phylogenetic relationships of *Eozostrodon, Kayentatherium*, and *Probainognathus* with respect to closest extant outgroup taxa. (After Gauthier et al. 1988; Wible 1991.) (b) Balancing-side jaw joint loadings mapped and optimized onto amniote phylogeny.

used to address conflicts in the way function, in the form of working- and balancing-side jaw-joint loading accompanying the slow-close/power stroke phase of the feeding cycle, can be identified in fossils.

Jaw-joint loadings are used as a functional character whose basis in homology across a range of taxa is unquestioned, although this measure can equally be seen as the static or dynamic consequences of the biomechanics of feeding. This latter interpretation likely will yield the same results, but I adopt the former for ease of discussion.

This discussion follows Crompton and Hylander (1986), who emphasized three therapsid taxa – *Probainognathus*, a tritylodontid (apparently *Kayentatherium*; viz. Sues 1986), and *Eozostrodon* – and their relevance to changes in jaw-joint loadings that accompany the origin of Mammalia. All are well studied in terms of jaw-joint construction and dental organization, as well as attachment sites and sizes of jaw muscles, and therefore all are amenable to biomechanical modeling. Although the positions of these taxa to other extinct cynodonts may vary from one study to another, there is consensus as to their phylogenetic position vis à vis extant taxa (Figure 3.12a). For example, Hopson and Barghusen (1986), Rowe (1988), and Wible (1991) all position *Eozostrodon* (as morganucodontids), *Kayentatherium* (as tritylodontids), and *Probainognathus* within Cynodontia as successively farther outgroup taxa from crown-group Mammalia. (For convenience, these three genera will be referred to as basal cynodonts, although there are several other taxa more basal to

these three; Hopson & Barghusen 1986.) Mammalia thus constitutes the extant sister-taxon to each of these extinct cynodonts. The remainder of the EPB is Reptilia, that is, the clade consisting of archosaurs (including birds), lepidosaurs, and turtles (Gauthier et al. 1988). In this way, these basal cynodonts fall within the EPB formed of Amniota (sensu Gauthier et al. 1988).

The number of these extant bracket taxa for which we have information on loadings on working- and balancing-side jaw joints is extremely small. For example, experimental strain-gauge and transducer studies have been carried out only on monkeys, dogs, and humans (Brehnan & Boyd 1979; Brehnan et al. 1981; Weijs 1981; Boyd et al. 1982; Hylander 1979a,b; Hylander, Johnson, & Crompton 1987; Dessem 1989). From bilateral electromyograms from jaw adducting muscles, it is possible to estimate relative working and balancing joint loadings (viz. Crompton & Hylander 1986); these are available for cats (Gorniak & Gans 1980), rabbits (Weijs & DeJongh 1977; Weijs & Dantuma 1981), ferrets (Dessem & Druzinsky 1992), insectivores (Oron & Crompton 1985), additional species of monkeys (Hylander 1979c), and opossums (Crompton & Hylander 1986; Thomason, Russell, & Morgeli 1991). I have surveyed other studies to similarly estimate relative jaw-joint loadings in other taxa, for example, *Varanus exanthematicus* (Smith 1982) and *Alligator mississippiensis* (Busbey 1989). In both cases, real jaw-joint loadings are obviously much more relevant to what follows, but I will go with what is available.

As noted by Crompton and Hylander (1986), the available sample of extant mammals appear to load both jaw joints in compression while the animal is biting at the front of the jaws and unilaterally toward the rear. In addition, reaction forces can be very high on the contralateral (balancing) side (as much as 50–75 percent of the bite force), while on the ipsilateral (working) side, the jaw joint is also subject to compression when the bite force involves the premolars or the first two molars. More distally, the working-side joint is subject to very little compression, no loading, or tension.

For Reptilia, the next extant outgroup to basal cynodonts with living members, the data base is exceedingly patchy. No work has been done on directly measuring jaw-joint loadings in any of the extant terminal taxa, and what is reported here is only a very tentative assessment of such loadings based on bilateral adductor electromyograms for *Alligator* and *Varanus*. According to the adductor electromyogram data reported for these two reptilians by Smith (1982) and Busbey (1989), muscle activity patterns (measured as amplitude, duration, and number of spikes) between working and balancing adductors appear to be very similar. This, together with the geometry of muscle force vectors, implies loading of the working-side and more especially the balancing-side jaw joint. Like the case among extant mammals, it seems reasonably likely that very high compressional loads act on both balancing- and working-side jaw joints during unilateral or bilateral biting along various positions of the tooth row.

Interpreted as homologies, these high jaw-joint loadings can be mapped and optimized to determine conditions at the bracket node for the cynodonts under consideration. In Mammalia, the extant sister taxon to the extinct cynodonts under consideration, there are no data on jaw joint loading for monotremes, and hence it is impossible to characterize this node. Moving upward, we encounter Theria. On the basis of information from carnivores, primates, rodents, rabbits, and insectivores, it appears that the placental portion of the clade is characterized plesiomorphically by unilateral biting, bilateral adductor muscle activity, and high compressional loading of the balancing-side joint during the slow close/power stroke phase of the feeding cycle. If conditions in opossums are plesiomorphic for Marsupialia, then we can draw down this same pattern of jaw loading to plesiomorphically include all of Theria.

In this phylogenetic analysis, it is obviously necessary to make comparisons with extant Mammalia, but at the same time these are not sufficient to establish the jaw-joint loading environment in the more primitive cynodont taxa. For this we must turn to the second extant outgroup, Reptilia. This clade also appears to be plesiomorphically characterized by bilateral adductor muscle activity and high compressional loading of the balancing-side joint during biting. This phylogenetic assessment, I point out again, is made on very slim data (only in a single lepidosaur and archosaur, and then without any experimental support) and may be easily overthrown with more detailed research.

Assuming that this distribution of function is true, relatively high levels of compressional loading on the balancing-side joint as a consequence of bilateral muscle activity is plesiomorphic at least to the level of Amniota (the EPB) and therefore to be expected in bracketed extinct taxa (i.e., in basal cynodonts) by parsimony alone (Figure 3.12b).

This phylogenetic assessment of high joint loadings contrasts with what has come out of modeling analyses of a host of extinct synapsids, from Sphenacodontidae through basal Mammalia (viz. Crompton 1963a,b; Barghusen & Hopson 1970; DeMar & Barghusen 1973; Bramble 1978). All tackle the problem of the reduced joint size and loadings during the shift from a quadrate-articular joint to one formed between the squamosal and dentary in basal cynodonts from the point of view of Newtonian – therefore, ahistorical – mechanics. In this context, I will discuss the modeling approach provided by Crompton and Hylander (1986) to directly assess the jaw mechanics in basal cynodonts.

The Crompton and Hylander model is based on free-body lever-link statics, whereby forces from the jaw musculature exert bite forces between occluding teeth as well as loadings at the two jaw joints. In simple terms, working-side adductor muscles exert forces at the bite point, as also do balancing-side muscles via force transfer across the mandibular symphysis to the bite point. At the same time, working- and balancing-side muscles also exert reaction forces at one or both jaw joints (viz. Hylander 1975, 1979b,c; Greaves 1978; Bramble 1978). Crompton and Hylander (1986) combined these relationships with information on muscle contraction vectors and cross-sectional areas for these adductor muscles to estimate the direction and relative bite-point forces and loadings at the jaw joints. Described solely as a network of input forces, lever-arms, and output forces, this model (Figure 3.13) was tested against known muscle-firing patterns and jaw-joint loading in macaques, opossums, and tenrecs and, in all three cases, was found to be in general agreement with experimental conditions for real joint loadings.

Modeling of jaw-joint loading for the three aforementioned basal cynodonts follows the same protocol as for extant taxa (Figure 3.14; details after Crompton & Hylander 1986). Working-side muscles are allowed to exert force directly on the bite point, while forces from balancing-side muscles are transferred to the

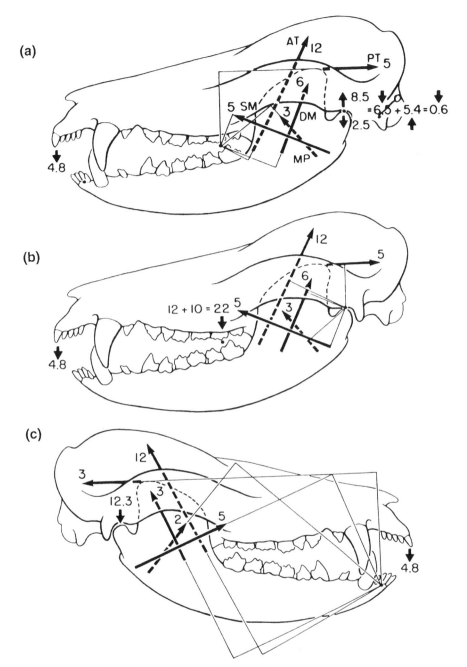

Figure 3.13. Crompton and Hylander (1986) model of mammalian jaw mechanics, using *Didelphis virginiana* as an example. Main adductor muscle vectors (AT, DM, MP, PT, SM) are assigned arbitrary values of contraction forces proportional to muscle masses. Resultant forces generated by these values are indicated for (a) the working-side jaw joint where working-side muscles generate a compressional force of 6.0 units and balancing-side muscles transfer through the symphysis a tension force of 5.4 (totaling to 0.6 units of compression), (b) the bite point where balancing-side and working-side forces add to 22 units, and (c) the balancing-side jaw joint where the balancing-side muscles provide 12.3 units of compression. (From Crompton & Hylander 1986; reprinted from *The Ecology and Biology of Mammal-like Reptiles*, ed. N. Hotton III et al. [Washington D.C.: Smithsonian Institution Press, 1987] by permission of the publisher. © 1987 Smithsonian Institution)

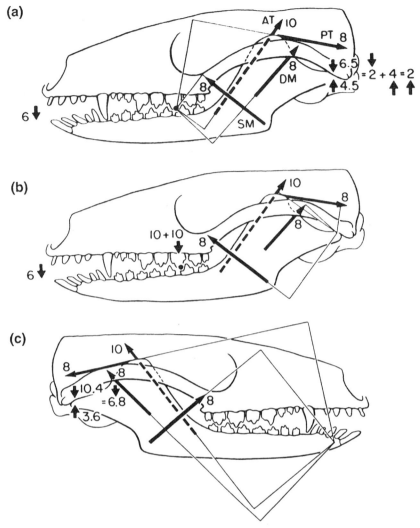

Figure 3.14. The Crompton and Hylander (1986) model applied to *Eozostrodon.* See Figure 3.13 for explanation. (From Crompton & Hylander 1986; reprinted from *The Ecology and Biology of Mammal-like Reptiles,* ed. N. Hotton III et al. [Washington D.C.: Smithsonian Institution Press, 1987] by permission of the publisher. © 1987 Smithsonian Institution)

bite point via the mandibular symphysis. Both jaw joints are then evaluated for the kind and magnitude of reaction forces they receive from bilaterally contracting adductor musculature. In all cases, the balancing jaw-joint in these basal cynodonts is loaded in compression, with levels ranging from 0 percent to 30 percent (relative to bite force), depending on bilateral or unilateral occlusion and the position of the bite point. Working-side joint loadings range from 29 percent tension to 10 percent compression. As noted by Crompton and Hylander (1986), these values are much lower than those found in extant primates, opossums, and tenrecs.

Herein lies the discordance: Phylogeny suggests relatively high compressive loads on the balancing-side joint in extinct, basal cynodonts, while modeling

suggests otherwise. At least one of these perspectives must be wrong (although in point of fact they may both be wrong). Without delving into which is right or wrong, if plesiomorphic function is given more weight than that determined through modeling, then the parameters of the model (or the modeling itself) may be called into question (Figure 3.15a). On the other hand, if it is possible to show that the pattern of low jaw-joint loadings discovered through modeling was more likely than the plesiomorphic high loadings of Amniota, then the functions identified for terminal taxa are not homologous (Figure 3.15b). The similarity in biomechanics among extant taxa may then be held to be convergently derived from a low-loaded jaw-joint condition (itself identified through modeling). This is probably the case

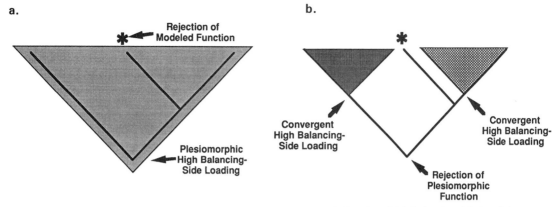

Figure 3.15. (a) Cladogram indicating strongly supported plesiomorphic function (high balancing-side jaw joint loading) and the overthrow of function determined through modeling. (b) Cladogram indicating strongly supported modeling of low balancing-side jaw joint loadings and the falsification of the functional homologs identified for terminal taxa.

for basal cynodonts (following the conclusions of Crompton & Hylander 1986), and is so as a consequence of the large amounts of extinction from ancestral Amniota to extant terminal taxa.

In this last example, I have outlined how modeling and phylogeny can bring different assessments of function in extinct organisms into focus. When extinct taxa are deeply nested members of a crown group, a great deal of information about wide-ranging extant terminal taxa is necessary to resolve the bracket node. These data may be hard to come by, but the efforts are worth it. For without both phylogeny and modeling, there can be no tension between approaches that, on the one hand, create functioning animals out of inanimate material and, on the other, employ extant organisms to phylogenetically cast their observable functions into the past.

ACKNOWLEDGMENTS

I thank all of the participants in this Symposium, as well as Bill Hylander, Arnold Kluge, Wolf Reif, Peter Dodson, Ron Heinrich, and Ralph Chapman for helpful information and comments as this study progressed. I thank Harry Greene for particular information on the distribution of constriction patterns among Alethinophidia and for his perspective on phylogeny, function, and behavior; Harold Bryant for discussion and for kindly providing me with a preprint of his and Tony Russell's paper on phylogeny and the inference of unpreserved attributes in extinct taxa; Blair Van Valkenburgh for a preprint of her paper on fossil vertebrates and ecomorphology of the past; and George Lauder for probing and prodding me over the meaning of function and behavior (and both together from a phylogenetic perspective). I especially thank Larry Witmer for discussions of phylogeny and its impact on our interpretation of extinct function, for his invention of the EPB approach, and for his help on the formative stages of this study. Larry Witmer, George Lauder, and Jeff Thomason read earlier drafts of this paper, and I thank them for their insightful comments. This work was supported in part by NSF grants DEB-7918490 and EAR-9004458.

REFERENCES

Alexander, R.M., Brandwood, A., Currey, J.D., & Jayes, A.S. 1984. Symmetry and precision of control of strength in limb bones of birds. *Journal of Zoology, London* 203, 135–143.

Anker, G.C. 1974. Morphology and kinetics of the head of the stickleback, *Gasterosteus aculeatus. Transactions of the Zoological Society of London* 32, 311–416.

Barghusen, H.R., & Hopson, J.A. 1970. Dentary-squamosal joint and the origin of mammals. *Science* 168, 573–575.

Barnes, L.G. 1979. Fossil enaliarctine pinnipeds (Mammalia: Otariidae) from Pyramid Hill, Kern County, California. *Contributions in Science, Natural History Museum of Los Angeles County* 318, 1–41.

Barnosky, A.D. 1981. A skeleton of *Mesoscalops* (Mammalia, Insectivora) from the Miocene Deep River Formation, Montana, and a review of the proscalopid moles: evolutionary, functional, and stratigraphic relationships. *Journal of Vertebrate Paleontology* 1, 285–339.

Berta, A. 1991. New *Enaliarctos** (Pinnipedimorpha) from the Oligocene and Miocene of Oregon and the role of "enaliarctids" in pinniped phylogeny. *Smithsonian Contributions to Paleobiology* 69, 1–33.

Biewener, A.A., Swartz, S.M., & Bertram, J.E.A. 1986. Bone remodeling during growth: dynamic strain equilibrium in the chick tibiotarsus. *Calcified Tissue International* 39, 390–395.

Bock, W.J., & Kummer, B. 1968. The avian mandible as a structural girder. *Journal of Biomechanics* 1, 89–96.

Boyd, R.L., Gibbs, C.H., Richmond, A.F., Laskin, J.L., & Brehnan, K. 1982. Temporomandibular joint forces in monkey measured with piezoelectric foil. *Journal of Dental Research* 61, 351.

Bramble, D.M. 1978. Origin of the mammalian feeding complex: models and mechanisms. *Paleobiology* 4, 271–301.

Braun, J., & Reif, W.-E. 1985. A survey of aquatic locomotion in fishes and tetrapods. *Neues Jahrbuch für Geologie und Paläontologie Abhandlungen* 169, 307–332.

Brehnan, K., & Boyd, R.L. 1979. Use of piezoelectric films to directly measure forces at the temporomandibular joint. *Journal of Dental Research* 58 (Special Issue A), 402.

Brehnan, K., Boyd, R.H., Laskin, J., Gibbs, C.H., & Mahan, P. 1981. Direct measurement of loads at the temporomandibular joint in *Macaca arctoides*. *Journal of Dental Research* 60, 1820–1824.

Bryant, H.N., & Russell, A.P. 1992. The role of phylogenetic analysis in the inference of unpreserved attributes of extinct taxa. *Philosophical Transactions of the Royal Society of London* B337, 405–418.

Busbey, A.B., III. 1989. Form and function of the feeding apparatus of *Alligator mississippiensis*. *Journal of Morphology* 202, 99–127.

Campbell, K.E., Jr., & Tonni, E.P. 1983. Size and locomotion in teratorns (Aves: Teratornithidae). *Auk* 100, 390–403.

Condon, K. 1987. A kinematic analysis of mesokinesis in the Nile monitor *(Varanus niloticus)*. *Experimental Biology* 47, 73–87.

Crompton, A.W. 1963a. On the lower jaw of *Diarthrognathus* and the origin of the mammalian lower jaw. *Zoological Society of London Proceedings, Series B* 108, 735–761.

Crompton, A.W. 1963b. The evolution of the mammalian jaw. *Evolution* 17, 431–439.

Crompton, A.W., & Attridge, J. 1986. Masticatory apparatus of the larger herbivores during Late Triassic and Early Jurassic times. In *The Beginning of the Age of Dinosaurs*, ed. K. Padian, pp. 223–236. Cambridge University Press.

Crompton, A.W., & Hylander, W.L. 1986. Changes in mandibular function following the acquisition of a dentary–squamosal jaw articulation. In *The Ecology and Biology of Mammal-like Reptiles*, ed. N. Hotton III., P.D. MacLean, J.J. Roth, & E.C. Roth, pp. 263–282. Washington, D.C.: Smithsonian Institution Press.

DeMar, R., & Barghusen, H.R. 1973. Mechanics and the evolution of the synapsid jaw. *Evolution* 26, 622–637.

Demes, B. 1984. Mechanical stresses at the primate skull base caused by the temporomandibular joint force. In *Food Acquisition and Processing in Primates*, ed. D.J. Chivers, B.A. Wood, & A. Bilsborough, pp. 407–413. New York: Plenum Press.

Demes, B., Preuschoft, H., & Wolff, J.E.A. 1984. Stress–strength relationships in the mandibles of hominoids. In *Food Acquisition and Processing in Primates*, ed. D.J. Chivers, B.A. Wood, & A. Bilsborough, pp. 369–390. New York: Plenum Press.

Dessem, D. 1989. Interactions between jaw-muscle recruitment and jaw-joint forces in *Canis familiaris*. *Journal of Anatomy* 164, 101–121.

Dessem, D., & Druzinsky, R.E. 1992. Jaw-muscle activity in ferrets, *Mustela putorius furo*. *Journal of Morphology* 213, 275–286.

Edwards, W.N. 1967. *The Early History of Palaeontology*. London: British Museum (Natural History).

Elshoud, G.C.A. 1980. APL and functional morphology. In *APL 80*, ed. G.A. van der Linden, pp. 175–181. New York: North-Holland Publishing Company (Elsevier).

Estes, R., & Price, L. 1973. Iguanid lizard from the Upper Cretaceous of Brazil. *Science* 180, 748–751.

Estes, R., Frazzetta, T.H., & Williams, E.E. 1970. Studies on the fossil snake *Dinilysia patagonica* Woodward. Part I. Cranial morphology. *Bulletin of the Museum of Comparative Zoology* 140, 25–74.

Estes, R., de Queiroz, K., & Gauthier, J. 1988. Phylogenetic relationships within Squamata. In *Phylogenetic Relationships of the Lizard Families*, ed. R. Estes & G. Pregill, pp. 119–218. Stanford: Stanford University Press.

Farris, J.S. 1970. Methods for computing Wagner trees. *Systematic Zoology* 9, 83–92.

Frazzetta, T.H. 1962. A functional consideration of cranial kinesis in lizards. *Journal of Morphology* 111, 287–319.

Fricke, H., Reinicke, O., Hofer, H., & Nachtigall, W. 1987. Locomotion of the coelacanth *Latimeria chalumnae* in its natural environment. *Nature* 329, 331–333.

Galton, P.M. 1974. The ornithischian dinosaur *Hypsilophodon* from the Wealden of the Isle of Wight. *Bulletin of the British Museum (Natural History) Geology* 25, 1–152.

Gauthier, J.A., Kluge, A.G., & Rowe, T. 1988. The early evolution of the Amniota. In *The Phylogeny and Classification of the Tetrapods*, Vol. 1, ed. M.J. Benton, pp. 103–155. Oxford: Oxford University Press.

Gingerich, P.D. 1971. Functional significance of mandibular translation in vertebrate jaw mechanics. *Postilla* 152, 1–10.

Goodship, A.E., Lanyon, L.E., & McFie, H. 1979 Functional adaptation of bone to increased stress. *Journal of Bone and Joint Surgery* 61A, 539–546.

Gorniak, G.C., & Gans, C. 1980. Quantitative assay of electromyograms during mastication in domestic cats (*Felis catus*). *Journal of Morphology* 163, 253–281.

Gorniak, G.C., Rosenberg, H.I., & Gans, C. 1982. Mastication in the tuatara, *Sphenodon punctatus* (Reptilia, Rhynchocephalia): structure and activity of the motor system. *Journal of Morphology* 171, 321–353.

Greaves, W.S. 1978. The jaw lever system in ungulates: a new model. *Journal of Zoology, London* 184, 271–285.

Greene, H.W., & Burghardt, G.M. 1978. Behavior and phylogeny: constriction in ancient and modern snakes. *Science* 200, 74–77.

Heinrich, R.E., Ruff, C.B., & Weishampel, D.B. in press. Femoral ontogeny and locomotor biomechanics of *Dryosaurus lettowvorbecki* (Dinosauria, Iguanodontia). *Zoological Journal of the Linnean Society*.

Hickman, C.S. 1988. Analysis of form and function in fossils. *American Zoologist* 28, 775–793.

Hildebrand, M. 1976. Analysis of tetrapod gaits: general considerations and symmetrical gaits. In *Neural Control of Locomotion* (*Advances in Behavioral Biology 18*), ed. R.M. Herman, S. Grillner, P.S.G. Stein, & D.G. Stuart, pp. 203–236. New York: Plenum Press.

Hopson, J.A., & Barghusen, H.R. 1986. An analysis of therapsid relationships. In *The Ecology and Biology of Mammal-like Reptiles*, ed. N. Hotton, III, P.D. MacLean, J.J. Roth, & E.C. Roth, pp. 83–106. Washington, D.C.: Smithsonian Institution Press.

Hylander, W.L. 1975. The human mandible: lever or link? *American Journal of Physical Anthropology* 43, 227–242.

Hylander, W.L. 1979a. Mandibular function in *Galago crassicaudatus* and *Macaca fascicularis*: an in vivo approach to stress analysis of the mandible. *Journal of Morphology* 159, 253–296.

Hylander, W.L. 1979b. An experimental analysis of temporomandibular joint reaction force in macaques. *American Journal of Physical Anthropology* 51, 433–456.

Hylander, W.L. 1979c. The functional significance of primate mandibular form. *Journal of Morphology* 160, 223–240.

Hylander, W.L., Johnson, K.R., & Crompton, A.W. 1987. Loading patterns and jaw movements during mastication in *Macaca fascicularis*: a bone-strain, electromyographic, and cineradiographic analysis. *American Journal of Physical Anthropology* 72, 287–314.

Kluge, A.G. 1991. Boine snake phylogeny and research cycles. *Museum of Zoology, University of Michigan, Miscellaneous Publications* 178, 1–58.

Kripp, D. von. 1933a. Der Oberschnabel-Mechanismus der Vögel. *Morphologisches Jahrbuch* 71, 469–544.

Kripp, D. von. 1933b. Die Kaubewegung und Lebensweise von Edmontosaurus spec. auf Grund der mechanisch-konstruktiven Analyse. *Palaeobiologica* 5, 409–421.

Lambe, L.M. 1920. The hadrosaur *Edmontosaurus* from the Upper Cretaceous of Alberta. *Canadian Geological Survey Memoir* 120, 1–79.

Lanyon, L.E. 1984. Functional strains as a determinant for bone remodeling. *Calcified Tissue International* 36, S56–S61.

Lanyon, L.E., Goodship, A.E., Pye, C., & McFie, H. 1982. Mechanically adaptive bone remodelling: a quantitative study on functional adaptation in the radius following ulna osteotomy in sheep. *Journal of Biomechanics* 15, 141–154.

Lanyon, L.E., & Rubin, C.T. 1985. Functional adaptation in skeletal structures. In *Functional Vertebrate Morphology*, ed. M. Hildebrand, D.M. Bramble, K.F. Liem, & D.B. Wake, pp. 1–25. Cambridge: Harvard University Press.

Lauder, G.V. 1981. Form and function: structural analysis in evolutionary morphology. *Paleobiology* 7, 430–442.

Lauder, G.V. 1990. Functional morphology and systematics: studying functional patterns in an historical context. *Annual Review of Ecology and Systematics* 21, 317–340.

Lauder, G.V. 1991. Biomechanics and evolution: integrating physical and historical biology in the study of complex systems. In *Biomechanics in Evolution*, ed. J.M.V. Rayner & R.J. Wooton, pp. 1–19. Cambridge University Press.

Lombard, R.E., & Wake, D.B. 1976. Tongue evolution in the lungless salamanders, Family Plethodontidae. I. Introduction, theory and a general model of dynamics. *Journal of Morphology* 148, 265–286.

Lombard, R.E., & Wake, D.B. 1977. Tongue evolution in the lungless salamanders, Family Plethodontidae. II. Function and evolutionary diversity. *Journal of Morphology* 153, 39–79.

Maddison, W.P., Donoghue, M.J., & Maddison, D.R. 1984. Outgroup analysis and parsimony. *Systematic Zoology* 33, 83–103.

McHenry, H.M. 1991. First steps? Analyses of the postcranium of early hominids. In *Origine(s) de la Bipédie chez les Hominidés*, ed. Y. Coppens, & B. Senut, pp. 133–141. Paris: Cahiers de Paléoanthropologie.

Nagurka, M.L., & Hayes, W.C. 1980. An interactive graphics package for calculating cross-sectional properties of complex shapes. *Journal of Biomechanics* 13, 419–451.

Nobiling, G. 1977. Die Biomechanik des Kieferapparates beim Stierkopfhai (*Heterodontus portusjacksoni* = *Heterodontus philippi*). *Advances in Anatomy, Embryology, and Cell Biology* 52, 1–52.

Nopsca, F. 1900. Dinosaurrieste aus Siebenbürgen. I. Schädel von *Limnosaurus transsylvanicus* nov. gen. et spec. *Denkschriften der Akademie der Wissenschaften, Wien* 68, 555–591.

Norman, D.B. 1984. On the cranial morphology and evolution of ornithopod dinosaurs. *Symposium of the Zoological Society of London* 52, 521–547.

Norman, D.B., & Weishampel, D.B. 1985. Ornithopod feeding mechanisms: their bearing on the evolution of herbivory. *American Naturalist* 126, 151–164.

Norman, D.B., & Weishampel, D.B. 1990. Iguanodontidae and related ornithopods. In *The Dinosauria,* ed. D.B. Weishampel, P. Dodson, & H. Osmólska, pp. 510–533. Berkeley: University of California Press.

Norman, D.B., & Weishampel, D.B. 1991. Feeding mechanisms in some small herbivorous dinosaurs: processes and patterns. In *Biomechanics in Evolution,* ed. J.M.V. Rayner & R.J. Wootton, pp. 161–181. Cambridge University Press.

Oron, U., & Crompton, A.W. 1985. A cineradiographic and electromyographic study of mastication in *Tenrec ecaudatus. Journal of Morphology* 185, 155–182.

Ostrom, J.H. 1961. Cranial morphology of the hadrosaurian dinosaurs of North America. *Bulletin of the American Museum of Natural History* 122, 33–186.

Otten, E. 1983. The jaw mechanism during growth of a generalized *Haplochromis* species: *H. elegans* Trewavas 1933 (Pisces, Cichlidae). *Netherlands Journal of Zoology* 33, 55–98.

Otten, E. 1991. The control of movements and forces during chewing. In *Feeding and the Texture of Food*, ed. J.F.V. Vincent & P.J. Lillford, pp. 123–141. Cambridge University Press.

Patterson, C. 1982. Morphological characters and homology. In *Problems of Phylogenetic Reconstruction*, ed. K.A. Joysey & A.E. Friday, pp. 21–74. New York: Academic Press.

Pauwels, F. 1980. *Biomechanics of the Locomotor Apparatus.* Berlin: Springer-Verlag.

Rieppel, O. 1988. A review of the origin of snakes. *Evolutionary Biology* 22, 37–130.

Rowe, T. 1988. Definition, diagnosis and origin of Mammalia. *Journal of Vertebrate Paleontology* 8, 241–264.

Rubin, C.T., & Lanyon, L.E. 1985. Regulation of bone mass by mechanical strain magnitude. *Calcified Tissue International* 37, 411–417.

Rudwick, M.J.S. 1964. The inference of function from structure in fossils. *British Journal of the Philosophy of Science* 15, 27–40.

Ruff, C.B., & Hayes, W.C. 1983. Cross-sectional geometry of Pecos Pueblo femora and tibiae – a biomechanical investigation: I. Method and general patterns of variation. *American Journal of Physical Anthropology* 60, 359–381.

Schwenk, K. 1988. Comparative morphology of the

lepidosaur tongue and its relevance to squamate phylogeny. In *Phylogenetic Relationships of the Lizard Families,* ed. R. Estes & G. Pregill, pp. 569–598. Stanford: Stanford University Press.

Schwenk, K., & Throckmorton, G.S. 1989. Functional and evolutionary morphology of lingual feeding in squamate reptiles: phylogenetics and kinematics. *Journal of Zoology, London* 219, 153–175.

Smith, K.K. 1982. An electromyographic study of the function of the jaw adducting muscles in *Varanus exanthematicus* (Varanidae). *Journal of Morphology* 173, 137–158.

Smith, K.K., & Hylander, W.L. 1985. Strain gauge measurement of mesokinetic movement in the lizard *Varanus exanthematicus. Journal of Experimental Biology* 114, 53–70.

Sues, H.-D. 1986. The skull and relationships of two tritylodontid synapsids from the lower Jurassic of western North America. *Bulletin of the Museum of Comparative Zoology, Harvard University* 151, 217–268.

Sues, H.-D., & Norman, D.B. 1990. Hypsilophodontidae, *Tenontosaurus,* Dryosauridae. In *The Dinosauria,* ed. D. B. Weishampel, P. Dodson, & H. Osmólska, pp. 498–509. Berkeley: University of California Press.

Swofford, D.L., & Maddison, W.P. 1987. Reconstructing ancestral character states under Wagner parsimony. *Mathematical Biosciences* 87, 199–229.

Thomason, J.J., & Russell, A.P. 1986. Mechanical factors in the evolution of the mammalian secondary palate: a theoretical analysis. *Journal of Morphology* 189, 199–213.

Thomason, J.J., Russell, A.P., & Morgeli, M. 1990. Forces of biting, body size, and masticatory muscle tension in the opossum *Didelphis virginiana. Canadian Journal of Zoology* 68, 318–324.

Timoschenko, S.P., & Gere, J.M. 1972. *Mechanics of Materials.* New York: Van Nostrand Reinhold.

Uicker, J.J., Jr., Denavit, J., & Hartenberg, R.S. 1964. An iterative method for the displacement analysis of spatial mechanisms. *Journal of Applied Mechanics (Series E)* 31, 309–314.

Van Valkenburgh, B. in press. Ecomorphological analysis of fossil vertebrates and their paleocommunities. In *Ecological Morphology: Integrative Organismic Biology,* ed. P.C. Wainwright & S. Reilly. Chicago: University of Chicago Press.

Versluys, J. 1923. Der Schädel des Skelettes von *Trachodon annectens* im Senckenberg-Museum. *Abhandlungen der Senckenbergische Naturforschenden Gesellschaft* 38, 1–19.

Weijs, W.A. 1981. Mechanical loading of the human jaw joint during unilateral biting. *Acta Morphologica Neerlando-Scandinavica* 19, 261–262.

Weijs, W.A., & Dantuma, R. 1981. Functional anatomy of the masticatory apparatus in the rabbit (*Oryctolagus cuniculus* L.). *Netherlands Journal of Zoology* 31, 99–147.

Weijs, W.A., & DeJongh, H.J. 1977. Strain in mandibular alveolar bone during mastication in the rabbit. *Archives of Oral Biology* 22, 667–675.

Weishampel, D.B. 1984a. Evolution of jaw mechanisms in ornithopod dinosaurs. *Advances in Anatomy, Embryology and Cell Biology* 87, 1–110.

Weishampel, D.B. 1984b. Interactions between Mesozoic plants and vertebrates: fructifications and seed predation. *Neues Jahrbuch für Geologie und Paläontologie Abhandlungen* 167, 224–250.

Weishampel, D.B. 1990. Dinosaurian distribution. In *The Dinosauria,* ed. D.B. Weishampel, P. Dodson, & H. Osmólska, pp. 63–139. Berkeley: University of California Press.

Weishampel, D.B. 1993. Beams and machines: modeling approaches to analysis of skull form and function. In *The Vertebrate Skull,* Vol. 3, ed. J. Hanken and B.K. Hall, pp. 303–344. Chicago: University of Chicago Press.

Weishampel, D.B., & Horner, J.R. 1990. Hadrosauridae. In *The Dinosauria,* ed. D.B. Weishampel, P. Dodson, & H. Osmólska, pp. 534–561. Berkeley: University of California Press.

Weishampel, D.B., & Norman, D.B. 1989. The evolution of occlusion and jaw mechanics in Late Paleozoic and Mesozoic herbivores. *Geological Society of America Special Paper* 238, 87–100.

Weishampel, D.B., Norman, D.B., & Grigorescu, D. in press. *Telmatosaurus transsylvanicus* from the Late Cretaceous of Romania: the most basal hadrosaurid dinosaur? *Palaeontology*

Weishampel, D.B., & Witmer, L.M. 1990. Heterodontosauridae. In *The Dinosauria,* ed. D.B. Weishampel, P. Dodson, & H. Osmólska, pp. 486–497. Berkeley: University of California Press.

Wible, J.R. 1991. Origin of Mammalia: the craniodental evidence reexamined. *Journal of Vertebrate Paleontology* 11, 1–28.

Wiley, E.O., Siegel-Causey, D., Brooks, D.R., & Funk, V.A. 1991. The Compleat Cladist. *University of Kansas Museum of Natural History Special Publication* 19, 1–158.

Witmer, L.M. 1992. Ontogeny, phylogeny, and air sacs: the importance of soft-tissue inferences in the interpretation of facial evolution in Archosauria. Unpublished Ph.D. dissertation, Johns Hopkins University.

Woo, S.-L.-Y., Kuei, S.C., Amiel, D., Gomez, M.A., Hayes, W.C., White, F.C., & Akeson, W.H. 1981. The effect of prolonged physical training on the properties of long bones: a study of Wolff's Law. *Journal of Bone and Joint Surgery* 63A, 780–787.

Woodward, A.S. 1901. On some extinct reptiles from Patagonia, of the genera *Miolania, Dinilysia,* and *Genyodectes. Proceedings of the Zoological Society of London* 1901, 169–184.

4

Masticatory function in nonmammalian cynodonts and early mammals

A.W. CROMPTON

ABSTRACT

Nontheriodont synapsids had massive quadrate–articular jaw joints capable of resisting vertical forces equal to those generated at the point of bite. In nonmammalian cynodonts and early mammals the quadrate and articular formed part or all of the jaw joint and conducted vibrations from a mandibular tympanic membrane to the stapes. In nonmammalian cynodonts the size of these bones and their ability to resist compressive forces was reduced, but compensated for by a reorganization of the adductor muscles. With the acquisition of a substantial squamosal–dentary articulation alongside the quadrate–articular joint, early mammals could again load the jaw joint.

The relative size of the jaw had a profound effect on the structure of the masticatory apparatus. When an opossum chews hard food, left and right adductor muscles are equally active, but the reaction force on the balancing side is greater than that on the active side. The ligaments of the symphysis transfer adductor force from the balancing to the active side, reducing compressive loading on the joint of the active side.

Adaptations of the masticatory apparatus in nonmammalian cynodonts to a small jaw joint are discussed. Herbivorous forms possessed bilateral postcanine occlusion. Most carnivorous forms lacked postcanine occlusion, restricting powerful biting to the incisor–canine region and reducing the relative size of the postcanines. Trithelodonts were the exception to this rule, as they had simple postcanine occlusion with a small squamosal–dentary contact. Early mammals were characterized by an enlarged squamosal–dentary articulation and unilateral occlusion.

Suckling in mammals partly depends upon the ability to form a seal between the tongue and the anterior region of the soft palate. They are brought into contact by tensing the veli palatini. The structure of the pterygoid bones in early mammals suggests that they possessed a tensor veli palatini muscle, supporting the view that early mammals suckled their young.

INTRODUCTION

The evolution of the mammalian jaw is one of the best documented examples of progressive morpho-logical change over a long period of time. Beginning with Permian sphenacodont pelycosaurs and ending with the Rhaeto–Liassic mammals, the dentary increased progressively in size (see Figure 4.1 for relationships of the animals discussed here). Meanwhile, the postdentary bones and the quadrate underwent a corresponding reduction (Crompton 1963a; Hopson 1966; Allin 1975; Crompton & Parker 1978; Allin & Hopson 1992). Allin (1975) claimed that the quadrate and articular of nonmammalian cynodonts not only formed the jaw joint, but also served to transmit vibrations from a tympanic membrane, supported by a reflected lamina of the angular, to the stapes. He argued that a reduction in size of the bones forming the jaw articulation improved their ability to serve as ear ossicles. Such reduction of the bones forming the jaw articulation did not occur in diapsids. In these forms the tympanic membrane developed behind the quadrate, and a single ossicle, the stapes, was present in the middle ear. In synapsids the tympanic membrane developed anterior to the jaw articulation. The progressive reduction in the size of the postdentary bones in noncynodont theriodonts (gorgonopsians and therocephalians) and nonmammalian cynodonts had a profound effect on the structure and function of the masticatory apparatus of different lineages.

In Figure 4.2 the jaw articulation of an early Triassic nonmammalian cynodont, *Thrinaxodon*, is compared with the jaw articulation and middle ear of a primitive mammal (*Morganucodon = Eozostrodon*) and an extant mammal (opossum) (Crompton & Jenkins 1979). The morphology and position of the elements indicate that the angular and tympanic, the articular and malleus, and the quadrate and incus are homologs. In nonmammalian cynodonts more advanced than *Thrinaxodon*, the dentary increased further in size. In some cynodonts and primitive mammals, the dentary contacted the squamosal to form a new temporomandibular joint alongside the

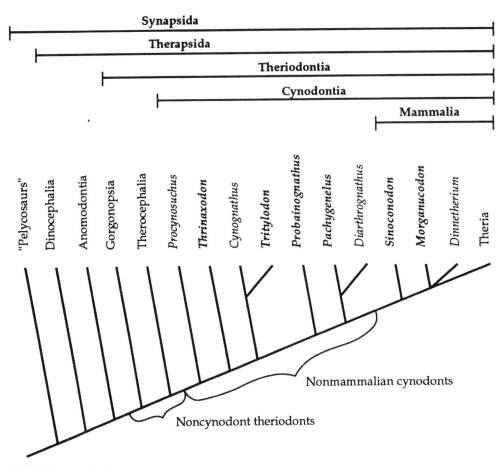

Figure 4.1. Cladogram showing a hypothesis of relationships of the taxa discussed in this chapter. (Modified from Allin & Hopson 1992.)

articular–quadrate joint. In several lineages of Mesozoic mammals the postdentary bones independently lost their contact with the dentary in the adult and formed a middle ear isolated from the dentary (Crompton & Sun 1985; Allin & Hopson 1992). The transition from a middle ear that formed the jaw articulation to one not involved in jaw suspension has been described and discussed in detail by Allin and Hopson (1992) and Rosowski (1992). This paper will consider the impact of a decrease in size of the articular–quadrate joint and the acquisition of a dentary–squamosal joint on the function of the masticatory system.

The first part of this paper will review the masticatory apparatus in four nonmammalian cynodonts – *Thrinaxodon, Probainognathus, Pachygenelus*, and *Tritylodon* – and a Liassic mammal, *Morganucodon* (Figure 4.3).

The two examples I present illustrate two primary difficulties in reconstructing the function of fossils that are not addressed by phylogenetic bracketing or similar techniques (Witmer, this volume; Weis-

hampel, this volume). The transition from a quadrate–articular to a dentary–squamosal jaw joint produced transitional forms whose jaw function appears to have differed from that in both of the extant bracketing taxa (mammals and "reptiles"). Similarly the timing of the acquisition of the soft palate cannot be assessed by reference to its presence or absence in living forms.

The data from which we infer functional details of the joint transition have three important features: (1) The osteology of the change is preserved in exquisite detail in numerous fossils; (2) the osteological changes are strongly indicative of the anatomical transition in jaw adducting musculature, and (3) in vivo experiments on the mechanics of biting in extant mammals suggest how the functional transition may have occurred. Sufficient relevant information is, therefore, available to support the robusticity of our inferences.

Inference on the timing of soft palate acquisition is much less robust. This example illustrates the classic problem of functional reconstruction in fossils, in

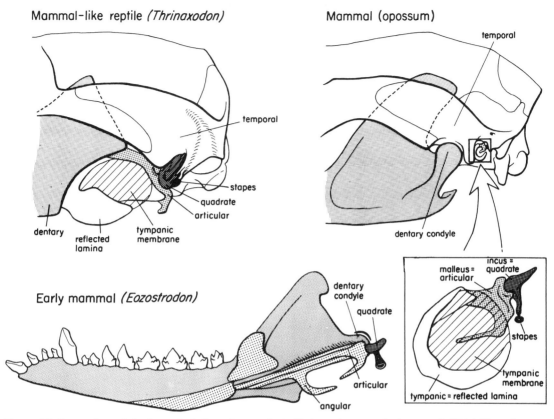

Figure 4.2. Comparison of the jaw articulation of a cynodont, *Thrinaxodon*; an extant mammal, *Didelphis;* and a Liassic mammal, *Morganucodon.* (After Crompton & Jenkins 1979.)

which as much information as possible is extracted from the bones but the question is not resolved unequivocally.

A brief account of some aspects of masticatory function in modern mammals, including discussion of the suckling mechanism, will follow. The discussion will address the function of the masticatory system of the four nonmammalian cynodonts and early mammals.

MATERIALS AND METHODS

Fossil material

Descriptions of the masticatory apparatus of *Thrinaxodon, Probainognathus, Pachygenelus, Tritylodon,* and *Morganucodon* are based upon material housed at the Museum of Comparative Zoology, Harvard University (MCZ), on loan from the Institute of Vertebrate Paleontology and Paleoanthropology in Beijing (IVPP) and the South African Museum (SAM), or upon published accounts.

Experiments

Two adult opossums (*Didelphis virginiana*) served as subjects for the study on mandibular bone strain and activity of adductor muscle during mastication. Six single-element strain gauges were placed on the ventral border of the mandible and electromyographic (EMG) electrodes were implanted within representative adductor muscles. Cineradiographic films were taken synchronously with the recordings of strain and EMG data. The results from the two experiments were nearly identical, and only the results from one are reported in this paper.

CINERADIOGRAPHIC PROCEDURES. In order to plot jaw movements, radio-opaque markers were placed in the upper and lower canines and third molars. The cineradiographic apparatus was the same as that described elsewhere (Oron & Crompton 1985; Thexton & Crompton 1989). Images were recorded on Kodak Plus-X film by an Eclair CV-16 cine camera operating at 85 to 100 frames per second. A total of 900 feet of film was exposed. Three sequences documenting chewing were chosen from each experiment for detailed analysis. Seven films were

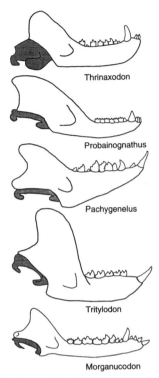

Thrinaxodon

Probainognathus

Pachygenelus

Tritylodon

Morganucodon

Figure 4.3. Outlines of the lower jaws in lateral view of the five nonmammalian cynodonts discussed in the text. The dentary is unshaded, the postdentary shaded. Not to scale.

taken in lateral projection and two in dorsoventral projection. Anterior views of the animals chewing were taken synchronously during cineradiographic filming with a Photosonics 16mm cine camera. Pulses corresponding to each frame of film from each camera were recorded on tape (see below).

ELECTROMYOGRAPHY AND STRAIN. Markers and EMG electrodes were inserted under halothane maintenance anesthesia administered in oxygen through a face mask. Induction was accomplished by Ketocet (Ketamine HCL, 35 mg/kg) in combination with Rompun (Xylazine HCL, 5 mg/kg). EMG electrodes were made from insulated Evanohm wires (size 002) under a dissecting microscope. The wire tips were bared for 1.0 mm and formed into a bipolar electrode of the type described by Basmajian and Steko (1962), then inserted with the aid of 25 gauge needles of varying lengths (16 to 38 mm). An incision was made in the ventral surface of the neck and above the sagittal crest. Muscles were exposed by blunt dissection. Electrodes were inserted along the line of muscle fibers. At the exit point, the electrode wires were sutured to the muscle. The wires of

the electrodes were passed through a subcutaneous tunnel to multipole connectors mounted between the scapulae. Fifteen electrodes were inserted but only twelve electrodes could be recorded simultaneously. A noninsulated subcutaneous wire, joined to a pin of the miniature connector, grounded each animal to the recording apparatus.

Electrodes were inserted in the right and left sides of the temporalis (anterior and posterior), deep masseter, superficial masseter, and the medial pterygoid (anterior and posterior). Six single-element strain gauges (WA-06-031-CF-120, Micromeasurements, Raleigh, North Carolina) were bonded to the ventral surface of both mandibles below the second premolar, first molar, and fourth molar on the left as well as below the second premolar and third molar and behind the fourth molar on the right (their positions are shown in Figure 4.13). Wires from the gauges were reinforced by attaching them to screws inserted immediately behind the gauges. The wires were also led subcutaneously to a miniature connector. The animals were allowed to recover at least 24 hours before attempting to record data.

RECORDING PROCEDURES. EMG signals from the muscles were recorded during filming. A 2-volt marker signal corresponding to each frame of film was generated by each cine camera. EMG activity was amplified ×200 to ×5000 (band width 100 Hz to 5 kHz and a 60-Hz filter) using Grass 511 differential preamplifiers and recorded at 15 ips on a 14-channel Bell & Howell CPR 4010 Datatape FM recorder together with the cine frame pulses. Each of the six strain-gauge elements formed one arm of a Wheatstone bridge. Bridge excitation was 2 volts. The voltage output from each element was first conditioned, then amplified (Measurements Group 2199 System), and then recorded with a seven-channel Bell & Howell CRP FM tape recorder together with the shutter pulse signals from each cine camera. The EMGs, strain, shutter pulses, and a synchronization signal were monitored during filming on Tektronix 5113N storage oscilloscopes.

In order to match the EMG and strain recordings with one another and to match these with individual frames of cine and cineradiographic film, several synchronizing pulses were recorded. When a circuit was closed, the voltage of the cineradiographic camera shutter pulse was reduced by half and that from the cine camera to zero. At the same time both the intensity of a 1000-Hz diode that marked the edge of the film in the cine camera and the voltage of the X-ray pulse of the cineradiographic equipment were reduced in intensity.

During the recording sessions (between 24 and 72 hours after surgery), the opossums were allowed to move freely within a Plexiglas box (30 by 60 cm) with an open top in which they were offered a variety of foods (chicken flesh, chicken wings, monkey chow, and pieces of apple).

DATA ANALYSIS. All the cineradiographic film was reviewed in detail, and only films taken in true lateral or dorsal aspect were chosen for measurement (i.e., with the cranium in the sagittal or horizontal plane as judged by superimposition of the shadows of bilateral features or symmetry of left and right sides). The positions of radio-opaque markers were digitized using a Vanguard stopframe projector (model M160W) coupled to a Graph/pen sonic digitizer (model C-P-6/50) and coordinates stored on Apple II microcomputer floppy disks.

EMG signals were passed through an analog-to-digital converter (Interactive Structures model AI13) attached to an Apple II microcomputer. The signals were full wave rectified, and integrated over a period of one frame of film (a period of either 10 or 11.8 milliseconds [ms], depending upon the camera speed). The resulting true voltage time integral was then reset to zero, the termination of the integration period being controlled by the recorded frame count pulse from the cineradiographic equipment. The 12-bit A/D converter sampled six channels of EMG data in the last 120 ms before the integrator was reset to zero. The integrated and digitized value for each EMG channel could then be related to positional data derived from matching film. During each feeding sequence, the amplification of the EMG signals from each muscle was held constant (feeding sessions lasted up to 30 minutes, during which time different foods were fed to the animals). Comparisons of the amplitude of EMG activity were made only with respect to such data, and no attempt was made to compare the precise absolute amplitude of the EMG signals at the electrodes when the signals had been recorded at different times or in different animals. Rectified and integrated EMGs and strain data were played out on an Okidata printer to the same length as the raw EMGs plotted by the Brush Gould recorder to check that the timing of activity was correct. When EMGs were played out on the Okidata or Hewlett-Packard plotter, the highest value for the EMGs of the sequence being plotted was set at 100 percent, and other values were scaled relative to this maximum value. EMG levels of a single muscle could, for example, be compared during the course of a single feeding session, as the active side shifted from left to right or vice versa. Comparisons between feeding sessions of the same or different animals were made only with respect to the timings of EMGs and the patterns of activity within each burst of activity.

ORAL FUNCTION IN EXTINCT AND EXTANT MAMMALS

Structure of the masticatory apparatus of selected cynodonts and early mammals

INFERRING MASTICATION IN *THRINAXODON*. In diapsids the lower jaw may be represented in two dimensions as a third-order lever. Because the insertion areas of the adductor muscles are closer to the jaw joint than the point of bite, the compressive forces generated are always higher at the joint than at the point of bite. With this arrangement of adductor muscles, reduction of the relative size of the jaw joint bones would not have been possible. Noncynodont theriodonts and nonmammalian cynodonts reduced the vertical forces acting through the joint by developing an ascending or coronoid process to the dentary and by modifying the direction of pull of the jaw adductors (Crompton 1963a; Bramble 1978; Crompton & Hylander 1986). In discussing the masticatory apparatus of each fossil, I will refer to muscles that have been reconstructed for them and reported previously. The reader is referred to the work, cited in each case, for the evidence on which each muscle was constructed. Reconstructions of synapsid jaw muscles are usually made both from osteological evidence on the fossils and by comparison with the conservative pattern of the masticatory musculature in extant mammals. The anterior fibers of the temporalis, for example, were nearly vertical to the longitudinal axis of the jaw (Figure 4.4), whereas the posteriors were more horizontal (based on the position of the coronoid process in the temporal fossa). In Permian nontherapsid synapsids (pelycosaurs), nontheriodont therapsids (deinocephalians), and noncynodont theriodonts (gorgonopsians and therocephalians), the posterior component of the temporalis appears to have been balanced by an anteriorly directed pterygoideus muscle that arose from the pterygoid bone and inserted on the angular–articular region of the postdentary rod (Figure 4.5a). In the nonmammalian cynodont the adductor mass was increased by adding a masseter muscle external to the coronoid process (Barghusen 1968; Crompton 1972a). In the early cynodont a small undivided masseter that inserted on the tip of the coronoid process was present. In more advanced cynodonts this muscle was divided to form deep and superficial portions. Within cynodont lineages both the pterygoideus muscle and the postdentary bones on which it inserted decreased in size, while the masseter increased in size. The superficial portion of this muscle counteracted the posteriorly directed component of force of the temporalis. Bramble (1978) proposed a bifulcral model to illustrate how changing muscle directions

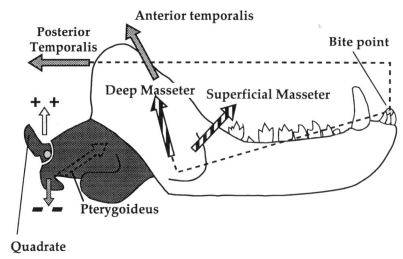

Figure 4.4. Direction of pull of the principal components of the jaw-closing muscles in *Thrinaxodon*. The moment arms of the posterior temporalis and superficial masseter around an incisal bite point are indicated. Activity in the superficial masseter (and the anterior temporalis and deep masseter) would place the jaw articulation under compression (+). The posterior temporalis acting alone would load the jaw in tension (−). In concert with the other muscles it would reduce the compression on the joint. Shading as in Fig. 4.3. (Redrawn after Bramble 1978.)

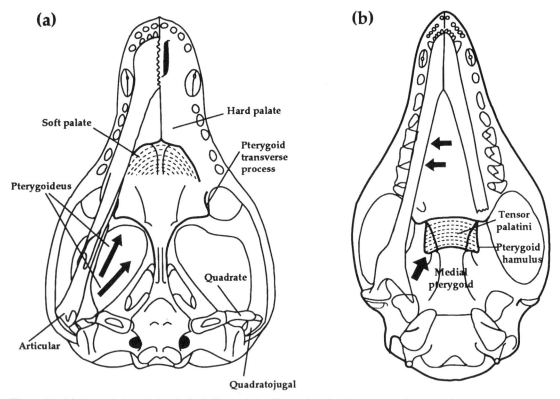

Figure 4.5. (a) Ventral view of the skull of *Thrinaxodon* illustrating the close contact between the transverse processes of the pterygoid and the medial surface of the lower jaw. (b) Ventral view of the skull of *Monodelphis* to illustrate the origin of the fibers of the *m. tensor veli palatini* within the anterior part of the soft palate. Note the distance separating the pterygoid hamulus from the inner surface of the lower jaw when the jaw is moved towards the side of dental occlusion (*small arrows*).

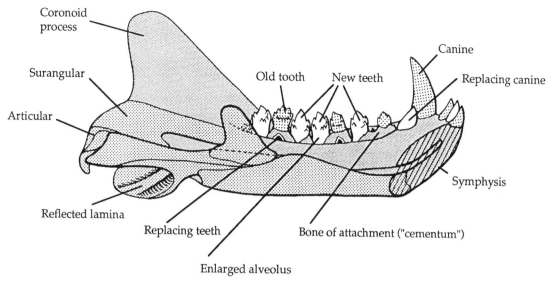

Figure 4.6. Medial view of lower jaw of *Thrinaxodon*. Postcanine tooth replacement is alternate. Fully erupted teeth are ankylosed to the dentary whereas erupting teeth are loosely held in an enlarged alveolus. (After Crompton & Parker 1978.)

could lead to a reduction of joint reaction forces. Figure 4.4 illustrates the orientations of the main components of the adductor muscles in *Thrinaxodon*. If moments were taken around an incisal bite point, contraction of the deep masseter, the superficial masseter, and the anterior temporalis would have generated a compressive force at the jaw joint, whereas contraction of the posterior temporalis would have decreased the level of the compressive force. In *Thrinaxodon*, when the jaws were closed, they fit snugly against the lateral surfaces of the transverse processes of the pterygoid bones (Figure 4.5a). These processes guided the lower jaw and restricted its movement to the vertical plane during the final phases of jaw closing. The lower postcanine teeth lay medial to the uppers, but the two sets did not contact one another. They served only to puncture food rather than both puncture and shear. In nearly all extant mammals the distance between the lower postcanine dentitions is also less than that between the uppers. Mammals differ from *Thrinaxodon* in that their lower postcanine teeth on the active side can be brought into contact with the uppers because the lower jaw can move transversely in the horizontal plane (Figure 4.5b). Lateral jaw movement in mammals is not restricted by the pterygoid bones. These are reduced in size to form the pterygoid hamuli, which are widely separated from the lower jaws.

In *Thrinaxodon*, tooth replacement was alternate and continued throughout life (Figure 4.6). A replacement tooth erupted into a greatly enlarged alveolus and was subsequently ankylosed (fused) to the jaw by the deposition of bone of attachment.

This was probably the equivalent of cementum in modern mammals. As a result, the periodontal space was almost completely obliterated. Figure 4.7a shows a transverse section through the mandible and a postcanine tooth of *Chiniquodon*, a carnivorous non-mammalian cynodont slightly more derived than *Thrinaxodon*. The "cementum" layer was thick. It filled most of the periodontal space and firmly united the tooth and alveolar bone. Replacement teeth initially developed medial to the functional tooth (Figure 4.6). As they enlarged they absorbed the surrounding bone and the root of the functional tooth, migrated laterally, and came to lie in a pocket below the crown of the functional tooth (Crompton, 1963b).

An interdigitating suture joined the hemimandibles and splenials at a massive symphysis in *Thrinaxodon* (Figure 4.5a,b). This suggests that while movement at the symphysis was possible, it was restricted.

Within the nonmammalian cynodont lineages later than *Thrinaxodon*, the size of the postdentary bones and the quadrate were further reduced. This occurred independently in forms adapted to different diets.

INFERRING MASTICATION IN *PROBAINOGNATHUS*. *Probainognathus* was a small carnivore from the middle Triassic about the same size as *Thrinaxodon*. The postdentary bones were considerably smaller, relative to skull length, than those of *Thrinaxodon* (Figure 4.8). An additional jaw articulation (Romer 1970; Crompton 1972a; Allin & Hopson 1992) was present between the surangular and the squamosal (Figure 4.9). These forms, therefore, possessed a double jaw articulation, one between the articular

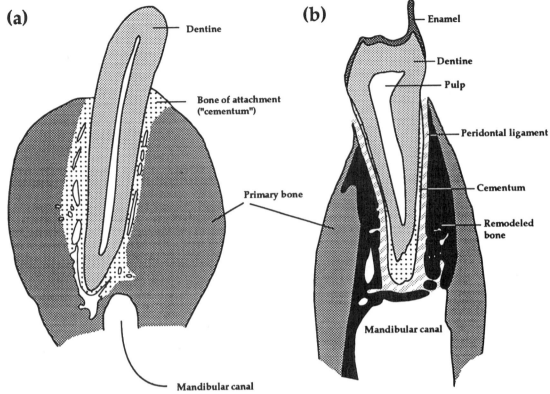

(a)

Dentine

Bone of attachment
("cementum")

Primary bone

Mandibular canal

(b)

Enamel

Dentine

Pulp

Peridontal ligament

Cementum

Remodeled
bone

Mandibular canal

Figure 4.7. Transverse section through lower jaw and postcanine of: (a) a Middle Triassic cynodont, *Chiniquodon* and (b) *Didelphis*. In *Chiniquodon*, the tooth is ankylosed to the primary bone surrounding the alveolus by bone of attachment ("cementum"). In *Didelphis*, the tooth is supported by a periodontal ligament and the alveolar bone is extensively remodeled (*black*).

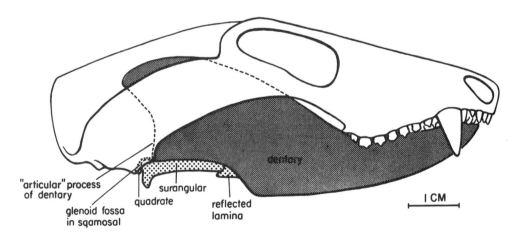

dentary

"articular" process
of dentary   surangular
quadrate
glenoid fossa   reflected
in sqamosal   lamina

I CM

Figure 4.8. Lateral view of the reconstructed skull of *Probainognathus*.

and quadrate and one between the surangular and squamosal. *Thrinaxodon* lacks the surangular–squamosal joint, but the ventral surface of the squamosal (Figure 4.9a; art.fl) closely approaches the surangular and the two were perhaps connected by a ligament. In preserved specimens of *Probainognathus*, the quadrate and postdentary bones are

usually either shifted from their natural position or lost altogether. This suggests that they were loosely joined to the skull and dentary.

The incisors and canines of *Probainognathus* were well developed, but the postcanines were relatively smaller than in *Thrinaxodon*. The upper and lower postcanines did not contact one another (Figure

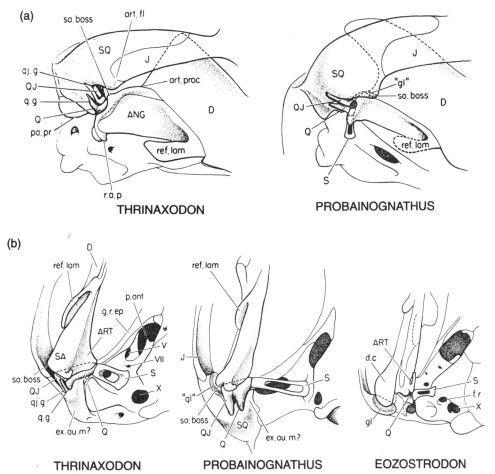

Figure 4.9. (a) Oblique lateral views of the jaw articulations in *Thrinaxodon* and *Probainognathus*. (b) Ventral views of the jaw articulations of *Thrinaxodon*, *Probainognathus*, and *Morganucodon* (=*Eozostrodon*). (After Crompton & Hylander 1986.) Abbreviations: ANG, angular; ART, articular; art.fl, articular; art.proc, articular process of dentary; D, dentary; dc, dentary condyle; ex.au.m?, possible position of external auditory meatus; fr, foramen rotundum; gl, glenoid; "gl," glenoid in squamosal that articulates with surangular boss; g.r.ep, quadrate ramus of epipterygoid; J, jugal; p.ant, pila antotica of petrosal; pa.pr, paroccipital process; q.g, groove for quadrate; QJ, quadrojugal; qj.g, groove in squamosal for quadratojugal; r.a.p., retroarticular process; ref.lam, reflected lamina of articular; S, stapes; sa.boss, surangular boss; SQ, squamosal; V, exit of trigeminal nerve; VII, primary exit of facial nerve; X, jugular foramen.

4.7b). Fully erupted postcanines were ankylosed to the jaw. *Probainognathus* probably used its canines and incisors for stabbing and tearing prey. The short postcanine row, at the most, played a minor role in processing food. The entire dentition in the specimen illustrated occupies about 35 percent of skull length. This compares with 54 percent in *Thrinaxodon*.

It is possible, based upon the skull and lower jaw morphology and the shape of the temporal opening, to reconstruct the adductor musculature of *Probainognathus*. Directions of pull and relative magnitude of the forces generated by different sections of the adductor complex (based upon cross-sectional area) can be estimated (Crompton & Hylander 1986). The space for adductor muscles in

Probainognathus was relatively larger than in *Thrinaxodon* suggesting an increase in the power of bite, despite the decrease in the relative size of the jaw joint bones. The posterior fibers of the temporalis had a ventrally directed component. If *Probainognathus* activated its adductor muscles bilaterally, it could have generated a powerful bite with the canines and incisors, without generating high reaction forces at the jaw joint (Crompton & Hylander 1986).

INFERRING MASTICATION IN THE TRITHELODONTS *PACHYGENELUS* **AND** *DIARTHROGNATHUS*. The Tritheledontidae are considered by Hopson and Barghusen (1986) to be the sister group of mammals. The best-known members of this family are

Pachygenelus and *Diarthrognathus*. These were small carnivores, the largest of which was slightly smaller than either *Thrinaxodon* or *Probainognathus*.

In both *Diarthrognathus* and *Pachygenelus*, the postdentary bones and quadrate were relatively smaller than in *Probainognathus*. The articular process of the dentary terminated in a rounded ridge with a pitted surface, suggesting that it was covered with cartilage (Crompton 1963b). It contacted the concave ventromedial surface of the squamosal (Allin & Hopson 1992). These forms lacked the squamosal glenoid and the surangular condyle of *Probainognathus* (Figure 4.9). Consequently, in *Pachygenelus* and *Diarthrognathus*, masticatory forces acting on the lower jaw could be transferred to the squamosal without subjecting the postdentary bones and the quadrate to high reaction forces.

As in *Thrinaxodon* and *Probainognathus*, tooth replacement as in *Pachygenelus* was alternate, and fully erupted teeth were firmly ankylosed to the maxilla and dentary. The postcanines developed wear facets on the labial surfaces of the lowers and lingual surfaces of the uppers. These wear facets produced sharp cutting edges along the apical border of the postcanines. As the postcanine teeth of *Pachygenelus* erupted, they immediately began to wear, against either fully or partially erupted teeth in the opposite jaw. Occlusion was probably bilateral. These forms retained well-developed transverse processes of the pterygoid that fitted snugly against the internal surface of the lower jaw and would have prevented lateral movements of the lower jaw. The ventral extensions of the pterygoid processes were separated by a narrow gap from the lower jaws. Relative to skull length, the area of the mandibular symphysis was reduced when compared to *Thrinaxodon* and *Probainognathus*. It was therefore probably fairly mobile. The wear facets on the lower postcanines were vertically oriented while those on the uppers tended to be more horizontal. This suggests that the mandible rotated during occlusion. The temporalis inserted mainly on the medial aspect of the dentary above the longitudinal jaw axis, whereas the masseter inserted on its external surface with some of its fibers inserting below the jaw axis. The medially directed components of the temporalis and masseter would, therefore, have tended to rotate the mandible around its longitudinal axis (Oron & Crompton 1985). The reduction in the postdentary bones was probably matched by a further reduction in the mass of the pterygoideus muscle. A mammalian medial pterygoid that could have controlled mandibular rotation was probably not present in the trithelodontids.

INFERRED MASTICATION IN TRITYLODONTIDS. In the herbivorous tritylodontids, the postdentary bones

and quadrate are relatively smaller than in *Thrinaxodon*. Like the Trithelodontidae, the tritylodontids lack a surangular–squamosal contact (Sues 1986). As all herbivorous and carnivorous cynodonts from the middle Triassic possess this contact, its loss is probably a derived feature of tritylodontids and trithelodontids. In tritylodontids, the jaw articulation lies exclusively between the quadrate and articular and there is no indication of a dentary–squamosal joint or ligament joining the two bones.

To break down plant material it is necessary to have lower postcanines that move in a horizontal or close to horizontal plane relative to the upper postcanines, and to have occlusal surfaces that support multiple shearing blades. In tritylodontids the upper and lower postcanines possessed a series of reciprocating blades and food was broken down as the lower jaws drew backward (Crompton 1972b). Occlusion was bilateral. In contrast to the forms discussed above, the postcanines were multirooted and appear to have been held in place by a periodontal ligament. The postcanine rows were parallel to one another. The transverse processes to the pterygoids were massive and prevented mediolateral movement of the lower jaw. The axis of the jaw joint was oblique to the longitudinal axis of the jaw. This orientation and the possibility of movement between the postdentary rod and the dentary on the one hand and the quadrate and the paroccipital process on the other, accommodated anteroposterior jaw movements in the horizontal plane. The mandibular symphysis lacked a heavily interdigitating suture and was probably mobile.

MASTICATION IN LIASSIC MAMMALS. The masticatory apparatus of early mammals such as *Morganucodon, Dinnetherium*, and *Sinoconodon* possessed several advanced features not seen in the nonmammalian cynodonts. *Sinoconodon* (Crompton & Sun 1985; Crompton & Luo in press) possessed a massive dentary condyle that articulated with the concave ventral surface of the squamosal, but lacked a well-defined glenoid on the squamosal. *Sinoconodon* had no occluding postcanines. These teeth were double rooted and held in place by a periodontal ligament.

In *Morganucodon* (Kermack, Mussett, & Rigney 1973), the dentary condyle was smaller than in *Sinoconodon*, and it articulated with a well-defined glenoid that had a ventrally facing articular facet (Figure 4.9b). The mandibular symphysis was slender and probably highly mobile. In contrast to forms such as *Probainognathus, Pachygenelus, Tritylodon*, and *Sinoconodon*, the postcanine row could be divided into a molar and premolar series (Figure 4.10). Complex occlusal wear facets developed between the molars of *Morganucodon* (Crompton & Jenkins 1968; Crompton 1974).

(a)

(c)

(b)

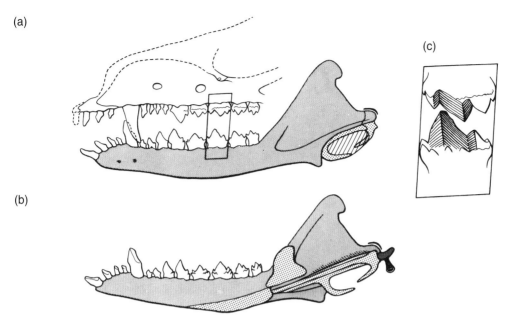

Figure 4.10. (a) External view of lower jaw and dentition of *Morganucodon*. (b) Internal view of lower jaw. (c) Occlusion between the internal surface of an upper molar and external (reversed) surface of a lower molar. Matching shearing surfaces indicated by cross-hatching. (After Crompton & Jenkins 1979.)

In early therian mammals such as the symmetrodonts and pantotheres, freshly erupted unworn teeth "fit" one another; the trigon of the uppers fit into an embrasure formed between the trigons of the lower molars and vice versa. Very little of the crown has to be worn away to produce accurately fitting shearing planes. This was not the case in Liassic mammals such as *Morganucodon* and *Kuehneotherium* or *Dinnetherium* (Crompton & Jenkins 1979; Jenkins, Crompton, & Downs 1982). When the molars of these forms erupted, they lacked matching morphologies on upper and lower molars. A large amount of the crown had to wear away to produce matching surfaces. The transverse processes of the pterygoid were reduced in size, and a small amount of mediolateral movement of the lower jaws became possible. Unilateral molar occlusion involved a combination of transverse movement and rotation of the mandible about its longitudinal axis. In *Morganucodon*, during jaw closure the lower molars moved from lateral to medial relative to the uppers, whereas in tritylodontids, the lower postcanines moved posteriorly relative to the uppers.

In *Morganucodon* the postdentary bones and the quadrate are loosely attached to the lower jaw and skull. They were probably better at conducting vibrations picked up by the tympanic membrane because the new temporomandibular joint had taken over the role of transferring masticatory forces to the skull.

A characteristic feature of mammals is that the inner ear is enclosed in a single ossification, the petrosal, rather than in several bones, as in the cynodonts (Kermack, Mussett, & Rigney 1981). A part of the petrosal, the promontorium, which houses the cochlear is visible in mammals on the ventral surface of the skull. The reorganization of the housing for the inner ear occurred with the acquisition of the new squamosal–dentary jaw articulation. The new joint freed the quadrate and articular from load bearing, and they were then free to specialize for sound conduction. The reorganization of the inner ear may be correlated with a more sensitive middle-ear mechanism.

A single replacement of the milk dentition by permanent incisors, canines, and premolars, and the presence of molars that are not replaced, has been linked inferentially with lactation and suckling (Hopson 1973; Pond 1977). A high percentage of total skull growth can occur during the suckling period. During this time, teeth are not required for the breakdown of food. Consequently, a single replacement of the entire milk dentition and the addition of molars is all that is necessary to match the dentition to skull growth that occurs after weaning. Nonavian diapsids that hatch from an egg must be able to break down food immediately after hatching. Many replacements of the dentition are necessary to accommodate skull growth that occurs between hatching and adulthood. There is no evidence that *Morganucodon* replaced its molars. If Hopson (1973) and Pond (1977) are correct, their argument suggests that *Morganucodon* and more advanced Liassic mammals suckled their young. The

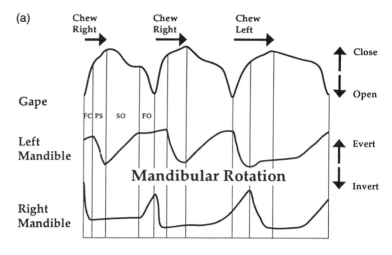

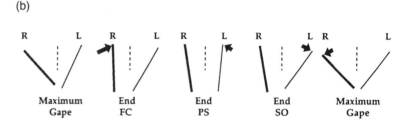

Figure 4.11. Rotation of mandible of *Didelphis* around its longitudinal axis. (a) Top trace gape profile during two right- and one left-sided chew. Each gape cycle is divided into four phases: fast close (FC), power stroke (PS), slow open (SO) and fast open (FO). Plots below the gape show the relative amount of inversion and eversion of the dentition during the phases of the jaw cycle. (b) Diagrammatic transverse section through the lower jaws to illustrate eversion and inversion of an active-side mandible (*right*) and balancing-side mandible (*left*).

situation in *Sinoconodon* is not clear. The postcanine row is not divided into premolars and molars, and postcanine teeth may have been added throughout life.

Aspects of oral function in extant mammals

MOLAR OCCLUSION.   Mastication is always unilateral in mammals. This has been documented in cineradiographic studies on opossums, tenrecs, armadillos, cats, dogs, shrews, and macaques and can readily be observed in other animals, especially the larger herbivores (Hiiemae 1978; Hylander 1979; Smith & Redford 1990; Oron & Crompton 1985). The active side jaw moves dorsomedially during occlusion. In mammals such as the opossum, the symphysis is mobile, and the mandible can rotate about its longitudinal axis. The dentition of the active side mandible is maximally everted at maximum gape (Figure 4.11a,b). During closure it inverts until both upper and lower dentitions contact the food (end fast close, FC). During the power stroke (PS) the active side jaw remains inverted and everts again during the last phase of opening (fast opening, FO). The balancing side mandible does not engage the food, and it only inverts during the power stroke, that is, later than does the active side jaw. Mammals control mandibular rotation with the aid of a power-

ful medial pterygoid muscle (Figure 4.5b). This muscle counteracts the tendency of the temporalis and fibers of the masseter, which insert on the ventral border of the dentary, to invert the dentition. The level of adductor activity during mastication depends in part on sensory information from the oral cavity on the position and consistency of the food (Thexton & Crompton 1989; Hiiemae & Crompton 1985; Hiiemae, Thexton, & Crompton 1992). For example, adductor activity levels are related to food hardness, and the timing of the jaw movements differs for hard and soft foods. Mechanoreceptors within the periodontal ligament monitor occlusal forces (Lavigne et al. 1987). These forces are transferred to the alveolar bone via the periodontal ligament, and in response, the alveolar bone in opossums and goats (D. Lieberman, personal communication) is extensively remodeled. Fluorescent dyes taken up by remodeling mandibular bone indicate that the alveolar bone remodels far more rapidly than does the remaining cortical bone (Figure 4.7b). In the cortical bone below the molars, the primary osteones and occasional secondary osteones (Haversian canals) are oriented parallel to the ventral surface of the mandible. In the vicinity of the premolars the primary osteones have a radial orientation. This may reflect the torque that is generated at the symphysis by mandibular rotation.

Activity levels of adductor muscles on the active

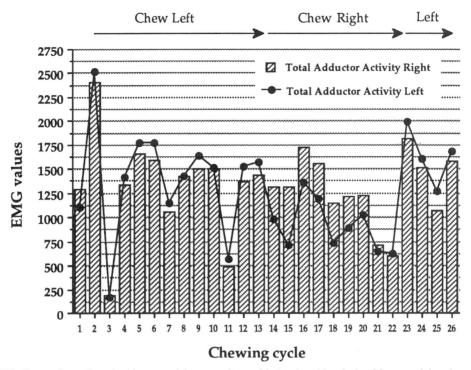

Figure 4.12. Comparison of total adductor activity on active and balancing sides during 26 sequential cycles while an opossum chewed a piece of bone. Active side shifted several times.

side when an animal chews soft food are higher than those on the balancing side (Hylander 1979). However, as harder foods such as bone are masticated, not only do the EMG levels of the active-side adductors increase, but those of the balancing side increase at a still greater rate until they match those of the active side (Crompton & Hylander 1986). Figure 4.12 shows a plot of the total activity of the principal jaw-closing muscles on the right and left sides while an opossum chewed a piece of bone. The muscles recorded from included: the anterior, middle, and posterior temporalis; the deep and superficial masseter, and the medial pterygoid. In the course of the 26 chewing cycles shown, the active side shifted several times. Activity on the right side of the jaw is represented as a bar graph and that of the left as a line graph. The total muscle activity levels do not change significantly as the active side shifts from left to right and vice versa, although there is a tendency for the active-side levels to be slightly higher than those of the balancing side.

TRANSFERENCE OF FORCE ACROSS A MOBILE SYMPHYSIS. The levels of activity in the muscles enable us to predict the loading on each hemimandible when an opossum chews unilaterally at the molars (Figure 4.13). The balancing-side musculature tends to adduct the symphysis against the reaction of the active mandible. The balancing mandible

is loaded in three points, bending down at the symphysis and joint, up at the muscle attachments, which will load the ventral border in compression (–). If it were assumed that the adducting force from the balancing side is transfered via the symphysis, the active-side mandible would be loaded at four points, bending up at the symphysis and muscle insertions and down at the bite point and condyles. This situation should load the ventral border in tension (+) between the bite point and symphysis, compression (–) between the condyle and muscles, and a summation of some compression and some tension between the bite point and muscles. Whether this sum results in net tension or net compression is difficult to predict.

These predictions were tested using single-element strain gauges positioned as in Figure 4.13, and were confirmed. The balancing mandible is loaded in compression ventrally; the active side is loaded in both tension and compression. In Figure 4.14 the strain levels recorded with single-element gauges on either side of the symphysis, through the same 26 cycles as in Figure 4.12, are compared. The strain on the balancing side is always compressive, that on the active side always tensile. The tensile strain on the active side is roughly equal in magnitude to the compressive strain on the balancing side. This means that all the force generated by the balancing-side musculature is transferred, via symphyseal ligaments,

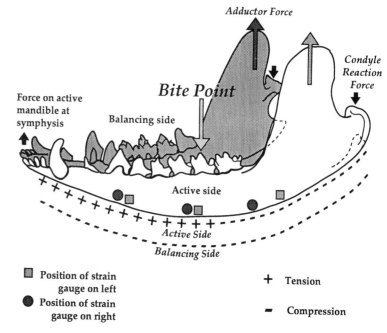

Figure 4.13. Diagram illustrating effect of active- and balancing-side jaw adductor musculature on loading of the active and balancing side temporomandibular joints. The position of single element gauges on the ventral border of the mandible are indicated. See text for explanation. Symbols: +, tensile strain; –, compression strain.

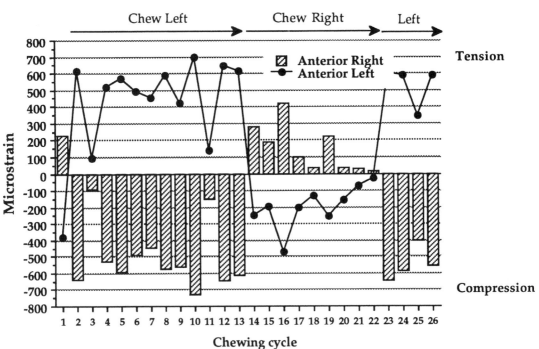

Figure 4.14. Amount of peak strain registered at the ventral surface of the left and right hemimandibles immediately behind the mandibular symphysis during the 26 cycles illustrated in Figure 4.12. Thus, these are strains from the anterior right (A.R.) and anterior left (A.L.) gauges. See text for explanation.

to the active side. The cruciate ligaments of the symphysis permit rotation of the two mandibles relative to one another, but do not prevent the transfer of vertically directed forces (Scapino 1981). The levels of strain recorded depend upon the position of the gauges on the ventral border of the mandible.

All the gauges situated below the teeth on the active side register tension when biting bone with the posterior molars. However, the gauge that lay behind the dentition (Figure 4.13; posterior gauge) always registers compression, whether it is on the active or the balancing side (Figure 4.15).

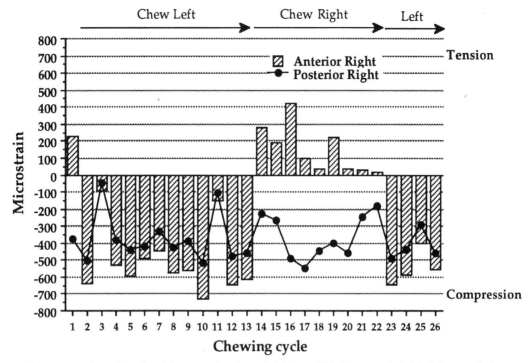

Figure 4.15. Comparison of strain of the ventral surface of right mandible immediately behind the symphysis (anterior right, A.R.) and on the ventral surface of the mandible behind the dentition (posterior right, P.R.). See Figure 4.13 for the position of the gauges. The ventral surface of the mandible behind the bite point always registers compression despite a shift of the chewing side. The 26 sequential chewing cycles are the same as those given in Figures 4.12 and 4.14.

In Figure 4.16 the relationship between the gape profile, strain levels and EMG levels are illustrated. Strain increases during closing and reaches its peak at the end of adductor activity. Strain profiles would match those of the EMGs, except in the opossum the EMGs precede the strain by about 50–75 ms.

JOINT REACTION FORCES. The distribution of forces around the mandible tends to reduce the vertical compressive force on the active-side jaw joint in two ways. The vertical component of the jaw-closing muscles is closer to the jaw joint than the symphysis (Figure 4.13). Therefore, if the level of adductor activity is the same on both sides, then on the balancing side, vertical loading at the joint is higher than at the symphysis. There is less vertical loading at the active-side joint because the bite point is closer to the jaw-closing muscles than to the symphysis. Vertical loading of the active-side joint is further reduced because of the transference of forces across the symphysis. The balancing-side force tends to elevate the active-side jaw at the symphysis, which tends to rock the active-side mandible about the bite point. This action reduces the compressive loading at the active-side joint. Thomason, Russell, and Morgeli (1990) have recorded bite forces on the anterior and posterior molars in the opossum. Based

upon these values they calculated joint reaction forces. Their assumption that balancing side force is transferred to the active side via a mobile symphysis has now been confirmed (see above). They concluded that the balancing-side jaw joint is under higher compressive loads than the active-side jaw joint, but that the discrepancy is less when biting with the anterior molars and premolars than with the posterior molars. When the posterior molars are involved in biting, the compressive force at the condyle could be as much as or greater than half the bite force.

If the adductor muscles are only active on the side on which the animal is chewing, it is theoretically possible, for reasons outlined in discussing Figure 4.3, to reduce the reaction forces, acting through the TMJ by differentially activating parts of the adductor complex. It is also theoretically possible to maintain low forces at the joint, if the adductor muscles on both sides are active to the same extent and occlusion is bilateral. Neither of these situations have been observed in mammals. Adductor muscles on both sides participate during powerful biting, and occlusion between the molars is always unilateral. With unilateral occlusion and high levels of activity in both the active and balancing side adductors, high compressive forces will always be generated at the

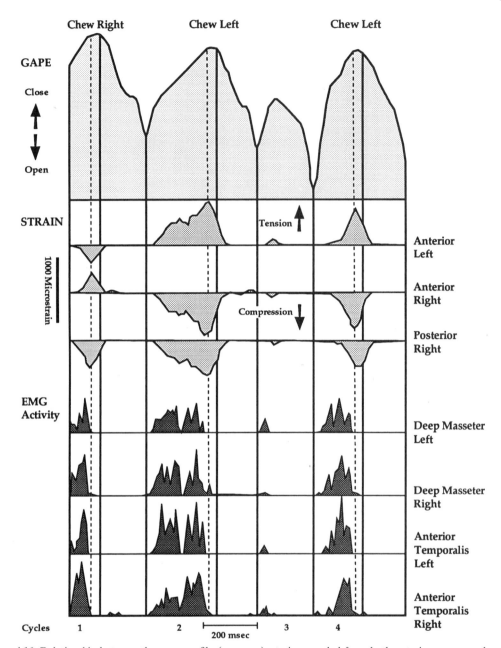

Figure 4.16. Relationship between the gape profile (*top trace*), strain recorded from both anterior gauges and one posterior gauge, and rectified and integrated EMGs of the masseter and temporalis on right and left sides. These are the first four cycles in Figures 4.12 and 4.14.

balancing-side condyle. In mammals the size of the mandibular condyles reflect the forces to which they are subjected (Crompton & Hylander 1986).

SUCKLING. The soft palate of mammals can be divided into anterior and posterior regions (Figure 4.17). The anterior region lies between the pterygoid hamuli (Figure 4.5b). The tensor veli palatini and palatoglossal muscles insert into this region. The posterior region of the soft palate that lies behind

the pterygoid hamuli surrounds the epiglottis and dorsal aspect of the larynx. The posterior region contains the palatopharyngeus muscle and is elevated by the levator veli palatini. These two parts of the soft palate help to form two seals: one at the fauces between the oral cavity and the oropharynx (seal #1) and the other between the oropharynx and the nasopharynx (seal #2) (Crompton 1989). The tongue and anterior part of the hard palate form a seal around the nipple. Several muscle groups work

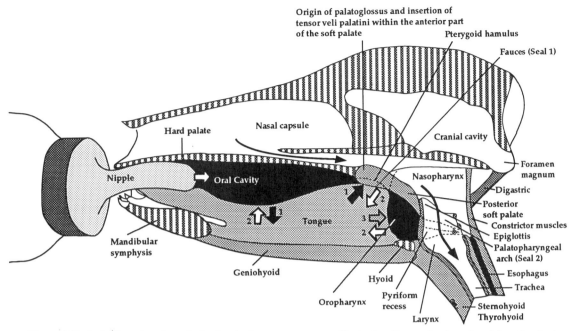

Figure 4.17. Sagittal section through the head of a young opossum illustrating the mechanism of suckling. See text for explanation.

together to form seal #1. The anterior portion of the soft palate is tensed by the tensor veli palatini muscle. The palatoglossal muscle can draw the tongue surface against the tensed anterior part of the soft palate, and the intrinsic tongue musculature can shape the tongue so that its dorsal surface is forced against the ventral surface of the anterior part of the soft palate. The tensor veli palatini originates on the side wall of the cranium close to the tympanic bone. It passes downward over the wall of the pharynx to the pterygoid hamulus where its muscle fibers become tendinous. They cross the trochlear ridge of the hamulus and radiate within the anterior part of the soft palate (Miller, Christensen, & Evans 1964). Barghusen (1986) has argued convincingly that part of the reptilian pterygoideus muscle is the homolog of the mammalian tensor veli palatini muscle. As the postdentary bones in advanced cynodonts decreased in size, so did the pterygoideus muscle. Part of the muscle shifted its attachment from the post-dentary rod to the alisphenoid and petrosal bones. Its origin migrated around the lateral edge of the pterygoid hamuli to invade the anterior part of the soft palate. Contraction of the palatopharyngeal muscle within the posterior portion of the soft palate maintains seal #2 by gripping the epiglottis and larynx.

Suckling can be broken down into three steps (German et al. 1992; Crompton, in preparation). During the first step (Figure 4.17; numbered arrows), the tongue is forced against the tensed anterior portion of the soft palate to form seal #1. The tongue behind the nipple and in front of seal #1 is depressed. This lowers intraoral pressure, and milk is drawn into the oral cavity. During the second step, seal #1 is broken (the soft palate is no longer tensed and the tongue is withdrawn from its contact with the soft palate). The tongue anterior to seal #1 rises, forcing the milk into the oropharynx. Seal #2 prevents milk from entering the nasopharynx. The oropharynx is enlarged by drawing the base of the tongue forward. The oropharynx is extended on either side of the larynx by the pyriform recesses (lateral food channels). These lie below the palatopharyngeal arch and terminate at the proximal end of the esophagus. Steps 1 and 2 are repeated several times until the oropharynx and pyriform recesses are distended with milk. During step 3, the tongue contacts the anterior part of the soft palate (seal #1), the posterior and near vertical surface of the tongue is drawn backward and this forces the milk in the oropharynx and pyriform recesses into the esophagus. This is accomplished without breaking seal #2 between the epiglottis and the soft palate. A patent airway (Figure 4.17; bold arrow) can be maintained during a liquid swallow in opossums (Crompton, in preparation). In opossums and all other mammals studied cineradiographically, seal #2 is broken when solid food is swallowed. The epiglottis and larynx are withdrawn from the palatopharyngeal arches; the posterior part of the soft palate contacts the pharyngeal wall to seal off the nasopharynx; and the bolus, en route to the esophagus, passes over the epiglottis, which is tilted over the larynx, and laterally to the larynx in the food channels. Olfactory sense is not impaired when food is chewed or transported to the oropharynx. The airway remains patent. Swallows break the epiglottal contact, but as they only take a fraction of a second, mastication does not interfere with olfaction (Negus 1949).

DISCUSSION

A probable explanation for the reduction of the post-dentary bones and quadrate within the nonmammalian cynodonts is that these bones were involved in conducting vibrations from the tympanic membrane to the stapes, as well as forming the jaw articulation. There was a selective advantage to reducing them in size. In nonmammalian cynodonts, the reduction of these two bones decreased the ability of the jaw joint to resist high reaction forces. But a progressive reduction in the size of the quadrate and articular was matched by an increase in the relative mass of the jaw-closing muscles. The temporal opening that housed the temporalis muscle became enlarged, and a masseter muscle differentiated from the adductor mass. Its origin was on the zygomatic arch and its insertion on the external surface of the dentary. This increase of muscle mass did not result in increased loading of the joint because a progressive reorganization of the jaw-closing muscles accompanied the reduction of the jaw-joint bones. The orientation of some of the fibers of the temporalis were horizontal, or close to horizontal, due to the presence of a tall coronoid or ascending process on the dentary and the elongation of the temporal opening. When active, these fibers would have reduced the vertical component of the joint reaction forces. The horizontal vector of the superficial masseter opposed the posteriorly directed horizontal component of the temporalis to prevent translation of the jaw.

The necessity of keeping joint reaction forces low placed constraints on the evolutionary potential of the masticatory apparatus. Forms such as *Probainognathus* appear to have restricted powerful biting to incisor and canine regions. In modern mammals adductor muscles on both sides are equally active during powerful bites. Hylander and Crompton (1986) reconstructed the direction of pull of the principal components of the adductor mass in *Probainognathus*. Based upon this reconstruction, they took moments around an incisal or canine bite point. They concluded that if adductors on both sides were equally active, powerful bites would only generate minor reaction forces at the jaw joint. If unilateral chewing had involved the postcanine dentition, re-

action forces at the balancing-side joint would have been high. But several features suggest that unilateral postcanine chewing did not take place: The jaw joint was too small to resist major forces; the large transverse processes of the pterygoids restricted jaw movements to the vertical plane; wear facets were not present on the postcanines; and the alveolar bone surrounding the postcanine teeth showed no sign of remodeling.

Carnivorous nonmammalian cynodonts, all the therocephalians but the bauriamorphs (Crompton 1962), and the gorgonopsians, are characterized by powerful incisors and canines, and by relatively weak postcanines. During prey capture, the canines had to resist struggling prey. It was essential that the adductor muscles be organized so that the bite forces did not generate high reaction forces at a reduced jaw joint. If these animals had large postcanines, and if they had been accustomed either to subduing or breaking down prey with these teeth, it would have been impossible to avoid unilateral loading of the mandible at all times. The inclusion of the jaw-joint bones in an ossicular chain between the tympanic membrane and stapes, and the extreme reduction of all three bones probably explain the small and nonoccluding postcanines in gorgonopsians, therocephalians, and most carnivorous nonmammalian cynodonts.

In "primitive" modern mammals such as the opossum, the situation is the reverse of that in *Probainognathus*. Lateral jaw movements are not restricted; the molars are large and occlude unilaterally; alveolar bone is, in response to biting forces, heavily remodeled; and high compressive forces are generated at the balancing-side jaw joint. In mammals, the medial pterygoid and masseter muscles form a sling. Differential contraction of the jaw-closing muscles controls the complex jaw movements that take place during occlusion. Precise control of the jaw is essential for the correct alignment of the lower molars relative to the uppers as they come into occlusion. Control of jaw movements is, in part, dependent on sensory feedback from the oral region. An important component of this feedback comes from mechanoreceptors within the periodontal ligament. Jaw movements in *Probainognathus* and more primitive carnivorous forms were simply orthal (vertically rotating) and guided by the transverse processes of the pterygoids. The pterygoideus muscle was reduced, and a mammalian medial pterygoid had probably not yet differentiated from the adductor mass. The jaw was not held in a sling of muscle.

Two types of occlusion developed within the nonmammalian cynodonts and Liassic mammals. (The haramyids are excluded from consideration because so little is known about the placement of postcanines in the jaw.) In *Pachygenelus*,

Diarthrognathus, and mammals such as *Morganucodon*, shearing planes of the upper postcanines always lay external to the matching planes on the lowers. In tritylodontids, on the other hand, the upper postcanines "straddled" the lowers so that two anteroposteriorly aligned rows of cusps on the lowers occluded between three rows of cusps on the uppers.

Pachygenelus represents an advance over the condition in *Probainognathus*. Jaw movements were still restricted to the vertical plane by the transverse processes of the pterygoids, but the postcanine rows occluded with one another. This was effected by increasing the transverse distance between the two rows of lower postcanines. *Pachygenelus* had a very simple form of shearing occlusion that did not involve complex movements of the lower jaw relative to the upper. Fully erupted teeth were fused to the jaws and therefore lacked periodontal ligaments. Occlusion was bilateral. This did not rule out the possibility of food particles being positioned unilaterally and the generation of high reaction forces at the contralateral jaw joint. *Pachygenelus* possessed a squamosal–dentary contact, and although this may not have formed a well-defined synovial joint, it would have helped to reduce the forces acting on the articular–quadrate joint. Occasional unilateral occlusion was less likely in *Probainognathus* because the postcanines did not occlude with one another.

All Liassic mammals are characterised by a load-bearing squamosal–dentary joint. The acquisition of this type of joint removed the constraint placed upon the masticatory system by a jaw joint that could not be subjected to major reaction forces. A load-bearing joint permitted unilateral chewing with the molars. Reaction forces at the joint are high when muscles on both sides are equally active. This is always the case when hard food is chewed. The molars were larger relative to jaw length, and positioned closer to the jaw joint. In *Morganucodon*, occlusion was still rather simple. The transverse processes of the pterygoids were reduced, but were large enough to constrain extensive transverse movement of the lower jaw. Occlusion involved a combination of mandibular rotation and minimal transverse movement.

Molar evolution during the Mesozoic is characterised by an increase in the freedom of movement of the lower jaw. The transverse processes of the pterygoids were withdrawn from their contact with the lower jaws. A medial pterygoid muscle, which inserted on the medial surface of the angle of the jaw, probably arose after the postdentary bones had lost their connection with the lower jaw in the adult. The muscular sling formed by the masseter and medial pterygoid is an important component of lateral jaw movements.

Sinoconodon possessed a more massive dentary condyle than *Morganucodon*, but lacked a well-defined glenoid facet on the squamosal. As in *Morganucodon*, this new joint permitted the quadrate and articular to function solely for the transmission of auditory vibrations. Improved auditory acuity in mammals may also have resulted from the enlargement of the promontorium to house the cochlear and the rest of the inner ear in a single ossification. It is interesting to note that a reorganization of the inner ear occurred when a new jaw articulation developed. In a juvenile specimen of *Pachygenelus*, the cochlear appears to be housed in a promontorium rather than within the basisphenoid (Hopson & Crompton, in preparation). That these forms had a squamosal–dentary contact or joint supports the correlation between the freeing of the quadrate and articular from jaw-joint function and the housing of the inner ear in a single ossification. *Sinoconodon* enigmatically lacked postcanine occlusion, despite the presence of a load-bearing jaw joint.

Tritylodontids retained a small quadrate–articular jaw joint. As all known Early to Middle Triassic cynodonts possessed a surangular–squamosal contact or connection, it is assumed that tritylodontids lost this feature, and never developed a squamosal–dentary joint. Without a load-bearing joint, how could they have developed complex occlusion? The answer probably lies in their bilateral occlusion. The postcanine rows were parallel and had interdigitating rows of cusps. The transverse processes of the pterygoid bones prevented lateral jaw movements. It is impossible to occlude the postcanine row of one side without occluding the other. The general form of the adductor muscles in tritylodontids can be reconstructed with a fair degree of confidence (Crompton & Hylander 1986). If the levels of adductor activity were equal on both sides of the head during mastication, it would have been possible to develop substantial bite forces between the postcanine rows without generating high reaction forces at the jaw joint. This would have been impossible if postcanine occlusion had been unilateral and the adductors of both sides were active. During occlusion the lower jaws were drawn posteriorly. Anteroposterior movements of the lower jaws during occlusion indicate differential contraction within the adductor mass and sensory feedback from the oral region. Tritylodontid postcanines are held in place by periodontal ligaments and, as in mammals, these were probably the site of mechanoreceptors, whose output could modify the activity pattern of the adductors.

Practically nothing is known about the structure of the soft palate, larynx, or hyoid complex in nonmammalian cynodonts. The posterior edge of the hard palate is arch shaped and was probably continued by a soft palate (Figure 4.5a; Barghusen

1968; Allin & Hopson 1992). In *Pachygenelus* and *Morganucodon*, the ventral tips of the transverse processes are withdrawn medially away from the lower jaw to create a narrow gap. This would have made it possible for the fibers of part of the reduced reptilian pterygoideus muscle to invade the soft palate and form the tensor veli palatini muscle. The palatoglossal muscle probably also invaded the soft palate at the same time. Both muscles are essential for the formation of a seal between the oral cavity and the oropharynx, which, in modern mammals, constitutes an essential component of the suckling mechanism. The possible presence of a tensor veli palatini muscle in *Pachygenelus* and *Morganucodon* suggests that they may have suckled their young. This alone is meager evidence for such a claim. It may, however, confirm the view, based upon dental evidence, that early mammals suckled their young (Hopson 1973).

REFERENCES

Allin, E.F. 1975. Evolution of the mammalian middle ear. *Journal of Morphology* 147, 403–438.

Allin, E.F., & Hopson, J.A. 1992. Evolution of the auditory system in Synapsida (Mammal-like reptiles and primitive mammals) as seen in the fossil record. In *The Evolutionary Biology of Hearing*, ed. D.B. Webster, R.A. Fay, & A.N. Popper, pp. 587–614. New York: Springer-Verlag.

Barghusen, H.R. 1968. The lower jaw of cynodonts (Reptilia, Therapsida) and the evolutionary origin of mammal-like adductor musculature. *Postilla* 116, 1–49.

Barghusen, H.R. 1986. On the evolutionary origin of the therian tensor veli palatini and tensor tympani muscles. In *The Ecology and Biology of Mammal-like Reptiles*, ed. N. Hotton III, P.D. MacLean, J.J. Roth, & E.C. Roth, pp. 263–282. Washington, D.C.: Smithsonian Press.

Basmajian, J.V., & Stecko, G. 1962. A new bipolar electrode for electromyography. *Journal of Applied Physiology* 17, 849.

Bramble, D.M. 1978. Origin of the mammalian feeding complex: models and mechanisms. *Paleobiology* 4, 271–301.

Crompton, A.W. 1962. On the dentition and tooth replacement in two bauriamorph reptiles. *Annals of the South African Museum* 46, 231–255.

Crompton, A.W. 1963a. On the lower jaw of *Diarthrognathus* and the origin of the mammalian jaw. *Proceedings of the Zoological Society of London* 140, 697–750.

Crompton, A.W. 1963b. Tooth replacement in the cynodont *Thrinaxodon liorhinus* Seeley. *Annals of the South African Museum* 46, 479–521.

Crompton, A.W. 1971. The origin of the tribosphenic molar. *Zoological Journal of the Linnean Society, (supplement 1)* 50, 65–88.

Crompton, A.W. 1972a. The evolution of the jaw articulation in cynodonts. In *Studies in Vertebrate Evolution*, ed. K.A. Joysey & T.S. Kemp, pp. 231–251. Edinburgh: Oliver & Boyd.

Crompton, A.W. 1972b. Postcanine occlusion in cynodonts and the origin of the tritylodontids. *Bulletin of the British Museum of Natural History (Geology)* 21, 21–71.

Crompton, A.W. 1974. The dentitions and relationships of the southern African mammals, *Erythrotherium parringtoni* and *Megazostrodon rudnerae*. *Bulletin of the British Museum of Natural History (Geology)* 24, 397–437.

Crompton, A.W. 1989. The evolution of mammalian mastication. In *Complex Organismal Functions: Integration and Evolution in Vertebrates*, ed. B.D. Wake & G. Roth, pp. 23–40. Berlin: John Wiley & Sons.

Crompton, A.W., & Hylander, W.L. 1986. Changes in mandibular function following the acquisition of a dentary–squamosal jaw articulation. In *The Ecology and Biology of Mammal-like Reptiles*, ed. N. Hotton III, P.D. MacLean, J.J. Roth, & E.C. Roth, pp. 263–282. Washington, D.C.: Smithsonian Press.

Crompton, A.W., & Jenkins, F.A., Jr. 1968. Molar occlusion in Late Triassic mammals. *Biological Reviews* 43, 427–458.

Crompton, A.W., & Jenkins, F.A., Jr. 1979. Origin of mammals. In *Mesozoic Mammals, the First Two-thirds of Mammalian History*, ed. J.A. Lillegraven, Z. Kielan-Jaworowoska, & W.A. Clemens, pp. 59–73. Berkeley: University of California Press.

Crompton, A.W., & Luo, Z. In press. Relationships of the Liassic mammals *Sinoconodon*, *Morganucodon oehleri* and *Dinnetherium*. In *Mammalian Phylogeny*, ed. F.S. Szalay, M.C. McKenna, & M.J. Novacek. New York: Springer-Verlag.

Crompton, A.W., & Parker, P. 1978. Evolution of the mammalian masticatory apparatus. *American Scientist* 66, 192–201.

Crompton, A.W., & Sun, A. 1985. Cranial structures and relationships of the Liassic mammal *Sinoconodon*. *Zoological Journal of the Linnaean Society* 85, 99–119.

German, R.Z., Crompton, A.W., Levitch, L.C., & Thexton, A.J. 1992. The mechanism of suckling in two species of infant mammal: miniature pigs and long-tailed macaques. *Journal of Experimental Zoology* 261, 322–330.

Hiiemae, K.M. 1978. Mammalian mastication: A review of activity of jaw muscles and the movements they produce in chewing. In *Development, Function and Evolution of Teeth*, ed. P.M. Butler & K.A. Joysey, pp. 359–398. London: Academic Press.

Hiiemae, K.M., & Crompton, A.W. 1985. Mastication, food transport and swallowing. In *Functional Vertebrate Morphology*, ed. M.E. Hildebrand, D.M. Bramble, K.L. Liem, & B.D. Wake, pp. 262–290. Cambridge: Harvard University Press.

Hiiemae, K.M., Thexton, A.J., & Crompton, A.W. 1992. The effect of food consistency on jaw profile in macaques. *Journal of Dental Research* 72, 744.

Hopson, J.A. 1966. The origin of the mammalian middle ear. *American Zoologist* 6, 437–450.

Hopson, J.A. 1973. Endothermy, small size, and the origin of mammalian reproduction. *American Naturalist* 107, 446–452.

Hopson, J.A., & Barghusen, H.R. 1986. An analysis of therapsid relationships. In *The Ecology and Biology of Mammal-like Reptiles*, ed. N. Hotton III, P.D. MacLean,

J.J. Roth, & E.C. Roth, pp. 83–106. Washington: Smithsonian Press.

Hylander, W.L. 1979. Mandibular function in *Galago crassicaudatus* and *Macaca fascicularis*: An in vivo approach to stress analysis of the mandible. *Journal of Morphology* 159, 253–296.

Jenkins, F.A., Jr., Crompton, A.W., & Downs, W.R. 1982. Mesozoic mammals from Arizona: New evidence on mammalian evolution. *Science* 222, 1233–1235.

Kermack, K.A., Mussett, F., & Rigney, H.W. 1973. The lower jaw of *Morganucodon*. *Zoological Journal of the Linnaean Society* 53, 87–175.

Kermack, K.A., Mussett, F., & Rigney, H.W. 1981. The skull of *Morganucodon*. *Zoological Journal of the Linnean Society* 71, 1–158.

Lavigne, G., Kim, J., Valiquette, S.C., & Lund, J.P. 1987. Evidence that periodontal pressoreceptors provide positive feedback to jaw closing muscles during changes in the pattern of mastication brought about by objects between the teeth. *Journal of Neurophysiology* 58, 342–358.

Miller, M.E., Christensen, G.C., & Evans, H.E. 1964. *Anatomy of the Dog*. Philadelphia: Saunders Co.

Negus, V.E. 1949. *The Comparative Anatomy and Physiology of the Larynx*. London: W. Heinmann Ltd.

Oron, U., & Crompton, A.W. 1985. A cineradiographic and electromyographic study of mastication in *Tenrec ecaudatus*. *Journal of Morphology* 185, 155–182.

Pond, C.M. 1977. The significance of lactation in the evolution of mammals. *Evolution* 31, 177–179.

Romer, A.S. 1970. The Chañares (Argentina) Triassic reptile fauna. VI. A chiniquodontid cynodont with an incipient squamosal–dentary articulation. *Breviora* (*Museum of Comparative Zoology, Harvard University*) 344, 1–8.

Rosowski, J.J. 1992. Hearing in transitional mammals: predictions from the middle-ear anatomy of extant mammals. In *The Evolutionary Biology of Hearing*, ed. D.B. Webster, R.A. Fay, & A.N. Popper, pp. 587–614. New York: Springer-Verlag.

Scapino, R. 1981. Morphological investigation into the function of the jaw symphysis in carnivorans. *Journal of Morphology* 167, 339–375.

Smith, K.K., & Redford, K.H. 1990. The anatomy and function of the feeding apparatus in two armadillos (Dasypoda): Anatomy is not destiny. *Journal of Zoology* 221, 27–47.

Sues, H.-D. 1986. The skull and relationships of two tritylodontid synapsids from the lower Jurassic of western North America. *Bulletin of the Museum of Comparative Zoology, Harvard University* 151, 217–268.

Thexton, A.J., & Crompton, A.W. 1989. Effect of sensory input from the tongue on jaw movement in normal feeding in the opossum. *Journal of Experimental Zoology* 250, 233–243.

Thomason, J.J., Russell, A.P., & Morgeli, M. 1990. Force of biting, body size and masticatory muscle tension in the opossum *Didelphis virginiana*. *Canadian Journal of Zoology* 68, 318–324.

5

Correlations between craniodental morphology and feeding behavior in ungulates: reciprocal illumination between living and fossil taxa

CHRISTINE M. JANIS

ABSTRACT

The diversity of living and fossil ungulates provides multiple examples of evolutionary adaptations to the challenge of herbivory. In this paper I review modern studies on ungulate craniodental anatomy in the historical context of conceptual and technological advances. The important influences on the modern studies include (1) the renaissance in biomechanical studies of form and function dating from around the middle of this century; (2) the emergence of new techniques such as scanning electron microscopy and electromyography; (3) the later rise of comparative method studies, aided by the prior accumulation of ecological data on living taxa; and (4) the appearance of the personal computer for easy and rapid data analysis. An understanding of the correlation between form and function in living ungulates of known ecology and behavior can be extended to the reconstruction of the dietary behavior and mode of mastication of extinct taxa. Studies of individual fossil taxa or lineages include both those with living representatives, such as horses and pigs, and those without living relatives or analogs, such as arsinotheres. Comparative method studies of a broad spectrum of living ungulates have provided anatomical correlates of precise herbivorous dietary categories such as grazer, browser, and mixed feeder. This information can not only provide evidence about the diets of individual extinct taxa, but can also be extended to the study of paleocommunities, including habitat reconstruction and environmental change over evolutionary time. Some outstanding problems are the difficulty in distinguishing mixed-feeding ungulates from the more specialized grazers and browsers, the precise functional reason for high-crowned cheek teeth, and the reason for the isometric scaling of dental dimensions. Although much of the flow of information has been from living taxa to extinct ones, the diversity of form in fossil taxa and the timing of events in the fossil record has resulted in the reinterpretation of hypotheses of form and function based on extant taxa alone.

INTRODUCTION

Ungulates, or hoofed mammals, are a highly diverse and successful group, comprising six living orders and a number of extinct ones. With the exception of whales (now technically classified with ungulates), living ungulates are mostly herbivorous, with a craniodental apparatus that has been highly modified for the mastication of a fibrous diet. Suoids (pigs and peccaries) are more omnivorous, but they have their own specialized craniodental morphology. Terrestrial ungulates provide a myriad of examples of parallel evolution of adaptations for eating herbage (whales and sea cows will not be considered in this chapter). While ungulates are not the only herbivorous mammals, their large size, morphological diversity, and extensive fossil record make them an appealing group for studying functional and ecological diversity within the general framework of adaptation to herbivory.

Herbivores in general among mammals have modified their skulls and teeth more than those with other diets. There are a number of advantages to herbivory: Food is usually abundant, and does not have to be pursued or subdued. However, the disadvantages lie in the fibrous nature of the herbage, which requires extensive mastication, and often some type of fermentation system within the gut to break down cellulose. The low protein content of herbage compared with animal matter means that this problem is compounded by the need to eat large quantities of the material. Thus there are considerable demands on the craniodental apparatus in ungulates, which have been "solved" in a variety of ways and which provide examples of evolutionary adaptations and convergence.

This chapter reviews the studies of the past few decades on the form and function of the ungulate craniodental apparatus. I initially take a historical perspective approach to show how changing ideas and the rise of new types of information and technology have influenced the type of work conducted. I present case studies on individual taxa, specific issues, and specific methodologies, and discuss how fossil ungulates can be used in paleocommunity studies for the understanding of past environments. Finally, I suggest avenues for future research, and discuss ways in which the study of fossil ungulates has influenced our understanding of living taxa.

HISTORICAL PERSPECTIVES

The rise of the study of biomechanical principles and the power of analogy

Although functional approaches to the study of form have existed throughout scientific history (see Fortelius 1990 for a discussion of the views of Aristotle and Cuvier on dental form and function), the types of study on functional morphology prevalent today stem from around the middle of this century. Studies of mammalian craniodental anatomy from earlier in the twentieth century were mainly descriptive (e.g., Loomis 1925), with the exception of works that linked the evolution of hypsodont (high-crowned) cheek teeth to the spread of grasslands (see later section on the rise of studies of dental wear and enamel structure), and occasional works like that of Gregory (1920) on the function of the preorbital fossae in fossil horses. However, it was not until the 1950s that workers generally saw the morphology of organisms as something that could be understood and modeled in terms of biomechanical principles.

Papers that summarized the trends of the early second half of this century, such as Bock and von Wahlert (1965) and Gould (1970), laudably stressed the "current revival" of morphological studies and the study of morphology in combination with function and ecology. Bock and von Walhert (1965) introduced the idea of a *form–function complex,* the study of an organism's adaptations in a context that relates its morphological features to its lifestyle. Gould (1970) discussed "a new science of form," whereby adaptations can be studied by quantitative methods, and complex form can be reduced to a set of "generating factors and causal inferences." Maglio (1972) also discussed how the study of the evolution of complex functional systems as a unit can provide a uniformitarian approach to the understanding of comparative anatomy.

Gould (1970) pointed out that the physical laws of the universe must affect all organisms in a similar fashion, and emphasized the effect of absolute body size on the ways these laws affect any particular organism. It follows that models based on these laws must apply to organisms in the past as well as organisms alive today. Thus the study of fossil taxa could move from simple observations, where the principle of homology was the guiding factor, to the application of biomechanical techniques and inferences,

where the principle of analogy (from the modern form to the fossil form) dominanted the argument. Organisms could be treated as machines, with function determined by principles of engineering, rather than by homology of structure. For example, all herbivorous mammals have a jaw joint that is offset from the level of the tooth row (see discussion later in this section). But this feature has clearly been evolved separately in different lineages, and so cannot be considered as strictly "homologous" (see also Gans 1985). However, an understanding of the biomechanical reasons for this structural change is independent of the issue of homology, and also enables the interpretation of the jaw morphology of certain herbivorous dinosaurs among the ornithopods, where the offsetting is clearly not homologous to the mammalian condition as it is in a different direction (below the level of the tooth row rather than above it as in mammals). Gould thus determined that adaptation could be studied and analyzed in a mechanical, experimental fashion, although he later came to criticize the naive application of adaptationist arguments (Gould & Lewontin 1979).

The demise of studies on functional morphology in vertebrate paleontology in the past couple of decades may be linked with the rise to predominance of cladistic methodologies in systematics, and the imperative to establish pattern before studying process (e.g., Lauder 1981). This could be interpreted as the re-emergence of the principle of homology over the principle of analogy. However, the "hardline" cladistic approach has been somewhat ameliorated in recent years (e.g., Lauder 1991, this volume), and it is certainly possible to study aspects of an extinct animal's biology in a "taxon-free" context (see Damuth 1992).

The semipopular paper by Maynard Smith and Savage (1959) on mammalian craniodental anatomy was a critical and influential study in the early part of these "revival years." In addition to presenting their own new ideas, they also summarized some earlier work, such as that of Becht (1953). They compared the skulls of carnivores and herbivores, and suggested biomechanical reasons for the characteristic differences in skull shape (see also a later review of these principles in Crompton & Hiiemae 1969). Figure 5.1 illustrates some of the salient differences between the skulls of carnivores and herbivores. Herbivores contrast with carnivores in the need for large muscles to chew food at the back of the jaw, in combination with a transverse jaw motion acting across ridged teeth. The muscles best suited for this action are the masseter and pterygoideus complexes. Herbivores have little need of the large temporalis of carnivores, which is advantageously positioned to resist the force at the front of the jaws produced by struggling prey. In herbivores, the small

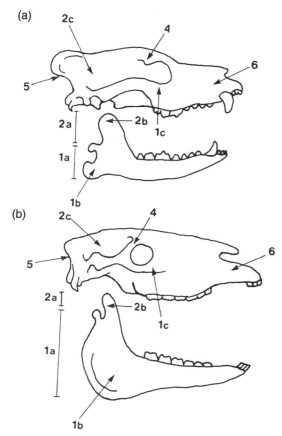

Figure 5.1. Schematic skulls illustrating the difference in craniodental design between (a) a carnivore and (b) a herbivore. Number 1 relates to the masseter muscle: 1a, larger moment arm for masseter in herbivore than in carnivore; 1b, larger area for insertion of masseter (larger angle of dentary) in herbivore; 1c, larger area for origin of masseter (anterior extension of zygomatic arch) in herbivore. Number 2 relates to the temporalis muscle: 2a, smaller moment arm for temporalis in herbivore than in carnivore; 2b, smaller area for insertion of temporalis (smaller coronoid process) in herbivore; 2c, smaller area for origin of temporalis (smaller temporal fossa and lack of sagittal and nuchal crests) in herbivore. Number 3 refers to the position of the jaw joint, which is set above level of tooth row in herbivores (allows for simultaneous occlusion of cheek teeth), and is on the level of the tooth row in carnivores. The postglenoid process, which prevents jaw dislocation due to action of a large temporalis, is present in the carnivore, absent in the herbivore. Number 4 refers to the postorbital bar, a force absorbing strut, which is present in the herbivore and absent in the carnivore. Number 5 indicates the nuchal area, which is large and elevated in the carnivore (for large neck muscles), and smaller in the herbivore. Number 6 describes the face: long face, with retracted nasals and postcanine diastema in the herbivore; shorter face with absence of a diastema in the carnivore.

temporalis plays a minor role on the balancing (nonchewing) side during occlusion to prevent dislocation of the jaw joint. (However Gingerich [1971, 1972] saw the temporalis of herbivores as important in forming a "link and lever" system with the lower jaw to resist the bite force at the teeth, and in producing orthal retraction during initial puncture-crushing of the food.)

Thus, in contrast with carnivores, herbivores have a large moment arm for the action of the masseter and pterygoid, reflected in the large angle of the dentary and the high position of the jaw joint, and also in the more pronounced zygomatic ridge for the origin of the masseter. They also have a small moment arm for the temporalis, reflected in the small coronoid process, and also in the smaller temporal area and reduced sagittal crest for the origin of this muscle. Note, however, that archaic Paleogene ungulates retained a larger temporal fossa (suggestive of a larger temporalis muscle) than most extant forms (Radinsky 1985), and among more modern ungulates a large temporal fossa is retained in oreodonts (Greaves 1973b), hyraxes (Janis 1983), and camelids (Turnbull 1970).

Turnbull (1970) studied the biomechanics of mammalian skull design and also the relative masses of the jaw elevator muscles. He showed that herbivores did indeed have a larger proportional mass of masseter and pterygoid complex muscles and a smaller-sized temporalis than carnivores. In a comparison of a horse (*Equus caballus*), a sheep (*Ovis aries*), and a deer (*Odocoileus virginianus*), he concluded that grazers (the two former species) have a proportionally larger masseter than browsers, and related this to their greater mastication of a more fibrous diet. However, Axmacher and Hofmann (1988) did not bear out this conclusion for browsing versus grazing ruminant artiodactyls, although there are changes in the arrangement of the muscles and hence in the morphology of the skull. Radinsky (1985) suggested that the predominance of the masseter and pterygoid muscle complexes in herbivores was also related to the need for precise jaw control, and Scapino (1972) linked the reduction of the temporalis in herbivores to a reduced and reoriented reaction force at the jaw joints.

Further studies of biomechanical principles in ungulate craniodental design were conducted by Greaves (and reviewed by him in this volume). For example, he pointed out (Greaves 1974) that the characteristic position of the jaw joint in herbivores, set above the level of the tooth row, may not be directly related to muscle moment arms as suggested by Maynard Smith and Savage (1959), but instead would allow for *simultaneous* (rather than sequential) occlusion of the cheek teeth. Moreover, the jaw-joint position in ungulates also allows for the same set of bilaterally symmetrical muscles to move the lower jaw in two different directions (Greaves 1980). The characteristic postorbital bar of ungulates may act as a strut to absorb forces generated by mastication of fibrous material (Greaves 1985), and the positioning of the cheek tooth row, where the distance from the jaw joint to the last molars approximates the length of the cheek tooth row, places the entire postcanine apparatus in the most mechanically advantageous position for the generation of force by the jaw elevator muscles (Greaves 1978a). Additionally, with the reduction in size of the temporalis in herbivorous ungulates, the resultant change in craniodental design reduces the distance between the jaw joint and the toothrow, along with reduction of the equivalent distance in the skull, increasing the resistance of the skull to torsion and allowing for the reduction of bone mass (Greaves 1991a).

Other characteristic features of the ungulate skull may also have functional/biomechanical explanations. The nuchal area is not greatly elevated: The expanded nuchal area of carnivores may relate to the presence of powerful neck muscles needed to resist struggling prey. The long face of ungulates, reflected by the postcanine diastema, may reflect the need for an area where the tongue can manipulate the food in mastication. Greaves (1991b) suggested that a long face might reflect increased space for larger nasal passages and an increased respiratory surface area in cursorial ungulates. Alternatively (or perhaps additionally), it seems to me that a long face may merely reflect the "need," in an animal that has modified its forelimbs for cursorial locomotion, to position the food-handling device (the incisors) at a distance from the eyes, to both protect the eyes and to permit the animal to focus on the food material (see also Greaves 1978a). (Note the relatively shorter face of kangaroos and herbivorous primates and rodents, mammals who can still use the hands for food manipulation.) The relatively small jaw condyles of herbivorous ungulates (in contrast to omnivores such as pigs) relate to the shift from orthal occlusion with heavy loading of the jaw joint to a more transverse motion, with a concomitant more posteriorly directed temporalis (Herring 1985).

Thus, the studies of biomechanical principles that underly the differing morphologies of the skulls of living mammals of known dietary type should enable the transfer of such principles to fossil animals. Certainly it should be possible to at least determine if a fossil mammal was a carnivore, an omnivore, or a herbivore. As grazing herbivores have a coarser diet than browsers, one might expect them to have more extreme modifications of the herbivore condition such as a larger mandibular angle (see Turnbull 1970; Axmacher & Hofmann 1988). Of course,

qualitative observations of this nature well predate the biomechanical studies of the 1960s and 1970s, but once such studies had been done it became possible to phrase such determinations within a rigorous, testable scientific format, and to test them experimentally. The biomechanical studies of this period also showed that craniodental anatomy could be designed for other functions than feeding. For example, bony features, such as flared mandibular angles and a low coronoid process, and muscular features, such as the orientation and pinnation of the superficial masseter muscle, are better related to the capacity for a wide gape during threat display than for reasons of diet in suoids such as peccaries and hippos (Herring 1975, 1980) and in hyraxes (Janis 1983). The legacy of the application of such biomechanical principles to fossil ungulates is reviewed in a later section.

A further development of this type of study, commencing primarily in the early 1970s, was the implementation of technology to study mammalian mastication in an experimental fashion with the techniques of X-ray cinefluoroscopy, electromyography, and strain-gauge analysis. These studies were largely pioneered by Crompton, Hiiemae, and colleagues (e.g., Crompton & Hiiemae 1970). Although such studies obviously cannot be directly applied to extinct organisms, they nevertheless can be used to test predictions derived from biomechanical analyses, and hence to increase confidence in the application of those predictions to fossil taxa. Initial studies were used to investigate the basic mammalian pattern in primitive mammals such as opossums, and were later applied extensively to primates (e.g., Kay & Hiiemae 1974). Kay and Hiiemae (1974) also confirmed the predictions of jaw movements in herbivores originally made by Butler (1952) and Mills (1955) based on studies of dental wear facets; that is, that dental occlusion occurs not only as the lower teeth move into centric occlusion with the uppers *(Phase I motion)*, but also in the initial stages of jaw opening as the teeth disengage *(Phase II motion)*. The first decade of such studies is reviewed by Hiiemae (1978).

Unfortunately, such studies have rarely been applied to ungulates, probably because the size of the experimental equipment is not conducive to investigating large animals, and ungulates are difficult and expensive to maintain in animal-care facilities. DeVree and Gans (1975) applied electromyography and motor analysis to the pygmy goat (*Capra hircus*), and confirmed the biomechanical predictions that mastication in ungulates was on one side of the jaw only at a time, and that the pattern of the jaw elevator muscles was asymmetrical. I used cinefluoroscopy and electromyography to examine mastication in the rock hyrax, *Procavia habessinica* (Janis 1979a), and verified many of the predictions of Maynard Smith and Savage (1959) (see also discussion in the next section).

The rise of studies of dental wear and enamel structure

One of the more obvious features of ungulates that eat more fibrous vegetation (primarily grazers) is their hypsodont, or high-crowned, cheek teeth. It has been repeatedly observed for more than a hundred years that horses and other ungulates have increased the height of their cheek teeth over evolutionary time, and that this event may be correlated with the spread of grassland habitats (e.g., Kowalevsky 1873–1874; Merriam 1916; Matthew 1926; Stirton 1947; Simpson 1951; Gregory 1971), an observation that persists in present-day studies (e.g., Webb 1977, 1978, 1983; Solounias & Dawson-Saunders 1988; Pascual & Ortiz Jaureguizar 1990). In a review of modes of predicting the diets of fossil mammals, Sanson (1991) pointed out that it is clear that the degree of hypsodonty and lophodonty (ridging) of the cheek teeth is related to the relative amount of fiber in the diet in a wide variety of herbivores, not just ungulates. Figure 5.2 summarizes these differences in ungulates in relation to diet. Such general aspects of the dentition will be covered further in a later section. Here I will review the emergence of more specific dental analyses that have developed over the past few decades. Dental-wear studies are exciting and important because they represent one of the few areas of study where there is direct morphological evidence of function in fossil taxa.

As previously discussed, Butler (1952) first made the observation that occlusal contact produces characteristic patterns of wear in mammalian teeth that can be homologized over a wide variety of taxa, and that such *wear facets* can be used to deduce the direction of jaw movements. Butler (1972) later defined various grades of occlusion in mammals: Herbivorous mammals were classed as the "grinding grade," where additional wear facets suggest a new Phase II pattern of jaw movement. Greaves (1973a) also showed that the difference in wear between the edges of the enamel ridges in lophed teeth could also determine the direction of jaw movement during occlusion.

Many initial studies of mammalian wear facets were on the evolution of the patterns of occlusion in primitive mammals (e.g., Crompton 1971), and the action of the jaw muscles in producing the appropriate occlusal movements was confirmed by experimental studies (e.g., Crompton & Hiiemae 1970; Kay & Hiiemae 1974). Further studies on primates led to the common assumption that all herbivores would

exhibit a large grinding type of Phase II movement in mastication, but cinefluoroscopic studies on the hyrax (Janis 1979a) showed that there was little in the way of a predicted Phase II shift in this small ungulate. Additionally, dental-wear studies on hyraxes showed a *decrease* in the extent of the Phase II with an increasingly fibrous diet in the three living genera; moderately large Phase II wear facets are apparent in the browsing *Dendrohyrax* and in the mixed feeding *Heterohyrax,* but only small Phase II facets are present in the grazing *Procavia.* A similar comparison of progressively hypsodont genera of fossil horses *Mesohippus, Parahippus,* and *Merychippus* (assuming that increasing hypsodonty reflects an increasingly fibrous diet; see Janis 1988 and Figure 5.3) showed a similar pattern of reduction, and eventual loss, of Phase II facets. Additionally, examination of dental wear in a wide variety of living ungulate species confirmed that Phase II wear was reduced or absent in those taxa with highly fibrous diets (Janis 1979b, 1990a). I thus concluded (Janis 1979a) that ungulates use their teeth to shred vegetation in a different fashion from the crushing and grinding type of motion seen in herbivorous primates.

Gross dental wear can also be related to diet. Biomechanical analysis of the way in which different food materials fracture can be related to tooth design (Lucas 1979; Lucas & Luke 1984), and herbivores show a strong correlation between the ratio of shearing to crushing wear on the teeth and the amount of fiber in the diet (more herbivorous taxa having a greater amount of shearing wear; see Janis 1990a). Because gross dental wear is cumulative, animals in the same relative wear stage must be selected for comparison (Janis 1990a; Sanson 1991). Most studies of the correlation of gross dental wear with diet, and its application to the determination of the diets of extinct mammals, have been qualitative in nature and applied to primates (e.g., Kay 1977; Maier 1984). I initially determined dietary correlates of qualitatively assessed dental wear in living ungulate taxa of known diet, and applied these to the examination the evolution of herbivory in ungulates (Janis 1979b). Later, I made a preliminary quantification of these patterns, and showed that there were demonstrable differences between groupings of taxa assigned to different dietary types (Janis 1990a). These differences were extended to a few example fossil species in this study showing, for instance, that the Late Eocene equid *Mesohippus* took a greater proportion of leaves in its diet than the more omnivorous Early Eocene *Hyracotherium.*

With the rise in scanning electron microscopy (SEM) studies in the 1960s, the 1970s saw such technological equipment become a common component of most departments of biology (at least in the

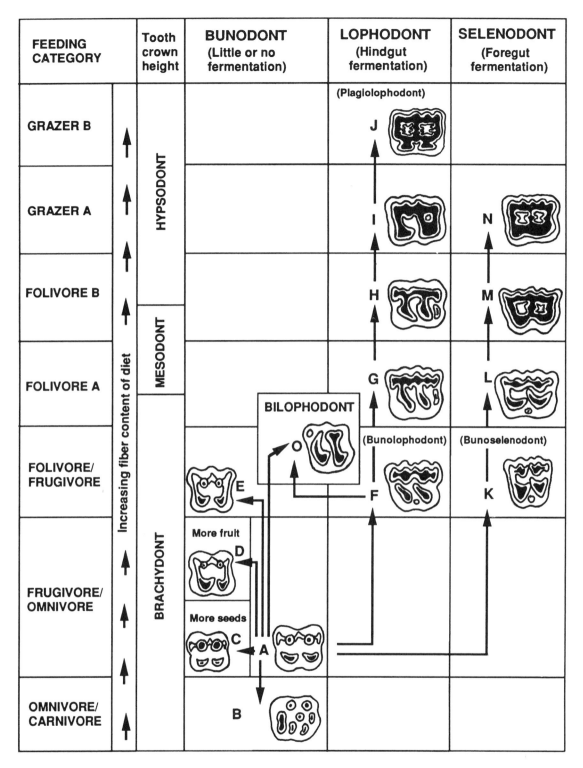

Figure 5.2. Diagram of changes in tooth-crown height and occlusal morphology with diet in herbivorous mammals. Sketches of teeth are approximations of living herbivores of known diet. All teeth represent upper left second molars (M^2) at the wear stage when the third molars are fully erupted but unworn or only slightly worn. Black areas represent exposed dentin. Abbreviations: A, generalized frugivore/omnivore (no specific living taxon); B, wild pig *Sus scrofa*; C, black-and-white colobus monkey *Colobus guereza*; D, red colobus monkey *Colobus badius*; E, black colobus monkey *Colobus satanus*; F, tree hyrax *Dendrohyrax dorsalis*; G, bush hyrax *Heterohyrax johnstoni*; H, rock hyrax *Procavia capensis*; I, white rhino *Ceratotherium simum*; J, common zebra *Equus burchelli*; K, lesser mouse deer *Tragulus javanicus*; L, white-tailed deer *Odocoileus virginianus*; M, impala *Aepyceros melampus*; N, wildebeest *Connochaetes taurinus*; O, lowland tapir *Tapirus bairdii*.

U.S.A.), and gross dental-wear studies were supplanted by SEM studies of microwear. The promise of such studies was that they could yield more precise and quantifiable correlates between diet and dental wear, and so provide a more "scientific" approach to inferring the diet of extinct animals. A pioneering paper by Walker, Hoeck, and Perez (1978) showed that the more browsing hyrax *Heterohyrax* could be distinguished from the grazing hyrax *Procavia* by the relative abundance of microscopic pits and scratches on the enamel (see also Rensberger 1978 for early microwear analyses). Microwear pits are more numerous in browsers, probably caused by crushing hard items such as stems, bark, and fruit, whereas scratches are more numerous in grazers, apparently caused by shearing silaceous grass blades (Solounias & Hayek 1993).

As with gross dental wear, most subsequent microwear studies have been on primates. However, Solounias has pioneered microwear studies of ungulates, classifying living ungulates into *grazer, browser,* and *mixed feeder* categories on the basis of their dental microwear with a fair degree of confidence (e.g., Solounias, Teaford, & Walker 1988; Solounias & Moelleken 1992a,b, in press; Solounias & Hayek 1993). Some interesting conclusions have resulted from applying this technique to fossil ungulates in comparison with the patterns seen in the living taxa. The Miocene giraffid *Samotherium boissieri* appears to have been a mixed feeder or a grazer rather than a browser like living giraffids (Solounias et al. 1988), and the diversity of the presumed grazing hipparionine equids of the late Cenozoic of Africa and Eurasia have been shown to have had a variety of feeding types; they were mainly mixed feeders rather than grazers as previously assumed (Hayek et al. 1992).

Yet despite the appeal of microwear in terms of scientific precision and quantification, some problems remain. The rate of dental wear can be extremely rapid (Skogland 1988), and dietary changes can result in rapid changes in microwear patterns (Walker et al. 1978), perhaps in as little as a matter of hours (Solounias, Fortelius, & Freeman, in preparation). Such rapid changes make the interpretation of the diet of a fossil animals subject to the "last supper phenomenon" (Fred Grine, personal communication): An animal that died at the end of a dry season might not have been consuming its usual diet immediately prior to death, and microwear patterns might be misleading. In contrast, gross dental wear, although providing less precision in quantification, is cumulative over a lifetime and would not be so biased by short term dietary changes. However, a problem here is that the comparison studies on living taxa are usually based on the gross wear patterns of lightly worn teeth (e.g., Janis 1979b, 1990a),

which may not be available in a fossil sample. Clearly, both gross dental wear and microwear may provide useful information about the diets of extinct species, but cautions and limitations apply in both cases.

The rise in SEM studies of mammalian enamel in the 1970s related not only to the patterns of microwear, but also to the study of the microstructure of the enamel itself (see Fortelius 1985 for review). Studies that relate specifically to the understanding of fossil ungulate dietary behavior concern the evolution of Hunter–Schreger (HS) bands, zones of decussating enamel prisms that appear to render the tooth less liable to fracture. Von Koenigswald, Rensberger, and Pfretzshner (1987) examined the teeth of a variety of Paleogene mammals, and showed that HS bands were present in all ungulates, including the most primitive artcocyonid "condylarths." They concluded that HS bands represent a fundamental adaptive feature of ungulates, relating to the demands of mastication of tougher diets. Fortelius (1984) discussed the vertical (rather than concentric) decussation of HS bands in living rhinocerotids, which wear to a pattern of fine transverse ridges or flutes along the enamel shearing ridges. Although this feature of rhinos had been known previously, it was unique among living mammals, and thus it was impossible to determine if it was adaptive or merely a unique peculiarity of the taxon (see Rensberger, this volume, for further discussion). Fortelius showed that a similar arrangement of HS bands occurred in a variety of extinct, large, lophodont ungulates or ungulatelike mammals, concluding that vertical HS bands must be an adaptive feature. They may be related to increased resistance to cracking, with the resultant enamel flutes perhaps acting as stress concentrators on the high-relief occlusal ridges of the cheek teeth (Boyde & Fortelius 1986).

A final area of dental studies is the unique work of Rensberger, who used a mixture of SEM technology and biomechanical modeling. While most of Rensberger's work is devoted to rodents, some papers are pertinent to ungulate dental evolution and diet. For example, the initial change from insectivory to herbivory in mammalian teeth was accompanied by a change in the surface orientation and length of the shearing surfaces, which resulted in a flatter occlusal surface and a reduced shearing mode, reflecting a change from food fracturing by shearing to the predominance of compression in trituration (Rensberger 1988). Among Paleogene ungulates, comparison of periptychid and phenacodontid "condylarths" showed that early members of both groups had cheek teeth suitable for the compression of food, but later forms diverged. Periptychids increased the size of their tooth cusps for greater compression, whereas phenacodontids converged their cusps into incipient lophs, a change that Rensberger (1986) interpreted as a pre-adaptation for the more oblique chewing mechanisms adopted by the later perissodactyls. Finally, studies of a sequence of Tertiary equids showed further changes in the edge alignment of the shearing surfaces and the flattening out of the occlusal plane. This reflected the loss of the original dual functions of shearing plus compression seen in earlier horses, with a change to a single-directional shearing function in *Merychippus,* interpreted as reflecting a shift to a diet composed primarily of tough, low-relief food items such as grass (Rensberger, Forstén, & Fortelius 1984).

The rise of ecological studies and the comparative method

The comparative method has been hailed primarily as an advance in statistical and methodological concepts in the late 1970s (see Clutton-Brock & Harvey 1979; Harvey & Pagel 1991). However, the history of how such studies came to such prominence is more complex. First, the 1980s saw the personal computer become commonplace (again, predominantly in affluent Western nations) and, together with it, user-friendly statistical software packages. Quantification problems that would have taken hours of labor in the preceding decades could now be solved rapidly and relatively easily. Second, and more importantly, precise behavioral data on mammals were now available for comparison with morphological data. The mammalian skulls and skeletons that repose in museums have been accumulated over the last century, but until there was an accumulation of behavioral and ecological data for use in comparative studies, their use was largely taxonomic. Contemporary comparative studies would be largely impossible without the vast gains in ecological and behavioral information that commenced in the 1960s.

The rise in the interest in scientific studies of animal ecology and behavior in the 1960s, combined with the affluence and growth of universities in Europe and North America during this time period, generated dozens of precise, long-term observational studies in the wild, including studies on ungulate diets. Many of these studies were thesis topics. When I was an undergraduate at the University of Cambridge in the early 1970s, a period of several years spent abroad was considered almost mandatory for a Ph.D. thesis in animal ecology or behavior. Funding for such trips did not appear to be of great concern. The trend was still evident when I returned to Cambridge as a Research Fellow in the early 1980s, although by then money seemed a little tighter, and more would-be ungulate ecologists were studying red deer on the Scottish Isle of Rhum than were studying impala in the Serengeti. Still, wildlife and con-

servation biology remain a priority in many countries, and data continue to accumulate, albeit at a slower rate than in the first flush of the golden days of academic expansion.

Papers providing precise data on ungulate diets first appeared during the early 1960s to early 1970s (e.g., Lamprey 1963; Gwynne & Bell 1968; Bell 1969; Schaller 1967, 1977; Hofmann 1973; Jarman 1974; Geist & Walther 1974). Hofmann and Stewart (1972) provided a quantitative definition of the dietary terms of *grazer* (more than 90 percent grass in the diet), *browser* (less than 10 percent grass in the diet), and *mixed feeder* (intermediate levels of grass), and Hofmann (1973) showed that these ecological categories correlated with stomach anatomy. Suddenly data were available to determine how diet might correlate with craniodental anatomy in a quantifiable, statistical fashion. The results of such work, in the late 1980s and early 1990s, are discussed in later sections on hypsodonty and comparative-method studies.

CASE STUDIES

Studies of individual taxa using biomechanical principles

The principles of biomechanical analysis have been applied in various ways to the understanding of functional craniodental adaptations of fossil ungulates, and can be conceptually divided into two groups: (1) studies that attempt to explain the adaptations of extant ungulates in a historical context, by looking at the evolution of the relevant characters in a sequence of fossil relatives, and (2) studies that attempt to identify and explain the adaptations in a particular extinct taxon that has no close living relatives for comparison.

ADAPTATIONS OF EXTANT UNGULATES PLACED IN HISTORICAL CONTEXT. Turnbull (1970) sought to understand the derivation of the peculiar tendinous insertion of the deep masseter muscle in a fossa on the mandible in the present-day horse, *Equus caballus.* Tracing the development of this fossa through a series of fossil horse skulls, he noted that it increased in size from the Late Eocene *Mesohippus* to the present-day *Equus.* The fossa also became more anteriorly situated, which he interpreted as placing the deep masseter in a position of greater mechanical advantage, now inserting further from the jaw joint. Turnbull recognised a distinct change in the size and position of the fossa between *Parahippus* and *Merychippus,* which he interpreted as being related to a switch from browsing to grazing (*Merychippus* was the first truly hypsodont equid). This change in the position of the deep masseter was also related to a decrease in the size of the temporalis muscle, as inferred from the size and the degree of cresting of the temporal fossa. This observation ties in rather nicely with the work of Rensberger et al. (1984) discussed earlier, showing a change in the mode of dental occlusion at the same point in time.

Radinsky (1983, 1984) also examined a series of fossil horse skulls to solve the controversy between two competing hypotheses for the reason for the elongate face of present-day horses, ontogenetic versus phylogenetic change. He noted that in hypsodont fossil horses (*Merychippus* and its descendants) the cheek-tooth battery was located relatively farther from the jaw joint, and that this also correlated with larger areas for both origin and insertion of the masseter and pterygoid muscles. Hypsodont species also had relatively longer jaws, a postorbital bar, wider occiputs, and larger, more posteriorly placed, orbits, but had a smaller temporal fossa and a lower coronoid process. Radinsky showed that preorbital lengthening over evolutionary time is not the same as the simple lengthening of the tooth-bearing portion of the face seen in the ontogeny of living horses, but rather is related to a ventral and foward displacement of the entire tooth row relative to the jaw joint and the orbit. This would allow for an increased moment arm for the masseter and pterygoid muscles, and allow space in front of the orbit for housing the long roots of the hypsodont upper cheek teeth.

Herring (1972) determined how differences in canine morphology in pigs and peccaries influenced other aspects of their cranial design. Present-day pigs have upwardly directed canines, whereas peccaries have downwardly directed upper canines that tightly interlock with the lowers. She noted that peccaries have well-developed pre- and postglenoid jaw processes, with the jaw joint positioned only slightly above the level of the cheek-tooth row, and an unenlarged coronoid process. She interpreted this morphology as enabling peccaries to transmit posteriorly directed bite forces to a postglenoid buttress, with the interlocking canines placing constraints on transverse shearing. Later Kiltie (1981) observed that peccaries used their posterior molars to crack open hard nuts and seeds, aided by thick enamel on the postcanine teeth and, via biomechanical analyses, concluded that the interlocking canines functioned to resist jaw-dislocating forces generated by such activities. In contrast to peccaries, pigs lack glenoid buttresses, and may have more complexly lophed teeth with a greater emphasis on transverse chewing.

Reviewing the fossil record, Herring (1972) noted that the peccary condition was the more primitive one, and it was not until the Late Miocene that pigs acquired their present-day morphology. The upturning of the canines would release the lower jaw from the orthal occlusion mandated by interlocking canines, and later pigs also reduced the postglenoid process, and acquired increasingly complex cheek teeth. At around the same time, peccaries became more derived in the secondary lowering of the jaw joint and developed a preglenoid process in addition to the original postglenoid one. Herring interpreted the different evolutionary patterns as being related to differences in social behavior. Peccaries are herd-forming and sexually monomorphic, whereas pigs are more solitary and sexually dimorphic; an initial modification of the canines in pigs for sexual display released them from their original interlocking role, enabling the ability for transverse jaw motion in later forms.

A final example in this section is that of Maglio (1972), who studied the unique craniodental system of proboscideans. He noted a trend in the evolution of present-day elephants for the cheek teeth to acquire a greater number of more tightly packed lophs, with a concomitant thinning of the occlusal enamel. Concurrent with this was the loss of the mandibular tusks, a more vertical orientation of the temporalis muscle, a forward displacement of the occipital region on the braincase, and the elevation of the parietal area. These skull changes all combined to shift the center of gravity of the skull backward, in correlation with a change in the masticatory cycle from a transverse power stroke to one where the lower jaw was swung forward into occlusion by the action of the temporalis muscle, and food was sheared isognathously (simultaneously on both sides of the jaw) along the transverse ridges of the cheek teeth. Maglio suggested that these unique elephantid features correlated with a rapid Plio–Pleistocene shift into a new adaptive feeding zone correlated with a grazing diet. The original transverse grinding/shearing action of the ancestral gomphothere masticatory apparatus was changed to one of horizontal shearing with forward jaw action and isognathous occlusion. Elephants also acquired their more hypsodont cheek teeth and horizontal mode of tooth replacement at this time. Increasing hypsodonty alone would not have facilitated this dietary shift because of limitations on the alveolar depth of the mandible.

ADAPTATIONS OF EXTINCT UNGULATES WITH NO CLOSE LIVING RELATIVES. Greaves (1973b) sought to explain the rather peculiar cranial anatomy of the North American Oligo–Miocene oreodont artiodactyls. In comparison to modern artiodactyls, oreodonts had selenodont cheek teeth like ruminants, in which the primary ridges are crescent-shaped and run in an anteroposterior direction across the teeth, rather than the straighter ridges with a primary buccolingual

direction of lophodont teeth (see Figure 5.2). In contrast with ruminants, oreodonts maintained the more primitive condition of a large temporal region and a short face with the virtual absence of a diastema, combined with a derived (but non-ruminantlike) condition of a strongly fused jaw symphysis. Greaves interpreted aspects of oreodont cranial anatomy as relating to the need to maintain a large gape with a short face, which meant that the leverage action of the masseter and temporalis muscles could not be improved (as in ruminants) by moving their origin, as this would limit the gape. He showed that the temporal fossa provides evidence for the presence of a large, divided temporalis muscle, the ventral portion of this muscle performing the role of the temporalis in ruminants, and the dorsal portion acting to supplement the role of the masseter.

I later pointed out that hyraxes, which are craniodentally very similar to oreodonts, despite having lophodont rather than selenodont cheek teeth, actually do have a divided temporalis muscle resembling the condition postulated by Greaves (Janis 1983). Hyraxes also have a wide-gape threat display, and both hyraxes and oreodonts have the additional modifications for this display described by Herring (1975, 1980) in peccaries and hippos. Greaves (1978b) later noted that large oreodonts had zygomatic arches that were widely and laterally expanded, which he interpreted as creating a shelf for the attachment of the ventral portion of the temporalis. The usual mode of increasing the area of origin for the temporalis is to increase the size of the sagittal and lambdoidal crests, a morphology often seen in larger mammals because size-scaling constraints dictate that larger animals should have proportionally larger muscles. However, this option would not be open to oreodonts because of the division of the temporalis into dorsal and ventral portions. (A brief examination of the fossil evidence also suggests that large hyracoids also look similar.)

Harris (1975) studied the peculiar deinothere proboscideans that lacked upper tusks but possessed recurved lower tusks. In comparison with other proboscideans, they had an increased area of origin for the temporalis, and their bilophodont tapirlike teeth suggested a vertical type of shearing occlusal motion. The tooth row was apparently bifunctional, with the more posterior molars offset laterally from the premolars. The premolars were apparently designed to crush and grind food, whereas the more highly lophed posterior molars acted to shear herbage with beveled, self-sharpening occlusal edges. The probable function of the tusks proved more elusive. Their rounded conical points suggested general-purpose tools, rather than instruments for digging. They may have served to clear superfluous vegetation

from a desired food source, or perhaps were used as a source of purchase for a trunk in food manipulation. Wear on the anteromedial surfaces of these tusks is also consistent with their use in stripping bark or other vegetation.

Wall (1980) studied the extinct early Tertiary amynodont rhinos. He showed that the skulls of the *Amynodontopsis–Cadurcodon* lineage could be compared with those of living tapirs. Features such as the posterior enlargement and expansion of the nasal incision, along with a cavity to house a laterally placed nasal diverticulum, and a large infraorbital foramen for sensory nerves returning from the anterior face, suggested the presence of a tapirlike proboscis. He also noted that similar modifications had taken place in other ungulate taxa, such as the oreodont *Brachycrus* and the litoptern *Macrauchenia*.

Joeckel (1990) studied the Oligo–Miocene entelodont artiodactyls, or "giant hogs," using both biomechanical modeling and comparisons with living taxa. Entelodonts had a strange mixture of primitive and derived features: They combined a large temporal fossa and an unreduced dental formula with the derived features of a laterally expanded zygomatic arch, a low coronoid process, a jaw joint at the level of the tooth row, a fused mandibular symphysis, and isognathous occlusion. Ontogenetic changes showed that with the increasing size of the temporal fossa, the coronoid apex of the lower jaw moved forward and a thickened postcoronoid bar was developed between the jaw condyle and the base of the coronoid process, a morphology unique to entelodonts. Joeckel concluded that entelodont jaws had a powerful and primarily orthal stroke, with incisors and canines acting to punch, grasp, and chop the food. The molars were relatively small, with apical wear on the premolars, and the cranial design reflected the capacity for a wide gape. He interpreted the entelodont craniodental design as consistent with the hypothesis of a catholic, omnivorous diet, possibly including scavenging of carrion.

Court (1992) examined the arsinotheres of the Late Eocene or Early Oligocene of Egypt, which resemble rhinoceroses but are more closely related to proboscideans. The molars had a peculiar dilambdodont morphology, with high buccal cusps and low lingual cusps, limiting the occlusal motion to a single-phase shearing action. However, the premolars were offset medially from the molars, with an occlusal morphology suggesting a more crushing and grinding function (analogous to the dentition of deinotheres). The mandibular condyle was uniquely double-faceted, interpreted as allowing for a bifunctional system of occlusion whereby the jaw joint could be positioned differently for shearing with the molars or crushing with the premolars. Arsinotheres also had a narrow muzzle, and Court interpreted the entire craniodental design as adapted to a highly specialized selective type of browsing diet.

The final example in this section is that of Lambert (1992), who studied the shovel-tusked gomphothere proboscideans *Amebelodon* and *Platybelodon*. Their peculiarly expanded lower jaws and tusks had previously been interpreted as a device for digging or for scooping up aquatic vegetation. By comparing the wear on the tusks of these extinct ungulates with that on the tusks of a variety of living mammals, Lambert concluded that the highly polished wear surfaces were not consistent with the previous hypotheses. SEM studies led him to conclude that *Platybelodon* probably used the lower tusks in combination with a long trunk to scythe through vegetation, whereas *Amebelodon* combined trunk and tusk use to scrape and strip branches. Note that the original reconstructions of these animals suggested only a short, stiff trunk, whereas Lambert's interpretation necessitates the presence of a long, flexible trunk.

Hypsodonty

The distinction between hypsodont and brachydont ungulates has long been noted, and hypsodonty is usually considered to be a functional adaptation to an abrasive diet. However, White (1949) considered hypsodonty, at least in Miocene equids, to be a reflection of high mineral intake combined with endocrine imbalance, depositing dental cement on the teeth as "waste material." Van Valen (1960) gave equal weight to hypsodonty as an adaptation to diet; it extended the life span of an animal on its existing diet.

It is worth reviewing the nature of abrasive diets and dental durability. Grass contains particles of silica, which are thought to act as abrasive agents "necessitating" the evolution of high-crowned teeth. The issue of hypsodonty, and of wear-resistant teeth in general, is actually more complex than one might expect. Janis and Fortelius (1988) reviewed the ways in which herbivorous mammals have rendered their dentitions more durable, and noted that a number of solutions are available, but that a number of constraints also apply. Bilophodont teeth cannot easily be made hypsodont, and only teeth with a flat occlusal morphology can be subjected to the type of molar progression seen in elephants and warthogs. Moreover, hypsodonty is not necessarily an intermediate step between brachydonty and hypselodonty (ever-growing cheek teeth) (contra Mones 1982). Hypselodonty is not a viable option for ungulates that need to preserve a complex occlusal morphology, as the central enamel fossae on the tooth surface must be laid down before eruption, and cannot be renewed once the original pattern has been worn away (Janis & Fortelius 1988). However the extinct rhinoceros *Elasmotherium* exhibits a unique solution to this problem, maintaining its complex occlusal pattern by infolding from the coronal enamel rim (see fig. 3 in Janis & Fortelius 1988).

The idea that "hypsodonty equals grazing" has been a truism for the past century. I decided to test this hypothesis by examining the correlation of hypsodonty with diet in living ungulates (Janis 1988), determining a *hypsodonty index* (see Van Valen 1960) of unworn M_2 height divided by the tooth width. The results showed that, although tooth height scaled isometrically with body mass (paralleling the scaling of other dental dimensions such as length and width; Fortelius 1985), the hypsodonty index showed no correlation with size. Thus hypsodonty indices can be used to compare taxa of differing body mass without being confounded by scaling effects. While the hypsodonty index could clearly distinguish grazers from browsers, no clear distinction could be made with mixed feeders. Mixed feeders in open habitats were not easily distinguished from grazers, and mixed feeders in closed habitats could not be distinguished from browsers. This problem is clearly exemplified by animals such as the pronghorn, *Antilocapra americana*, which is a highly hypsodont, open-habitat animal, which takes only around 12 percent of grass in its diet (Hansen & Clark 1977). Additionally, *fresh grass grazers*, feeding near water, are less hypsodont than regular grazers, and *high level browsers*, feeding on tree leaves rather than on near-ground shrubs and herbs, are less hypsodont than regular browsers (Janis 1988). I concluded, therefore, that habitat and feeding style were at least as important as diet in determining the level of hypsodonty, and that dust and grit accumulating on low level dicotyledonous plants or on grasses in dry environments were as least as likely candidates for the evolution of wear-resistant teeth as the silica in the grasses.

Further work, with a more restricted set of ungulate taxa for which information was available on the percentage of grass in the diet, does show a correlation of hypsodonty index with dietary fiber (Figure 5.3). Although the regression line is significantly different from zero ($p < 0.001$), the amount of grass in the diet only explains about 40 percent of the variation in hypsodonty index ($r^2 = 0.39$), and this is confounded by the fact that all grazers are found in open habitats. Thus, the hypsodonty index alone cannot be used to determine if an extinct ungulate was a true grazer, although it may be a good indicator of an open-habitat environmental preference.

The idea also exists that the hypsodonty index has generally increased in ungulates over time,

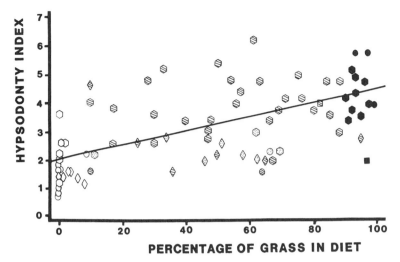

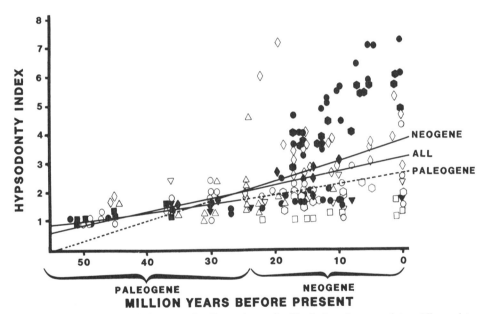

Figure 5.3. Correlation between hypsodonty index and percentage of grass in the diet. Slope of line = 0.025, r^2 = 0.39. Slope significantly different from zero ($p < 0.001$). Key to taxa: hexagons, bovids; diamonds, other selenodont artiodactyls; squares, suoid artiodactyls; circles, perissodactyls and hyraxes. Key to feeding strategies: filled (solid) symbols, grazers; hatched symbols, mixed feeders in open habitats; dotted symbols, mixed feeders in closed habitats; open symbols, browsers.

Figure 5.4. Changes in hypsodonty index over the Cenozoic era for North American ungulates. All ungulates: slope of line = 0.05, r^2 = 0.23. Paleogene ungulates: slope of line = 0.03, r^2 = 0.28. Neogene ungulates: slope of line = 0.07, r^2 = 0.09. Slopes of lines significantly different from zero for all ungulates and Neogene ungulates. Key to taxa: filled circles, equids; open circles, rhinos and tapirs; filled squares, brontotheres; open squares, peccaries; filled diamonds, protoceratids; open diamonds, camelids; open triangles, oreodonts; open inverted triangles, traguloids; filled inverted triangles, moschids; open hexagons, dromomerycids and cervids; filled hexagons, antilocaprids.

correlated with the spread of grasslands in the Neogene (e.g., Simpson 1951; Gregory 1971; Webb 1977). While a plot of hypsodonty index against time for North American ungulates seems to confirm this hypothesis (Figure 5.4), the real picture is actually more complicated. If the taxa making up this pattern are partitioned into their constituent families, then a significant increase overtime ($p < 0.01$) is seen only in the equids, the rhinos, and the antilocaprine antilocaprids. If the equids are examined more closely (Figure 5.5), it can be seen that

this trend is only true for the subfamily Equinae (*Merychippus* and its descendants); taxa in the subfamilies Hyracotheriinae and Anchitheriinae show no such trend, even though anchitheriines persisted until around 8 MYA (millions of years ago). Thus, the story of increasing hypsodonty over time, at least in North America, is really the story of the appearance of more hypsodont taxa at around 18 MYA and 10 MYA, and such taxa do not necessarily subsequently further increase their level of hypsodonty over time.

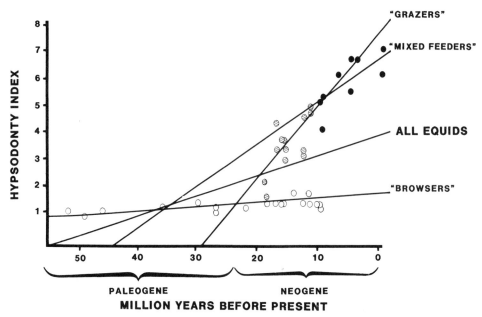

Figure 5.5. Changes in hypsodonty index over the Cenozoic era for North American equids. All equids: slope of line = 0.08, r^2 = 0.31. Browsers (open circles): slope of line = 0.03, r^2 = 0.15. Mixed feeders (hatched circles): slope of line = 0.28, r^2 = 0.42. Grazers (solid circles): slope of line = 0.16, r^2 = 0.57. Slopes significantly different from zero in the case of all equids, mixed feeding equids, and grazing equids.

Comparative method studies

Most comparative method studies have focused on two aspects of ungulate craniodental characters: incisor morphology and muzzle size (shape or width of the premaxillary region). Boué (1970) made an initial qualitative comparison of incisor morphology in ungulates, and showed that grazers tend to have incisors that are all the same size, whereas browsers have central incisors that are conspicuously larger than the more lateral ones. This was confirmed later in quantitative studies (Gordon & Illius 1988; Janis & Ehrhardt 1988). A cautionary note was introduced by McKenzie (1990), who suggested that the incisor difference could be related to grooming functions. Gordon and Illius (1988) also showed a difference in the curvature of the incisor arcade between browsers and grazers: Whereas grazers tend to have a flatter and wider mouth than browsers in general, with increasing body size grazers reduce the curvature of the incisor arcade, whereas browsers increase the curvature. They interpreted these findings to reflect decreasing selectivity in large grazers, but increasing selectivity in large browsers.

Janis and Ehrhardt (1988) also examined muzzle size, using the ratio of the width of the muzzle at the premaxillary/maxillary boundary versus the width of the palate at the level of M^2 to determine a measure of relative muzzle width. We showed that grazers have broader muzzles than browsers, again probably reflecting differences in food selectivity, but the

narrowest muzzles belonged to the mixed feeders in open habitats. We also introduced a note of caution in showing that, although perissodactyls and artiodactyls followed similar trends, the absolute values were different. For example, grazing horses are actually less selective than grazing bovids (see Bell 1969), and yet they have absolutely narrower muzzles (although proportionally wider than those of other, nongrazing perissodactyls). A different type of quantitative study of muzzle shape was performed by Solounias et al. (1988) and Solounias and Moelleken (1993), using shape analysis of premaxilla proportions to show that grazers had square-shaped muzzles and browsers had pointed muzzles; mixed feeders were intermediate and sometimes had a club-shaped muzzle. Solounias et al. (1988) used premaxillary shape analysis in combination with dental microwear to reach the conclusion that the Miocene giraffid *Samotherium boissieri* was probably a mixed feeder or a grazer. Solounias and Moelleken (1993) examined a broad variety of both Old World and North American extinct ruminants and concluded, for example, that the extinct protoceratid *Synthetoceras* was probably a grazer rather than a browser as previously supposed (e.g., Webb 1983).

Studies considering a large number of craniodental variables in correlation with diet have been performed by Solounias and Dawson-Saunders (1988) and myself (Janis 1990b). I included 136 species of ungulates and 24 craniodental variables, looking at

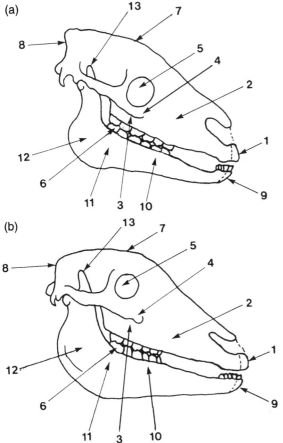

Figure 5.6. Schematic skulls illustrating the differences between (a) a browser and (b) a grazer. (Derived from information in Solounias and Dawson-Saunders [1988] and Janis [1990b].) Number 1 indicates a broad muzzle in the grazer versus a narrower muzzle in the browser. Number 2 indicates a high facial region above premolar row in the browser (for insertion of lip musculature) versus a shallower face in the grazer. Number 3 refers to the large area of the posterior maxilla (for origin of the deep masseter muscle) in the grazer versus the smaller area in the browser. Number 4 refers to the large anterior extension of the zygomatic arch, with a scar for the insertion of the superficial masseter above M^1 in the grazer (ruminants only), versus a shorter zygomatic arch and the absence of a prominent scar in the browser. Number 5 refers to the posteriorly situated orbit, at or behind level of M^3 in the grazer, versus an orbit sited at level of M^2 in the browser. Number 6 refers to the size of the posterior molars: M3 larger than M2 in the grazer (ruminants only), M2 and M3 the same size in the browser. Number 7 refers to the length of the basicranium and the basicranial angle: greater length of basicranium and less acute basicranial angle (less flexion of face on basicranium) in the browser than in the grazer. Number 8 refers to the occipital region: higher occipital region in the browser than in the grazer. Number 9 refers to the size of the incisors: Lateral incisors smaller than central incisors in the browser, same size as central incisors in the grazer. Number 10

refers to the length of the premolar row: long premolar row and large premolars in the browser, short premolar row and smaller premolars in the grazer (ruminants only). Number 11 refers to the depth of the lower jaw: jaw shallow beneath M_2 in the browser, jaw deep beneath M_2 in the grazer (to house roots of hypsodont cheek teeth). Number 12 refers to the angle of the dentary: angle larger and more convex in the grazer than in the browser (for insertion of the masseter muscle). Number 13 refers to the length of the coronoid process: longer coronoid process in the grazer than in the browser. Note that in addition the skull and lower jaw of the grazer is longer than in the browser.

the residuals of the distribution of taxa of different feeding types around a regression line computed against body mass. Solounias and Dawson-Saunders included 27 species of ruminant artiodactyls and 13 variables. They ranked living ungulates as high, intermediate or low for values of these variables, and then applied a Kruskal–Wallis test to show significant differences between the average total scores for the feeding categories of browser, mixed-feeder, and grazer. The application of their findings to fossil ungulates is discussed later in this section, and in the next. Figure 5.6 summarizes the results of these studies, and shows that there are clear, quantifiable differences in the skulls between browsers and grazers. However, it is harder to distinguish mixed feeders, as they typically tend to have intermediate values between browsers and grazers. Although mixed feeders may comprise a real group in ecological terms (see Hofmann 1973), they do not fall out as a discrete group on morphological criteria, although they may be distinguishable by certain features (see discussion below in this section). A similar problem exists when trying to separate mixed feeders from other types by dental microwear (Solounias et al. 1988; Solounias & Moelleken 1992a), although gross dental-wear patterns may show clearer discrimination (Janis 1990a).

Other problems result from the fact that artiodactyls and perissodactyls may follow similar trends for most craniodental features, but they do not have identical measurements, as discussed above for muzzle width. (Separate regression lines for selenodont artiodactyls and perissodactyls plus hyracoids were computed in Janis [1990b]. See also Damuth [1990] for similar observations on molar dimensions.) Additionally, different types of ungulates may vary in the pattern of changes in morphological traits with dietary type. Selenodont artiodactyls usually shorten the length of the premolar row with increasing hypsodonty, whereas perissodactyls plus hyracoids show a slight increase in its length. Greaves (1991b) suggested that anterior premolar loss occurs in long-faced ruminants

Table 5.1. *Results of discriminant analyses of living ungulates grouped according to dietary type*

	Categories of classification			
	Correct %	Grazers #	Mixed feeders #	Browsers #
Selenodont artiodactyls, perissodactyls, and hyraxes (*N* = 127)				
Grazers (25)	72	18	7	0
Mixed feeders (69)	65	7	45	17
Browsers (33)	90	0	3	30
Selenodont artiodactyls only (*N* = 108)				
Grazers (17)	82	14	3	0
Mixed Feeders (65)	62	9	40	16
Browsers (26)	92	0	2	24
Perissodactyls and hyraxes only (*N* = 19)				
Grazers (8)	100	8	0	0
Mixed feeders (4)	50	0	2	2
Browsers (7)	57	0	3	4

Notes: Grouping was done using the morphological features of hypsodonty index (unworn M_2 height/M_3 width), relative premolar row length (premolar row length/molar row length) and relative muzzle width (muzzle width/palatal width).

because teeth placed in this region cannot optimally transmit bite force to the posterior portion of the skull, and torsion is poorly resisted at this anterior point with a long-faced skull design. However, his suggestion does not explain why horses, which are not only similarly long-faced but also similarly narrow-jawed (Fortelius 1985), retain large anterior premolars.

The differences in premolar row lengths may instead be related to differences in feeding strategies in correlation with differences in digestive physiology: foregut fermentation in the former group and hindgut fermentation in the latter. Work in progress by myself and Emily Constable (an undergraduate at Brown University) shows that horses chew their food more on initial ingestion than do cows, and that horses also increase the amount they chew their food with increasing levels of fiber in the diet, while cows do not (Janis & Constable 1993). If these differences in craniodental morphology can be shown to correlate with differences in feeding behavior, then it may prove possible in the future to distinguish foregut from hindgut fermenters in the fossil record. This might be especially useful for ungulate lineages that have no living representatives, such as oreodonts and the South American notoungulates and litopterns.

A problem in the application of the results of such studies to fossil ungulates is that the full range of craniodental variables is rarely available for extinct taxa. However, it is possible to make a fair distinction between feeding types using just three variables: hypsodonty index, relative length of the premolar row (compared with the molar row), and relative muzzle width (defined here as in comparison with palatal width, but molar row length could also be used to provide a relative measure as it correlates well with body mass; see Janis 1990c). Hypsodonty index and premolar row length have the advantage of being fairly readily available in fossil samples, and premolar row length and muzzle width are among the few variables that can be used to distinguish mixed feeders from other feeding types. Browsers have a low hypsodonty index, a long premolar row, and a medium muzzle width. Grazers can be distinguished from browsers by having a greater hypsodonty index, a shorter premolar row in ruminants (slightly longer in hindgut fermenters), and a broader muzzle. Mixed feeders cannot be distinguished from grazers by their hypsodonty index, but can be distinguished by moderate-to-high levels of hypsdonty in combination with a short premolar row (in ruminants) and a narrow muzzle.

Table 5.1 shows the results of multivariate discriminant analysis with these three variables on 127 species of living ungulates (excluding the suoid species considered in Janis 1990b). Results are poor if all ungulates are lumped together, but improve if they are segregated by digestive physiology. Note that grazers are never classified as browsers, and vice versa, but the problem again resides with the mixed feeders, although the misclassifications are "explainable." That is, a mixed feeder misclassified as a grazer is one taking a relatively high percentage of grass in the diet and so on. However, it has been possible to place fossil taxa in this discriminant matrix, with some interesting results. For example, North American hipparionine horses, usually assumed to be grazers, fall out as having a variety of feeding styles, comprising both mixed feeders and grazers. This parallels the conclusions of Hayek et al. (1992) on the diets of the Old World hipparionines, based on dental microwear studies. The protoceratid *Synthetoceras* is classified as a grazer, paralleling the conclusions of Solounias and Moelleken (in press) from premaxillary shape analyses. This research has not yet been published in full, but some preliminary results were summarized in Janis (1989). This technique was used to determine the feeding style of the equids shown in Figure 5.5, and the feeding style of the fossil ungulates used in Janis, Gordon, and Illius (in press), as discussed in the following section. The applications of the findings of Solounias and Dawson-Saunders (1988) to fossil taxa are also discussed below.

FOSSIL UNGULATES AND PALEOCOMMUNITY RECONSTRUCTIONS

If one can determine the diet of a fossil taxon, it should be possible to use the distribution of the feeding types in a paleocommunity to elucidate past vegetational patterns. The pioneering work of Andrews and colleagues (e.g., Andrews, Lord, & Evans 1979) has shown that present-day habitats can be distinguished by the diversity of species composition as related to taxonomic order, body size distribution, and locomotor and feeding habitats, and that fossil communities can be compared quantitatively with living ones to see which vegetational type is the best fit. The role of ungulates in such studies is mainly limited to broad dietary comparisons. For example, the Miocene Paşalar fauna from Turkey is interpreted as a subtropical semideciduous forest, in part because of the high diversity of browsing and frugivorous ungulates, and the low diversity of grazers (Andrews 1990). A finer subdivision of the feeding strategies of ungulates might allow for more precise determinations. This approach has been used for carnivores by Van Valkenburgh for both trophic

categories (1988) and locomotor categories (1985). However, carnivores comprise a single order with a greater uniformity of morphology than is found across the diversity of ungulates, and so it has proved easier to isolate a few key characters that can determine dietary type in this group than has proved possible for ungulates to date.

Many previous studies have used general estimates of hypsodonty to indicate paleoenvironmental change. For example, the level of hypsodonty of the endemic ungulates indicate that grassy savannas were probably present in South America by the Early Miocene (Webb 1978; Pascual & Ortiz Jaureguizar 1990), but did not appear in North America until the Late Miocene (Webb 1977, 1983). However, the problems of using the hypsodonty index to distinguish between grasslands and open habitats have already been discussed. I used the hypsodonty index in a rather different manner (Janis 1984). I plotted hypsodonty index against M^2 area (as an estimate of body mass) and compared the fauna of the present-day East African savannas with that of the Late Miocene of North America. The patterns were very different; notably most large hypsodont taxa in Africa were ruminant artiodactyl species, whereas in the North American Miocene their place was taken by a variety of equids. I concluded that this represented a fundamental difference in the types of savanna habitat, with the American Miocene habitat representing a drier, less-productive environment more conducive to hindgut fermenting grazers than to ruminants (Janis 1976).

This predominance of equid grazers in the North American Neogene faunas was examined in a more quantitative fashion in Janis et al. (in press), although our particular focus was on the browsing ungulates. Here I worked with Iain Gordon and Andrew Illius on a model (Illius & Gordon 1992) that can be used to predict the percent daily energy needs for a particular type of ungulate (ruminant or hindgut fermenter) on a herbage of particular potential digestibility (grass or browse), and can be used to determine the body mass "cut off" limit for sustenance on any one type of forage. We ran this model "in reverse" to examine changes in the diversity of browsing ungulates (equids and ruminants) during the North American Miocene. We used the measurements described previously to determine which of the ungulates were true browsers, and determined their body masses from scaling algorithms (Janis 1990c). The body masses of the smallest ungulates in the faunas were then used to determine the probable minimum quality of available browse from Early, Middle, and Late Miocene horizons. The results we obtained were consistent with a declining quality of browse forage through the Miocene epoch, with the diversification of the browsing dromomerycid ruminants forcing the browsing equids to adopt larger body sizes to avoid competition.

Vrba used the distinction between browsing and grazing ungulates to examine African Plio–Pleistocene climatic changes. For example, she (Vrba 1980) showed that the distribution of living bovids can define African habitats into open or closed, and moist or arid. She also argued (e.g., Vrba 1985) that changes in the diversity and abundance of bovid ecological types mark a widespread environmental change to open grasslands across sub-Saharan Africa around 2.5 MYA, with a further change at around 1.8 MYA. However, while she (Vrba 1980) mentioned morphological features that might be used to determine feeding types in fossil bovids, she did not quantify them. Instead her assignations of feeding types are based on taxonomic affinities, which at best would not serve as accurate indicators of diet prior to the Plio–Pleistocene.

An example of how a more precise analysis of feeding types is important for determining more ancient paleohabitats is provided by Solounias and Dawson-Saunders (1988). They reconstructed the feeding habits of the bovids at the Late Miocene Greek sites of Pikermi and Samos, using the comparison with living taxa previously described. These sites had previously been considered to be savannas, based on the presence of bovids, equids, and hyaenids – taxa that might inhabit savannas today had they not undergone considerable evolutionary change in the past 10 million years. They showed that, of a wide diversity of fossil bovids present, only one might be classified as a grazer, with the rest being mixed feeders and browsers. The assemblage showed no resemblance to a modern East African savanna, and bore the greatest resemblance, out of a variety of living faunas, to the forest woodland ruminants of India and South East Asia, a conclusion that was also supported by paleobotanical evidence (Ioakim & Solounias 1985).

Other studies have concentrated on various aspects of dental morphology to explain patterns of ungulate evolution in the context of environmental change. Collinson and Hooker (1987, 1991) showed that the general degree of lophing of the teeth of ungulates in Europe increased during the late Middle and Late Eocene, suggesting a more fibrous diet that was also correlated with paleobotanical changes accompanying the reduction of tropical floral elements. I showed similar dietary changes in North American ungulates at this time by examination of patterns of gross dental wear (Janis 1979b; see also Wing & Tiffney 1987; Stucky 1990).

The studies described so far show broad congruence between evidence from ungulate paleofaunas and paleobotany. However, inferences of paleohabitats from paleopedological data (fossil soils)

appear to contradict the ungulate evidence. Retallack (1983, 1992a) interpreted the Early Oligocene White River Badlands as being a type of savanna environment, asserting that large areas of open grassland were apparent in North America by the end of the Oligocene. He interpreted the brachydont ungulates in the Early Oligocene as animals that were not "optimally adapted to their environment" (Retallack 1983). Although there were some highly hypsodont mammals by the Late Oligocene of North America, such as the hyraxlike leptauchiniine oreodonts and the gazellelike stenomyline camelids, these taxa had very narrow muzzles suggestive of a mixed-feeding rather than a grazing diet. The oreodont *Leptauchenia* had auditory region specializations suggesting adaptations to arid conditions (Joeckel 1992), but aridity does not necessarily signify a savannalike habitat. Qualitative studies on ungulate diversity and morphology in general (Janis 1982), and quantitative studies on carnivore locomotor diversity (Van Valkenburgh 1985), both support the idea of a predominantly woodland habitat in the North American Oligocene.

The issue of the absence of hypsodonty is an interesting issue in this case. While hypsodont animals may not necessarily signify grassland, I maintain that it would be impossible to have a grassland environment without the presence of hypsodont taxa. It might be possible for animals to be "suboptimally adapted" in certain circumstances; the polar bear, as a clumsy swimmer new to the aquatic environment, might be able to keep afloat both literally and figuratively. But because grit or silica in the diet in an open grassland would rapidly abrade the teeth, there would be high selection for hypsodonty; an ungulate with a brachydont dentition might wear its teeth down to stubs before reaching sexual maturity, and it is a fairly simple developmental procedure to make a tooth more highly crowned (see Fortelius 1985; Janis & Fortelius 1988). Additionally, if the Early Oligocene ungulates were indeed suboptimally adapted to their environment, one might expect to see a large proportion of individuals with highly worn teeth, which is not apparent in the faunal assemblage (John Damuth, personal communication). Note that Leopold, Liu, and Clay-Poole (1992) provide an alternative explanation of the Oligocene environment from paleopedological data, suggesting a dense packed woody type of "savanna" without grasses, and claim that grassy savannas were not prominent in North America until the Late Miocene. Retallack (Retallack, Dugas, & Bestland 1990; Retallack 1992b) has also claimed the presence of a grassy savanna at the Middle Miocene Fort Ternan site in Africa, but evidence from both carbon isotopes (Cerling et al. 1991) and bovid locomotor adaptations (Kappleman 1991) suggest a more wooded environment (see Cerling et al. 1992 for review).

CONCLUSIONS

Future directions: promises and problems

What types of future studies might be applied to ungulate craniodental anatomy? For a start, it is encouraging to see a recent flourishing of "classical" types of studies on fossil ungulates (e.g., Joeckel 1990; Court 1992; Lambert 1992), and I hope that this trend continues. The use of ungulates in paleo-community studies would be greatly strengthened by the determination of a few simple, frequently preserved, morphological features that could determine diet in a quantitative fashion, as determined by Van Valkenburgh (1988) for carnivores. A combination of hypsodonty index and gross dental wear might be appropriate, although methods must be sought to determine such indices in worn teeth, as these frequently represent the only available fossil data. As previously discussed, work in progress may allow for a quantitative assessment of the digestive physiology of extinct ungulates, but further research is needed on the biomechanical reasons for differences in craniodental design between living foregut and hindgut fermenters.

Further investigation is needed into the true functional reasons for hypsodonty. It is clear (see Figure 5.3) that the correlation between hypsodonty index and the amount of grass in the diet leaves much of the variance unexplained, and thus that silica in the grass cannot be the sole causal agent of dental abrasion. My previously discussed suggestion, of abrasion caused by the accumulation of dust and grit on the food surface, is intuitively appealing and is supported by certain data (Janis 1988). Yet the case is far from proven, and this idea need further investigation (although I admit that I can think of no easy way to frame this question as a testable hypothesis).

Another outstanding problem is the reason for isometric (rather than allometric) scaling of dental dimensions, which defies the "commonsensical" prediction that dental dimensions should scale with metabolic rate, that is, both variables scaling with body mass (M) as $M^{0.75}$ (see discussions in Fortelius 1985, 1988, 1990). Fortelius (1987, 1988) has devised an ingenious explanation for this phenomenon, relating to the negative scaling of chewing rate (as $M^{-0.25}$), suggesting that isometric scaling of occlusal area in combination with negative scaling of chewing rate results in true metabolic scaling ($M^{1.0} \times M^{-0.25} = M^{0.75}$). He also suggested that the volume between the occluding teeth might be the true

determinant of dental dimensions, which would be expected to scale isometrically (volume comminuted per chew versus the volume function of body mass).

However, Fortelius admitted (1987, personal communication) that this suggestion does not explain the isometric scaling of tooth crown height. If a large and a small mammal have similar diets, then because of metabolic scaling a large mammal would consume relatively less food per lifetime than a small one, and so tooth height should scale as $M^{0.75}$ rather than as $M^{1.0}$. Does the same food induce less tissue loss per chew in a small mammal than a large one, perhaps because of lower absolute forces resulting in a lower probability of cracking or other damage? Do larger ungulates consume proportionally more fibrous food even if the diets are similar, perhaps because the selection of choice dietary items is more difficult at a larger size? Or could it be related to the hypothesis of "wear particles" distributed within the food (Fortelius 1987), rather than the food being continuously abrasive, so that the thickness of the food between the teeth (which scales isometrically) is the determining factor of the amount of abrasion experienced?

Reciprocal illuminations

Much of the flow of information of ungulate craniodental anatomy is from the living taxa to the fossils. It is rare to find direct evidence of the diet of an extinct species (but see Voorhies & Thomasson 1979; von Koenigswald & Schaarschmidt 1983). But if the morphology of living species can be correlated with their diets, then these correlations can be applied to determine the probable diets of extinct taxa. The value of knowing the diets of fossil ungulates relates not just to curiosity about the lifestyle of a particular extinct animal, but also to the use of paleocommunities to reconstruct past vegetation, and hence to the examination of environmental change over time. Such studies are important not only as a backdrop to evolutionary patterns, but may also relate to issues such as global warming in the present day.

However, there are a number of instances where the study of the fossil taxa themselves have illuminated our understanding of living ungulates. For example, a number of studies on fossil horses (Turnbull 1970; Radinsky 1983, 1984; Rensberger et al. 1984) all show that the transition to the modern type of equid craniodental anatomy took place with the evolution of the genus *Merychippus*, thus implicating the adoption of a more fibrous diet as the key feature behind the morphological adaptations. Studies on extinct taxa without close living relatives can provide examples of adaptive types that are lacking today, and hence can expand our knowledge of constraints and possibilities in design. For example, the bifunctional dentitions of deinotheres (Harris 1975) and arsinotheres (Court 1992) allowed for both crushing and shearing within different parts of the tooth row, and may represent a specialized type of browsing diet probably not seen today.

Studies on fossil ungulates have also illuminated a broader-scale understanding of evolutionary strategies. A previously discussed example is the distribution in fossil taxa of enamel containing vertically oriented Hunter–Schreger bands (Fortelius 1984), which suggested that this enamel structure was functionally related to tooth design rather than to a peculiarity of the living family Rhinocerotidae. A further example is my own work on the digestive systems of artiodactyls and perissodactyls (Janis 1976), which to my great surprise became a "classic" in the world of animal science (see Duncan et al. 1990). Perissodactyls, with their hindgut mode of fermentation, had long been considered inferior to ruminant artiodactyls with their complex rumen, or foregut fermentation, an idea originating with Kowalevsky (1873–1874). I argued for these different types of digestive systems as representing "equal and opposite" strategies, with hindgut fermenters having an advantage over ruminants in certain instances. Much of the paper summarized the ecological and physiological literature, and the same general conclusions could be reached without recourse to the fossil record (see Foose 1982). However, I consider that my study of fossil evidence provided a unique solution to this evolutionary problem. The transition from omnivory to herbivory in these two orders, as evidenced by patterns of dental wear, occurred at a larger body size in the ruminants than in the perissodactyls, and I argued that the body size at the time of the adoption of a more fibrous diet was the key feature in determining which type of digestive strategy was developed.

In conclusion, the craniodental adaptations of ungulates are interesting not merely as systems of mechanical design, but because the diversity of ungulates can illustrate how different solutions can be provided to the evolutionary problem of adopting a herbivorous diet. The extension of the correlation of diet with morphology to fossil ungulates is interesting not merely in the reconstruction of the ecology and behavior of extinct taxa, but in how these taxa can be then used in paleocommunity studies to track environmental changes over evolutionary time. And the study of the fossil record of ungulates is not merely interesting for phylogenetic histories, but can be used to challenge assumptions about the conditions, adaptive or otherwise, evident in living taxa.

ACKNOWLEDGMENTS

This paper was greatly improved by comments from fellow members of the Valio Armas Korvenkontio Unit of Dental Anatomy in Relation to Evolutionary Theory: John Damuth, Mikael Fortelius, and Nikos Solounias. This paper is dedicated to the memory of Joe, a hyrax, who taught both me and the editor of this volume much of what we understand about ungulate design.

REFERENCES

Andrews, P. 1990. Palaeoecology of the Miocene fauna from Paşalar, Turkey. *Journal of Human Evolution* 19, 569–582.

Andrews, P., Lord, J., & Evans, E.M.N. 1979. Patterns of ecological diversity in fossil and modern mammalian faunas. *Biological Journal of the Linnean Society* 11, 177–205.

Axmacher, H., & Hofmann, R.R. 1988. Morphological characteristics of the masseter muscle of 22 ruminant species. *Journal of Zoology, London* 251, 463–473.

Becht, G. 1953. Comparative biologic–anatomical researches on mastication in some mammals, I and II. *Proceedings Konenklijke Nederlandse Akademie van Wetenshappen, Ser. C* 56, 508–527.

Bell, R.H.V. 1969. The use of the herb layer by grazing ungulates in the Serengeti. In *Animal Populations in Relation to Their Food Resource,* ed. A. Watson, pp. 111–128. Symposium of the British Ecological Society. Oxford: Blackwells.

Bock, W.J., & von Wahlert, G. 1965. Adaptation and the form–function complex. *Evolution* 19, 269–299.

Boué, C. 1970. Morphologie fonctionnelle des dents labials chez les ruminants. *Mammalia* 34, 696–771.

Boyde, A., & Fortelius, M. 1986. Development, structure and function of rhinoceros enamel. *Zoological Journal of the Linnean Society* 87, 181–214.

Butler, P.M. 1952. The milk molars of the Perissodactyla, with remarks on molar occlusion. *Proceedings of the Zoological Society of London* 121, 777–817.

Butler, P.M. 1972. Some functional aspects of molar evolution. *Evolution* 26, 474–483.

Cerling, T.E., Kappleman, J., Quade, J., Ambrose, S.H., Sikes, N.E., & Andrews, P. 1992. Reply to comments on the paleoenvironment of *Kenyapithecus* at Fort Ternan. *Journal of Human Evolution* 23, 371–377.

Cerling, T.E., Quade, J., Ambrose, S.H., & Sikes, N.E. 1991. Fossil soils, grasses, and carbon isotopes from Fort Ternan, Kenya: grassland or woodland? *Journal of Human Evolution* 21, 295–306.

Clutton-Brock T.H., & Harvey, P.H. 1979. Comparison and adaptation. *Proceedings of the Royal Society of London* B205, 547–565.

Collinson, M.E., & Hooker, J.J. 1987. Vegetational and mammalian faunal changes in the Early Tertiary of southern England. In *The Origin of Angiosperms and Their Biological Consequences,* ed. E.M. Fries, W.G. Chaloner, & P.R. Crane, pp. 259–303. Cambridge University Press.

Collinson, M.E., & Hooker, J.J. 1991. Fossil evidence of interaction between plants and plant-eating mammals. *Philosophical Transactions of the Royal Society, London, Series B* 333, 197–208.

Court, N. 1992. A unique form of dental bilophodonty and a functional interpretation of peculiarities in the masticatory system of *Arsinotherium* (Mammalia, Embrithopoda). *Historical Biology* 6, 91–111.

Crompton, A.W. 1971. The origin of the tribosphenic molar. In *Early Mammals*, ed. D.M. Kermack & K. Kermack. *Journal of the Linnean Society London (Zoology) Supplement 1*, 65–87.

Crompton, A.W., & Hiiemae, K. 1969. How mammalian molar teeth work. *Discovery* 5, 21–47.

Crompton, A.W., & Hiiemae, K. 1970. Functional occlusion and mandibular movements during occlusion in the American opossum *Didelphis marsupialis* L. *Zoological Journal of the Linnean Society* 49, 21–47.

Damuth, J.D. 1990. Problems in estimating body masses of archaic ungulates using dental measurements. In *Body Size Determination in Mammalian Paleobiology*, ed. J. Damuth & B.J. MacFadden, pp. 229–253. Cambridge University Press.

Damuth, J.D. 1992. Taxon-free characterization of animal communities. In *Terrestrial Ecosystems through Time*, ed. A.K. Behrensmeyer, J.D. Damuth, W.A. DiMichele, R. Potts, H.-D. Sues, & S.L. Wing, pp. 183–203. Chicago: University of Chicago Press.

DeVree, F., & Gans, C. 1975. Mastication in pygmy goats, *Capra hircus*. *Annals of the Royal Society of Belgium* 105, 255–306.

Duncan, P., Foose, T.J., Gordon, I.J., Gakahu, G.C., & Lloyd, M. 1990. Comparative nutrient extraction by grazing bovids and equids: a test of the nutritional models of equid/bovid competition and coexistence. *Oecologia* 84, 411–418.

Foose, T.J. 1982. Trophic Strategies of Ruminant versus Nonruminant Ungulates. Unpublished Ph.D. thesis, University of Chicago.

Fortelius, M. 1984. Vertical decussation of enamel prisms in lophodont ungulates. In *Tooth Enamel IV*, ed. R.W. Fearnhead & S. Suga, pp. 427–431. New York: Elsevier.

Fortelius, M. 1985. Ungulate cheek teeth: developmental, functional, and evolutionary interrelations. *Acta Zoologica Fennica* 180, 1–76.

Fortelius, M. 1987. A note on the scaling of dental wear. *Evolutionary Theory* 8, 73–75.

Fortelius, M. 1988. Isometric scaling of mammalian cheek teeth is also true metabolic scaling. In *Teeth Revisited*, ed. D.E. Russell, J.P. Santoro & D. Sigogneau-Russell. *Mémoires du Muséum national d'Histoire naturelle (C)* 53, 459–462.

Fortelius, M. 1990. The mammalian dentition: a "tangled" view. *Netherlands Journal of Zoology* 40, 312–328.

Gans, C. 1985. Differences and similarities: comparative methods in mastication. *American Zoologist* 25, 291–301.

Geist, V., & Walther, F., ed. 1974. *The Behaviour of Ungulates and its Relation to Management*. Morges, Switzerland: I.U.C.N. Publication No. 24.

Gingerich, P.D. 1971. Functional significance of mandibular translation in vertebrate jaw mechanics. *Postilla* 152, 1–10.

Gingerich, P.D. 1972. Molar occlusion and jaw mechanics of the Eocene primate *Adapis*. *American Journal of Physical Anthropology* 36, 359–368.

Gordon, I.J., & Illius, A.W. 1988. Incisor arcade structure and diet selection in ruminants. *Functional Ecology* 2, 15–22.

Gould, S.J. 1970. Evolutionary paleontology and the science of form. *Earth-Science Reviews* 6, 77–119.

Gould, S.J., & Lewontin, R.C. 1979. The spandrels of San Marco and the Panglossian paradigm: a critique of the adaptationist programme. *Proceedings of the Royal Society of London* B205, 581–598.

Greaves, W.S. 1973a. The inference of jaw motion from tooth wear facets. *Journal of Paleontology* 47, 1000–1001.

Greaves, W.S. 1973b. Evolution of the merycoidodont masticatory apparatus (Mammalia, Artiodactyla). *Evolution* 26, 659–667.

Greaves, W.S. 1974. Functional implications of mammalian jaw joint position. *forma et functio* 7, 363–376.

Greaves, W.S. 1978a. The jaw lever system in ungulates: a new model. *Journal of Zoology, London* 184, 271–285.

Greaves, W.S. 1978b. The posterior zygomatic root in oreodonts. *Journal of Paleontology* 52, 740–743.

Greaves, W.S. 1980. The mammalian jaw mechanism – the high glenoid cavity. *American Naturalist* 116, 432–440.

Greaves, W.S. 1985. The mammalian postorbital bar as a torsion-resisting helical strut. *Journal of Zoology, London* 207, 125–136.

Greaves, W.S. 1991a. The orientation of the force of the jaw muscles and the length of the mandible in mammals. *Zoological Journal of the Linnean Society* 102, 367–374.

Greaves, W.S. 1991b. A relationship between premolar loss and jaw elongation in selenodont artiodactyls. *Zoological Journal of the Linnean Society* 101, 121–129.

Gregory, J.T. 1971. Speculations on the significance of fossil vertebrates for the antiquity of the Great Plains of North America. *Abhandlung Hessisches Landesamtes Bodenforschung (Heinz Tobien Festschrift)* 60, 64–72.

Gregory, W.K. 1920. Studies in comparative myology and osteology, No. V. – On the anatomy of the preorbital fossa of Equidae and other ungulates. *Bulletin of the American Museum of Natural History* XLII, 265–283.

Gwynne, M.D., & Bell, R.H.V. 1968. Selection of vegetation components by grazing ungulates in the Serengeti National Park. *Nature* 220, 390–393.

Hansen, R.M., & Clark, R.C. 1977. Food of elk and other ungulates at low elevations in northwest Colorado. *Journal of Wildlife Management* 41, 76–80.

Harris, J.M. 1975. Evolution of feeding mechanisms in the family Deinotheriidae (Mammalia: Proboscidea). *Zoological Journal of the Linnean Society* 56, 331–362.

Harvey, P.H., & Pagel, M.D. 1991. *The Comparative Method in Evolutionary Biology*. Oxford: Oxford University Press.

Hayek, L.-A., Bernor, R.L., Solounias, N., & Steigerwald, P. 1992. Preliminary studies of hipparionine diet as measured by tooth wear. *Annals Zoologici Fennici* 28, 187–200.

Herring, S.W. 1972. The role of canine morphology in the evolutionary divergence of pigs and peccaries. *Journal of Mammalogy* 53, 501–512.

Herring, S.W. 1975. Adaptations for gape in the hippopotamus and its relatives. *forma et functio* 8, 85–110.

Herring, S.W. 1980. Functional design of cranial muscles: comparative and physiological studies in pigs. *American Zoologist* 20, 283–293.

Herring, S.W. 1985. Morphological correlates of masticatory patterns in pigs and peccaries. *Journal of Mammalogy* 66, 603–617.

Hiiemae, K. 1978. Mammalian mastication: a review of the activity of jaw muscles and the movements they produce in chewing. In *Development, Function and Evolution of Teeth,* ed. P.M. Butler & K. Joysey, pp. 359–398. London: Academic Press.

Hofmann, R.R. 1973. *The Ruminant Stomach.* East African Monographs in Biology, No. 2. Nairobi: East African Literature Bureau.

Hofmann, R.R., & Stewart, D.R.M. 1972. Grazer or browser: a classification based on the stomach-structure and feeding habits of East African ruminants. *Mammalia* 36, 226–240.

Illius, A.W., & Gordon, I.J. 1992. Modelling the nutritional ecology of ungulate herbivores: evolution of body size and competitive interactions. *Oecologica* 89, 428–434.

Ioakim, C., & Solounias, N. 1985. A radiometrically dated pollen flora from the Upper Miocene of Samos island, Greece. *Revue de Micropaleontologie* 28, 197–204.

Janis, C.M. 1976. The evolutionary strategy of the Equidae, and the origin of rumen and cecal digestion. *Evolution* 30, 757–774.

Janis, C.M. 1979a. Mastication in the hyrax and its relevance to ungulate dental evolution. *Paleobiology* 5, 50–59.

Janis, C.M. 1979b. Aspects of the Evolution of Herbivory in Ungulate Mammals. Unpublished Ph.D. thesis, Harvard University.

Janis, C.M. 1982. Evolution of horns in ungulates: ecology and paleoecology. *Biological Reviews* 57, 261–318.

Janis, C.M. 1983. Muscles of the masticatory apparatus in two genera of hyraces (*Procavia* and *Heterohyrax*). *Journal of Morphology* 176, 61–87.

Janis, C.M. 1984. The use of fossil ungulate communities as indicators of climate and environment. In *Fossils and Climate,* ed. P.J. Brenchley, pp. 85–104. Chichester: John Wiley and Sons.

Janis, C.M. 1988. An estimation of tooth volume and hypsodonty indices in ungulate mammals, and the correlation of these factors with dietary preferences. In *Teeth Revisited,* ed. D.E. Russell, J.P. Santoro, & D. Sigogneau-Russell. *Mémoires de Muséum national d'Histoire naturelle (C)* 53, 367–387.

Janis, C.M. 1989. Estimation of diets in fossil ungulates. *Journal of Vertebrate Paleontology* 9, 27A.

Janis, C.M. 1990a. The correlation between diet and dental wear in herbivorous mammals, and its relationship to the determination of diets of extinct species. In *Paleobiological Evidence for Rates of Coevolution and Behavioral Evolution,* ed. A.J. Boucot, pp. 241–259. New York: Elsevier.

Janis, C.M. 1990b. Correlation of cranial and dental variables with dietary preferences: a comparison of macropodoid and ungulate mammals. *Memoirs of the Queensland Museum* 28, 349–366.

Janis, C.M. 1990c. Correlation of cranial and dental variables with body size in ungulates and macropodoids. In *Body Size Determination in Mammalian Paleobiology,* ed. J. Damuth & B.J. MacFadden, pp. 255–299. Cambridge University Press.

Janis, C.M., & Constable, E. 1993. Can ungulate craniodental features determine digestive physiology? *Journal of Vertebrate Paleontology* 13, abstract.

Janis, C.M., & Ehrhardt, D. 1988. Correlation of relative muzzle width and relative incisor width with dietary preference in ungulates. *Zoological Journal of the Linnean Society* 92, 267–284.

Janis, C.M., & Fortelius, M. 1988. On the means whereby mammals achieve increased functional durability of their dentitions, with special reference to limiting factors. *Biological Reviews* 63, 197–230.

Janis, C.M., Gordon, I.J., & Illius, A.W. In press. Modelling equid/ruminant competition in the fossil record. *Historical Biology.*

Jarman, P.J. 1974. The social organisation of antelopes in relation to their ecology. *Behaviour* 48, 213–267.

Joeckel, R.M. 1990. A functional interpretation of the masticatory system of entelodonts. *Paleobiology* 16, 459–482.

Joeckel, R.M. 1992. Comparative anatomy and functions of the leptaucheniine oreodont middle ear. *Journal of Vertebrate Paleontology* 12, 505–523.

Kappleman, J. 1991. The paleoenvironment of *Kenyapithecus* at Fort Ternan. *Journal of Human Evolution* 20, 95–129.

Kay, R.F. 1977. The evolution of molar occlusion in the Cercopithecidae and early catarrhines. *American Journal of Physical Anthropology* 46, 227–256.

Kay, R.F., & Hiiemae, K. 1974. Jaw movement and tooth use in recent and fossil primates. *American Journal of Physical Anthropology* 40, 227–256.

Kiltie, R.A. 1981. The function of interlocking canines in rain forest peccaries (Tayassuidae). *Journal of Mammalogy* 62, 459–469.

Koenigswald, W. von, & Schaarschmidt, F. 1983. Ein Urpferd aus Messel, das Weinbeeren frass. *Natur und Museum* 113, 79–84.

Koenigswald, W. von, Rensberger, J.M., & Pfretzschner, H.-U. 1987. Changes in the tooth enamel of early Paleocene mammals allowing increased diet diversity. *Nature* 328, 150–152.

Kowalevsky, W. 1873–1874. Monographie der battung *Anchitherium* Cuvier und Versuch einer naturlichen classificationder fossilen Hufthiere. *Palaeontographica* 22, 131–385.

Lambert, W.D. 1992. The feeding habits of the shovel-tusked gomphotheres: evidence from tusk wear patterns. *Paleobiology* 18, 132–147.

Lamprey, H.F. 1963. Ecological separation of the large mammal species in the Tarangire Game Reserve, Tanganyika. *East African Wildlife Journal* 1, 63–92.

Lauder, G.V. 1981. Form and function: structural analysis in evolutionary morphology. *Paleobiology* 7, 430–442.

Lauder, G.V. 1991. Biomechanics and evolution: integrating physical and historical biology in the study of complex systems. In *Biomechanics in Evolution,* ed. J.M.V. Rayner & R.J. Wootton, pp. 1–19. Cambridge University Press.

Leopold, E.B., Liu, G., & Clay-Poole, S. 1992. Low-biomass vegetation in the Oligocene. In *Eocene-Oligocene Climatic and Biotic Evolution,* ed. D.R. Prothero & W.A. Berggren, pp. 399–420. Princeton: Princeton University Press.

Loomis, F.B. 1925. Dentition of artiodactyls. *Bulletin of the Geological Society of America* 36, 583–604.

Lucas, P.W. 1979. The dental-dietary adaptations of mammals. *Neues Jarhbuch für Geologie und Paleontologie* 8, 486–512.

Lucas, P.W., & Luke, D.A. 1984. Chewing it over: basic principles of food breakdown. In *Food Acquisition and Processing in Primates*, ed. D.J. Chivers, B.A. Wood, & A. Bilsborough, pp. 283–302. New York: Plenum Press.

Maglio, V.J. 1972. Evolution of mastication in the Elephantidae. *Evolution* 26, 638–658.

Maier, W. 1984. Tooth morphology and dietary specialisation. In *Food Acquisition and Processing in Primates*, ed. D.J. Chivers, B.A. Wood, & A. Bilsborough, pp. 303–330. New York: Plenum Press.

Maynard Smith, J., & Savage, R.J.G. 1959. The mechanics of mammalian jaws. *School Science Review* 141, 289–301.

Matthew, W.D. 1926. The evolution of the horse. A record and its interpretation. *Quarterly Review of Biology* 1, 139–185.

McKenzie, A.A. 1990. The ruminant dental grooming apparatus. *Zoological Journal of the Linnean Society* 99, 117–128.

Merriam, J.C. 1916. Tertiary vertebrate fauna from the Cedar Mountain region of western Nevada. *University of California Publications, Bulletin of the Department of Geological Sciences* 9, 161–198.

Mills, J.R.E. 1955. Ideal dental occlusion in the primates. *Dental Practitioner* 6, 47–61.

Mones, A. 1982. An equivocal nomenclature: what means hypsodonty? *Palaeontologisches Zeitschrift* 56, 107–111.

Pascual, R., & Ortiz Jaureguizar, E.O. 1990. Evolving climate and mammal faunas in Cenozoic South America. *Journal of Human Evolution* 19, 23–60.

Radinsky, L.B. 1983. Allometry and reorganisation in horse skull proportions. *Science* 221, 1189–1191.

Radinsky, L.B. 1984. Ontogeny and phylogeny in horse skull evolution. *Evolution* 38, 1–15.

Radinsky, L.B. 1985. Patterns in the evolution of ungulate jaw shape. *American Zoologist* 25, 303–314.

Rensberger, J.M. 1978. Scanning electron microscopy of wear and occlusal events in some small herbivores. In *Development, Function, and Evolution of Teeth*, ed. P.M. Butler & K.A. Joysey, pp. 415–438. London: Academic Press.

Rensberger, J.M. 1986. Early chewing mechanisms in mammalian herbivores. *Paleobiology* 12, 474–494.

Rensberger, J.M. 1988. The transition from insectivory to herbivory in mammalian teeth. In *Teeth Revisited*, ed. D.E. Russell, J.P. Santoro & D. Sigogneau-Russell. *Mémoires du Muséum national d'Histoire naturelle (C)* 53, 351–365.

Rensberger, J.M., Forstén, A., & Fortelius, M. 1984. Functional evolution of the cheek tooth pattern and chewing direction in Tertiary horses. *Paleobiology* 10, 439–452.

Retallack, G.J. 1983. A paleopedological approach to the interpretation of terrestrial sedimentary rocks: the mid-Tertiary fossil soils of Badlands National Park, South Dakota. *Geological Society of America, Bulletin* 94, 823–840.

Retallack, G.J. 1992a. Paleosoils and changes in climate and vegetation across the Eocene/Oligocene boundary. In *Eocene-Oligocene Climatic and Biotic Evolution*, ed. D.R. Prothero & W.A. Berggren, pp. 382–398. Princeton: Princeton University Press.

Retallack, G.J. 1992b. Middle Miocene fossil plants from Fort Ternan (Kenya) and evaluation of African grasslands. *Paleobiology* 18, 383–400.

Retallack, G.J., Dugas, D.P., & Bestland, E.A. 1990. Fossil soils and grasses of a Middle Miocene East African grassland. *Science* 247, 1325–1328.

Sanson, G.D. 1991. Predicting the diet of fossil mammals. In *Vertebrate Paleontology of Australasia*, ed. P. Vickers-Rich, J.M. Monaghan, R.F. Baird, & T.M. Rich, pp. 201–228. Melbourne: Pioneer Design Studios, Monash University Publications.

Scapino, R.P. 1972. Adaptive radiation of mammalian jaws. In *Morphology of the Maxillo-Mandibular Apparatus*, ed. G.H. Schumacher, pp. 33–39. Leipzig: G. Thieme.

Schaller, G.B. 1967. *The Deer and the Tiger*. Chicago: Chicago University Press.

Schaller, G.B. 1977. *Mountain Monarchs*. Chicago: Chicago University Press.

Simpson, G.G. 1951. *Horses*. New York: Columbia University Press.

Skogland, T. 1988. Tooth wear by food limitation and its life history consequences in wild reindeer. *Oikos* 51, 238–242.

Solounias, N., & Dawson-Saunders, B. 1988. Dietary adaptations and paleoecology of the Late Miocene ruminants from Pikermi and Samos in Greece. *Palaeogeography, Palaeoclimatology, Palaeoecology* 65, 149–172.

Solounias, N., Fortelius, M., C.A., & Freeman, P. In press. Molar wear rates in ruminants: a new approach. *Annales Zooligici Fennici*.

Solounias, N., & Hayek, L.-A.C. 1993. New methods of tooth microwear and application to dietary determination of two extinct antelopes. *Journal of Zoology, London* 229, 421–445.

Solounias, N., & Moelleken, S.M.C. 1992a. Dietary interpretation of *Eotragus sansaniensis* (Mammalia, Ruminantia): tooth microwear analysis. *Journal of Vertebrate Paleontology* 12, 113–121.

Solounias, N., & Moelleken, S.M.C. 1992b. Dietary adaptation of two Miocene goat ancestors and evolutionary implications. *Geobios* 6, 797–809.

Solounias, N., & Moelleken, S.M.C. 1993c. Dietary adaptations of some extinct ruminants determined by premaxillary shape. *Journal of Mammalogy* 74, 1059–1074.

Solounias, N., & Moelleken, S.M.C. In press. Dietary differences between two archaic ruminant species from Sansan, France. *Historical Biology*.

Solounias, N., Teaford, M., & Walker, A. 1988. Interpreting the diet of extinct ruminants: the case of a non-browsing giraffid. *Paleobiology* 14, 287–300.

Stirton, R.A. 1947. Observations on evolutionary rates in hypsodonty. *Evolution* 1, 32–41.

Stucky, R.K. 1990. Evolution of land mammal diversity in North America during the Cenozoic. In *Current Mammalogy*, vol. 2, ed. H.H. Genoways, pp. 375–432. New York: Plenum Press.

Turnbull, W.D. 1970. Mammalian masticatory apparatus. *Fieldiana: Geology* 18, 149–356.

Van Valen, L. 1960. A functional index of hypsodonty. *Evolution* 14, 531–532.

Van Valkenburgh, B. 1985. Locomotor diversity within past and present guilds of large predatory mammals. *Paleobiology* 11, 406–428.

Van Valkenburgh, B. 1988. Trophic diversity within past and present guilds of large predatory mammals. *Paleobiology* 14, 155–173.

Voorhies, M.R., & Thomasson, J.R. 1979. Fossil grass Anthoecia within Miocene rhinoceros skeletons: diet in an extinct species. *Science* 206, 331–333.

Vrba, E.S. 1980. The significance of bovid remains as indicators of environment and predation patterns. In *Fossils in the Making*, ed. A.K. Behrensmeyer & A.K. Hill, pp. 247–271. Chicago: University of Chicago Press.

Vrba, E.S. 1985. African Bovidae: evolutionary events since the Miocene. *South African Journal of Science* 81, 263–266.

Walker, A., Hoeck, H.N., & Perez, L. 1978. Microwear of mammalian teeth as indicators of diet. *Science* 201, 908–910.

Wall, W.P. 1980. Cranial evidence for a proboscis in *Cadurcodon* and a review of snout structure in the family Amynodontidae (Perissodactyla, Rhinocerotoidea). *Journal of Paleontology* 54, 968–977.

Webb, S.D. 1977. A history of savanna vertebrates in the New World. Part I: North America. *Annual Review of Ecology and Systematics* 8, 355–380.

Webb, S.D. 1978. A history of savanna vertebrates in the New World. Part II: South America and the Great Interchange. *Annual Review of Ecology and Systematics* 9, 393–426.

Webb, S.D. 1983. The rise and fall of the Late Miocene ungulate fauna in North America. In *Coevolution*, ed. M.H. Nitecki, pp. 267–306. Chicago: University of Chicago Press.

White, T.E. 1949. The endocrine glands and evolution. No. 2. The appearance of large amounts of cement on the teeth of horses. *Journal of the Washington Academy of Sciences* 39, 329–335.

Wing, S.L., & Tiffney, B.H. 1987. The reciprocal interaction of angiosperm evolution and tetrapod herbivory. *Review of Palaeobotany and Palynology* 50, 179–210.

6

Functional predictions from theoretical models of the skull and jaws in reptiles and mammals

WALTER S. GREAVES

ABSTRACT

A group of simple mechanical models provides hypotheses that can explain and in some cases have been used to predict a number of morphological conditions found in typical mammalian, and some reptilian, skulls and jaws. These models suggest that the biological or natural design of these structures is stereotyped at a basic level. They also assume that skulls and jaws function in a nearly optimal manner and use a minimum amount of tissue when possible. The models generate the following results:

1. Measurements along the jaw are best taken perpendicular to the vector of the resultant force of the jaw muscles.
2. This vector divides the jaw in the ratio of 3:7 which maximizes the *average* bite force.
3. The resultant acts inside a triangle of support formed by the two jaw joints and the bite point which precludes dislocation of these joints.
4. The postorbital bar if present forms an angle of 45° with the resultant vector to optimally resist torsion of the skull.
5. In animals that bite strongly at anterior teeth there is a minimum jaw width relative to jaw length below which torsion cannot be optimally resisted.
6. The condyle is high above the lower tooth row in mammalian herbivores because the glenoid cavity lies above the upper tooth row.
7. The teeth in a geometrically defined region at the rear of the tooth row exert subequal bite forces that approach a maximum magnitude.
8. Jaw elongation results in the positioning of additional cheek teeth in this region of potentially maximum force at the rear of the tooth row.
9. A long diastema is simply a by-product of jaw elongation.
10. Since the jaw joint is high in mammals when the temporalis muscle is relatively small, the distance between the jaw joint and the tooth row is minimized.

INTRODUCTION

One approach to explaining the natural design of the head in fossil tetrapods is to study preserved hard structures from a mechanical and functional perspective. Analogical models can be used to construct hypothetical explanations for the bony apparatus in question. Living animals that serve as modern analogs for fossils not only provide information but eventually can be used to test the resulting ideas. Parts of organisms are appropriately modeled as machines or subassemblies of machines because bones and other structures of the head must adhere to the same mechanical rules that govern any other physical entity, albeit with a series of special limitations because they are or have been living structures.

Two important additional assumptions simplify this general approach to an explanation of form: (1) Structures such as the skull are imagined to operate in an optimal manner to the extent possible (see discussion in Thomason 1991), and (2) they are assumed to be constructed with only enough metabolically expensive tissue to allow adequate performance. This is not to suggest that animals carry out their functions in a technically optimal way. They obviously do not. Nor does this view imply that safety factors are unnecessary or that there is no redundancy. However, these simplifying assumptions are important because they make for tractable models (Beatty 1980).

Unlike the engineer who provides solutions to problems, the approach utilized here to model animal structure is basically the reverse (Laithwaite 1977). The skull and jaw are taken to represent the embodiment of a series of "solutions" to adaptive "problems" faced by an animal. Our task is then to work backward in an attempt to determine what the problems were that particular structures "solved." This attempt is complicated by the need to disentangle a single solution from the many and because the meaning of many solutions is not obvious.

The analysis of idealized models produces less than perfect predictions about both living and fossil forms because these analogical devices are designed to create *simplified* descriptions of real animals. Models are by definition technically *incorrect*. In constructing a model of a specific part of a fossil organism, assumptions are sometimes made that can be supported only marginally by measurements on analogous living animals. But a model's strength lies in this simplicity. These heuristic devices are at once tractable and allow stepwise improvement; a simple, albeit imperfect, but reasonably well-understood model paves the way for other representations that more closely approach reality. If part of a living animal does indeed function like an idealized machine, if most members of the group have this structure, and if predicted changes are found in a documented series of transitional stages, then the importance of this function to the animals may be inferred.

The models described below are biophysical analyses (Lauder 1991, this volume) and as such are only a part of our search for an explanation of fossil structure. Careful observations or measurements of fossil as well as living forms (where they are often called "experiments") and real experiments using living analogs (observation both before and after a change) are vitally important because they can test the predictions made using a model. Just as modeling and other forms of hypothesizing are essential if data are to be synthesized, so the gathering of appropriate data is required if models are to be properly tested.

Yet modeling, experimenting, and observing are very different activities that individuals may have difficulty combining successfully in a single investigation. Pleas for completeness aside, these activities are probably best undertaken, without apology, as individual studies. Team efforts may be appropriate to avoid the addition of trivial and self-evident hypotheses to good observational and experimental work.

A simple and generally accepted model can be improved in a number of ways. In the body of this paper three approaches will be discussed with examples. First, errors or incorrect interpretations that may have halted progress can be rectified. Second, the original model may be accepted without radical change in its basic tenets while further work extends the model and may result in a serious break with older ideas. Finally, an entirely new view may be constructed that relies only upon some aspects of the original model.

INCORRECT INTERPRETATION OF ACCEPTED MODELS

Some functional studies are confusing or provide only a partial answer. The approach might be reasonable but the analysis is either faulty or incomplete. As originally analyzed, the position of the condyle, or lower half of the jaw joint, in mammalian jaws exemplifies this kind of difficulty. Apparently the geometry of this situation had not been carefully studied. Furthermore, the use of a pair of scissors (a common and seemingly well-understood tool) as an analogy may have added an intuitive logic to the explanation that masked the flaws or missing observations in the analysis.

The lower jaws of many mammals can generally be divided into two basic groups according to the location of the articular condyle of the lower jaw relative to the tooth row (Becht 1953; Davis 1964; de Wolff Exalto 1951; Gregory 1951; Maynard Smith & Savage 1959; Moss 1968). In carnivores the articular condyle is close to the level of the lower teeth so that a straight line intersecting most of these teeth also intersects the condyle. Herbivores, on the other hand, have a condyle that is some distance above a line passing through the lower tooth row.

The most influential explanation stated that scissorlike cutting, as found in many carnivores, required a low condyle (Colbert 1945; Gregory 1951; Davis 1964). The second part of this explanation claimed that simultaneous occlusion (where at least all the cheek teeth occluded at approximately the same time because the upper and lower rows are parallel when the teeth meet) was present in herbivores. This kind of occlusion was said to occur only when the condyle was some distance above the teeth.

This explanation was derived from the claim that the pivot point, or rivet, in a pair of scissors is generally in line with the cutting edges. This geometry, it was claimed, constrains the system so that the edges of the two blades, in passing each other, form an acute angle familiar to anyone who has used a pair of shears. Carnivore jaws generally form an acute angle with each other, hence the analogy with scissors. Simultaneous meeting of the cutting blades was apparently thought to occur only in modified scissors such as certain garden snips.

Simultaneous occlusion (in lateral but not in dorsal view) of all or most of the teeth in the upper and lower tooth rows is clearly important in herbivores (Gregory 1951; Shute 1954; Tattersall 1972). After the lower jaw closes, it moves medially and raised enamel ridges on opposing teeth move, as small cutting edges, relative to one another (Rensberger 1973). The direction of jaw movement can be directly observed in living forms. It had also been inferred in fossils by studying the placement of wear facets (Butler 1952; Crompton & Sita-Lumsden 1970) and their topography (Greaves 1973b; Rensberger 1973; Costa & Greaves 1981). Yet, a careful study of the geometric variables that determine exactly how tooth rows, or scissor blades, meet each other was lacking. The major flaw in previous analyses was the presumption that in a pair of scissors the rivet and the edges of the cutting blades lie on the same line. They do not. Once this was recognized the geometry of the system was quickly elucidated. It was the reverse of the previously assumed situation.

The way the blades meet each other is indeed governed by the relationship between the edges and the rivet (Figure 6.1). But, in a common pair of scissors, the rivet is actually below the level of the cutting edge of one blade and above that of the other (Figure 6.1a). If the edges are parallel when they meet, like knife blades placed on edge on opposite surfaces of a glass plate, the distances between the rivet and the blade edges will be equal and in the same direction relative to the rivet (Figure 6.1b,c). That is, both blades must be the same distance either above or below the rivet. If the distances are different, or if one blade is above and the other below the rivet (regardless of the magnitude of the distances), the blades will meet at an angle as in a pair of common scissors (Figure 6.1a).

Clarifying the relationship between the condyle and the *lower tooth row* calls attention to the relationship between the upper half of the jaw joint, the glenoid cavity, and the *upper tooth row* (cf. DeMar & Barghusen 1973). The glenoid cavity is almost always above the level of the upper tooth row in mammals (Greaves 1974). If simultaneous occlusion is required in herbivores, then the lower jaw must match the upper (Figure 6.1b). Since the glenoid is above the upper row, the condyle must be above the lower row, and by an equal amount (Greaves 1974). An understanding of how tooth rows meet each other thus provides a reason for the high condyle in herbivores. The reason the glenoid is high in herbivores has not been satisfactorily explained (contra Greaves 1980) but may simply reflect the more dorsal positioning of the new dentary–squamosal joint that formed at the origin of mammals.

Simultaneous occlusion is not required in carnivores, and a low condyle accompanied by a high glenoid causes the upper and lower tooth rows to meet at an angle during jaw closing. Yet in carnivores simultaneous occlusion again in lateral view is sometimes important. For example, the large molars in dog jaws are essentially in line with the jaw joint and occlude at about the same time. Because the

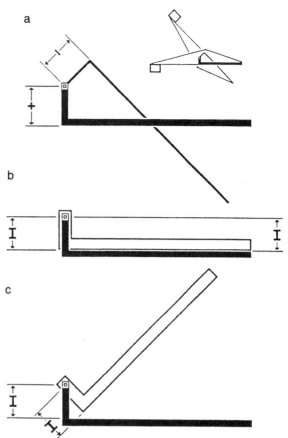

Figure 6.1. (a) Relationship between the rivet and the edges of the blades of scissors. Rivet is above (+) heavy line representing the upper blade and below (−) light line representing the lower blade. Inset is a diagrammatic representation of an entire pair of scissors. (b) Blades (or tooth rows) are parallel when they meet when the distances (I) between the blades and the rivet (or joint) are equal, and both are either above or below the rivet. (c) Same blades as in (b) drawn in the open position.

tooth row is convex downward, the more anterior teeth do not occlude simultaneously with the molars.

The assumption that a high condyle moved up in order to improve the leverage of some of the jaw muscles is an entirely different kind of analytical error (cf. Crompton & Hiiemae 1969). The condyle's location (or the joint's location) can be defined by its position relative to the tooth row. A muscle's lever or moment arm is the distance between the action line of the *muscle* and the condyle (or joint). Therefore, the position of the tooth row relative to the condyle does not affect the muscle's moment arm. Stating that the distance between the condyle and the tooth row changed says nothing necessarily about whether or not the condyle moved relative to a muscle (Greaves 1974). Furthermore, most jaw joints probably did not move relative to the skull (Greaves 1974). Even if some joints did move, as apparently was the case in saber-toothed cats, the leverage of at least one jaw elevating muscle must necessarily be degraded and the leverage of the teeth must also change because both are now farther from the fulcrum (see Bryant & Russell, this volume).

The above examples of incorrect assumptions produced a long period of stasis in our understanding. This often occurs when something is wrong with our way of looking at a problem. Once the assumptions are corrected progress can be rapid and new problems are often suggested: For instance, the high glenoid in mammals rather than the high condyle still requires an explanation.

EXTENDING ACCEPTED MODELS

In essence the masticatory apparatus is a relatively straightforward mechanical system with many similarities between reptiles and mammals. In an unadorned approximation, two bars are hinged and brought together by a series of muscles. Most early workers used this model and considered only the joint muscles and teeth on one side of the head (but see Gysi 1921). Most were concerned with such things as changes in lever arms of muscle or teeth, the magnitude of the input muscle force, bite forces at individual teeth, and the shapes of the bony levers themselves. In a lever system like this the largest bite force is exerted at the most posterior biting position; all the other locations (teeth) are farther from the joint and have smaller bite forces (Figure 6.2a).

Yet, this lever model is unable to explain a number of features commonly found in reptilian and mammalian jaws. Teeth are not found far to the rear of the jaw, which is apparently the "best" location. Furthermore, an intriguing geometric relationship present in many mammals is that the length of the molar row (sometimes including the premolars) is approximately equal to the distance between the joint and the last molar (Stöckmann 1975, 1979; Greaves 1978), and regions without teeth (diastemata) are often present (Ardran, Kemp, & Ride 1958; Landry 1970).

Location of the teeth

A study of the location of the teeth along the jaw may be examined as an important example of an extension of an accepted model. The jaw muscles will be considered first because there is a relationship between the locations of these muscles and the tooth row. Jaw muscles are present in a variety of

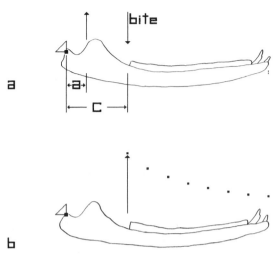

Figure 6.2. (a) Lower jaw represented as a simple lever. Up-arrow indicates the resultant force of the jaw muscles. Bite force increases as the ratio a/c increases, implying an advantage to moving the teeth posteriorly. (b) Jaw represented with a more accurate position for the muscle resultant. Biting on teeth located behind this resultant would tend to dislocate the jaw joint. Dotted line indicates relative bite force along the tooth row.

sizes, orientations, and patterns of pinnation. Some major groups of mammals can even be defined in part by the characteristics of their jaw muscles (Turnbull 1970). These muscles are generally concentrated at the posterior end of the jaw near the jaw joints in reptiles and mammals, suggesting that this location is significant.

At least two reasons for the posterior positioning of jaw elevating muscles are perhaps obvious: (1) Gape of the jaws might be restricted by the amount of stretch available to muscles (cf. Herring & Herring 1974; Herring 1975), and (2) teeth are expected to be located in front of the resultant force of these muscles to avoid disarticulating the jaw joint. A tooth behind the resultant (Figure 6.2b) would act as a fulcrum as the jaw closes, and the jaw would rotate around it and move the posterior end down, thus disarticulating the joint (Gysi 1921; Bramble 1978; Greaves 1978; Druzinsky & Greaves 1979). Such rotation could be resisted by tensed ligaments of the joint capsule. But as joints are compression-resisting devices, a jaw that is repeatedly disarticulated is unlikely. Disarticulation of the jaw joint is essentially eliminated if all the teeth lie anterior to the resultant force of the muscles.

However, the actual location of the resultant force of the jaw muscles has yet to be measured. Reasonably careful estimates, from dissections (my own and those of students in a long-running graduate course) of a number of different mammals, place the resultant at about one-third of the length of the jaw from the jaw joint. This position conflicts with one's intuitive grasp of jaw structure; many would predict that the location of the resultant would be different in different animals (cf. Gans & de Vree 1987; Gans 1988). Yet these careful estimates of the resultant's position match the location of the most posterior tooth; this tooth often lies just in front of this one-third location, tending to confirm the idea that the avoidance of dislocation is significant.

An important reason for examining this estimated location of the resultant force and the posterior tooth is worth repeating. This relationship is seen over and over again, suggesting that this location is important (Greaves 1978, 1982, 1983, 1985b; Walker 1978; Weijs 1980; Weijs & Dantuma 1981). A notable simplification in the discussion below, allowing for a tractable model, is the assumption that all the jaw-closing muscles act both maximally and at the same time.

Accepting the possibility that the resultant of the jaw-muscle force almost always acts at approximately one-third of the way along the jaw and immediately behind the last tooth suggests that other positions are either not as good or are precluded for some reason. A reasonable approach would be to search for a mechanical optimum by analyzing a number of different theoretical positions of the resultant in front of and behind the one-third location. If an optimum were found, this would be evidence that selection favors a particular resultant location or tooth-row position.

An obvious approach is simply to plot relative bite forces along the entire tooth row for different hypothetical locations of the resultant force of the muscles, because the resultant's distance from the fulcrum affects the forces at the teeth. This is true for both a bar pivoted at one end or a triangular plate that more realistically represents the roughly triangular jaws of many mammals (Greaves 1988a).

In these experimental models the teeth have to be anterior to the resultant to avoid disarticulation. Therefore, a fixed jaw length means that the tooth row would be shorter for more anterior resultant locations (Figure 6.2b). Conversely, more posterior locations allow for potentially longer tooth rows.

Plotting relative bite force along the tooth row for a single resultant location results in a rectangular hyperbola that ascends at the rear of the jaw, and descends at the front (Figure 6.2b). Using different theoretical anteroposterior resultant locations changes the length of the tooth row and generates a family of such curves. At one extreme a long tooth row is paired with relatively low bite forces and at the other a short row has higher bite forces. Calculating an *average* bite force for the entire tooth row for each hypothetical resultant position is one way to evaluate different tooth-row lengths.

This requires the summation of all possible bite forces that theoretically could be applied along the tooth row and the division of this sum by the jaw length. (This sum is represented by the area under each curve, thus requiring the integration of each curve's equation.) These averages can be plotted against the corresponding location of the most posterior tooth along the row and, thus, the muscle resultant. This gives a curve with average bite forces that are low for more anterior and posterior resultant positions but are higher as you approach the one-third position. The curve opens downward with a maximum value at 0.30 of jaw length (Greaves 1988a). The maximization of average bite force suggests a reason or explanation for this 0.30 resultant/posterior tooth position. It also implies that this location is important to animals. From the point of view of maximizing average bite force, this is the "best" way to locate the tooth row along the jaw (cf. Parrington 1934) and is an explanation for why teeth do not typically extend all the way back to the jaw joint (cf. Gans & de Vree 1987; Gans 1988).

Analyzing reptilian jaws in the same way gives a different "best" location for the resultant and the posterior end of the tooth row in many of these animals (20 percent, rather than 30 percent, of the way along the jaw). This is mainly due to a difference in jaw shape in dorsal view; the reptilian jaw is pentagonal instead of triangular. The changes necessary to transform the reptilian into the mammalian model are just those changes seen in the fossils that span the reptile/mammal transition (see Crompton, this volume). The importance of this observation is that it tends to support the idea that these relationships were critical to these reptiles as well as to early mammals (Greaves 1988a).

The geometry that limits the length of the molar row

Biting and the mastication of food are clearly important functions of jaws. Relatively large and efficiently produced bite forces probably are also important, especially at the molars. According to a lever model, bite force can be increased in two basic ways: (1) by increasing muscle or input force and (2) by improving the leverage of the system by a change in the relative positions of muscles, teeth, or joints. The discussion in the previous section strongly suggests that the relative positions of structures are already optimal and are thus unlikely to change.

Input force can be increased by adding muscle and/or by including the force due to activity of the muscles on both sides of the head (Gysi 1921). Many animals chew on only one side at a time. If forces from the muscles on the nonbiting side are to be transmitted to the biting side, the mandibular symphysis must be able to transmit strong muscle force from the nonchewing side since this is the only path this force can take (e.g., Rigler & Mlinšek 1968; Weijs 1981). There are advantages to an unfused mandibular symphysis in mammals (Scapino 1965, 1981; Mills 1966; Hiiemae & Ardran 1968; Kallen & Gans 1972; Weijs & Dantuma 1981) including fine alignment of carnassials, fitting teeth to food objects, and absorbing occlusal shock. Most workers feel, however, that a fused symphysis is less likely to fail when muscle forces from the nonchewing side are transmitted across the symphysis (Du Brul & Sicher 1954; Hylander 1975, 1979, 1984, 1985b; Beecher 1977, 1979; Weijs 1981). An unfused or patent symphysis functions well in selenodont artiodactyls (Greaves 1978) as well as in some carnivores (Dessem 1985a,b, 1989). Fused symphyses are uncommon among mammals and have significant costs (Greaves 1988b), for example, the large amount of bone needed to buttress this region against the bending and shearing strain that accompanies mastication.

In any event, the muscles on both sides of the head are active as can be confirmed by palpating the jaw muscles on a chewing person or an animal, and as indicated by a number of studies (Leibman & Kussick 1965; Møller 1966; Kallen & Gans 1972; Herring & Scapino 1973; de Vree & Gans 1976; Hylander & Bays 1978; Hylander 1979, 1985a, b; Gorniak & Gans 1980; Weijs & Dantuma 1981; Dessem 1985a, 1989). If only the muscles on one side are considered, a simple bar representing the lower jaw, and a muscle force in the second dimension, are adequate. A V-shaped bar or a triangular plate is required if two jaw joints, one limited region of tooth occlusion, and muscles from both sides acting in the third dimension, are involved (Nagel & Sears 1958; Piffault & Duhamel 1963; Walker 1976, 1978; Greaves 1978; Smith 1978).

The forces resisted at the three angles of the triangular plate (the joints and bite point) depend directly on the location of the resultant force of the jaw muscles. Joint dislocation is precluded, and the forces at the joints and bite point will change, if the resultant moves *inside* the triangle of support. This is equivalent to the two-dimensional case where the resultant is behind the teeth, and the bite point moves forward and backward along the tooth row (e.g., Greaves 1978; Druzinsky & Greaves 1979; Dessem 1985a,b 1989).

Once muscle activity on both sides of the head is included, the model of the masticatory apparatus significantly changes; force distribution as well as increased intensity becomes an issue. Consider a dorsal view of a *modified* triangular lower jaw where the molar rows are now parallel, and lines passing through these rows intersect their respective jaw

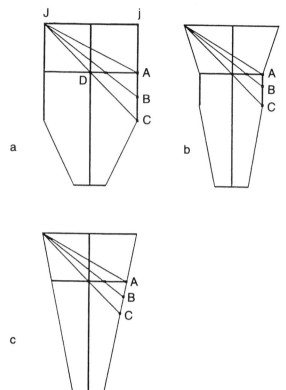

Figure 6.3. (a) Dorsal view of a stylized lower jaw where the posterior tooth rows are parallel and in line with the jaw joints. A, B, and C are representative bite points and D is a point on the midline. J and j indicate the nonchewing or balancing-side jaw joint and the chewing or working-side joint, respectively. (b) Similar drawing where the tooth-rows are still parallel but closer to the midline. (c) Depiction of tooth rows converging toward the front of the jaw.

joints (Figure 6.3a). (This is similar in some ways to the reptilian jaw but is, nevertheless, quite different.) A line from the balancing side joint (J) can be drawn through any bite point (e.g., A, B, or C) on the biting side. Each of these lines ($\overline{JA}$, $\overline{JB}$, or $\overline{JC}$) represents the jaw levers for these bite points in this idealized example. Given this system, the calculated magnitude of the maximum possible bite force is always half of the input muscle force (Greaves 1978).

This result is not intuitively obvious and may appear to be anomalous until the geometry of Figure 6.3 is examined closely. A series of right triangles can be defined by the two jaw joints (J and j) and one of the many bite points (e.g., A, B, or C). This series of triangles (e.g., JjC, JjB) has a common base ($\overline{Jj}$) and the sides perpendicular to the base all lie along line $\overline{jC}$. Lines parallel to $\overline{jC}$ divide the base and every hypotenuse (e.g., $\overline{JB}$) in the same ratio. For example, every hypotenuse is bisected by the midline.

Moreover, different muscle activity patterns on

each side of the head place the muscle resultant at different mediolateral positions. For example, equal activity with 10 units of force on each side produces a resultant that is large (20 units) and lies on the midline. Unequal muscle activity where, say, the left side exerts 10 units of force while the right side is silent, gives a resultant located on line $\overline{jC}$ that is only half as large. Intermediate magnitudes and locations result from other muscle activity patterns, but all resultants with the same magnitude lie on a single line parallel to $\overline{jC}$ and thus divide each hypotenuse in the same ratio. For a given bite point, say B, the largest possible bite force is exerted when the resultant muscle force acts on the hypotenuse ($\overline{JB}$) of the triangle defined by that bite point; the hypotenuse is the jaw lever and is the line joining the bite point, the muscle resultant, and the joint. Thus the maximum bite force at the end of every jaw lever is equal because equal resultant forces act at the same relative positions on all the jaw levers; equal forces act on lever systems with the same mechanical advantage.

Muscle resultants are largest at the midline and smaller laterally (see above). Because the jaw levers for particular bite points (say, $\overline{JC}$ and C, respectively) lie at an angle to the midline, smaller resultants are farther from the fulcrum (J) and closer to the bite point (C) while larger resultants act closer to the fulcrum and farther from C. Larger input muscle forces will necessarily be paired with less favorable lever systems and vice versa. Thus, the maximum output bite forces are always equal, even when the resultants have *different* magnitudes and act at *different* locations along each jaw lever. An alternate method of calculating bite forces (Weijs 1981) gives the same result.

If the joints are in line with parallel tooth rows in dorsal view (Figure 6.3a), these two variables balance each other so that the *maximum* bite force is always half of the input muscle force. If the tooth rows were still parallel but closer to, or farther from, the midline (Figure 6.3b) the maximum bite forces would be some other fraction of the input muscle force because the position of the tooth row affects the length of the output lever (Greaves 1978). If the tooth rows are not parallel (Figure 6.3c) the maximum bite forces will only approach equality.

Confining the muscle resultant to an anteroposterior position that is located 30 percent of the way along the jaw places limits on the above analysis. If line $\overline{DA}$ is 30 percent of the way along the jaw, variable bilateral muscle activity can position the resultant muscle force at any point on this line. The jaw levers from the left joint (e.g., $\overline{JA}$, $\overline{JB}$, and $\overline{JC}$) that intersect line $\overline{DA}$ can intersect the tooth row only from point A to point C. The resultant muscle force can be applied on the left side at A and move to D if the muscles vary their activity appropriately. As the resultant moves from A to D the same bite force (half the muscle force) will be exerted first at A and then at each more anterior bite point in turn until point C is reached.

Bite forces in front of this region decrease rapidly because the efficiency of the lever system decreases with each more anterior tooth position (the muscle resultant force cannot increase beyond its maximum magnitude). The geometry of this simplified system dictates that the length of this region ($\overline{AC}$) along the molar tooth row where the bite force is half the muscle force is equal to the distance from the jaw joint to the last tooth ($\overline{jA}$). This relationship has long been evident (Greaves 1973a, 1978; Stöckmann 1975, 1979).

This *idealized* model assumes that the jaw joint on the nonchewing side (J) resists all the joint reaction forces, a condition that is only approached in an animal. Joint forces in the initial subjects of this analysis, the selenodont artiodactyls (e.g., cattle, antelope, sheep), have not yet been measured, but in monkeys these forces are larger at the joint on the nonchewing side (Brehnan et al. 1981; Hylander 1985a).

Essentially all of the modern selenodont artiodactyls have subparallel cheek-tooth rows, and their cheek teeth are located in the region of maximum bite force, implying that this geometry is important in this group. As indicated, if the tooth rows are not parallel but rather converge toward the anterior end of the jaw (Figure 6.3c), the relative bite forces at the posterior teeth will only approach equality but the molars in most mammalian jaws will be located in this region ($\overline{AC}$) of largest bite forces (Greaves 1978). Moreover, the teeth in this region generally have a large occlusal area, which is smaller in more anterior teeth. A change in width is often abrupt in mammals and is expected if the force per unit area of a tooth is roughly constant, and if bite force along the posterior region is large but decreases rapidly anterior to point C (Figure 6.3).

In carnivores, carnassial and molariform teeth are in the region of high force. This region of high force is shorter than in artiodactyls ($\overline{AC} \neq \overline{jA}$; see Figure 6.3c) because the tooth rows in these animals are not parallel but converge anteriorly. The carnassial, at the anterior end of this region, is approximately half way along the jaw, as opposed to 60 percent of the way, as is point C (Radinsky 1981; Greaves 1983, 1985b). This region of high force is apparently absent in many reptiles as it probably also was in the earliest mammals (Greaves 1988a). It developed later in some reptiles and most mammals and requires variable jaw-muscle activity patterns as described above (Greaves 1978; and references in Druzinsky & Greaves 1979).

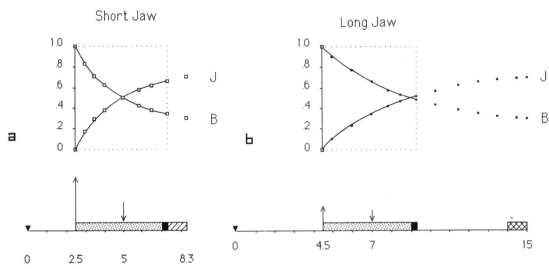

Figure 6.4. Relative bite force (B) and joint reaction force (J) versus biting position along the tooth row based on a two-dimensional model for (a) a short jaw, 8.3 units long and (b) a long jaw, 15 units long. Symbols: inverted triangles, jaw joints; filled rectangles, first premolars; stippled rectangles, molars and remaining premolars; hatched rectangles, incisors and canines. Upper boxes isolate areas above and below the curves in the region of the cheek teeth. Arrows pointing up represent the resultant force of the muscles which is 30 percent of jaw length from the jaw joint in each case (2.5 and 4.5 units, respectively). Arrows pointing down represent sample bite points. The respective bite and joint force curves have the same shape because they are reciprocals of each other and can be superimposed by rotating one or the other curve around a horizontal axis. The area above the bite force curve is, therefore, the same as the area under the curve for joint force.

Jaw elongation and the diastema

Another concern is the presence in some animals of a long *diastema,* a region where teeth are absent. While the presence of this region may have some functional significance, clearly no bite forces are applied here (Ardran et al. 1958; Landry 1970). Following the previous analysis, a reasonable assumption is that selection would not favor locating important teeth in a region where bite forces are too low (anterior to point C in Figure 6.3). Teeth are metabolically expensive and would be better placed in a region ($\overline{AC}$) of higher force. This is an intuitively satisfying partial explanation for the absence of teeth anterior to the equivalent of point C in selenodont artiodactyls. But how might this be accomplished? The following analysis (as well as another in the section on torsion resistance) bears on this problem.

The resultant force of the jaw muscles in two jaws that differ only in length is absolutely farther from the jaw joints in the longer jaw because 30 percent of a longer jaw is a greater distance than 30 percent of a shorter jaw. Being farther away, the joints will resist less reaction force in the longer jaw. Accepting the simplification that muscle forces produce only joint and bite forces, then the bite forces in the longer jaw must necessarily be larger if the joint forces are smaller.

In Figure 6.4 relative bite forces (B) and reaction forces at the joint (J) are compared for different biting positions along the tooth rows of a short and a long jaw modeled in two dimensions. The lengths of the jaws (8.3 and 15 units, respectively) are arbitrary but the ratio between them insures that, when tooth size is held constant, the cheek teeth will just fit into the region of high force ($\overline{AC}$ in Figure 6.3a) in the long jaw (see below and the fuller discussion in the next section). At any position along the tooth row the sum of relative biting and joint forces equals one and the *areas* under the curves represent the sum of all possible forces. For biting on the cheek teeth, the sum of biting and joint reaction forces is represented by the dotted rectangles. Each of these areas divided by the length of the cheek-tooth row gives the average muscle, bite, and joint forces, respectively. The fractions of the average muscle force that are applied at the teeth and the jaw joint can be calculated using these averages. In the case of the short jaw in Figure 6.4, 54 percent of the average input muscle force acts at the teeth while 46 percent is wasted as reaction force at the jaw joint (efficiency = 0.54). The respective percentages for the long jaw are 69 percent and 31 percent (efficiency = 0.69). Thus, in this example, an additional 15 percent of the average input force in the longer jaw is used productively at the teeth rather than being wasted at the jaw joint.

The three-dimensional model discussed in the previous section suggests, however, that the maximum bite force (half the maximum input muscle force) is already being exerted at a region at the rear of the tooth row. Increasing jaw length while holding the geometry constant increases the distance ($\overline{jA}$) in Figure 6.3 as well as the length of the region of large forces ($\overline{AC}$). The tooth row is held constant in this example. Thus, the molar row is absolutely the same length in both jaws but *relatively shorter* in the long jaw and thus occupies less of distance $\overline{AC}$. It follows that more teeth (premolars) will now lie in the region ($\overline{AC}$) of maximum force. Thus, the increased bite force in the longer jaw is exerted at the anterior end of a longer region of maximum force. The maximum bite force has *not* been increased, rather the bite forces at some teeth have been increased because they now lie in the region of maximum force. This can be seen in Figure 6.5 where the three-dimensional model is assumed, and relative joint and bite forces are plotted against position along the tooth row in short, intermediate, and long jaws. A region of high and equal bite force is obvious at the rear of each tooth row; its presence was discussed in the previous section. This region increases in length as the jaw lengthens, until all the cheek teeth are included. Thus, beginning at the rear of the premolar row, the bite force at each premolar is increased in turn with increasing jaw length until all these teeth exert the maximum bite force. A diastema is absent initially (left panel) but is present in the other panels where it is longest in the longer jaw.

The average bite force (the area under the curve divided by the length of the cheek-tooth row) increases about 9 percent from the shorter to the longer jaw. However, a greater fraction of the muscle force is exerted at the teeth, and so the efficiency rises from 0.52 in the shortest jaw to 0.59 in the intermediate case, to 0.65 in the long jaw with the first premolar present, and to 0.67 without the premolar as is the case in modern selenodont artiodactyls. The respective average joint forces are 0.48, 0.41, 0.35, and 0.33. Thus 15 percent more of the average muscle force is usefully exerted at the teeth in long jaws because less is being wasted at the two jaw joints.

The apparent lengthening of the jaw in selenodont artiodactyls (Greaves 1991a) increased the length of the region of high bite force, thus including in this region the premolars as well as the molars. The three-dimensional model implies that selection will favor locating all the cheek teeth in a region of high force. The analyses in Figure 6.5 and in the next section indicate that lengthening the jaw did just that. But jaw lengthening where everything else remains the same implies a *relatively shorter* tooth row and, thus, a diastema, or a longer diastema, as a by-product.

Similar changes apparently also took place in other mammals such as horses (Granger 1908).

Lengthening the jaw and thus increasing the bite force at the premolar teeth is not expected in most mammals. In carnivores, for example, the carnassial and molar teeth are already at optimal locations. The bite force at the canine region important in these animals is slightly lower, and bite force remains unchanged at the anterior incisors (Greaves 1983).

NEW MODELS

Torsion resistance

The discussion so far has dealt mainly with the positions or locations of input and output forces that are associated with the masticatory apparatus. These forces also influence skull buttressing and general overall shape variables of the skull and jaws, such as length and width, because a structure's ability to resist these kinds of forces dictates the limits on the general morphology of the skull.

The masticatory forces as described above in the model of the selenodont artiodactyl jaw mechanism have a twisting effect on the skull (Greaves 1985a). The jaw adductor muscles exert a downward force on the skull and, by way of the lower jaw, upward forces at the joints and bite points. Large forces are exerted on the upper teeth on only one side because these animals chew on one side of the head at a time. The upward joint forces are larger at the joint on the nonchewing side of the head (see above and Hylander 1985a). This asymmetrical force regime tends to twist the teeth and the rostrum in one direction, and the jaw joint and the braincase in the other, around an axis located approximately at the midline of the palate.

The orbital region is located at the junction of the braincase and the rostrum. Thus, other structures that enable this weaker area to resist torsion are expected. The bony postorbital bar that completes the orbit at the rear seems to function in this way (Greaves 1985a).

When a skull or any structure is twisted, the maximum tensile and compressive shear strains are located along helical paths (Figure 6.6) that wind around the skull at an angle of 45° to the twisting forces (Perry & Lissner 1962). Strains can be resisted in an optimal manner by strengthening the bone along these helices in preference to other regions of the skull.

Strengthening could take a number of different forms. In essence, the skull is a tubular shell, and this shape is naturally strong in torsion. If additional strengthening were required, regions of curved or folded bone could be situated along these helical

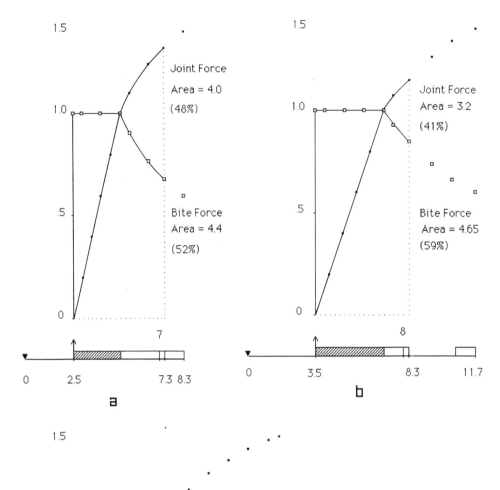

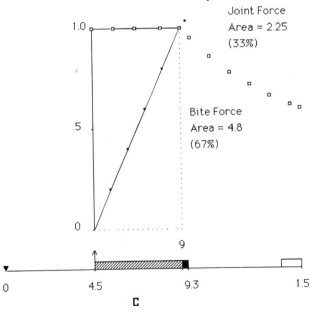

Figure 6.5. Bite and joint force curves utilizing the three-dimensional model. (a) Curves for a short (8.3 units long) jaw. (b) Curves for a jaw of intermediate length (11.7 units long). (c) Curves for a long (15 units long) jaw. Shaded portion of the rectangles representing teeth indicate the region of maximum force. Other symbols as in Figure 6.4.

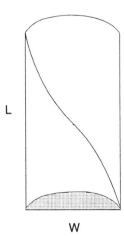

L

W

Figure 6.6. Drawing of a half cylinder showing a 45° helix on the curved surface that intersects the plane surface at two points. Width is the diameter of the cylinder. Length is the distance parallel to the axis of the cylinder from one intersection of the helix to the other.

paths. Bone may also be thickened, or additional struts of bone may even be found in places where the skull shell is interrupted. In all of these cases, unless the modified bone lies along one of the helices that describes the location of the maximum strains, the modification will not have its greatest strengthening effect. No matter how it is strengthened, a helical tract must also intersect both the rostrum and the braincase if torsion between these two structures is to be resisted. Finally, for maximum effect this helical tract should be on the surface of the skull so as to be as far from the axis of twisting as possible.

One example of a torsion-resisting strut may be the postorbital bar (Greaves 1985a). Postorbital bars connect the front and back halves of the skull by spanning the weak orbital regions where the shell of the skull is interrupted. These bars are as far from the axis of twisting as possible and, perhaps most telling, are coincident with segments of the 45° helices that can be traced around the skull and that map the location of the maximum shear strains. A postorbital bar is not present in the majority of mammals. The component of muscle force tending to twist most skulls is small because of the orientation of the muscle resultant. Only in selenodont artiodactyls, most primates, horses, and a few other mammals is a large component of muscle force oriented so as to twist the skull strongly and thus require a postorbital bar (Greaves 1985a). Simple working models tend to confirm these predictions and also implicate the zygomatic arches. Nevertheless, strains in the postorbital bars of skulls of selenodont artiodactyls have not yet been measured.

Measured strains in various parts of some primate skulls seem to conform to predictions but the data have been variously interpreted (Russell 1986; Oyen 1987; Ravosa 1988, 1991a,b; Hylander, Picq, & Johnson 1991a,b; Oyen & Tsay 1991).

Accepting that the orbit in selenodont artiodactyls is a weak region in the skull, one could reasonably expect a preorbital as well as a postorbital bar. There is no preorbital bar because the skull is complete anterior to the orbit. However, curved and folded bone is found along a 45° helix in this region in selenodont artiodactyls. Moreover, a line of thickened bone was observed in a single, frontally sectioned goat skull. A critical observation is that an imagined helix from the most anterior cheek tooth, the second premolar, winds around the skull past the anterosuperior edge of the biting side orbit and intersects the balancing joint (Greaves 1991a). The length *(L)* from the joint to the premolar and the width *(W)* of the skull are in the ratio: $L/W = \pi/2$. This means that torsion due to biting at the second premolar can be adequately resisted with a minimum structure. The first premolar is absent in most selenodont artiodactyls. If it were present, the 45° helix emanating from it would be too far forward, and it would not intersect the jaw joint. That helix could not resist torsion between the front and back of the skull because it does not join these two regions. The skull would not be a minimum structure because additional bony buttressing would be required. Thus, one expects the first premolar to be missing in these animals because, as jaw length in selenodont artiodactyls increased, this tooth in effect moved anterior to the postulated buttressing helix (Greaves 1991a). These considerations recall the previous discussion of the diastema. Jaw lengthening includes the premolars in a region of high bite force and produces a long diastema as a by-product. This would presumably continue until all the cheek teeth were included in this region. However, if torsion could no longer be resisted with a skull of minimum structure, tooth loss and/or the cessation of jaw lengthening would be expected.

GENERAL SKULL AND JAW SHAPE. The previous section emphasized strain-resisting struts and thickenings. Limiting the general shape of the skull, so that muscle, joint, and tooth forces are restricted to particular locations where they can be resisted in an optimal manner, also seems important.

Large canine teeth are important to carnivores. During canine biting both canines may encounter either soft tissue or bone, or one canine may meet bone while the other does not. In the latter case, the reaction force at one canine will approach zero, whereas the force at the other will be large. Moreover, the force at the diagonally opposite jaw joint

Figure 6.7. Jaw length plotted against jaw width for a sample of living and fossil Carnivora where most of the genera are represented. Analysis predicts that the length-to-width ratio of the jaws will approach but should not exceed π/2 (1.57) when selection favors long narrow jaws. The line (W = 0.64L) indicates this limiting ratio. Ratios smaller than this will be evident when selection favors large bite forces. In general canids and viverrids closely approach this limit; felids and mustelids and ursids are farthest away; and the remaining families are intermediately positioned.

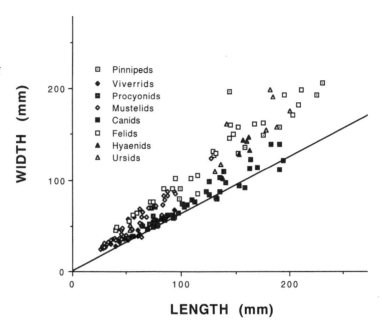

will be larger than that at the other joint if the jaw muscle activity on each side of the head is approximately equal. A force regime like this will twist the rostrum and the braincase relative to one another.

A carnivore skull can be modeled as a right circular half cylinder with the plane surface approximating the palate (Greaves & Covey 1992). A 45° helix that winds around the curved surface will intersect both edges of the plane surface. The distance L along the long axis of the cylinder, from one of these intersections to the other, varies with the diameter or width W of the cylinder such that $L/W = \pi/2$ (Figure 6.6). Twisting forces are best resisted if they act at each end of this helix because a minimum amount of extra bone if needed can be confined to this area. If L/W for the jaw does equal $\pi/2$ then both the canine and the diagonally opposite joint will be intersected by a single 45° helix that winds around the surface of the skull (Figure 6.6). Maximum shear strains will be located along this helix during canine biting. Torsion can be resisted with a minimum amount of bone if strengthening devices or extra bone are concentrated along this helix. Therefore, skull width is not expected to be narrower than length divided by π/2 or, alternatively, length longer than width times π/2 (Greaves & Covey 1992). If it were, helices emanating from the joint and the bite point would not be coincident but would pass each other on the surface of the skull with an interval between them. Neither helix would connect both the front and rear units of the skull, and extra bone would be required to buttress the skull against twisting; a minimum structure for torsion resistance would be precluded.

Relatively greater widths (or lesser lengths) are

expected to be common. The helices will again pass each other but this time *both* will connect the anterior and posterior regions of the skull. Presumably the shear strains will be shared between the helices so that a relatively smaller amount of strengthening will be required for torsion resistance. Thus, accepting that selection favors a minimum skull structure to resist torsion, the minimum width relative to jaw length of the jaw is limited (but the maximum width is not), and jaws wider than this minimum will be more torsion resistant (Greaves & Covey 1992). (If jaw width is given, the longest jaw length is limited, but the shortest is not.)

Length is plotted against width in Figure 6.7 for a large sample of carnivores representing all the families. Points representing canids (dogs) and viverrids (mongooses, etc.), where long jaws seem important, cluster around the limiting line *(W = 0.64L)* as predicted. Felids (cats), mustelids (weasels, etc.), and ursids (bears) are located away from this line, while the remaining carnivores are found in intermediate positions. Thus, while canid skulls are minimum structures that resist torsion very well, felid skulls should be even more resistant to torsion. This is reasonable since greater width gives room for relatively more musculature. This results in larger bite forces and, thus, greater torsional forces (Greaves 1985b). Furthermore, the surface of the rostrum of, say, the felid skull is farther away from the twisting axis, which also increases resistance to torsion in skulls of this general shape.

The same relationship between jaw length and width is also found in many rodents even though their skulls and jaws are significantly different from those of carnivores. This suggests a similar functional

Functional modeling of the skull

explanation for this basic skull relationship. During incisal, rather than canine, biting in these animals, the magnitude of the twisting forces is smaller because the difference in reaction force at the two jaw joints is not very great since the incisors are located near the midline. Yet, virtually continuous incisor chewing and sharpening may make a skull with a minimum structure even more important than in the carnivores where larger, but intermittent, twisting forces are present. The same L/W relationship is seen in many primates, and the 45° helices intersect the brow ridges as they pass around the orbits. These paths of the helices may help to interpret the primate data mentioned in the previous section.

Orientation of the muscle resultant

In the previous analysis – suggesting that the last tooth and, thus, the jaw muscle resultant, should be positioned at 30 percent of the way along the jaw – the teeth and the joints were assumed to lie in the same plane, and the muscle resultant was perpendicular to that plane (Greaves 1988a). Thus, distances along the jaw were the same as distances from the resultant vector. If the resultant vector is not perpendicular to the occlusal plane (or if the jaw joints are not in the plane of occlusion), we might consider components of the resultant both in and perpendicular to the plane. Another approach is to treat jaw muscles like any other muscles in the body and measure their moment arms (the perpendicular dropped from, say, a joint to the action line of the muscle). If a bony element and a moment arm happen to be coincident, they will be the same length; if they are not coincident the bone will be longer. From this point of view jaw length, should not be measured *along* the horizontal jaw ramus but rather perpendicular to the resultant vector. Vertical and horizontal components of the resultant force of the jaw muscles and "planes" of occlusion are constructs that are not always helpful (but see Spencer 1991).

The skulls of carnivores need to resist large forces directed anteriorly and ventrally, produced by struggling prey. Carnivores tend to have a large temporalis muscle and, therefore, a dorsal and more posterior orientation of the muscle resultant, opposite to the force from the prey (Maynard Smith & Savage 1959; Gingerich 1971). The resultant force in animals which do not have to struggle with prey tends to be oriented more anteriorly because the masseter and internal pterygoid muscles taken together are larger than the temporalis. Radinsky (1985) found that this latter condition arose independently many times. This recurrent acquisition suggests that advantages accrue to animals whose muscle resultant force has an anterior as well as a dorsal orientation.

Radinsky (1985) suggested that this arrangement

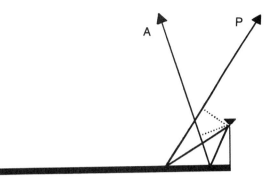

Figure 6.8. Jaw with a posteriorly directed muscle resultant (P) and an anteriorly directed resultant (A). Each resultant is the same perpendicular distance (*dotted lines*) from the inverted triangle representing the jaw joint. The distances from the joint to the point where each resultant intersects the tooth row are not the same (*solid lines*).

allowed for finer control of the lower jaw, and Scapino (1972) pointed out that it reduced and reoriented the joint reaction force. The length of that part of the skull and jaw behind the teeth also varies with the orientation of the resultant force (Greaves 1991b). This posterior region of the jaw is longer for posterior, and shorter for anterior, orientations of the muscle force (Figure 6.8). A posteriorly oriented resultant vector with a given moment arm will intersect the jaw at some distance from the jaw joint, and the teeth will be positioned anterior to that point. A resultant with a more anterior orientation, but with the same moment arm, will meet the jaw at a point much closer to the joint. This is significant because (1) less bone tissue will be needed to construct both the skull and the jaw, and (2) the lower jaw as well as the relatively weak orbital region of the skull, being reduced in length, are stronger and thus better able to resist torsion as well as bending at this critical region. This explanation predicts the following. Unless there is a clear reason for a posteriorly oriented resultant force – for example, carnivores with large caniniform teeth and large temporalis muscles to resist struggling prey (Maynard Smith & Savage 1959) and many primates with large food handling incisors (Hiiemae & Kay 1972) – the resultant force of the jaw muscles will pull up and *forward* (Greaves 1991b).

The idea that the muscle resultant should act 30 percent of the way along the jaw (see above and Greaves 1988a) can be superimposed on this analysis where the perpendicular distance from the joint to the resultant vector is held constant. Since the resultant divides jaw length in the ratio of 3:7, the region anterior to the resultant must also remain the same. These distances in front and in back of the resultant are projected as a unit onto the bony

jaw. They are projected absolutely closer to, or farther from, the joint depending on the orientation of
the resultant. Thus, if the mechanical advantage of
the resultant force is held constant (constant distance from joint to resultant), anterior reorientation
of the resultant reduces the absolute distance from
the joint to the last tooth, even though the distances
perpendicular to the resultant remain the same.

An alternative to holding constant the perpendicular from the joint to the resultant is to maintain
the distance from the joint to the most posterior
tooth. The resultant vector will simply rotate around
its point of application to change its orientation
(Greaves 1991b). As the vector rotates from a posterior to an anterior orientation, the perpendicular
from the joint to the vector will increase to a maximum, whereas the direct distance from the joint to
the last tooth remains the same. If the jaw joint is
above the level of the tooth row, the resultant will
have an anterior orientation when this maximum is
reached – again explaining the early acquisition of a
high joint (DeMar & Barghusen 1973).

Since the vector is always 30 percent of the way
along the jaw, if the distance from the joint to the
vector increases, then the length of the entire jaw
must also increase. Thus, the anterior, tooth-bearing
region of the jaw increases in length as the resultant
force becomes first vertical and then more anteriorly
oriented until a maximum length is reached. This
reasoning suggests that jaw elongation and muscle
resultant reorientation in selenodont artiodactyls
could have taken place at the same time. Thus, the
increase in the length of the region of high and
subequal bite force at the rear of the tooth row (that
increased the bite force at the premolars), the formation of a diastema, and the reorientation of the
resultant may all have taken place simultaneously,
with only the anterior teeth needing to move physically as the diastema formed.

The jaw joint in reptiles is often below rather than
above the tooth row as in mammals. Moreover, the
resultant vector of the adductor musculature is usually oriented dorsally and posteriorly. Yet, the above
analysis applies here also because an anterior resultant with a joint above the teeth is geometrically
equivalent to a posterior resultant with a joint below the teeth.

CONSOLIDATION AND SYNTHESIS

The models reviewed above are examples of what
Lauder (1991, this volume) calls biophysical analyses.
As such they are mainly ahistorical studies. Nevertheless, these analyses complement and are consistent with each other (i.e., different analyses give the
same answer to a given question), and a summary

statement might be clearer and more revealing if
these explanations were gathered together and presented in the form of an interrelated scenario. This
creates a story with a generality and even an inevi-
tability that is surely artifactual. But this approach
has the advantage that it may better draw attention
to some of the major physical imperatives that seem
to have constrained the natural design of jaws and
skulls.

Reptilian and mammalian jaws are quite simple
systems in which the efficient use of muscle forces
seems to be important. The largest average bite force
for the shape of their jaws was present in the reptilian ancestors of mammals. In mammals an even
larger average bite force resulted from a change in
jaw shape and muscle repositioning, and is correlated with the beginning of increased interaction
between upper and lower teeth. Thus, during the
transition from reptiles to mammals, the pentagonally
shaped jaws in cynodonts had the resultant muscle
force located 20 percent of the way along the jaw.
This system had changed in the earliest mammals so
that the jaw had a triangular shape with the muscle
resultant located at 30 percent of the jaw length from
the jaw joints. The tooth row necessarily shortened
from the rear during this transition, which precluded
disarticulation of the joints by positioning the resultant muscle force within the triangle of support
formed by the joints and the bite points. The tooth
row subsequently lengthened, again at the rear, and
the position of the resultant force was maintained
within the triangle of support by differential activity
of the muscles on both sides of the head.

Differential muscle activity also made possible
maximum and subequal bite forces in this new posterior region. The teeth in this region became large
and molariform, or carnassial, in many mammal lineages, coincident with an abrupt transition to smaller
anterior teeth. This general plan of the jaws has been
preserved essentially intact in those mammals where
a large, posteriorly directed temporalis muscle is
present.

Under some conditions canine biting twists the
front and back portions of the skull relative to one
another in carnivores and, presumably, in any animal biting with large anterior teeth. This torsion can
be resisted by a skull using a minimum amount of
bone if the length of the jaw divided by the width is
equal to $\pi/2$ as is the case in dogs, mongooses, and
related forms. Jaws narrower than this are possible
but are predicted to be absent because more bone
tissue than necessary would be required to resist
torsion. This condition is not found in either fossil
or recent carnivores, nor in most other animals.
Wider jaws, as in cats and bears, allow for more jaw
musculature and higher bite forces, although the
resulting higher torsion is more efficiently resisted.

These relationships are predicted for other animals with large caniniform teeth and also for animals like rodents, lagomorphs, and primates where large forces act at anterior incisors.

Animals with no need for large canines and the large temporalis muscle that usually goes with them (e.g., most herbivores) can take advantage of an additional method of limiting the amount of unnecessary bone in the skull and jaws and at the same time strengthen these structures. A relatively small temporalis shifts the muscle resultant to a more anterodorsal orientation. Provided that the jaw joints lie above the upper tooth row, an anterodorsally oriented resultant means that the part of the jaw behind the teeth will be as short as possible.

In selenodont artiodactyls and some other herbivores, the jaw apparently elongated whereas the teeth remained essentially the same size. This increased the number of teeth in the region of high and sub-equal bite force at the rear of the tooth row, and a long diastema probably developed as a by-product of this elongation. Jaw elongation was "arrested" because torsion, induced by biting at the anterior premolars, could no longer be resisted with a skull of minimum bony structure. Jaw elongation is not expected to occur in animals where canines (or tusks) apply large forces, because the carnassials are already located in a region of maximum bite force, the bite force at the anterior incisors is not changed, and the bite force at the posterior incisors and canines decreases slightly.

If we accept that large bite forces, optimality, and minimum structure are factors that have played an important role in the natural design of the skull and jaws, the scenario above implies that the morphology of the skulls and jaws of mammals, many reptiles, and perhaps other groups as well share at a basic level a series of common biological or natural design features.

ACKNOWLEDGMENTS

M.L. Greaves and D.G. Covey kindly read the manuscript and made many helpful suggestions. Many of the measurements of carnivores were taken by D.G. Covey who also contributed to the analysis of torsion in carnivore skulls. M.L. Greaves assisted with the figures.

REFERENCES

Ardran, G.M., Kemp, F.H., & Ride, W.D.L. 1958. A radiographic analysis of mastication and swallowing in the domestic rabbit *Oryctolagus cuniculus. Proceedings of the Zoological Society of London* 130, 257–274.

Beatty, J. 1980. Optimal design models and the strategy of model building in evolutionary biology. *Philosophy of Science* 47, 532–561.

Becht, G. 1953. Comparative biologic-anatomical researches on mastication in some mammals, I & II. *Proceedings of the Koninklijke Akademie van Wetenschappen, Amsterdam, Series C* 56, 508–527.

Beecher, R.M. 1977. Function and fusion at the mandibular symphysis. *American Journal of Physical Anthropology* 47, 325–336.

Beecher, R.M. 1979. Functional significance of the mandibular symphysis. *Journal of Morphology* 159, 117–130.

Bramble, D.M. 1978. Origin of the mammalian feeding complex: models and mechanisms. *Paleobiology* 4, 271–301.

Brehnan, K., Boyd, R.L., Laskin, J., Gibbs, C.H., & Mahan, P. 1981. Direct measurement of loads at the temporomandibular joint in *Macaca arctoides. Journal of Dental Research* 60, 1820–1824.

Butler, P.M. 1952. The milkmolars of Perissodactyla, with remarks on molar occlusion. *Proceedings of the Zoological Society of London* 121, 777–817.

Colbert, E.H. 1945. *The Dinosaur Book.* New York: McGraw-Hill.

Costa, R.L., & Greaves, W.S. 1981. Experimentally produced tooth wear facets and the direction of jaw motion. *Journal of Paleontology* 55, 635–638.

Crompton, A.W., & Hiiemae, K.M. 1969. How mammalian molar teeth work. *Discovery* 5, 23–34.

Crompton, A.W., & Sita-Lumsden, A. 1970. Functional significance of the therian molar pattern. *Nature* 227, 197–199.

Davis, D.D. 1964. The giant panda: a morphological study of evolutionary mechanisms. *Fieldiana: Zoology Memoirs* 3, 13–39.

DeMar, R., & Barghusen, H.R. 1973. Mechanics and the evolution of the synapsid jaw. *Evolution* 26, 622–637.

Dessem, D.A. 1985a. Interactions Between Muscle Activity and Jaw Joint Forces. Unpublished Ph.D. Thesis, University of Illinois at Chicago.

Dessem, D.A. 1985b. The transmission of muscle force across the unfused symphysis in mammalian carnivores. *Fortschritte der Zoologie* 30, 289–291.

Dessem, D.A. 1989. Interactions between jaw muscle recruitment and jaw joint forces in *Canis familiaris. Journal of Anatomy* 164, 101–121.

de Vree, F., & Gans, C. 1976. Mastication in pygmy goats (*Capra hircus*). *Annales Societe Royale Zoologique Belgique* 105, 255–306.

Druzinsky R.E., & Greaves, W.S. 1979. A model to explain the posterior limit of the bite point in reptiles. *Journal of Morphology* 160, 165–168.

Du Brul, E.L., & Sicher, H. 1954. *The Adaptive Chin.* Springfield, Ill.: Charles C. Thomas.

Gans, C. 1988. Muscle insertions do not incur mechanical advantage. *Acta Zoologica Cracoviensia* 31, 615–624.

Gans, C., & de Vree, F. 1987. Functional bases of fiber length and angulation in muscle. *Journal of Morphology* 192, 63–85.

Gingerich, P.D. 1971. Functional significance of mandibular translation in vertebrate jaw mechanics. *Postilla* 152, 1–10.

Gorniak, G.C., & Gans, C. 1980. Quantitative assay of electromyograms during mastication in domestic cats (*Felis catus*). *Journal of Morphology* 163, 253–281.

Granger, W. 1908. A revision of the American Eocene horses. *Bulletin of the American Museum of Natural History* 24, 221–264.

Greaves, W.S. 1973a. Evolution of the merycoidodont masticatory apparatus (Mammalia, Artiodactyla). *Evolution* 26, 659–667.

Greaves, W.S. 1973b. The inference of jaw motion from tooth wear facets. *Journal of Paleontology* 47, 1000–1001.

Greaves, W.S. 1974. Functional implications of mammalian jaw joint position. *forma et functio* 7, 363–376.

Greaves, W.S. 1978. The jaw lever system in ungulates: a new model. *Journal of Zoology, London* 184, 271–285.

Greaves, W.S. 1980. The mammalian jaw mechanism – the high glenoid cavity. *American Naturalist* 116, 432–440.

Greaves, W.S. 1982. A mechanical limitation on the position of the jaw muscles of mammals: the one-third rule. *Journal of Mammalogy* 63, 261–266.

Greaves, W.S. 1983. A functional analysis of carnassial biting. *Biological Journal of the Linnean Society* 20, 353–363.

Greaves, W.S. 1985a. The mammalian postorbital bar as a torsion-resisting helical strut. *Journal of Zoology, London* 207, 125–136.

Greaves, W.S. 1985b. The generalized carnivore jaw. *Zoological Journal of the Linnean Society* 85, 267–274.

Greaves, W.S. 1988a. The maximum average bite force for a given jaw length. *Journal of Zoology, London* 214, 295–306.

Greaves, W.S. 1988b. A functional consequence of an ossified mandibular symphysis. *American Journal of Physical Anthropology* 77, 53–56.

Greaves, W.S. 1991a. A relationship between premolar loss and jaw elongation in selenodont artiodactyls. *Zoological Journal of the Linnean Society* 101, 121–129.

Greaves, W.S. 1991b. The orientation of the force of the jaw muscles and the length of the mandible in mammals. *Zoological Journal of the Linnean Society* 102, 367–374.

Greaves, W.S., & Covey, D.G. 1992. Torsion of the skull due to canine biting and the dimensions of the jaw in carnivores. *Journal of Vertebrate Paleontology* 12, 31A.

Gregory, W.K. 1951. *Evolution Emerging*, 2 vols. New York: Macmillan.

Gysi, A. 1921. Studies of the leverage problem of the mandible. *Dental Digest* 27, 74–84, 144–150, 203–208.

Herring, S.W. 1975. Adaptations for gape in the hippopotamus and its relatives. *forma et functio* 8, 85–100.

Herring, S.W., & Herring, S.E. 1974. The superficial masseter and gape in mammals. *American Naturalist* 108, 561–576.

Herring, S.W., & Scapino, R.P. 1973. Physiology of feeding in miniature pigs. *Journal of Morphology* 141, 427–460.

Hiiemae, K.M., & Ardran, G.M. 1968. A cinefluorographic study of mandibular movement during feeding in the rat *(Rattus norvegicus)*. *Journal of Zoology, London* 154, 139–154.

Hiiemae, K.M., & Kay, R.F. 1972. Trends in the evolution of primate mastication. *Nature* 240, 486–487.

Hylander, W.L. 1975. The human mandible: lever or link? *American Journal of Physical Anthropology* 43, 227–242.

Hylander, W.L. 1979. An experimental analysis of temporomandibular joint reaction force in macaques. *American Journal of Physical Anthropology* 51, 433–456.

Hylander, W.L. 1984. Stress and strain in the mandibular symphysis of primates: a test of competing hypotheses. *American Journal of Physical Anthropology* 64, 1–46.

Hylander, W.L. 1985a. Mandibular function and temporomandibular joint loading. In *Developmental Aspects of Temporomandibular Joint Disorders,* ed. D.S. Carlson, J.A. McNamara, & K.A. Ribbens., pp. 19–35. *Monograph 16, Center for Human Growth and Development,* Ann Arbor: University of Michigan.

Hylander, W.L. 1985b. Mandibular function and biomechanical stress and scaling. *American Zoologist* 25, 315–330.

Hylander, W.L., & Bays, R. 1978. An in vivo strain-gauge analysis of the squamosal–dentary joint reaction force during mastication and incisal biting in *Macaca mulatta* and *Macaca fascicularis*. *Archives of Oral Biology* 24, 689–697.

Hylander, W.L., Picq, P.G., & Johnson K.R. 1991a. Function of the supraorbital region of primates. *Archives of Oral Biology* 36, 273–281.

Hylander, W.L., Picq, P.G., & Johnson, K.R. 1991b. Masticatory-stress hypotheses and the supraorbital region of primates. *American Journal of Physical Anthropology* 86, 1–36.

Kallen, F.C., & Gans, C. 1972. Mastication in the little brown bat, *Myotis lucifugus*. *Journal of Morphology* 136, 385–420.

Laithwaite, E.R. 1977. Biological analogues in engineering practice. *Interdisciplinary Science Reviews* 2, 100–108.

Landry, S.O., Jr. 1970. The Rodentia as omnivores. *Quarterly Review of Biology* 45, 351–372.

Lauder, G.V. 1991. Biomechanics and evolution: integrating physical and historical biology in the study of complex systems. In *Biomechanics in Evolution,* ed. J.M.V. Rayner & R.J. Wootton, pp. 1–19. Cambridge University Press.

Leibman, F.M., & Kussick, L. 1965. An electromyographic analysis of masticatory muscle imbalance with relation to skeletal growth in dogs. *Journal of Dental Research* 44, 768–774.

Maynard Smith, J., & Savage, R.J.G. 1959. The mechanics of mammalian jaws. *School Science Review* 40, 289–301.

Mills, J.R.E. 1966. The functional occlusion of the teeth of Insectivora. *Journal of the Linnean Society, London* 46, 1–25.

Møller, E. 1966. The chewing apparatus. *Acta Physiologica Scandinavica* 69 (supplement), 1–229.

Moss, M.L. 1968. Functional cranial analysis of mammalian mandibular ramal morphology. *Acta Anatomica* 71, 423–447.

Nagel, R.J., & Sears, V.H. 1958. *Dental Prosthetics*. St. Louis: C.V. Mosby Co.

Oyen, O.J. 1987. Bone strain in the orbital region of growing vervet monkeys. *American Journal of Physical Anthropology* 72, 39–40.

Oyen, O.J., & Tsay, T.P. 1991. A biomechanical analysis of craniofacial form and bite force. *American Journal of Dentofacial Orthopaedics* 99, 298–309.

Parrington, F.R. 1934. On the cynodont genus *Galesaurus*, with a note on the functional significance of the changes in the evolution of the theriodont skull. *Annals and Magazine of Natural History* 13, 38–67.

Perry, C.C., & Lissner, H.R. 1962. *The Strain Gage Primer*, 2nd ed. New York: McGraw-Hill.

Piffault, C., & Duhamel, J. 1963. A propos des réactions subies par les condyles maxillaires à l'équilibre. *Revue d'Odonto-Stomatologie* 21, 13–17.

Radinsky, L.B. 1981. Evolution of skull shape in carnivores. I. Representative modern carnivores. *Biological Journal of the Linnean Society* 15, 369–388.

Radinsky, L.B. 1985. Patterns in the evolution of ungulate jaw shape. *American Zoologist* 25, 303–314.

Ravosa, M.J. 1988. Browridge development in Cercopithecidae: A test of two models. *American Journal of Physical Anthropology* 76, 535–555.

Ravosa, M.J. 1991a. Ontogenetic perspective on mechanical and non-mechanical models of primate circumorbital morphology. *American Journal of Physical Anthropology* 85, 95–112.

Ravosa, M.J. 1991b. Interspecific perspective on mechanical and non-mechanical models of primate circumorbital morphology. *American Journal of Physical Anthropology* 86, 369–396.

Rensberger, J.M. 1973. An occlusion model for mastication and dental wear in herbivorous mammals. *Journal of Paleontology* 47, 515–528.

Rigler, L., & Mlinšek, B. 1968. Die Symphyse der Mandibula beim Rinde. *Anatomischer Anzeiger* 122, 293–314.

Russell, M.D. 1986. In vivo testing and refinement of the frame model of the craniofacial skeleton. *American Journal of Physical Anthropology* 69, 259.

Scapino, R.P. 1965. The third joint of the canine jaw. *Journal of Morphology* 116, 23–50.

Scapino, R.P. 1972. Adaptive radiation of mammalian jaws. In *Morphology of the Maxillomandibular Apparatus,* ed. G.H. Schumacher, pp. 33–39. *IXth International Congress of Anatomists.* Leipzig: G. Thieme.

Scapino, R.P. 1981. Morphological investigation into functions of the jaw symphysis in carnivorans. *Journal of Morphology* 167, 339–375.

Shute, C.C.D. 1954. Function of the jaws in mammals. *Journal of Anatomy* 88, 565.

Smith, R.J. 1978. Mandibular biomechanics and temporomandibular joint function in primates. *American Journal of Physical Anthropology* 49, 341–350.

Spencer, M. 1991. Primate masticatory system configuration and function: a three-dimensional biomechanical approach. Unpublished Ph.D. proposal, Doctoral Program in Anthropological Sciences, State University of New York, Stony Brook.

Stöckmann, W. 1975. Die Form der Mandibel Afrikanischer Bovidae (Mammalia) und ihre Beeinflussung durch die Ernährung. Unpublished Ph.D. Dissertation, University of Hamburg.

Stöckmann, W. 1979. Formenunterschiede der Mandibel Africanischer Bovidae (Mammalia) und ihre Beziehungen zur Zusammensetzung der Nahrung. *Zoologische Jahrbücher für Systematik, Oekologie und Geographie der Tiere* 106, 344–373.

Tattersall, I. 1972. The functional significance of airorhynchy in *Megaladapis. Folia Primatologica* 18, 20–26.

Thomason, J.J. 1991. Cranial strength in relation to estimated biting forces in some mammals. *Canadian Journal of Zoology* 69, 2326–2333.

Turnbull, W.D. 1970. Mammalian masticatory apparatus. *Fieldiana: Geology* 18, 147–356.

Walker, A. 1976. A 3–dimensional analysis of the mechanics of mastication. *American Journal of Physical Anthropology* 44, 213.

Walker, A. 1978. Functional anatomy of oral tissues: mastication and deglutition. In *Textbook of Oral Biology,* ed. J.H. Shaw et al., pp. 277–296. Philadelphia: W.B. Saunders.

Weijs, W.A. 1980. Biomechanical models and the analysis of form: a study of the mammalian masticatory apparatus. *American Zoologist* 20, 707–719.

Weijs, W.A. 1981. Mechanical loading of the human jaw joint during unilateral biting. *Acta Morphologica Neerlando-Scandinavica* 19, 261–262.

Weijs, W.A., & Dantuma, R. 1981. Functional anatomy of the masticatory apparatus in the rabbit (*Oryctolagus cuniculus* L.). *Netherlands Journal of Zoology* 31, 99–147.

Wolff, Exalto A.E. de. 1951. On differences in the lower jaw of animalivorous and herbivorous mammals, I and II. *Proceedings of the Koninklijke Nederlandse Akademie van Wetenschappen, Amsterdam, Series C* 54, 237–246, 405–410.

7

Carnassial functioning in nimravid and felid sabertooths: theoretical basis and robustness of inferences

HAROLD N. BRYANT and ANTHONY P.
RUSSELL

ABSTRACT

The carnassials of sabertoothed carnivorans (some Felidae and most Nimravidae) are placed more posteriorly, and experienced higher rates of attritional wear than those of most other carnivorans. The correlation of these attributes with the sabertooth morphotype suggests some functional association. Greaves's (1983) model for carnassial functioning in carnivorans indicates that the posteriorly placed carnassials of sabertooths maximized the available leverage, but did not compensate for the alleged smaller size, of the jaw adductors. Among the Nimravidae, carnassial wear in *Dinictis* was similar to that of most extant carnivorans, whereas that in *Hoplophoneus* was much greater. In the latter, carnassial rotation, especially of P^4, was required to maintain carnassial occlusion throughout the life of the animal. Heavy wear and carnassial rotation are associated not with the geometry of jaw closure, which is the same in the two genera, but with differences in dental morphology. This explanation may not apply to heavy wear in other sabertooths, suggesting the absence of a common explanation for the heavy wear to the carnassials in sabertooths. Valid inferences of functional or other unpreserved attributes in fossils are either deductions based on phylogenetic relationships or extrapolatory interpretations of the known features of the fossil. Functional inferences about extinct taxa rely heavily on studies of extant taxa and must entail greater uncertainty. Nonetheless, the study of fossils can test generalizations based on extant taxa and broadens our view of the functional morphology of vertebrates.

THE CARNASSIALS OF SABERTOOTHED CARNIVORANS

The Order Carnivora (Mammalia) is characterized by the localization of shearing occlusion in the cheek teeth to the most posterior upper premolar and the most anterior lower molar, the carnassials P^4 and M_1. Cats (Felidae and Nimravidae) have reduced or

lost the most anterior grasping premolars, lost the crushing molars, and further emphasized the shearing function of the carnassials. Both the Felidae and Nimravidae include major radiations of sabertoothed cats in which the upper canine is elongated and laterally compressed, a morphology that contrasts with the conical-toothed canines of extant felids. Sabertoothed canines correlate with a series of cranial and mandibular features that are usually interpreted as modifications for increased maximum gape (e.g., Emerson & Radinsky 1980), which was necessary to allow sufficient clearance between the sabers and the lower canines. The large upper canines and these associated modifications have been the primary focus of most functional studies on sabertooths (e.g., Matthew 1901, 1910; Bohlin 1940; Simpson 1941; Akersten 1985; Van Valkenburgh & Ruff 1987; Bryant 1990). Nonetheless, sabertooths also display consistent differences from other carnivorans in the morphology and functioning of the carnassials, and these warrant detailed study.

Some early studies suggested that carnassial functioning in sabertooths was less well developed than that of extant carnivorans. Sinclair and Jepsen (1927) concluded that the extreme development of the sabers in the nimravid *Hoplophoneus sicarius* was associated with marked reduction in the shearing function of the cheek teeth. They concluded that this reflected a stabbing rather than a biting mode of attack. Marinelli (1938; quoted from Miller 1969) stated that the carnassials in the felid *Smilodon* provided "an inferior meat cutting mechanism." Bohlin (1940) argued that the sabers compensated for the reduced shearing function of the carnassials; he proposed that the sabers were used to slice strips of meat from the carcass.

Most authors, however, have concluded that the carnassials of sabertoothed carnivores were equally, or more, efficient shearing devices, as compared to those of extant carnivorans. Merriam and Stock (1932; p. 48) described the upper carnassial of *Smilodon* as "a beautifully developed cutting tooth" and suggested that the scissorlike action of the carnassials helped to enlarge the wound generated by the stabbing sabers. Miller (1969) concluded that *Smilodon*'s long, narrow, high-crowned P^4 was an ideal shearing device, and that the carnassials were more efficient cutting teeth than those of various extant carnivorans. In various sabertooth lineages the carnassials are enlarged and more specialized, when compared to those of other carnivorans, whereas other portions of the postcanine dentition are reduced (Turnbull 1978). Sabertooths provide some of the most extreme examples of carnassial development (Martin 1980); in the nimravid *Barbourofelis fricki* the P^4 is effectively the only functional upper cheek tooth and is a 6-cm long shearing blade which occludes with the overlapping P_4 and M_1. Martin (1984) noted the parallel tendency of the barbourofeline nimravids and smilodontine felids to develop enlarged carnassials in conjunction with increased length of the upper canines. Together with extreme carnassial development, many sabertooths display much heavier wear to the carnassials than is characteristic of extant carnivorans. Although the elaboration and marked wear of the carnassial system seem indicative of large bite forces, most recent discussions of carnassial functioning in sabertooths have concluded that cranial modifications that allow for increased gape would have reduced the force of the jaw adductors. Compensatory modifications have been considered necessary for the maintenance of a strong bite.

In this chapter we (1) review the arguments regarding the bite force at the carnassials and possible compensatory mechanisms, together with an analysis of the posterior placement of the carnassials within the context of Greaves's (1983) model for carnassial functioning, and (2) consider functional rationales for the heavy carnassial wear in sabertooths, with particular reference to a comparison between the nimravids *Dinictis* and *Hoplophoneus*. These analyses illustrate the applicability of a proposed methodology for the inference of unpreserved attributes in fossils (Bryant & Russell 1992) to the study of functional morphology in fossil vertebrates. An understanding of the appropriate theoretical framework for the inference of function in fossils allows one to determine the limits and uncertainties of such analyses.

ACRONYMS. AMNH, American Museum of Natural History, New York; F:AM, Frick American Mammals, AMNH, New York; PU, Princeton University Collection; YU, Yale University, New Haven; SDSM, South Dakota School of Mines, Rapid City; UCMZ, University of Calgary Museum of Zoology, Calgary.

THE THEORETICAL BASIS AND ROBUSTNESS OF FUNCTIONAL INFERENCES IN FOSSIL VERTEBRATES

Because a fossil vertebrate usually consists of only some portion of the skeleton, functional interpretations necessarily entail the inference of unpreserved attributes. Many functional inferences entail considerable extrapolation because they depend on previous inferences regarding unpreserved portions of the skeleton and soft anatomy. Particular functional inferences may be specifically phylogenetic or extrapolatory (Bryant & Russell 1992), or may involve a complex mixture of both. See Lauder's and Witmer's contributions to this volume for other categorizations of functional inferences in fossils.

Phylogenetic inference

Phylogenetic inference involves the transfer of known attributes in other taxa to the fossil taxon based on an assumed hypothesis of phylogenetic relationships between the fossil taxon and two or more other taxa. In the absence of contradictory information, one assumes that the fossil taxon shares the synapomorphies that diagnose the clades to which it belongs. The unequivocal transfer of an attribute to a fossil taxon necessitates information regarding not only the appropriate sister taxon but also one or more additional taxa. This comparative context provides the basis for inferring that the feature in question occurred in the common ancestor of the fossil taxon and its nearest extant relatives (Figure 7.1; fossil-extant node). The feature is then transferred to the fossil. The consistency of various aspects of the jaw apparatus among all extant carnivorans provides a phylogenetic rationale for inferring the same structure and functioning in nimravids. The robustness of phylogenetic inferences depends on the support for the cladistic relationships involving the fossil and entails the assumption that the fossil retained the features attributed to the fossil-extant node.

Recent revision of the systematics of catlike carnivorans has changed the comparative framework for phylogenetic inferences in some sabertoothed carnivorans. Traditionally all sabertoothed carnivorans were included in the Felidae, and the single origin of the morphotype was implicitly assumed (e.g., Simpson 1945). Within this systematic context extant felids provided the most appropriate phylogenetic context for inferences regarding unpreserved

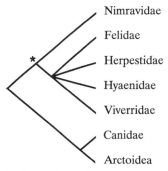

Nimravidae
Felidae
Herpestidae
Hyaenidae
Viverridae
Canidae
Arctoidea

Figure 7.1. Phylogenetic relationships among the Nimravidae and extant carnivoran taxa (after Bryant 1991 and Flynn et al. 1988). The Nimravidae is the sister taxon of the extant Feliformia (Felidae, Herpestidae, Hyaenidae, Viverridae); Canidae and Arctoidea = the Caniformia. Of the genera referred to in the text, *Barbourofelis, Dinictis, Eusmilus, Hoplophoneus,* and *Nimravus* are nimravids; *Homotherium, Machairodus, Metailurus, Nimravides,* and *Smilodon* are felids (see Table 7.1). Phylogenetic inferences in fossils entail the transfer of attributes from the fossil-extant node (*) to the fossil. Reconstruction of the fossil-extant node requires information regarding not only extant feliforms but also at least the next most closely related group, the Caniformia. (See Bryant & Russell [1992] for detailed explanation and discussion.)

features in all sabertoothed carnivorans. However, detailed anatomical study of the basicranium and auditory region of the Eocene to Miocene radiation of sabertoothed carnivorans resulted in the referral of these taxa to a separate family, the Nimravidae (Tedford 1978; Baskin 1981; Neff 1983; Hunt 1987). The phylogenetic relationships of this group have been controversial (Flynn, Neff, & Tedford 1988) but cladistic analysis suggests that the Nimravidae is the sister taxon to the extant Feliformia (Figure 7.1; Bryant 1991).

The phylogenetic distance between the Felidae and Nimravidae invalidates the direct transfer of felid features to nimravids on phylogenetic grounds. If the Nimravidae is the sister group to the extant Feliformia, phylogenetic inferences regarding nimravid features require information regarding attributes throughout the Feliformia and in the Caniformia. The phylogenetic distance between felid and nimravid sabertooths also casts doubt on the appropriateness of using the felid *Smilodon*, the best-studied sabertooth carnivoran, as a model for all sabertooths. Equivalence in functional systems among different clades of sabertooths must be demonstrated, not assumed. For example, the present analysis argues for a common explanation for the posterior position of the carnassials in sabertooths, but suggests that extreme wear of the carnassials may have different causal explanations in different sabertooth clades.

Extrapolatory inference

Extrapolatory inferences entail the analysis of the known features of the fossil within the context of assumed biological generalizations which provide a basis for the inference of unpreserved attributes. Such generalizations are akin to nomological statements regarding the relationship among variables (Bock 1989). Given the assumption that the generalization is valid for a particular group of organisms, the preservation of one variable in the fossil allows the extrapolation of particular unpreserved variables. For example, given an established association between dental morphology and diet (shearing surfaces and carnivory), the preservation of the dental features in the fossil provides the basis for inferences regarding diet or feeding behaviors. Extrapolatory approaches can be divided into (1) form–function correlation and (2) biomechanical design analysis (Hopson & Radinsky 1980).

In form–function correlation generalizations are based on observed associations among particular morphologies, functions, or behaviors at a particular level of systematic generality (e.g., skeletomuscular systems in vertebrates). Model taxa often act as analogs for the inference of unknown attributes in fossils. Form–function correlations are usually restricted to a particular, if broad, phylogenetic context. Because functional attributes are directly observable only in extant organisms, functional inferences in fossils rely heavily on the study of extant organisms. Inferences regarding muscle performance in fossil taxa are based on correlations in extant taxa between particular functional attributes and skeletal features. The force a muscle can generate is related to its size or physiological cross-sectional area; muscle size is often correlated at least to some degree with the size and nature of its attachment sites (but see Bryant & Seymour 1990). Such correlations have been the basis for inferring reduced power of the temporalis in sabertooths from the muscle's apparently smaller area of origin (temporal fossa on the skull) or insertion (small coronoid process on the mandible). The robustness of form–function correlations depends on the testing and corroboration of the generalization that forms the basis for the inference and the applicability of the correlation to the taxon in question.

Biomechanical design analysis infers function or other attributes through the study of the fossil from a biomechanical or engineering perspective. There is no direct use of analogs, and only an extremely broad phylogenetic framework may be applicable. The analysis of carnassial or other dental functioning involves the application of engineering principles to occlusion and mastication (e.g., Rensberger 1973; Osborn & Lumsden 1978). In this study analysis of the factors that influence the wear to the cutting surfaces of a guillotine (Atkins & Mai 1979) provides the basis for correlating morphological differences between *Dinictis* and *Hoplophoneus* with differences in carnassial wear. Biomechanical design analysis often entails assumptions that are based in turn on phylogenetic inference or form–function correlations. Greaves's (1983) model for carnassial functioning in carnivorans (see the section below in which we use the position of the carnassials in sabertooths to test his model) entailed assumptions regarding reaction forces at the craniomandibular joint and the role of the balancing-side musculature that were based, to some degree, on studies of other mammals that were assumed to apply to carnivorans. The theoretical basis behind biomechanical approaches is often well founded but the applicability of a particular approach to the biological system being studied is often uncertain. For example, there has been considerable debate regarding the most applicable lever model for the study of jaw mechanics (e.g., Turnbull 1970; Bramble 1978; Greaves 1983).

Functional inferences in sabertoothed carnivorans

The analysis of jaw mechanics and carnassial functioning in sabertoothed carnivorans entails a complex mix of phylogenetic and extrapolatory inferences. Considerable detail regarding muscle architecture and ligament systems can be inferred phylogenetically from the known morphology of extant carnivorans. Form–function correlations include inferences regarding lower-jaw movement from symphyseal and craniomandibular joint morphology, and dental functioning from wear facets. Because sabertooths have no extant functional analogs, many of the functional inferences associated directly with the morphotype are based on biomechanical design analysis. Many of these inferences involve the application of lever models to jaw mechanics (e.g., inference of poor leverage in the temporalis from the small coronoid process, improvement in leverage through the posteriorly positioned carnassials). Such inferences may still find support, however, in form–function correlations in extant taxa with similar design requirements. Information regarding the capacity for stretching in the muscles of extant mammals, together with form–function correlations in taxa with large gapes such as the hippopotamus (Herring 1975), support the view that the reduced coronoid process on the mandible in sabertoothed carnivorans is a modification for increased gape (e.g., Emerson & Radinsky 1980).

Functional inferences in extinct taxa entail a high degree of uncertainty because they tend to be predicated on previous inferences that also involve inherent uncertainty, or that may not be fully testable. Relative determinations of muscle force through lever models must assume comparable muscular architecture and performance in the taxa being compared. These assumptions cannot be tested, and any comparative inferences must carry a proviso of equivalency in those attributes. Inferences in fossils often involve possibilities rather than solid conclusions.

THE FUNCTIONAL MORPHOLOGY OF THE CARNIVORAN CARNASSIAL SYSTEM

Mechanical processing by the teeth of vertebrate muscle, which is primarily soft and compliant but contains tougher connective tissue components, requires sharp blades (Lucas & Lake 1984). In carnivorans this function is carried out primarily by the carnassials, P^4 and M_1. Shearing occlusion involves the metacrista on P^4 (formed by the anterior paracone and posterior metastyle; Figure 7.2a; pa, pm) and the paracristid on M_1 (formed by the anterior paraconid and posterior protoconid; Figure 7.2a; pad, prd). Initial contact tends to occur at the two ends of the occluding blades; as the teeth pass each other the two points of contact converge at the central notches (Figure 7.2a). The cutting edges are reciprocally curved (Figure 7.2b,c) so that the entire blade surfaces are never in contact at any one instant, a morphology that also characterizes the shearing crests on the molars of primitive therian mammals (Crompton & Hiiemae 1970; Kay & Hiiemae 1974). In their most specialized form, which characterizes most sabertooths, the blades of the carnassials are aligned parallel to the tooth rows, and other functions of these teeth are reduced or absent. Carnassial design effectively promotes both retention and compression, requirements for the effective division of food items (Osborn & Lumsden 1978).

Retention of food

Retention mechanisms prevent food items from slipping anteroposteriorly and maintain close apposition between occluding carnassials so that food items do not become jammed between the blades, forcing them apart. The pointed premolars and the notches on the carnassials (Figures 7.2 & 7.3) prevent anteroposterior movement of the food item. Carnassial contact can be maintained by either muscular force or autocclusal mechanisms in which the dental morphology itself promotes proper occlusion. Those portions of the jaw adductor musculature that have a significant laterally directed

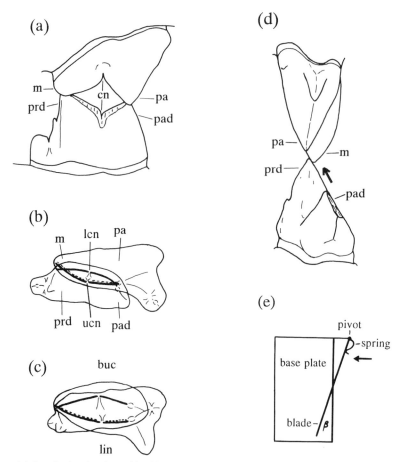

Figure 7.2. Carnassial functioning in nimravids. (a) Lateral view of right carnassials (P^4/M_1) at the beginning of occlusion in *Dinictis*. (b) Occlusal view of teeth in (a) to illustrate the positional relationship of the reciprocally concave blade edges (metacrista on P^4 marked with dashed line); paracone and the posterior end of metastyle on P^4 are buccal to paraconid and protoconid, respectively, but carnassial notch on P^4 (ucn) is lingual to that on M_1 (lcn). (c) Occlusal view of the carnassials in *Hoplophoneus* at initial contact between the metastyle and the protoconid (as in [d]; metacrista on P^4 marked with dashed line as in [b]); unlike *Dinictis*, paracone on P^4 is lingual to paraconid on M_1 and the space between the notches is larger. (d) Anterior view of the carnassials of *Hoplophoneus* at initial contact between metastyle and protoconid (perpendicular to reciprocal wear facets); paracone and paraconid have yet to occlude and paracone is lingual to the paraconid. Arrow indicates the direction of movement of M_1 relative to P^4. (e) Diagrammatic representation of a guillotine, such as a paper cutter, consisting of a base plate against which a blade cuts. When blade is fully or partly open, as illustrated, pivot orients blade at an angle β to edge of base plate. The spring supplies the sideways force (*arrow*) that holds the blade against the base plate. Wear at the cutting point is a function of the size of the angle, and the force supplied by the spring. See text for comparison to occluding carnassials. Abbreviations: buc, buccal; cn, carnassial notch; lcn, lower carnassial notch; lin, lingual; m, metastyle; pa, paracone; pad, paraconid; prd, protoconid; ucn, upper carnassial notch.

component of force (e.g., the masseter) have been considered essential for effective carnassial contact (Kurtén 1952; Scapino 1965). However, the morphology of P^4 and M_1 provides an autocclusal mechanism that, in the presence of a food item with sufficient resistance, holds the teeth together. As P^4 and M^1 approach each other, the food item will be forced into the space between the notches on the two teeth (Figures 7.2a & 7.4a). Because of the buccal excavation associated with the notch on P^4 and the lingual excavation associated with the notch on M_1,

as the teeth move closer together a turning moment is applied to food, forcing it against the buccal surface of the notch on P^4 and the lingual surface of the notch on M_1 (Figure 7.4b; fig. 1 in Mellett 1981). As long as the notch on M_1 is buccal to the notch on P^4 when the ends of the blades begin to overlap, the buccally oriented force applied by the food to M_1 will bring the carnassial blades into proper alignment. Precise alignment of the carnassials through muscular or other mechanisms is not necessary for effective shearing. Although Mellett (1981) empha-

(a) (b)

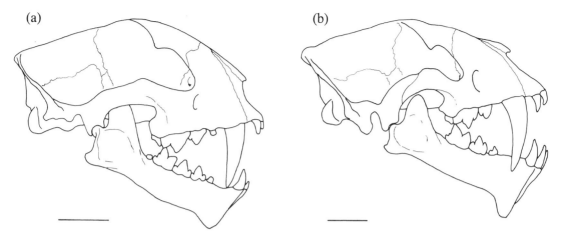

Figure 7.3. Right lateral views of skull and mandible of the nimravids (a) *Dinictis* (PU 13587) and (b) *Hoplophoneus* (PU 12953) shown at the beginning of carnassial occlusion. (Redrawn and modified from Scott & Jepsen [1936].) Scale bars = 30 mm.

sized the role of the reciprocally curved blades (Figure 7.2b) in this autocclusal mechanism, the excavation of the carnassials at the notches has the primary role in applying the turning moment to the food that brings the blade edges together.

Compression of food

The food item will be compressed as the diamond-shaped space formed by the notches of the occluding carnassials becomes smaller as the teeth pass each other (Figure 7.2a). A compliant material, such as flesh, will first conform to the opening between the carnassials; then as the opening diminishes most of the material between the notches will be forced against, and cut by, the blade edges before tooth-on-tooth contact occurs (Abler 1992). Although the morphology of the teeth suggests that occlusal forces are concentrated at the small areas of actual dental contact (Savage 1977), Abler's (1992) analysis of how the food is actually divided indicates that cutting forces are more diffuse.

Dental wear: attrition and abrasion

Osborn and Lumsden (1978) characterized two types of dental wear. *Attrition* results from tooth-on-tooth contact; it usually produces a sharp enamel edge and a flat wear facet that passes smoothly from enamel to dentine. *Abrasion* results from contact between the tooth and the food item; tooth edges become blunt and the dentine wears faster than the enamel, forming smoothly concave surfaces. The shearing facets that develop on the faces of the carnassials in carnivorans are produced by attrition as the blade edges pass each other. However, the tooth-food contact described by Abler (1992) could potentially

abrade the blade edges. The sharp enamel edges of the carnassials of most carnivorans suggest that given a predominantly meat diet the rate of attrition is greater than that of abrasion. Other diets can result in predominantly abrasive wear (e.g., ferrets fed on dry pellets, Berkovitz & Poole [1977]).

Rensberger (1973) identified various factors that affect the rate of dental attrition. Of these, the duration of occlusal contact during the chewing cycle and the pressure between surfaces are probably most amenable to comparative functional analyses in fossils because they should vary with differences in dental morphology, occlusion, and jaw mechanics. Factors influencing the pressure forcing the carnassial blades together can be identified by considering an analogous system comprised of a guillotine blade that is spring-loaded against a base plate so that the blade cuts in a skew fashion (Figure 7.2e). The pressure forcing the base plate and blade together is directly proportional to the skew angle and the sideways force imparted by the stiffness of the spring (Atkins & Mai 1979). The skew angle is analogous to the angle between the occluding carnassial blade edges (Figure 7.2b,c), and the sideways force is imparted by trapped food (Figure 7.4) and possibly muscular contraction.

BITE FORCE AT THE CARNASSIALS IN SABERTOOTHS

The necessary relationship between the extreme length of saberlike upper canines and a series of cranial modifications that allow for an increased gape has long been recognized (e.g., Pomel 1843). Emerson and Radinsky (1980) related the following features to increased gape: the more ventrally

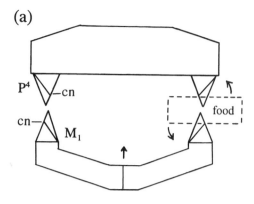

(a)

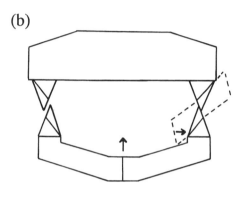

(b)

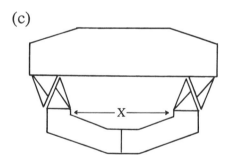

(c)

Figure 7.4. Diagrammatic representation of the positional relationships among skull, carnassials, and lower jaw in cross section at various stages of jaw closure in *Hoplophoneus*. Carnassials are represented by triangles (see Fig. 7.2d); larger triangles are the cross sections at the posterior ends of the blades and smaller internal triangles are the cross sections at the notches. The difference in area represents the excavation associated with the notch. The working side is on the right. (a) Position just before carnassial occlusion. Blade edges need not be aligned precisely; food (*dashed outline*) has been forced into the notches and will rotate as indicated by the arrows. (b) Position at initial blade contact. Food has been forced against buccal surface of P^4 and lingual surface of M_1; force against the latter moves the dentary buccally bringing the blades into occlusion. Carnassials of some carnivorans (e.g., extant felids) have more vertically oriented blade surfaces but positional relationships associated with the autocclusal

positioned craniomandibular joint, the smaller postglenoid process, the anteroposteriorly narrower temporal fossae, the upward rotation of the facial portion of the skull, and the shortened coronoid process and laterally displaced angular process on the lower jaw. The general consensus is that some of these modifications had a negative effect on the bite force at the carnassials.

Matthew (1910) argued that the ventrally positioned craniomandibular joint increased the length of the temporalis, and the small coronoid process (Figure 7.3b) prevented overstretching of this muscle during the large excursions of the mandible associated with an increased gape. The smaller coronoid process reduced the leverage of the temporalis, and the size and position of the zygomatic arch and masseteric fossa resulted in a weak masseter with poor leverage. This muscular arrangement might suggest a quick but weak closure of the jaws and reduced occlusal force at the shearing cheek teeth (as suggested for the marsupial sabertooth, *Thylacosmilus*, by Churcher 1985).

Kurtén (1952) was the first to provide a detailed analysis of the effects of the cranial anatomy of sabertooths on the functioning of the carnassials. He argued that the modifications for increased gape had three negative effects on the bite strength at the carnassials in *Smilodon*.

First, Kurtén argued that the maximum torque of the jaw adductors occurred at much larger gapes than those associated with carnassial occlusion. He assumed that maximum force output occurred at the midpoint of contraction; because of the larger gape in *Smilodon*, there would have been greater disparity between the point of maximal force output and the point of carnassial occlusion than in extant felids. MacKenna and Turker (1978) have since demonstrated, however, that, at least in cats, maximum force output occurs at close to maximum gape; this result invalidates the primary assumption of Kurtén's argument.

Second, Kurtén predicted reduced torque for the temporalis from the lesser strength and shorter moment arm of this muscle. The first was inferred from the relatively smaller temporal fossa, and the second was inferred from the small coronoid process on the mandible. Kurtén judged that the more posterior position of the carnassial would have compensated for the reduced moment arm of the

mechanism are the same. (c) Position at jaw closure. M_1 moved dorsomedially during occlusion; if autocclusion on the balancing side also governs mandibular movement (see text), the distance between the dentaries (X) must be smaller at jaw closure than during carnassial occlusion as in (a) and (b).

temporalis, and that the efficiency of the temporalis should be increased by the more dorsal position of the temporal fossa and the lowered mandibular fossa that gave the muscle a more vertical orientation. Nonetheless, Kurtén concluded that the force at the carnassials was less than that in extant felids.

Third, Kurtén argued that the role of the masseter in carnassial occlusion was reduced. He reasoned that, because the skull is relatively narrower across the zygomatic arches than it is in extant felids, the masseter would have been oriented in a more vertical plane and less obliquely than it is in extant felids; this increased the torque of the muscle in closing the jaw, compensating to some degree for the weaker temporalis. However, because Kurtén assumed that the lateral component of the masseter's action was essential for effective carnassial occlusion, the more vertical orientation of the muscle suggested reduced carnassial functioning. He concluded, as had Bohlin (1940), that the sabers, rather than the carnassials, were probably used in processing the carcass.

Martin (1980) disagreed with Kurtén's (1952) argument that the temporalis of sabertooths was relatively weak, but agreed that the mechanical advantage of all the jaw adductors was reduced by their more posterior insertions. Martin concluded instead that the muscle attachment sites, and hence the musculature, were actually relatively larger in sabertooths than in conical-toothed felids. He agreed with Kurtén that the more posterior position of the carnassials compensated for the reduced mechanical advantage of the musculature. Martin concluded, as had Miller (1969), that there was no basis for assertions that sabertooths had weak bites.

Emerson and Radinsky (1980) demonstrated that most of the cranial modifications, and the suggested ramifications for carnassial functioning, that occur in *Smilodon* are characteristic of most sabertoothed carnivorans. Although the smaller temporal and masseteric fossae suggested that both of these muscles were relatively smaller in sabertooths than in similarly sized extant felids, they argued that compensatory mechanisms and additional circumstantial evidence indicated that bite force at the carnassials was not necessarily lower. These authors also noted that recent research on muscle physiology and functioning suggests that accurate estimates of bite strength in fossil taxa are not possible because of the unavailability of information on detailed muscle architecture and performance.

Emerson and Radinsky (1980) concluded that the efficiency of the temporalis was increased by the lowering of the glenoid fossa, the shortening of the temporal fossa, and the upward rotation of the facial region; these modifications caused the temporal fibers to be oriented more nearly perpendicular to the upper tooth row. The distance from the condyle to the tip of the coronoid process was considered the best measure of the effective moment arm of the temporalis. Although this moment arm is shorter in sabertooths than in extant felids, the difference is less than that of previous comparisons that were based on the height of the coronoid process. Emerson and Radinsky demonstrated that Merriam and Stock's (1932) and Kurtén's (1952) observations of the more posteriorly positioned carnassials in *Smilodon* are applicable to sabertooths in general. The distance from the condyle to M_1 is inversely correlated with the size of the upper canine. Emerson and Radinsky argued that the more posteriorly placed carnassials increased the mechanical advantage of the adductor musculature, especially the temporalis; a given unit of muscular force is expected to produce more force at the carnassials because of the shorter output lever.

The shorter zygomatic arch and masseteric fossa of sabertooths suggested a relatively smaller masseter. However, Emerson and Radinsky (1980) argued against the necessity of Kurtén's (1952) conclusion that the masseter had a more vertical orientation; the shorter distance from the condyle to the ventral margin of the mandible may have compensated for the narrowness of the skull at the zygomatic arches, thus maintaining the more oblique orientation of the muscle. Emerson and Radinsky (1980) also argued that the relative similarity in carnassial size and mandible depth and cross section in sabertooths and extant felids suggested that bite forces were probably comparable. The extremely large upper carnassials, deep mandible, and relatively large moment arm of the temporalis in *Barbourofelis* suggested an even more powerful bite than that of extant felids.

Van Valkenburgh and Ruff (1987) considered the strength of saberlike canines to bending about their mediolateral axis as a reflection of large bite forces. They concluded that bite force at the canines equaled or exceeded that of extant felids. The high bending strength of the mandibular corpora of *Smilodon* also seems indicative of powerful bite forces at the anterior dentition (Biknevicius & Van Valkenburgh 1991). The relevance of these results to bite forces at the carnassials is uncertain.

Recent functional analyses of sabertoothed carnivorans all indicate that despite morphology that should have produced lower bite forces at the carnassials, compensatory devices resulted in forces at least comparable to those in extant carnivorans. These inferences are based, in part, on traditional lever models that consider only the dentary on the working side of the jaw (e.g., Kurtén 1952; Maynard Smith & Savage 1959; Bramble 1978); the craniomandibular joint of that dentary is considered to be the force-bearing fulcrum. These models

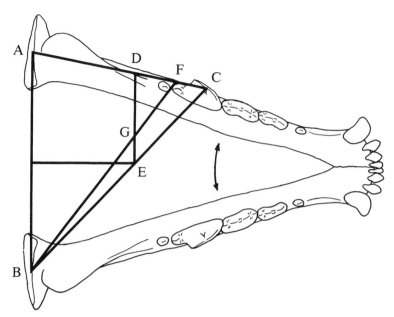

Figure 7.5. Dorsal view of lower jaw of *Dinictis* (F:AM 62148). The triangular plate in Greaves's model is delimited by the positions of joints (A, B) and the working-side carnassial (C). If the muscle resultant is located 60 percent of the distance from A to C (as in extant carnivorans) it will lie along $\overline{DE}$. F indicates position of carnassial in *Smilodon.* See text for details. Arrows indicate relative movement between dentaries in the frontal plane.

implicitly assume that all the force at the occluding carnassials is supplied by the musculature on the working side. These assumptions are inconsistent with recent studies on the jaw mechanics of various mammals (e.g., Hylander 1979; Gorniak & Gans 1980; Weijs & Dantuma 1981; Gorniak 1985; De Gueldre & De Vree 1990), suggesting that alternative approaches that consider the dentaries and musculature on both the working and balancing sides are more appropriate.

The position of the carnassials in sabertooths: a test of Greaves's model

Greaves (1978, this volume) developed an alternative model of the mammalian jaw mechanism that assumes that musculature on both the working and balancing sides of the jaw contribute to occlusal force and that both craniomandibular joints are subjected to reaction forces. Greaves (1983, 1985) applied the model to carnassial functioning in carnivores; the relatively consistent location of the lower carnassial at the midpoint of the lower jaw in extant carnivorans (Radinsky 1981) was explained as the position that permits a reasonably wide gape but also maximizes the bite force. The model predicts a constant geometry among the locations of the jaw joints, the carnassial, and the muscle resultant in that the latter is expected to be located 60 percent of the distance from the condyle to the carnassial. Other geometries reduce the available bite force at the carnassials.

The relatively posteriorly positioned carnassials of sabertoothed carnivorans (Emerson & Radinsky 1980) contrast with Radinsky's (1981) generalization that carnivoran carnassials tend to be placed at the

midpoint of the lower jaw. More posterior carnassials have generally been viewed as a compensatory mechanism for one or both of the poorer leverage and purportedly reduced size of the jaw adductors, especially the temporalis. The more posterior position of the carnassials in sabertooths provides a test of Greaves's model as an adequate representation of the carnivoran carnassial system. In turn, Greaves's model may provide a basis for choosing between the alternative explanations for the position of the carnassials in sabertooths.

Greaves's (1983) model (Figure 7.5) considers the lower jaw at carnassial occlusion as a triangular plate; the three corners of the triangle are the bite point (C) and the centers of cranial contact at the craniomandibular joints (A, B). The forces of the adductor musculature on both sides of the head are resolved into a single resultant oriented perpendicularly to the surface of this plate. The resultant generates torque on the lower jaw about an axis running through the jaw joints. Because the craniomandibular joints in carnivorans prohibit most movements other than this orthal rotation, the resultant represents the major vector in the system. The position of the resultant is determined by the attachment sites of the adductors and the relative force output of the musculature on the two sides of the head. Greaves assumed that the resultant would be located within the triangular plate because this provides a stable system in which forces acting at the joints are compressive. If the resultant were outside the triangle, the mandible would tend to rotate around the effective lever arm and dislocate the working-side joint, producing an unstable and inefficient system. When the resultant is located 60

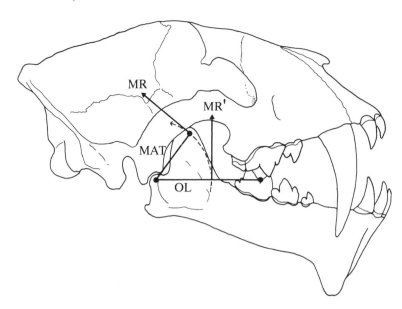

Figure 7.6. Right lateral view of skull and mandible of *Hoplophoneus* (after Scott & Jepsen 1936) illustrating input and output levers and muscle resultants for carnassial occlusion due to jaw rotation through contraction of the temporalis (see text for discussion and Table 7.1). Abbreviations: MAT, moment arm of temporalis for orthal rotation of mandible at mandibular fossae (input lever); MR, muscle resultant for orthal rotation (perpendicular to MAT); MR', effective muscle vector perpendicular to the output lever (muscle resultant in Greaves's model); OL, output lever. Dashed line projects length of MAT into plane of OL.

percent of the distance from the joints to the carnassial (along line $\overline{DE}$ in Figure 7.5), muscles on both the working and balancing sides can contribute fully to force generation; the resultant will lie along the midline (at E), and the effective lever arm will run from the carnassial through the resultant to the balancing-side joint (Figure 7.5; line $\overline{CEB}$).

If the carnassial is located more anteriorly, bite force drops because of poorer leverage. If the carnassial is located more posteriorly, leverage is improved but bite force again drops because the contribution of the balancing-side musculature must be drastically reduced in order to keep the resultant within the triangular plate (fig. 2 in Greaves 1983). All else being equal, the more posterior carnassial in sabertooths is inconsistent with Greaves's model. Assume that the musculature on each side of the head delivers 10 units of force and that 100 percent of the force generated by the balancing-side musculature is transferred to the working side. If the resultant is 60 percent of the distance from the condyles to the carnassial, the full 20 units of force can be delivered because the resultant is located within the support triangle (Figure 7.5: E). The bite force at the carnassial will be 12 units (60 percent of 20). Consider the situation in the sabertoothed felid *Smilodon* in which the carnassials are located only 41 percent of the distance from the condyle to the anterior end of the jaw (Figure 7.5; F, where $\overline{AF}$ = 82 percent of $\overline{AC}$). This position would improve the leverage on the working side so that the 10 units of muscular force would yield 7.3 units at the carnassial ($10 \times 60/82$). However, the output of the balancing-side musculature must be reduced to 53 percent to yield a resultant at G which is within the triangle of support (which is now ABF). At this output the

balancing side produces only 3.9 units of force at the carnassial. Because of the large drop in usable force from the balancing side, the total bite force is reduced to 11.2 units (7.3 + 3.9) despite the improvement in leverage.

For the position of the carnassials in sabertooths to be consistent with Greaves's model, the muscle resultant must also have been located more posteriorly. The exact position of the muscle resultant is difficult to predict, even in extant taxa, and can only be approximated in fossils. The distance from the muscle resultant to the joint is determined by the lengths of the moment arms and the relative force outputs of the individual jaw adductors. The lengths of the moment arms, especially that of the temporalis, suggest that the muscle resultant must have been located more posteriorly in sabertooths. The moment arm of the temporalis (Figure 7.6; MAT) extends from the center of mandibular rotation at the joint to the center of its insertion near the tip of the coronoid process. Thus the length of MAT is related directly to the height of the coronoid process; a tall coronoid process is correlated with a long MAT. Because the coronoid process is reduced in sabertooths, MAT is shorter and the resultant of the temporalis was in a more posterior position.

The leverage of the temporalis can be determined by dividing MAT into the output lever, the distance from the joint to the carnassial (Figure 7.6; OL; COM1 of Emerson & Radinsky 1980). Table 7.1 gives the values for MAT, OL, and the leverage of the temporalis in a variety of felid and nimravid sabertooths. These values are compared with averages for extant members of four carnivoran families (calculated from Radinsky 1981). The leverage in the felid sabertooths is similar to that in extant felids

Table 7.1. *M_1 position in sabertoothed carnivorans*

	JL	MAT	OL	OL/JL	MAT/OL	OL^E	MAT/OL^E
Nimravidae							
Dinictis	11.48	2.68	5.74	0.50	0.467	5.97	0.449
Hoplophoneus	12.15	2.72	5.56	0.458	0.489	6.32	0.430
"*Eusmilus*"	17.67	3.90	7.54	0.423	0.517	9.19	0.424
Barbourofelis morrisi	17.0	4.38	7.15	0.421	0.613	8.84	0.495
Barbourofelis fricki	25.7	6.51	10.69	0.416	0.609	13.36	0.487
Felidae							
Machairodus	26.0	5.0	10.73	0.413	0.466	13.52	0.370
Homotherium	23.5	5.2	9.8	0.417	0.531	12.22	0.426
Smilodon	20.55	4.16	8.44	0.411	0.493	10.69	0.389
Extant sp.		2.54	5.09		0.498		
Viverridae		1.49	3.21		0.464		
Canidae		2.46	5.63		0.437		
Mustelidae		1.55	2.55		0.608		

Notes: Lengths are in centimeters; means of one to 12 specimens (fossils) or one specimen of each of 14 to 17 species (extant families). "*Eusmilus*" = *Hoplophoneus sicarius*. Abbreviations: JL, lower jaw length, measured from the back of the condyle to the front of the median incisor alveolus; MAT, moment arm of the temporalis; OL, output lever (condyle to carnassial); OL/JL, relative position of the carnassial; MAT/OL, mechanical advantage of the temporalis; OL^E, expected output lever based on Radinsky (1981) = 0.52 × (JL). JL, MAT, and OL from Emerson & Radinsky (1980) and Radinsky (1981).

and viverrids. If the carnassial were in the "expected," more anterior position seen in most extant carnivorans (OL^E), the leverage would be considerably lower. The temporalis of most nimravid sabertooths has similar leverage to that of sabertooth and extant felids and viverrids. The most appropriate comparison for nimravids among extant carnivorans is less obvious, but the similarity to felids may reflect the general similarity in cranial proportions. Thus, given that the resultant of the temporalis is located more posteriorly in sabertooths, the more posterior carnassial maintains the leverage at similar values to that in extant carnivorans, especially felids and viverrids.

The leverage of the temporalis in *Barbourofelis* is unusually high (Table 7.1). MAT is relatively large in this genus because, despite its extremely reduced height, the coronoid process is displaced anteriorly (Emerson & Radinsky 1980). Nonetheless, *Barbourofelis* may still be consistent with Greaves's model. The elongated P^4 in *Barbourofelis* is considerably longer than M_1 and occludes with both M_1 and P$_4$. Martin (1980) suggested that the region of overlap between P$_4$ and M_1 was the functional carnassial notch in this genus. The notch on M_1 is extremely shallow and is inappropriate for the measurement of OL; the output lever extends farther forward, reducing the leverage to a value that is probably more similar to that of other sabertooths.

The lever arms for the other adductors may also have been shifted posteriorly but their positions are more difficult to estimate. Nonetheless, the temporalis is the largest adductor, and the more posterior position of its resultant almost certainly shifted the overall resultant posteriorly. The degree of consistency in the leverage of the temporalis (Table 7.1) suggests that the more posterior carnassial maintains the geometry among the positions of the joints, the muscle resultant, and the carnassial that is predicted by Greaves's model. The maintenance of this geometry, despite the changes to cranial proportions that occur in sabertoothed carnivorans, provides strong corroboration for the general applicability of Greaves's model and the importance of carnassial position to the jaw mechanics of carnivorans. This geometry may be a synapomorphy for the Carnivora.

Given the applicability of Greaves's model to sabertoothed carnivorans, the explanations for the posterior position of the carnassials can be reevaluated. The posterior carnassials have been considered a compensatory mechanism for either the poorer leverage or the reduced size of the jaw adductors, especially the temporalis. Greaves's model indicates that the posterior carnassials can compensate only for the potential reduction in leverage. The posterior carnassial reduces the length of the output lever to match the reduction to the input lever associated with the lowered coronoid process. This is consistent with Kurtén's (1952) observation that the ratio of the input lever to output lever is the same in *Smilodon* and the less-derived sabertoothed felid

Metailurus. However, Gans (1988) argued that the position of muscle insertions does not incur mechanical advantage. From this perspective, the more posterior carnassial would not be a compensatory device for reduced mechanical advantage of the temporalis, but rather would simply reflect a repositioning of the tooth to maximize the available leverage associated with the more posteriorly placed muscle resultant. In any case, Greaves's model predicts that the more posterior carnassial cannot compensate for the reduction in the size of the musculature. This suggests that if sabertoothed felids and nimravids did have a smaller volume of adductor musculature for their size, the potential maximum bite force at the carnassials, and at other points along the tooth row, may have been lower (see Greaves 1985). Given that the relationship between muscle volume and bite force requires information on muscle performance and architecture that is not available for fossils, the prediction of actual bite forces is not feasible.

CARNASSIAL WEAR IN SABERTOOTHED CARNIVORANS

Heavy wear of the carnassials has been reported in various sabertoothed carnivorans. The lingual surface of the P^4 of *Smilodon* was subjected to heavy shearing wear that exposed the pulp cavity in old individuals (Miller 1969), and heavy wear also occurred to the deciduous carnassials in this genus (Turnbull 1978). Sinclair and Jepsen (1927) described the vertical grooving of the shearing surfaces of both the upper and lower carnassials of the nimravid *Hoplophoneus sicarius*, a feature that is indicative of heavy attritional wear. In many sabertooths the carnassials retained their shearing edge by the rapid loss of the enamel on the working surfaces, creating an enamel–dentin interface which was self-sharpening but was subject to extremely heavy attrition (Martin 1980). The microwear on the M_1 of *Smilodon* is distinct from that of various extant carnivorans in having narrow, long scratches and an extremely low pit frequency (Van Valkenburgh, Teaford, & Walker 1990). These features most resemble those of the cheetah, which consumes the least bone among the extant carnivorans studied. The authors suggested that *Smilodon* actively avoided contact with bone, possibly to avoid potential breakage to the sabers, which were vulnerable to bending about their mesiodistal axis (Van Valkenburgh & Ruff 1987).

Heavy carnassial wear occurred in the sabertoothed felids *Nimravides*, *Machairodus*, *Homotherium*, and *Smilodon*, as well as in most of the Nimravidae. Although *Nimravus* has few sabertooth features, M_1 was subjected to considerable attrition (fig. 2 in Toohey 1959). The strong correlation

between saberlike upper canines and heavy, usually attritional, wear to the carnassials in both felids and nimravids suggests that some functional connection existed between these features. One might postulate that changes in diet, the geometry of the chewing cycle, or autocclusal mechanisms (Mellett 1985) associated with the sabertoothed canines may have led to increased wear on the carnassials.

In the more derived nimravid sabertooths, such as *Hoplophoneus* and *Barbourofelis*, extreme attrition was accompanied by rotation of one or both carnassials around the mesiodistal axis of the tooth. Hough (1949) noted that in *Hoplophoneus* P^4 rotated inward and M_1 rotated outward with age; and suggested that the resultant wear to the teeth was self-sharpening. Baskin (1981) described the marked buccal rotation of M_1 and the less extensive lingual rotation of P^4 in *Barbourofelis lovei*; in old individuals the wear surface on M_1 is almost parallel to the base of the crown. Baskin suggested that the close proximity of the sabers to the extremely large ventral flange at the anterior end of the mandible in *Barbourofelis* (Figure 7.3b) restricted the mediolateral movement of the lower jaw, which allows the carnassials to come into proper occlusion in extant carnivorans; carnassial rotation was considered necessary to maintain carnassial contact.

Baskin's (1981) functional explanation for carnassial rotation in *Barbourofelis* resembles that of Mellett (1969, 1977) for the carnassial rotation in the creodont *Hyaenodon*. In some species of this genus, the carnassials M^1 and M^2 rotate lingually around their mesiodistal axes, so that in old individuals the roots are rotated almost 90°, the crowns are completely obliterated, and wear facets occur on the roots. Mellett associated the rotation with a high rate of attrition resulting from the prolonged contact between the upper and lower carnassials. Mellett judged that the increased lateral excursion of the mandible necessary for the maintenance of carnassial contact as the surfaces wore was precluded by the fused mandibular symphysis. Lingual rotation of the upper carnassials maintained the original intercarnassial distance as the teeth wore down. Mellett (1977) suggested that carnassial rotation was not necessary in carnivorans because of the greater potential excursion of the dentaries provided by the patent mandibular symphysis. Baskin's (1981) analysis of *Barbourofelis* suggests that some sabertooths may be exceptions to this generalization.

The closely related nimravids *Dinictis* and *Hoplophoneus* (Figure 7.3) provide an appropriate comparison for analysis of the functional morphology associated with differential wear and carnassial rotation. These genera differ primarily in the degree of sabertooth morphology and both demonstrate marked carnassial development (Figures 7.2 & 7.3).

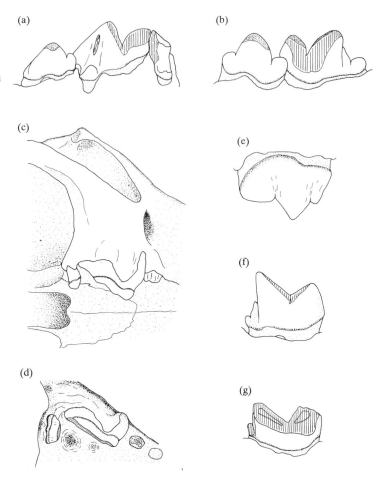

Figure 7.7. Carnassial wear in *Dinictis* and *Hoplophoneus*. Attrition facets are marked by parallel lines; abrasion facets in (a) and (b) are heavily stippled. (a) Lingual view of P^3, P^4, and M^1 in *Dinictis* (F:AM 62050); note abrasion to apex of P^3. (b) Buccal view of worn P_4 and M_1 in *Dinictis* (based on F:AM 62047); note abrasion to P_4 and blade edges on M_1. (c) Right ventrolateral view of skull of an old individual of *Hoplophoneus* (AMNH 11858) showing carnassial rotation and extreme wear to P^4 (compare with [e]). (d) Ventral view of P^4, M^1, and alveoli of P^3 in AMNH 11858 (as in [c]) showing oblique orientation of wear facet on P^4 relative to surface of palate and curvature of anterobuccal root. (e) Buccal view of an unworn P^4 of *Hoplophoneus* (AMNH 1792, reversed; compare with [c] and [d]). (f) Buccal view of an almost unworn right M_1 of *Hoplophoneus* (AMNH 5338; compare with [g]). (g) Buccal view of a heavily worn right M_1 of *Hoplophoneus* (F:AM 62107; compare with [f]).

Nonetheless, wear of the carnassials differs considerably. In *Dinictis* carnassial wear is more typical of that of most extant carnivorans; attritional loss of enamel occurred from the working surfaces of P^4 and M_1 but the crowns of these teeth remained largely intact, even in old individuals (Figure 7.7a,b). In *Hoplophoneus* the attritional wear is extreme; P^4 erupted normally but subsequently rotated lingually, as in the molars of *Hyaenodon*, so that in old individuals the shearing surface is almost parallel to the base of the crown, and most of the crown is obliterated (Figure 7.7c,d). Of the possible morphological correlates that might be associated with a functional explanation for this difference in wear, the following seemed most amenable to study in these fossil taxa: (1) differences in the geometry of jaw movement and (2) differences in dental morphology and occlusion. The differences were predicted to be related in some way to the size of the sabers.

The geometry of jaw movement in *Dinictis* and *Hoplophoneus*

METHODS. The geometry among the two dentaries and the skull during various stages of the chewing cycle were inferred in *Dinictis* and *Hoplophoneus* from the study of associated skulls and lower jaws, including casts with bisected mandibles (*Dinictis*: F:AM 62148, SDSM 2663; *Hoplophoneus*: AMNH 11858, F:AM 69417, SDSM 2544). Comparisons with *Smilodon* were based on a cast of a skull and associated mandible. A survey of the Frick and American Museum collections of *Dinictis* and *Hoplophoneus* provided additional data regarding positional relationships during various stages of the chewing cycle and information regarding dental wear and carnassial rotation.

ANALYSIS. The palates of *Dinictis* and *Hoplophoneus* are shaped differently from those of extant felids (Figure 7.8). The anterior portion of the palate is relatively narrow, especially in *Hoplophoneus*, with roughly parallel margins. Posterior to P^2 the palate widens considerably. In felids the margins of the palate are straighter, with a constant taper, and the palate is relatively wider at the canines. The shape of the nimravid palate, especially its narrowness at P^3, places constraints on the movement of the dentaries during the chewing cycle. The geometry of

(a)   **(b)**

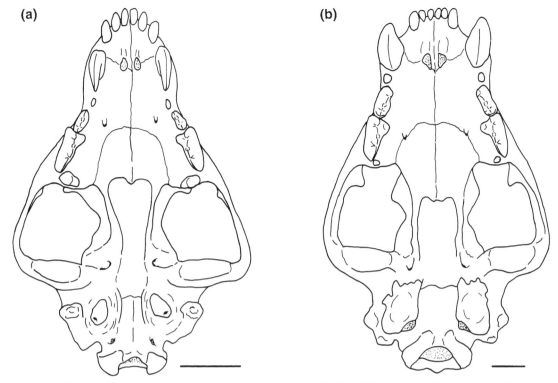

Figure 7.8. Ventral views of skulls of (a) nimravid *Hoplophoneus* (SDSM 2544) and (b) felid *Panthera* (UCMZ 1975.2). See text for description. Scale bars = 30 mm.

the chewing cycle in *Dinictis* and *Hoplophoneus* differs significantly from that of various extant carnivorans (Scapino 1965; Gorniak & Gans 1980).

During carnassial occlusion in both *Dinictis* and *Hoplophoneus*, the mandibular condyle on the working side is more or less centered in the mandibular fossa, with the medial margin of the condyle usually just lateral to the medial margin of the postglenoid process. The posterior portions of the upper and lower cheek tooth rows are almost parallel (Figure 7.9a). The narrowness of the palate at P^3 necessitates considerable medial movement of the dentary to position the lower premolars to the lingual side of the upper tooth row at complete jaw closure. At complete jaw closure the mesiodistal axes of the collateral P^4 and M_1 and the rest of the posterior portions of the upper and lower tooth rows are no longer parallel (Figure 7.9b), and the posterior ends of the carnassial blades are well separated. The articular surfaces of the mandibular condyles and the mandibular fossae are no longer aligned (Figure 7.9b). Most specimens are too distorted to allow accurate measurement, but in *H. primaevus* the condyle projects up to 5 mm medial to the inner margin of the postglenoid process; in *Dinictis* the medial projection is probably somewhat less. These observations indicate that during jaw closure involving carnassial occlusion the working dentary moves

medially, as in other carnivorans, but in doing so it rotates in the frontal plane about a dorsoventral axis in the incisor region (Figure 7.9) so that the amount of medial translation by the dentary is minimal anteriorly and increases posteriorly.

Medial movement of the dentary was governed primarily by an autocclusal mechanism involving P^4 and M_1. In *Dinictis*, as in other carnivorans, carnassial occlusion caused the dentary to move mediad relative to P^4 as the reciprocally curved blades slid past each other (Figure 7.2b). The oblique orientation of the cutting edge on P^4 relative to the tooth row (Figure 7.9) also tended to force the dentary posteriorly. Because the postglenoid process prevented significant posterior movement, a mediad turning moment was generated at the anterior carnassial cusps, resulting in the early loss of contact at the posterior ends of the blades. Most specimens of *Dinictis* have a distinct wear facet on the anterolingual surface of P^4 that extends from high on the flank of the paracone toward the protocone (Figure 7.7a). The paraconid slid dorsolingually along this sloped surface during occlusion. In *Hoplophoneus* the anterolingual surface of the paracone is less sloped, and no marked wear facet developed. Medial movement of the dentary was generated by the marked angle between the edges of the occluding blades (Figure 7.2c). As the wear facets developed, they

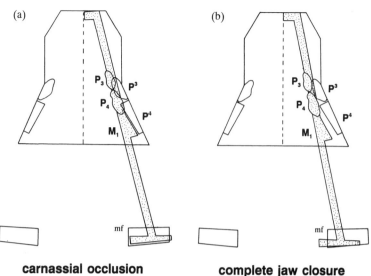

Figure 7.9. Diagrammatic representation of positional relationships between palate, mandibular fossae, and dentary on working side at (a) carnassial occlusion and (b) complete jaw closure in nimravid *Dinictis*. Midline of palate is indicated by dashed line; dentary is stippled. Dentary rotates in the frontal plane about a dorsoventral axis in the incisor region during jaw closure (see text and Fig. 7.4). Abbreviation: mf, mandibular fossa.

carnassial occlusion   **complete jaw closure**

acquired a considerably oblique orientation relative to the vertical plane (Figure 7.2d), producing a significant posteromedially directed force to M_1 as occlusion proceeded. As in *Dinictis* the postglenoid process prevented posterior movement of the dentary, resulting in its medial rotation in the frontal plane. In both genera the rotation generated by carnassial occlusion was sufficient to move the dentary on the working side into its most medial position at complete jaw closure.

Because the dorsoventral axis of rotation of the working dentary in the frontal plane is located in the incisor region, substantial mediolateral movement posteriorly was not precluded by the close apposition between the sabers and the lateral surface of the dentaries anteriorly. The symphysis remained unfused in all nimravids. The symphyseal surface has raised projections and deep pits that mesh intimately with the matching surface on the complementary symphyseal surface. Series of projections and pits extend posteroventrally, dorsally and posterodorsally from the anterior margin. This morphology most closely resembles Scapino's (1981; fig. 3) Class III carnivoran symphysis type that occurs in extant ursids and large felids. Scapino (1981) argued that Class III symphyses are extremely stiff and allow little relative movement between dentaries. If the nimravid symphysis was rigid, the spread between the dentaries must have been no greater than that which occurred at full jaw closure. The shape of the palate and the position of the mandibular fossae do not preclude the rotation of a rigid mandible about a point in the incisor region; however, other aspects of the functional system suggest that the symphysis was not rigid and allowed relative movement between the dentaries in the frontal plane.

If the symphysis was rigid and the distance

between the condyles was the same at carnassial occlusion as at complete jaw closure, the condyle on the balancing-side dentary would have projected medially of the glenoid surface by almost a centimeter at the initiation of carnassial occlusion in both genera. This degree of medial excursion seems unlikely based on (1) craniomandibular joint morphology in extant carnivorans, (2) the shape of the articular surfaces of the craniomandibular joint in *Dinictis* and *Hoplophoneus*, and (3) inferences regarding carnassial functioning. Various ligaments are essential to the integrity of the craniomandibular joint in extant carnivorans (Scapino 1965; Gorniak & Gans 1980) and limit mediolateral movement of the condyle. A ligamentous joint capsule with the elasticity or "slack" to allow up to a centimeter of medial movement would presumably not have provided the support that would seem to have been necessary, especially in *Hoplophoneus* in which the pre- and postglenoid processes are reduced, during the marked mandibular excursions associated with a wide gape. In both *Dinictis* and *Hoplophoneus* the articular surfaces on the condyle and the glenoid surface are best developed medially. Given the medial projection of the balancing-side condyle at carnassial occlusion that would be necessitated by a rigid symphysis, most of the articular surfaces would not have been in contact. This poor fit at the craniomandibular joint on the balancing side at carnassial occlusion seems especially unlikely in light of Greaves's model (Figure 7.5) and studies in extant mammals (see De Gueldre & De Vree 1990) that suggest that the balancing-side craniomandibular joint bears the larger percentage of the reaction forces associated with occlusion between the posterior portions of the tooth rows.

The spread between the dentaries at complete

jaw closure is too narrow to allow the simultaneous optimal fit of the articular surfaces at both craniomandibular joints; the maintenance of this geometry during biting with the canines is unknown in extant carnivorans and seems extremely unlikely. If recent arguments suggesting that the mandible acted as a brace against the prey during canine biting in *Smilodon* (Akersten 1985) are applicable to nimravid sabertooths, one would infer that simultaneous centering of both condyles at the mandibular fossae was an important component of that functional system.

The morphology of the symphyseal surfaces in *Dinictis* and *Hoplophoneus* is compatible with movement in the frontal plane. In both genera this movement would have been facilitated by the anteroposterior narrowness of the symphysis (Figure 7.5). In addition, unlike large extant felids in which the anterior and posterior margins of the symphysis intermesh intimately simultaneously, on F:AM 62147 (*Dinictis*) tight meshing of the interdigitating surfaces anteriorly results in a significant gap posteriorly, suggesting that one dentary moved relative to the other by rotation around an anterior dorsoventral axis.

The distance or spread between the dentaries in the frontal plane was probably at a minimum at complete jaw closure and was greater with increasing gape (Figure 7.4). Occlusion between the carnassials was important, if not essential, for the medial movement of the dentaries during jaw closure (Figure 7.4c). The paracone on P^4 seems especially important as an occlusal guide for this movement. This suggests that occlusion between both pairs of carnassials, at least at the anterior cusps, occurred during jaw closure in these genera, even when carnassial biting did not occur. Jaw opening would have resulted in a spreading of the dentaries, possibly through some sort of "spring-loading" in the symphyseal region.

The inference that relative movement between the dentaries occurred in the frontal plane in both *Dinictis* and *Hoplophoneus* is a probability statement based on interpretations of the morphology of these taxa, together with phylogenetic inferences based on the morphology of extant carnivorans. The inference that the working mandible rotated about a dorsoventral axis in the incisor region is strongly supported by dental wear facets and specimens that illustrate the positional relationships in the system at complete jaw closure. However, the known morphology does not preclude the possibility of an immobile symphysis and, although the hypothesis of mobility is based on assumptions that appear sound given the functional morphology of extant carnivorans, such inferences must entail inherent uncertainty.

A rationale for carnassial wear and rotation in *Hoplophoneus*

The geometry of jaw closure is the same in *Dinictis* and *Hoplophoneus* and does not explain the differences in wear to the carnassials. The greater wear and necessary rotation of the carnassials in *Hoplophoneus*, as compared to *Dinictis*, can instead be explained as a consequence of differences in dental morphology, but only within the context of the geometry of jaw closure, and the autocclusal role of the carnassials therein, described above.

In *Dinictis* the shearing blades on the carnassials are oriented somewhat obliquely to the margin of the palate and the axis of the lower jaw. Wear includes both attrition and abrasion (Figure 7.7a,b). The lingual surface of P^4 and the buccal surface of M_1 developed the attrition facets that are typical of carnivorans. On P^4 this wear is most pronounced on the posterior portion of the paracone and the metastyle, from which most of the enamel is often worn away. On many specimens the carnassial notches have lost their slitlike shape with only minor wear and are broad and rounded from abrasion. In some specimens abrasion also occurred to the apex of the paracone and to the cutting edge. Abrasion of M_1 occurred to the protoconid and to the blade edge.

In *Hoplophoneus* the shearing surfaces are oriented more in line with the margin of the palate and the axis of the lower jaw. Wear was dominated by attrition. Wear developed first immediately above the cutting edge on P^4 and below that on M_1, from which the enamel was quickly removed; less extreme wear occurred to the sides of the carnassial notches. P^4 began to rotate lingually about its mesiodistal axis before much of its crown was obliterated. Rotation was generated primarily by the continued eruption and inturning of the anterobuccal root, but also involved the posterior root at advanced wear stages. Subsequent wear to the shearing surface exposed the original pulp cavities, and in the oldest individuals most of the crown of P^4 is worn away (Figure 7.7c,d). Rotation maintained the oblique orientation of the wear facets to the surface of the palate that developed with initial attrition (Figure 7.2d). Wear to M_1 was less extreme (Figure 7.7g), and its buccal rotation was only moderate. Attrition was most pronounced on the paraconid, and the pulp cavities became exposed at advanced wear stages.

Differences in carnassial morphology and functioning in *Hoplophoneus*, as compared to *Dinictis*, probably increased the occlusal pressure between the carnassials, resulting in increased dental attrition. The mediolateral distance between the carnassial notches is greater in *Hoplophoneus* than in *Dinictis* because of differences in the orientations of the blade edges

(Figure 7.2b,c). In *Dinictis* the shape of the blade edges on P⁴ and M₁ closely resembles that of most extant carnivorans; they are shallow, reciprocal curves of similar radius, with a relatively small internotch distance (Figure 7.2b). In *Hoplophoneus* the cutting edge on P⁴ is straighter than the cutting edge on M₁, which is strongly curved (Figure 7.2c). At initial contact between the posterior cusps the mesiodistal axes of the teeth were not parallel and, unlike *Dinictis*, the paracone was positioned lingually to the paraconid (Figure 7.2c,d). As a result, the angle between the occluding blades and the distance between the notches are larger. Based on an analogous functional system consisting of a guillotine blade that is spring-loaded against a base plate (Figure 7.2e; Atkins & Mai 1979), both of these differences increased the pressure between the occluding blades. The larger angle between the blade edges increased the wear to occluding surfaces as the blades passed each other. Secondly, the larger internotch distance would have trapped a larger amount of food between the carnassials, increasing the turning moment generated by the food and, as a consequence, the appositional forces between the shearing surfaces. This effect was accentuated by the initial position of the paracone, which was lingual to the paraconid. Attrition in *Hoplophoneus* was increased further upon the development of oblique shearing surfaces because the distance over which the tooth surfaces were in contact would be increased. Thus, differences in wear in *Dinictis* and *Hoplophoneus* relate, at least in part, to differences in dental morphology. Attrition occurred more rapidly in *Hoplophoneus* and obliterated any abrasion.

Because of the more posterior position of the carnassials in *Hoplophoneus*, the posterior cusps began to occlude well before the anterior cusps (Figure 7.2d). This also occurred in at least some *Dinictis* (Figure 7.3a), but to a lesser extent. Contact between the anterior cusps generated a turning motion in the dentary, as described above, resulting in the loss of contact between the posterior cusps. This might suggest that occlusal forces were more concentrated in *Hoplophoneus*, initially between the posterior cusps and later between the anterior cusps, increasing wear at each. However, Abler's (1992) analysis suggests that occlusal forces at the carnassials were more diffuse. Wear is greater to the anterior cusps because contact at these surfaces was maintained longer, and masticatory forces were concentrated at these surfaces as the dentary rotated during the later stages of occlusion. The development of oblique wear facets on these surfaces increased the occlusal pressure between the teeth.

Heavy attrition to the carnassials results in potential changes to the positional relationships between the occluding teeth. In extant felids heavy attrition in older individuals increases the intercarnassial distance and necessitates greater lateral excursion by the mandible to bring the carnassials into occlusion; the orientation of the shearing surfaces is also altered. Carnassial rotation is a compensatory mechanism that can preserve initial occlusal relationships in the face of extreme attrition and probably increases the functional life of the teeth. The positional relationship between the carnassials was essential to jaw closure in *Dinictis* and *Hoplophoneus* and, given the extreme attrition that occurred in *Hoplophoneus*, rotation of P⁴ was necessary for the maintenance of the autocclusal mechanism between the carnassials. The rotation of P⁴ maintained throughout life the positional relationship between the carnassials that occurs in young individuals with newly erupted teeth and in *Dinictis*. Rotation also preserved the lingual position of the paracone, relative to the paraconid, at initial carnassial contact; this positional relationship may have been especially important in old individuals in which the original interlocking mechanism provided by the carnassial notches was degraded. At advanced wear stages in *Hoplophoneus*, the functioning of the carnassials was extremely scissorlike. This explanation for carnassial rotation in *Hoplophoneus* is predicated in part on previous inferences regarding the geometry of jaw closure in nimravids. Given the inherent uncertainties in that analysis, considerable uncertainty must be associated with this further extrapolation.

Applicability to other sabertoothed carnivorans

Although the correlation between sabertoothed upper canines and the development of heavy attritional wear to the carnassials is suggestive of some causal or functional relationship between these features, study of the nimravids *Dinictis* and *Hoplophoneus* does not provide an explanation that seems applicable to other sabertooths. *Dinictis* and *Hoplophoneus* share a particular geometry of jaw closure, and differences in carnassial wear seem to be associated with differences in the detailed morphology of the carnassials rather than differences associated with the degree of hypsodonty of the sabers. Different functional explanations for extreme attritional wear and carnassial rotation probably apply to other nimravids. The shape of the palate and the morphology of the carnassials in *Barbourofelis* differ considerably from those of *Hoplophoneus*, suggesting different explanations for wear and carnassial rotation. The shape of the palate of the felid *Smilodon* resembles that of other large felids, and explanations for heavy wear in this genus may involve functional morphology that is particular to felids. Nonetheless, the strong correlation between sabertoothed morphology and heavy carnassial wear does suggest some general functional explanation. Detailed studies of carnassial functioning in other sabertoothed carnivorans may help to resolve this apparent paradox.

THE INTERPLAY BETWEEN NEONTOLOGICAL AND PALEONTOLOGICAL STUDIES

Functional analyses in fossil taxa are necessarily dependent upon neontological studies. All reconstructions of unpreserved attributes in extinct organisms through phylogenetic inferences and form–function correlations rely on knowledge of extant organisms. The role of neontological information may be less obvious in biomechanical design analysis, but these inferences are often second order deductions that are based on previous inferences. Biomechanical inferences regarding the force generated by the temporalis in sabertooths take as given the inference that the temporalis existed and had particular attachments in the fossil taxon in question.

Nonetheless, the flow of information between neontological and paleontological study is not unidirectional. Fossil taxa broaden our view of the constraints associated with the vertebrate body plan and can provide test cases for hypotheses based on extant taxa. Greaves's (1983) model for carnassial functioning in extant carnivorans provided an explanation for the relatively consistent position of the lower carnassial at the midpoint of the mandible (Radinsky 1981). Sabertoothed carnivorans present a situation not seen in extant carnivorans in which the carnassial is located more posteriorly. However, because the muscle resultant for the jaw adductors is also located more posteriorly, the geometry among the carnassial, the craniomandibular joint, and the muscle resultant predicted by Greaves's model is maintained, providing corroboration of the model and its general applicability. In other words, morphologies that are restricted to fossil taxa can provide the basis for tests of models that were predicated on the study of extant taxa. Fossil taxa may also provide exceptions to generalizations based on extant taxa that may be the basis for form–function correlations. Although Scapino's (1981) studies of the jaw symphysis of extant carnivorans suggested that highly rugose symphyseal surfaces are associated with stiff symphyses, study of the jaw mechanics of *Dinictis* and *Hoplophoneus* indicates that the generalization does not hold for at least these taxa.

CLOSING STATEMENTS

Functional inferences in extinct organisms must entail greater levels of uncertainty than those in extant taxa. Phylogenetic and extrapolatory inferences rely on assumptions that are often unverifiable and therefore entail significant uncertainty. As a result, functional inferences in fossils should probably be considered plausibility statements; the assumptions upon which they are based should be explicit. Through careful stepwise analysis, with due regard to the assumptions and inherent limitations of particular approaches, the uncertainty and limitations associated with the inference of the functional morphology of extinct vertebrates can be identified, and unnecessary speculation can be reduced. Certain inferences about fossils can be robust, and extinct taxa broaden our view of the functional morphology of vertebrates and may provide tests of generalizations based on the extant taxa.

ACKNOWLEDGMENTS

J. Alexander, P. Bjork, R.H. Tedford, and J. Whitmore provided or facilitated the loan of and access to specimens in their care. B. Naylor of the Royal Tyrrell Museum facilitated the casting of specimens; C. Coy and J. McCabe provided their technical expertise. C.A. Shaw of the Page Museum provided casts of an associated skull and mandible of *Smilodon*. Blaire Van Valkenburgh provided a detailed review of the manuscript leading to marked improvements and clarifications. C. Bryant helped with the editing of the original manuscript. Financial support was provided by Natural Sciences and Engineering Research Council of Canada Grant A9745 to A.P.R.

REFERENCES

Abler, W.L. 1992. The serrated teeth of tyrannosaurid dinosaurs and biting structures in other animals. *Paleobiology* 18, 161–183.

Akersten, W.A. 1985. Canine function in *Smilodon* (Mammalia; Felidae; Machairodontinae). *Contributions to Science, Los Angeles County Museum of Natural History* 356, 1–22.

Atkins, A.G., & Mai, Y.W. 1979. On the guillotining of materials. *Journal of Materials Science* 14, 2747–2754.

Baskin, J.A. 1981. *Barbourofelis* (Nimravidae) and *Nimravides* (Felidae) with a description of two new species from the late Miocene of Florida. *Journal of Mammalogy* 62, 122–139.

Berkovitz, B.K.B., & Poole, D.F.G. 1977. Attrition of the teeth in ferrets. *Journal of Zoology, London* 183, 411–418.

Biknevicius, A.R., & Van Valkenburgh, B. 1991. Feeding behaviors of *Smilodon*. *American Zoologist* 31, 54A.

Bock, W.J. 1989. Principles of biological comparison. *Acta Morphologica Neerlando-Scandinavica* 27, 17–32.

Bohlin, B. 1940. Food habits of the machairodonts, with special regard to *Smilodon*. *Bulletin of the Geological Institute, Uppsala* 28, 156–174.

Bramble, D.M. 1978. Origin of the mammalian feeding complex: models and mechanisms. *Paleobiology* 4, 271–301.

Bryant, H.N. 1990. Implications of the dental eruption sequence in *Barbourofelis* (Carnivora, Nimravidae) for the function of upper canines and the duration of parental care in sabretoothed carnivores. *Journal of Zoology, London* 222, 585–590.

Bryant, H.N. 1991. Phylogenetic relationships and systematics of the Nimravidae (Carnivora). *Journal of Mammalogy* 72, 56–78.

Bryant, H.N., & Russell, A.P. 1992. The role of phylogenetic analysis in the inference of unpreserved attributes of extinct taxa. *Philosophical Transactions, Royal Society of London B337*, 405–418.

Bryant, H.N., & Seymour, K.L. 1990. Observations and comments on the reliability of muscle reconstruction in fossil vertebrates. *Journal of Morphology* 206, 109–117.

Churcher, C.S. 1985. Dental functional morphology in the marsupial sabre-tooth *Thylacosmilus atrox* (Thylacosmilidae) compared to that of felid sabre-tooths. *Australian Mammalogy* 8, 201–220.

Crompton, A.W., & Hiiemae, K.M. 1970. Functional occlusion and mandibular movements during occlusion in the American opossum, *Didelphis marsupialis* L. *Zoological Journal of the Linnean Society* 49, 21–47.

De Gueldre, G., & De Vree, F. 1990. Biomechanics of the masticatory apparatus of *Pteropus giganteus* (Megachiroptera). *Journal of Zoology, London* 220, 311–332.

Emerson, S.B., & Radinsky, L. 1980. Functional analysis of sabretooth cranial morphology. *Paleobiology* 6, 295–312.

Flynn. J.J., Neff, N.A., & Tedford, R.H. 1988. Phylogeny of the Carnivora. In *The Phylogeny and Classification of the Tetrapods, Volume 2: Mammals,* ed. M.J. Benton, Systematics Association Special Volume 35B, pp. 73–116. Oxford: Clarendon Press.

Gans, C. 1988. Muscle insertions do not incur mechanical advantage. *Acta Zoologica Cracoviensia* 31, 615–624.

Gorniak, G.C., 1985. Trends in the actions of mammalian masticatory muscles. *American Zoologist* 25, 331–337.

Gorniak, G.C., & Gans, C. 1980. Quantitative assay of electromyograms during mastication in domestic cats (*Felis catus*). *Journal of Morphology* 163, 253–281.

Greaves, W.S. 1978. The jaw lever system in ungulates: a new model. *Journal of Zoology, London* 184, 271–285.

Greaves, W.S. 1983. A functional analysis of carnassial biting. *Biological Journal of the Linnean Society* 20, 353–363.

Greaves, W.S. 1985. The generalized carnivore jaw. *Zoological Journal of the Linnean Society* 85, 267–274.

Herring, S.W. 1975. Adaptations for gape in the hippopotamus and its relatives. *forma et functio* 8, 85–100.

Hopson, J.A., & Radinsky, L.B. 1980. Vertebrate paleontology: new approaches and new insights. *Paleobiology* 6, 250–270.

Hough, J. 1949. The habits and adaptation of the Oligocene saber tooth carnivore, *Hoplophoneus*. *U.S. Geological Survey Professional Paper* 221-H, 125–137.

Hunt, R.M. 1987. Evolution of the aeluroid Carnivora: significance of auditory structure in the nimravid cat *Dinictis*. *American Museum Novitates* 2886, 1–74.

Hylander, W.L. 1979. An experimental analysis of temporomandibular joint reaction forces in macaques. *American Journal of Physical Anthropology* 51, 433–456.

Kay, R.F., & Hiiemae, K.M. 1974. Jaw movement and tooth use in recent and fossil primates. *American Journal of Physical Anthropology* 40, 227–256.

Kurtén, B. 1952. The Chinese *Hipparion* fauna. *Societas Scientiarum Fennica, Commentationes Biologicae* 13, 1–82.

Lucas, P.W., & Lake, D.A. 1984. Chewing it over: basic principles of food breakdown. In *Food Acquistion and Processing in Primates,* ed. D.J. Chivers, B.A. Wood, & A. Bilsborough, pp. 283–301. New York: Plenum Press.

MacKenna, B.R., & Turker, K. 1978. Twitch tension in the jaw muscles of the cat at various degrees of mouth opening. *Archives of Oral Biology* 23, 917–920.

Marinelli, W. 1938. Der Schädel von *Smilodon*, nach der Funktion des Kieferapparates analysiert. *Palaeobiologica* 6, 246–272.

Martin, L.D. 1980. Functional morphology and the evolution of cats. *Transactions of the Nebraska Academy of Sciences* 8, 141–154.

Martin, L.D. 1984. Phyletic trends and evolutionary rates. In *Contributions in Quaternary Vertebrate Paleontology: a Volume in Memorial to John E. Guilday,* ed. H.H. Genoways & M.R. Dawson, pp. 526–538. Pittsburgh: Carnegie Museum of Natural History, Special Publication 8.

Matthew, W.D. 1901. Fossil mammals of the Tertiary of northeastern Colorado. *Memoirs of the American Museum of Natural History* 1, 353–447.

Matthew, W.D. 1910. The phylogeny of the Felidae. *Bulletin of the American Museum of Natural History* 28, 289–310.

Maynard Smith, J., & Savage, R.J.G. 1959. The mechanics of mammalian jaws. *The School Science Review* 40, 289–301.

Mellett, J.S. 1969. Carnassial rotation in a fossil carnivore. *The American Midland Naturalist* 82, 287–289.

Mellett, J.S. 1977. Paleobiology of North American *Hyaenodon* (Mammalia, Creodonta). *Contributions to Vertebrate Evolution* 1, 1–134.

Mellett, J.S. 1981. Mammalian carnassial function and the "Every effect." *Journal of Mammalogy* 62, 164–166.

Mellett, J.S. 1985. Autocclusal mechanisms in the carnivore dentition. *Australian Mammalogy* 8, 233–238.

Merriam, J.C., & Stock, C. 1932. The Felidae of Rancho la Brea. *Carnegie Institute Publication* 422, 3–231.

Miller, G.J. 1969. A new hypothesis to explain the method of food ingestion used by *Smilodon californicus* Bovard. *Tebiwa* 12, 9–19.

Neff, N.A. 1983. The Basicranial Anatomy of the Nimravidae (Mammalia: Carnivora): Character Analyses and Phylogenetic Inferences. Unpublished Ph.D. dissertation, City University of New York.

Osborn, J.W., & Lumsden, A.G.S. 1978. An alternative to "thegosis" and a re-examination of the ways in which mammalian molars work. *Neues Jahrbuch für Geologie und Paläontologie. Abhandlungen* 156, 371–392.

Pomel, M. 1843. Notice sur les carnassiers à canines comprimées et tranchantes, trouvées dans les alluvions du val d'Arno et de l'Auvergne. *Bulletin de la Société Géologique de France* 14, 29–38.

Radinsky, L.B. 1981. Evolution of skull shape in carnivores. 1. Representative modern carnivores. *Biological Journal of the Linnean Society* 15, 369–388.

Rensberger, J.M. 1973. An occlusion model for mastication and dental wear in herbivorous mammals. *Journal of Paleontology* 47, 515–528.

Savage, R.J.G. 1977. Evolution in carnivorous mammals. *Palaeontology* 20, 237–271.

Scapino, R. 1965. The third joint of the canine jaw. *Journal of Morphology* 116, 23–50.

Scapino, R.S. 1981. Morphological investigation into functions of the jaw symphysis in carnivorans. *Journal of Morphology* 167, 339–375.

Scott, W.B., & Jepsen, G.L. 1936. The mammalian fauna of the White River Oligocene. Part 1 – Insectivora and Carnivora. *Transactions of the American Philosophical Society* 28, 1–153.

Simpson, G.G. 1941. The function of sabre-like canines in carnivorous mammals. *American Museum Novitates* 1130, 1–12.

Simpson, G.G. 1945. The principles of classification and a classification of mammals. *Bulletin of the American Museum of Natural History* 85, 1–350.

Sinclair, W.J., & Jepsen, G.L. 1927. The skull of *Eusmilus, Proceedings of the American Philosophical Society* 66, 391–407.

Tedford, R.H. 1978. History of dogs and cats, a view from the fossil record. In *Nutrition and Management of Dogs and Cats*, Chapter M23. St. Louis: Ralston Purina Co.

Toohey, L. 1959. The species of *Nimravus* (Carnivora, Felidae). *Bulletin of the American Museum of Natural History* 118, 71–112.

Turnbull, W.D. 1970. Mammalian masticatory apparatus. *Fieldiana Zoology* 18, 147–356.

Turnbull, W.D. 1978. Another look at dental specialization in the extinct sabre-toothed marsupial, *Thylacosmilus*, compared with its placental counterparts. In *Development, Function and Evolution of Teeth*, ed. K.A. Joysey, pp. 388–414. London: Academic Press.

Van Valkenburgh, B., & Ruff, C.B. 1987. Canine tooth strength and killing behaviour in large carnivores. *Journal of Zoology, London* 212, 379–397.

Van Valkenburgh, B., Teaford, M.F., & Walker, A. 1990. Molar microwear and diet in large carnivores: inferences concerning diet in the sabretooth cat, *Smilodon fatalis. Journal of Zoology, London* 222, 319–340.

Weijs, W.A., & Dantuma, R. 1981. Functional anatomy of the masticatory apparatus in the rabbit (*Oryctolagus cuniculus* L.). *Netherlands Journal of Zoology* 31, 99–147.

8

The artificial generation of wear patterns on tooth models as a means to infer mandibular movement during feeding in mammals

VIRGINIA L. NAPLES

ABSTRACT

Teeth are the most commonly preserved elements in the fossil record, and many phylogenies are based mostly or entirely upon their interpretation. Comparison with living animals is the most common method of reconstructing dental structure and function for fossils, although simple models of teeth have also been used. This study demonstrates a new method of modeling dental structure and function, using the teeth of the extant tree sloths *Bradypus* and *Choloepus* as examples. In contrast to most mammals, tree sloths lack deciduous teeth and enamel, and the ever-growing teeth erupt in juveniles as simple, vertically oriented cones, which only differentiate into anterior chisel-shaped or caniniform and molariform teeth in adults. A diastema in *Choloepus* develops by relatively greater growth of the anterior mandible and maxilla. Surface relief on adult teeth – central basins and sharp-edged facets – forms by wear. To determine how these features develop, plaster teeth with softer chalk cores, simulating the relatively hard, outer and soft, inner dentin of sloth teeth, were manufactured.

Model teeth were mounted in wooden blocks at the spacing, size, and angulation found in different growth stages of the two sloth genera. The teeth could be repositioned during the experiment to simulate ontogenetic changes in dental form. The blocks themselves were put in a jig, which simulated masticatory movements (previously determined from cineradiography) or specific components of the whole masticatory cycle. Wear patterns generated on the model teeth closely followed those seen in vivo.

The first features to form in model teeth of both sloth genera were basins. Sharp-edged wear facets appeared second, "cusps" appeared in young adults and increased in prominence gradually as the teeth reoriented with facial growth. The angle at which the teeth grew determined the pattern of wear-facet formation. In *Bradypus,* pointed molariform tooth "cusps" began to form only when growth angles were sufficiently tilted to permit intercuspation of a maxillary tooth between anterior and posterior mandibular teeth. In *Choloepus* "cusps" formed gradually in young adults and sharpened in older adults as the teeth reoriented with greater growth of the anterior facial region. Formation of molariform tooth wear-features was independent

of formation of sharp, angled facets on the caniniform teeth, as the two tooth types did not occlude simultaneously at any growth stage. Caniniform tooth wear surfaces were formed by thegosis, the sharpening of a single tooth-shearing surface by repetitive movement, against the corresponding surface of another tooth.

The technique described here could be used to replicate wear stages in other mammals, including fossils, and thereby reduce the confounding effects of dental wear in mammalian systematics.

INTRODUCTION

History of development of methods of analysis

Teeth of fossil mammals are the most commonly preserved element in the fossil record, and dental characteristics have been strongly weighted in mammalian systematics. Many researchers have noted that wear alters features on these teeth – for example, changing size and shape of crowns, ridges, basins and other dental structures – but they have not had many methods to predict the effects of wear. The observed dissimilarities in teeth at different stages of wear produce two problems that confound the study of tooth characters. First, teeth worn to a different degree may show great variation in crown features, which occurred during the life of the animal. In addition, parts of the crowns of fossil teeth may have been broken off or worn away, either from antemortem or postmortem damage. Both cases cause difficulties in systematics, where great significance is sometimes afforded to tiny dental features, and also to functional studies. This paper primarily addresses the first problem, variation in crown features with wear. Often, fossil teeth are found in isolation from other skeletal elements and may be damaged, or have portions missing and therefore cannot easily be homologized with other fossil material. The effects of incomplete data on parameters such as tooth size, shape and appearance of occlusal features, and age of the animal at death may lead to a significant misunderstanding of structure, function, and classification of fossils. Second, when only a few fossil teeth represent a taxon, some wear stages will be absent, and not all teeth belonging to that taxon may be easily homologized with those previously known. Therefore, it is desirable to establish methods for predicting how factors of growth, wear, and damage change the appearance and therefore the interpretation of a fossil specimen.

In addition to having tooth specimens, to resolve the problem described above, it is essential to have one or more hypotheses of how these teeth were moved past one another, as could be derived from observations of living animals. Studies of functional morphology have contributed greatly to this understanding, as masticatory movement patterns studied in living animals using such techniques as cineradiography (Hiiemae 1978; Naples 1982, 1985) show how animals move their jaws to produce the chewing movements that affect tooth structure. A combination of knowledge of how living animals chew, together with the effects the action has on the dentition, provides paleontologists a model independent of phylogenetic constraints to interpret masticatory function in fossil animals. Functional studies provide a separate data set because they do not depend on the interpretation of phylogenies, nor are they limited by predictions of how evolutionary changes will affect structures of either closely or distantly related groups. Therefore, functional models of systems that share similar structural constraints can provide explanations for structures in both closely and distantly related groups of animals.

One approach to resolve both phylogenetic and functional problems has been to compare teeth of fossil taxa with those of related extant taxa. This notion depends upon the belief that functional interpretations of structures in living animals can reasonably be applied to similar structures in fossils, that is, dental characters of a certain shape will behave in a predictable manner when interacting, regardless of the taxon of the animal. As these predictions are based upon physical principles governing the behavior of dental materials and mandibular movement patterns, it is reasonable to consider these assumptions valid. This method is useful for comparing ancestor–descendant lineages, because it is reasonable to assume that environmental and dietary influences and mandibular function were more similar for closely related groups than for animals unrelated to living forms, or for animals that have dental structures not shown by living taxa. Studies of diverse problems such as evolution of elephant masticatory mechanisms (Maglio 1973); structure, function,

and phylogeny of Mesozoic mammals (Crompton & Jenkins 1978, 1979; Kron 1979; Cassiliano & Clemens 1979; Kraus 1979; Bown & Kraus 1979; Kielan-Jaworowska, Eaton, & Bown 1979; Clemens 1979; Kielan-Jaworowska, Bown & Lillegraven 1979); prosimian primate phylogeny (Walker 1978); ungulate evolution (Thewissen & Domning 1992); mammalian origins (Wible 1991); and many other aspects of paleontology and neontology have depended greatly upon a comparison of fossil and Recent dental data, and are useful primarily in interpreting fossil material based upon conclusions from Recent animals. However, this approach is less reliable the more distant the relationship between the Recent animals and the fossil group to which the conclusions are extrapolated. Last, this method cannot be used when the structures under study in the fossils have no Recent animal model available to be used as a comparison. Many of the speculations, such as those describing the function of the large cutting teeth of multituberculates (Krause 1982) and sabertooth cats (Martin 1980), although based upon the best evidence available, cannot be tested in this manner.

A second approach to understanding and predicting dental changes and function in both fossil and living animals includes development of models to test theories of how wear features form and change with time. Some workers have developed models to explain certain aspects of wear-feature production (Greaves 1973; Rensberger 1973; Ryan 1979; Costa & Greaves 1981), but none of these methods have attempted to reproduce structures that resemble real teeth, nor were they derived from structures designed to function as do real teeth. Although one advantage of constructing physical models is that they can be used to interpret data from both fossil and living animals, they do not show the actual changes expected in teeth of real animals under changing conditions.

A new method for predicting and interpreting dental wear patterns

Experimental replication on tooth models of wear features similar to those caused by mandibular movement on natural teeth is a new tool in the field of functional morphology and vertebrate paleontology. This approach requires accurate modeling of (1) the occlusal features of individual teeth, using materials of similar relative hardness to those of the originals; (2) spacing and orientation of the model teeth in the mandible and maxilla, and (3) the appearance of teeth at some (but not necessarily all) stages of wear. The pattern of mandibular movement, if known, can make replication of wear features on artificial teeth a simple process of fine tuning the movements of the model. If the pattern of mandibular movement is unknown, the model can be used to test different movement patterns and to determine which is required to produce teeth worn in the manner observed in either fossil or living animals. This new technique has the capacity to increase greatly the amount and types of information attainable from fossils, including isolated tooth specimens. Specifically, wearing model teeth can show changes that occurred in the teeth of a single taxon during the lifetime of an individual. This information would allow identification of teeth belonging to different age classes of the same taxon, something not always possible with previous techniques. The method of simulating wear features was tested using models of sloth teeth which are ideal for this purpose because they are simple in comparison to those of many other mammals. Also, the spacing, orientation in the jaw, and mandibular movement patterns have been determined cineradiographically for some sloths (Naples 1982, 1985, 1986). In comparison to members of most mammalian taxa, which have teeth differentiated into incisors, canines, premolars, and molars, the tooth number in the tree sloths *Bradypus* and *Choloepus* is reduced, and the teeth are differentiated into only two types (Figure 8.1; Illiger 1811; Parker 1885; Winge, 1941). In adult *Bradypus* the anterior teeth are chisel-shaped, whereas more posterior teeth are simple cylinders bearing complicated patterns of facets on the occlusal surfaces (Sicher 1944; Naples 1982). The anterior teeth in *Choloepus* are large, self-sharpening, triangular, and separated from the cheek-tooth row by a diastema (Parker 1885; Winge 1941). The cheek teeth in this genus are simple cylinders which bear a complicated set of wear facets in adults (Scott 1937; Romer 1966). All teeth in tree sloths grow throughout the life of the individual. Sloths lack a deciduous dentition, and at eruption the anterior chisel or caniniform teeth and all cheek teeth are conical and show few surface features (Naples 1981, 1982). The diastema between the caniniform teeth and the cheek teeth in *Choloepus* is absent at birth, but appears with differential growth of the mandible and maxilla (i.e., greater anteriorly), and is prominent in adults (Figure 8.2; Naples 1982).

In all of the living Xenarthra (the group to which sloths belong) and in most of the extinct forms, the teeth lack enamel (Winge 1941). Sloths compensate for this lack by having a hard outer dentine shell with a softer dentine core. Because hard and soft dentins wear differently, as do enamel and dentin, the all-dentin teeth in the tree sloths wear like the enamel and dentin teeth of other mammals (Naples 1982). In contrast to other mammals, wherein the teeth show the adult pattern of tooth features upon eruption, dental features such as basins and "cusps" form in tree sloths with wear, appearing and changing gradually with the differential growth of the mandibles and maxillae.

Tooth characteristics formed by wear have been previously modeled and produced experimentally upon real teeth (Ryan 1979). Wear features have also been produced in artificial teeth composed of hard outer aluminum tubing shells to represent enamel and soft plaster cores to represent dentin by Costa and Greaves (1981). Wear features were produced on these composite teeth by moving a lower tooth against a fixed upper tooth along a repetitive path using a motor-driven turntable rotating in a single direction. To date, the only study to combine these experimental techniques was Naples (1990) wherein chalk and plaster model teeth were constructed and worn to illustrate the manner of "cusp" formation in teeth of the extinct ground sloth *Nothrotheriops shastensis*. Production of replicates of worn *Bradypus* and *Choloepus* teeth requires an exact simulation of tooth shape, spacing, orientation in the alveolus, and the mandibular movement pattern used by these sloths. The pattern of mandibular movement was determined from analyses of cineradiographic films and videotapes of live tree sloths chewing a variety of foods, in addition to observation of wear facets on all maxillary and mandibular teeth in sloths at all stages of growth (Figure 8.3; Naples 1982).

The masticatory stroke is anterolingually directed in most mammals (Hiiemae 1978). In addition to direct visual observation or frame-by-frame analysis of motion picture, videotape, or cineradiographic films, the general trend of the masticatory stroke has been determined for several mammals by examination of the orientation of wear striations across the occlusal surfaces of teeth. Representatives of the following groups have been studied: rodents (Hiiemae & Ardran 1968; Rensberger 1973, 1975); perissodactyls (Rensberger & Koenigswald 1980); selenodont artiodactyls (Every & Kuhne 1971; Greaves 1973); primates (Gordon 1982; Gingerich 1971; Kay & Hiiemae 1974); and marsupials (Crompton & Hiiemae 1970; Hiiemae & Crompton 1971). The direction of the masticatory stroke can be demonstrated precisely by examining the transitions between enamel and dentin in most mammals, and the hard and soft dentins in sloths. Several authors (Greaves 1973; Rensberger 1973; Costa & Greaves 1981) have demonstrated, using models, that the transition from the hard to soft material at the leading edge of a tooth forms a smooth, flush interface (Figure 8.4a). On the edge that trails during the masticatory stroke the transition from soft to hard material forms a noticeable step (Figure 8.4b). As this method has produced results in agreement with direct observation and analysis of masticatory movement patterns of living animals, it has been used to

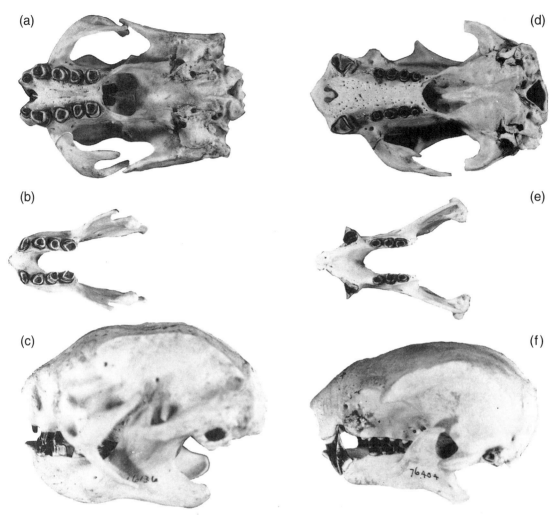

Figure 8.1. Occlusal views of (a) the adult palate and (b) mandible, and (c) lateral views of the maxillary and mandibular dentition of *Bradypus* aligned as at the end of the masticatory stroke. Parts (d-f) are corresponding views of *Choloepus*.

determine the direction of the masticatory stroke in extinct sloths (Naples 1987, 1990). However, teeth in all stages of wear are not typically available for study in extinct populations. It has not been possible, therefore, in studies of real teeth to demonstrate step-by-step how the small, newly erupted conical sloth teeth are transformed into teeth with sharp-edged cusps, facets, and distinct basins formed entirely through wear (Naples 1981, 1982). The present study undertakes to demonstrate how such a transition occurs by producing model sloth teeth correctly spaced in artificial maxillae and mandibles and oriented in positions equivalent to juvenile-through-adult wear stages, and by manipulating characteristics of the masticatory stroke to duplicate all stages in the development of the wear patterns seen in *Bradypus* and *Choloepus*.

EXPERIMENTALLY MODELING TOOTH WEAR FEATURES IN SLOTHS

Tooth wear and skull growth in living sloths

I examined the skulls and dentitions of 74 specimens of *Bradypus* sp. housed in the American Museum of Natural History (listed in Appendix 1, Naples 1982). These specimens were divided into three age classes, based upon skull length, tooth-row length (maxillary and/or mandibular), bone density, rugosity of cranial processes, and degree of sutural closure. These categories are estimates only; no specimens of known age were available. Small skulls were categorized as juvenile; adult-sized skulls lacking marked rugosities were considered young

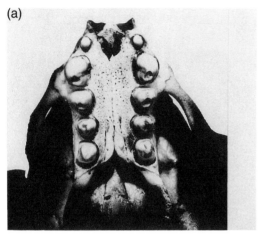

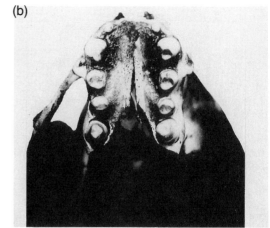

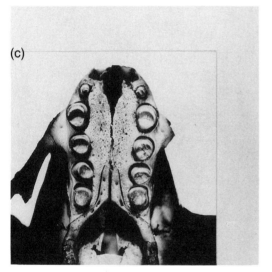

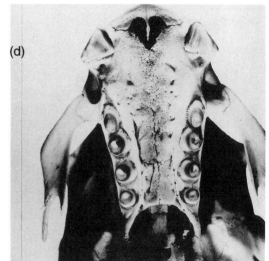

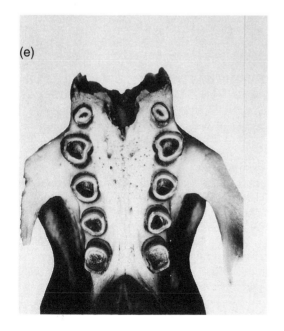

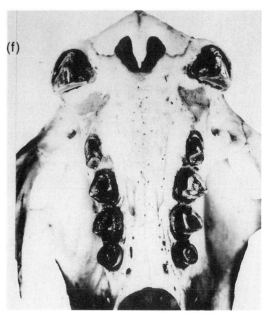

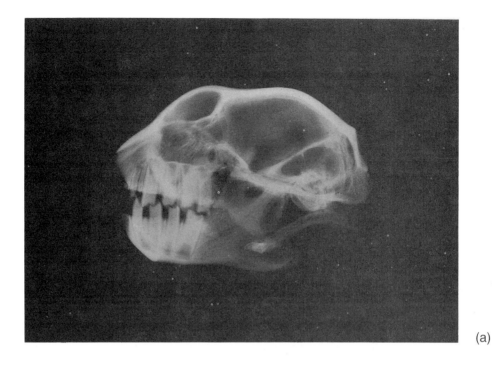

(a)

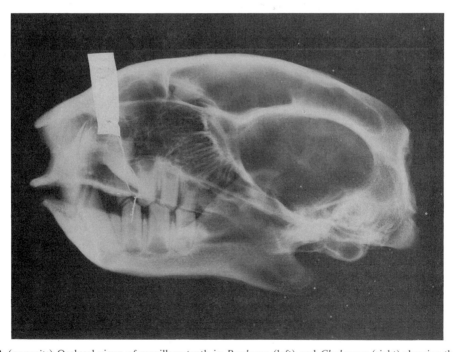

(b)

Figure 8.2. (*opposite*) Occlusal views of maxillary teeth in *Bradypus* (left) and *Choloepus* (right) showing three age classes into which specimens have been divided in this study. Parts (a-d) are juveniles, and (e) and (f) are adult sloths.

Figure 8.3. (*above*) Radiographs of (a) *Bradypus* and (b) *Choloepus* showing dentitions almost at full occlusion at end of masticatory stroke.

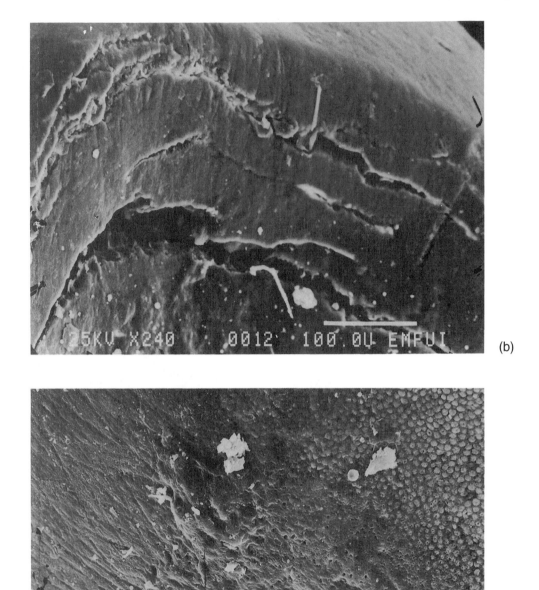

Figure 8.4. (a—lower panel) Scanning electron microscope (SEM) showing smooth, even transition area between hard, outer dentin (on lower left) and softer, inner dentin, on upper right, forming the basin at the leading edge of upper left third molariform tooth in *Bradypus*. Curved gouges worn into tooth reflect the anteromedial direction of jaw movement. (b) stepped transition area between softer, inner dentin (lower side) and harder, outer dentin in trailing edge (upper side) of same upper third molariform tooth in *Bradypus*. The anteromedially directed masticatory stroke would start at the lower right of (a) and continue across the tooth, in a vertical direction in (b). Scale bar = 100 µm.

adults; and adult-sized skulls with prominent rugosties were considered old adults. The illustrations showing the changes in tooth orientation from juvenile to adult were based upon the categories listed above (Figure 8.2a–c). The spacing and the angles at which the tooth models were embedded in the model jaws were determined, based upon observations of the typical orientations of teeth in specimens sorted into juvenile and adult categories.

Characteristics and types of teeth in *Choloepus* sp. were determined from observation of 98 specimens of skulls, mandibles, and dentitions housed in the American Museum of Natural History, and 13 specimens housed in the Field Museum of Natural History (listed in Appendix 1, Naples 1982). As no specimens of known age were available, animals were assigned to one of three age classes as was described for specimens of the genus *Bradypus* (Figure 8.2d–f).

Generating wear patterns on artificial teeth

To determine the details of all stages of the development of wear features on the teeth of *Bradypus* and *Choloepus,* model teeth were fabricated and subjected to wear using several patterns of masticatory movements. Plaster was used to simulate the hard outer dentin, and was layered around cores of colored chalk that represented the soft inner dentin of erupting molariform teeth in juvenile sloths. Prior to being dipped in plaster, the chalk cores were carved and sanded to the conical shape of the teeth of juvenile sloths. When the plaster outer shells were of equivalent thickness to the hard outer dentin in real teeth, they were sanded to the uniform rounded shape seen in the juvenile teeth. The hard outer dentin layer in the teeth of *Bradypus* differs from that of *Choloepus* because, in the former, the hard outer dentin shell is markedly thinner than in the latter.

Anterior chisel-shaped teeth *(Bradypus)* and large triangular-shaped caniniform teeth *(Choloepus)* were hand carved from chalk blanks of smaller and larger size. Plaster was layered over the chalk centers, and when dried to sufficient hardness, sanded into uniform rounded conical shapes. To emphasize the differences in wear facets caused by specific teeth, different colored chalk cores were used for each of the four upper and three lower molariform teeth. In this manner, wear into the colored core of any tooth could be uniquely identified, either on the tooth itself, or on the plaster shell or chalk core of the other tooth or teeth with which it occluded.

A mechanized apparatus was constructed to facilitate replication of the masticatory stroke in young, juvenile and adult sloths, and to produce replicate sets of worn teeth rapidly and efficiently (Figure 8.5). A motor-driven arm moved the model mandibular tooth row anteroposteriorly. Guides on the edges of the horizontal plate on which the mandibular model was mounted were able to alter the movement pattern from directly anteroposterior to mediolateral and any pattern between these directions of motion. The mandibular and maxillary models consisted of tooth models held in "alveoli" drilled into wooden blocks. The "maxillary" blocks were clamped to a supporting arm, which could be raised or lowered and held stationary throughout the experiments. The "mandibular" blocks were moved repeatedly along the same path at the correct distance from the maxilla to simulate tooth wear caused during mastication. The path of the mandibular tooth model remained identical for each test because it was rigidly attached to the supporting arm that moved along a guide directing the path of the plate. At the end of each cycle, the teeth disengaged, and the mandible was returned to the starting point automatically by completing a revolution of the controlling arm connected to the drive shaft of an electric motor which moved the plate upon which the mandibular model was mounted. As the teeth wore, the maxillary block was lowered by sliding it downward along an adjustable arm to ensure continued engagement of the teeth. This method ensured that similar amounts of tooth–tooth contact were maintained, and simulated growth of both the upper and lower model teeth.

Sets of two anterior chisel-shaped or caniniform and seven molariform (four maxillary and three mandibular) model teeth were used, for each sloth genus. The teeth were reoriented during the experiment, simulating the change in the angle of the teeth in the alveoli with growth. First, the teeth were placed in alveoli slightly angled, as they are in small juvenile sloths. Second, after the typical amount of wear seen on juvenile teeth appeared, the teeth were placed in alveoli angled as in large adults. In both cases, the teeth were spaced as in the observed juvenile and adult sloths. At all stages of wear, the teeth of both sloths occluded with the maxillary teeth a half-tooth length anterior to the mandibular teeth. Such spacing permits these sloths to have four maxillary molariform but only three mandibular molariform teeth. The small upper fourth molariform teeth occlude only with the posterior halves of the lower third molariform teeth. In some tree sloth specimens, the lower third molariform teeth are larger than in others. A trend to have an increased-size third molariform tooth is even more marked in such extinct sloth genera as *Glossotherium* (Naples 1989).

There was a slightly larger space between the anterior chisel-shaped teeth and the molariform-tooth row in adult *Bradypus* than between teeth of the molariform row, but with some anteroposterior

Figure 8.5. Lateral view of a mechanical apparatus designed to study wear in models of sloth teeth. Mandible is propelled by a constant speed motor, and the masticatory orbit is determined by a template that can be altered to produce movements ranging from directly anterior-posterior to completely mediolateral.

movement of the mandible, it was still possible for the anterior chisel-shaped teeth to intercuspate with the anterior molariform teeth. Spacing of the anterior caniniform teeth of *Choloepus* differed more with respect to the cheek-tooth row than for any other tooth type at all stages of either growth or wear. In *Choloepus* the growth of the anterior maxilla and mandible is so great that the caniniform teeth rapidly become isolated from all possibility of intercuspation with any of the molariform teeth (Figure 8.2d–f). Therefore, separate experimental chewing cycles were conducted with all sets of teeth to occlude the anterior teeth.

Five sets of teeth in each of the orientations described above were occluded according to one of four patterns of mandibular movement. The first experimental masticatory stroke was directly posterior to anterior, while the second was directly labial to lingual. The last two masticatory strokes used were anterolingual, but with the emphasis on the anterior component of motion in one, and on the lingual component of motion in the other.

Scanning electron micrographs revealed the details of wear-feature development on occlusal surfaces of real juvenile and adult sloth teeth. The teeth were magnified from 43 to 330 times to show the

orientation, depth, and shape of striations formed by wear. These observations were used to help determine the four patterns of mandibular movement tested in this study.

COMPARISON OF WEAR FEATURES ON ARTIFICIAL AND REAL SLOTH TEETH

The generation of wear features on the anterior chisel-shaped teeth of *Bradypus* and the caniniform-shaped teeth of *Choloepus* differed from wear-feature development on the molariform teeth of both sloths. Therefore, these development patterns will be discussed separately after treatment of the wear changes observed in the molariform teeth of both sloths.

In all of the experimental masticatory stroke directions tested, the model teeth showed less of a difference of depth between the hard outer plaster and the inner chalk core than is seen between the hard outer dentin and soft inner dentin core in real teeth. However, the experimental apparatus was only able to test the formation of wear patterns from tooth–tooth contact, and unable to assess the role of tooth–food–tooth contact in the formation of wear features. It has been demonstrated that a bolus of food may remain within a concavity in the occlusal surface of a tooth for more than one masticatory stroke, and this prolonged contact can cause significant abrasion of the surface in addition to that caused by tooth–tooth contact (Hiiemae & Ardran 1968; Hiiemae 1978; Greaves 1973; Costa & Greaves 1981; Gordon 1982). The same pattern of basin formation was obtained with all of the masticatory power-stroke orientations used in models of adult sloth teeth.

In model teeth of juveniles of both sloth genera, the maxillary or mandibular tooth of an occluding pair became concave, whereas its opposite remained slightly convex, somewhat flattened, or slightly concave. This pattern of uneven, somewhat irregular wear is also observed in the real teeth of juvenile sloths of both genera (Figure 8.6). Not until the sloths become larger juveniles or young adults do the wear patterns on all the teeth become more regular and typical of those seen in fully adult specimens. Both natural and model teeth that had flattened or slightly convex wear patterns showed no other features of interest in the ontogeny of wear patterns. However, in both real and model teeth showing concavities, raised ridges remained along the tooth margins parallel to the orientation of the masticatory stroke (Figure 8.7). The direction of the experimental masticatory stroke used with the tooth models was solely responsible for determining the location of the ridges and whether or not they were symmetrical. It is likely that ridge formation, location, and orientation was determined by similar masticatory movements of real teeth.

As the initial break through the harder outer dentin and early formation of basins began in the conical tips of the teeth in young sloths, it was clear that there was a large relative difference in resistance to wear between the outer harder and softer inner layers of dentin. The wear structures formed on the artificial teeth clearly showed that once the harder outer layer was breached, the softer core material began to erode much more quickly. These relative differences of rate at which tooth features appeared and changed with wear provided for (1) an initial period of slow wear (prior to wearing through of the harder outer dentin), (2) an intermediate period of very rapid wear, and therefore, a rapid change in tooth shape (once the hard outer layer was breached) and (3) a final, slower period of steady wear as the animal reached full adult size (corresponding with the wearing away of the smaller diameter, conical tips of the teeth). This change allowed the tooth surfaces in contact to be approximately the same diameter, or larger than the diameter of the roots of the ever-growing teeth.

Effects of masticatory stroke orientation on model sloth teeth

The four different masticatory-stroke directions tested in these experiments demonstrated that observations from films of live sloths chewing provided sufficient initial information to determine the pattern of mandibular movement and thus wear-feature formation. Neither of the masticatory strokes solely in anterior or labial directions produced any wear features similar to those seen on real sloth teeth. When the upper and lower teeth were aligned along a single vertical axis, both anterior and labial directions of mandibular movement produced flattened occlusal surfaces on all teeth. These orientations of masticatory stroke did result in formation of some basins and ridges on the teeth when the vertical axes were displaced by a half-tooth length, as is seen in juvenile tree sloths, but the wear features produced did not resemble those of real teeth. As these patterns did not resemble any stages of wear seen on teeth of live sloths, neither the anterior only nor the labial only masticatory strokes were tested with teeth oriented in the oblique patterns seen in the maxillae and mandibles of large juvenile and adult *Bradypus* and *Choloepus*.

Only juvenile sloths of both genera have teeth oriented approximately vertically in the alveoli. As the animals grow, the teeth become increasingly more obliquely oriented (Figure 8.3), although these orientations differ along the length of the tooth rows as well as between the two sloth genera. In *Bradypus* the maxillary teeth tilt inward slightly, while the mandibular teeth angle inward to a slightly greater degree. In *Choloepus* both tooth rows tilt inward to a greater degree than is exhibited in *Bradypus*, particularly in the mandibular row. In both sloths, the maxillary teeth are spaced a half-tooth anterior to the mandibular teeth, and these features, together with the direction of masticatory stroke determine the pattern of wear-feature production on the teeth of each sloth genus.

Molariform tooth wear-feature formation

The spacing of the molariform teeth differs only slightly from small juveniles to mature adults in both *Bradypus* and *Choloepus*. However, because only the tips of the vertically oriented conical teeth are in occlusion in small juveniles, the effective areas of contact are more widely spaced than in larger juveniles and adult sloths. The small size of the conical tooth tips in juvenile sloths results in the restricion of the wearing surfaces of each tooth to interaction with the single opposing upper or lower tooth (Figure 8.8). This is the only type of tooth–tooth contact in juvenile sloths, and results first in the formation

(a)

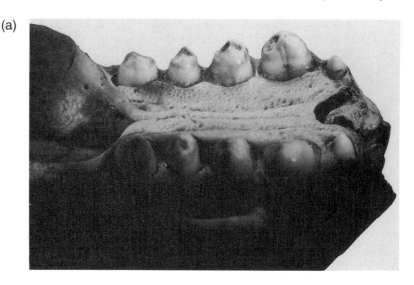

(b)

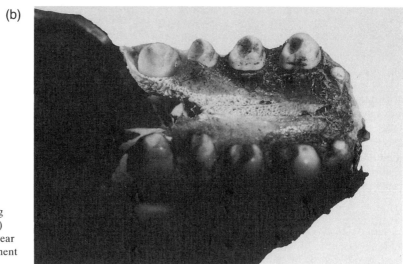

Figure 8.6. Photographs in oblique ventral view of young juvenile (a) *Bradypus* and (b) *Choloepus* to show uneven wear patterns and pitting development on the teeth.

of basins in the center of the teeth. Initially, the tooth tips wear slowly; however, once the hard outer dentin (or plaster in the models) has been abraded away, the basins enlarge and deepen rapidly. In larger juvenile and adult sloths, tooth-wear surfaces enlarge so greatly that the slightly tilted teeth in the mandibles and maxillae eventually occlude with more than one tooth throughout the masticatory stroke. The second set of wear features formed in both sloth genera are raised ridges surrounding the central basins. At this stage of tooth wear, the ridges are neither sharp-edged nor separated from the central tooth basins by steps. In some specimens, both of model and real teeth, striations indicating the orientation of mandibular movement were visible. The ridges formed both because the more resistant outer shells were worn more slowly than the softer centers and because the outer ridges received less tooth–

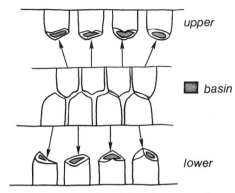

Figure 8.7. Depiction of model teeth worn to show ridges formed at edges of central basin.

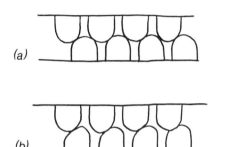

Figure 8.8. Depiction of model teeth, showing that only tips of teeth intercuspate in young juvenile sloths. (a) *Bradypus* and (b) *Choloepus.*

tooth contact, being engaged with teeth of the opposite occlusal surface only at either the start or the end of the intercuspation stroke.

The third and final type of major feature to appear are the sharp-edged wear facets in real teeth of both sloths. These features arise as the marginal ridges acquire ever steeper sides as the tooth basins deepen. At this time, the smooth, flush transitions between leading edges and stepped transitions to the trailing edges also appear. All maxillary and mandibular cheek teeth (except the small, upper fourth molariform teeth) occlude with two of their opposites (Figure 8.3). Thus, both an anterior and a posterior slope is formed on each tooth except the upper fourth molariform tooth. As wear increases, each of the "cusps" becomes more pointed and sharp-edged. In both sloths, the direction of the masticatory stroke is anterolingual (Naples 1982), but in *Bradypus* the anterior component of motion is emphasized while the lingual component is dominant in *Choloepus.* A structural difference that also influences the appearance of the worn adult teeth in the two sloths is that the harder, outer dentin shell is thinner in *Bradypus* than in *Choloepus.* The more anteriorly directed masticatory stroke in *Bradypus* brings the tooth ridges of the opposing occlusal surfaces into contact for a greater proportion of the

masticatory stroke, thus subjecting them to a greater amount of wear than is experienced by those of *Choloepus.* As these ridges are also thinner than those of *Choloepus,* they are less able to resist either tooth–tooth or tooth–food–tooth abrasion, and so are less likely to be able to retain as steeply angled cusps.

Formation of tooth wear features on anterior chisel-shaped teeth

In *Bradypus* the anterior chisel-shaped teeth acquire some of their wear features in a manner similar to those of the molariform teeth, although in adults the posterior faces of the lower anterior teeth only occlude at the start of intercuspation with the lower first molariform teeth. This interaction forms the basins on the posterior faces of the lower anterior chisel-shaped teeth. Although these teeth only intercuspate at the beginning of the masticatory stroke, there is no gap in time between the finish of intercuspation of these teeth and the initiation of intercuspation of the molariform teeth in most sloths; in some animals there may be partial overlap of the time that these teeth intercuspate. From manipulation of dried specimens and observations of film of some live *Bradypus* chewing, however, I have found that it is not always possible for both types of teeth to be in occlusion simultaneously.

Toward the end of the masticatory stroke, the upper anterior chisel-shaped teeth occlude with the lower anterior chisel-shaped teeth, often forming a distinct groove in the anterior hard dentin face of the lowers. This action also wears the surfaces of the small peg-shaped upper anterior chisel-shaped teeth to a flattened face, worn down more on the lingual than the labial side. Some of these teeth show development of a small basin, as was demonstrated by the model, which formed whenever a resistant ridge wore across a surface with a softer center. The anterior teeth in sloths are primarily used for biting, rather than chewing, and therefore this basin in real sloth teeth may have been enhanced by abrasion against food as well as the lower anterior chisel-shaped tooth.

Formation of tooth wear features on caniniform teeth

In *Choloepus,* the anterior portions of the maxillae and mandibles in young juveniles grow more rapidly than the rest of the face, and the caniniform teeth lose the ability to occlude with the molariform teeth very quickly, if ever this occurs at all. From this stage of growth onward, occlusion of the caniniform teeth is independent of molariform tooth occlusion, and, as the caniniform teeth elongate, they

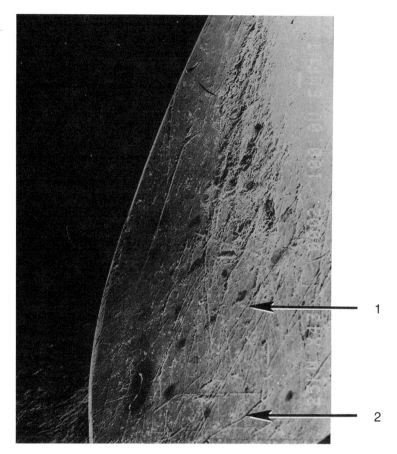

Figure 8.9. Scanning electron micrograph of anteriorly oriented occlusal face of a lower right caniniform tooth of *Choloepus* (labial is to left; lingual is to right). Gouges on surface occur in two main orientations of mandibular movement: (1) opening and closing of jaw (= longitudinally along the surface), and (2) medial movement of jaw that typically occurs at end of masticatory stroke (= anteromedially) which appears as curved scratches. Scale bar = 100 μm.

intercuspate when the mandible is opened to such an angle that the shorter molariform teeth have not yet neared intercuspation. The mandible in *Choloepus* shows more mediolateral movement than does that of *Bradypus*. This observation is supported by the crescent-shaped glenoid fossa in which the mandibular condyle rests in *Choloepus*, a shape that would allow the labiolingually widened and curved mandibular condyle on one side to slide anteriorly. The mental symphysis in sloths is fused; therefore, anterior movement of the labial surface of one mandibular condyle would require the mandible to swing medially, and the opposite condyle to move posteriorly on the labial side. The morphology of the condyles and glenoid fossae in *Choloepus* clearly demonstrate the capability to move in this fashion. In *Bradypus*, in contrast, the glenoid fossa permits mostly anteroposterior movement of the mandible (Naples 1982). The ability, in *Choloepus*, to generate a relatively greater amount of labiolingual mandibular movement also correlates with the ability to move the lower caniniform teeth into occlusion at a separate time in the masticatory stroke from when the molariform teeth occlude. From manipulation of dried specimens as well as observa-

tions of live sloths and analysis of cineradiographic film of sloths chewing, it is clear that in some specimens the caniniform teeth cannot be in occlusion at the same time as the molariform teeth. Therefore, the single occlusal face (posterior on upper and anterior on lower caniniform teeth) is created by wear of only these teeth against each other, or by abrasion from the objects these sloths bite. The orientation of wear striations on these teeth are either longitudinal scratches aligned along the orientation of the masticatory stroke, or shorter, curved grooves which are across the occlusal face, reflecting the anteromedial component of the masticatory stroke in *Choloepus* (Figure 8.9).

DISCUSSION

Wearing away of the smaller conical tips of the teeth alone would result in a larger surface area contact, but because the teeth also become increasingly more obliquely oriented in their alveoli by differential growth of the mandibles and maxillae, the surface area exposed for contact is even larger in larger and older animals. Tooth–food–tooth contact is responsible for additional deepening and definition of wear features on sloth teeth, as has been seen in other animals. The present set of experiments tested only the effects of tooth–tooth contact, which probably explains the somewhat reduced level of differentiation between the components of the wear features. Tooth–food–tooth contact as well as abrasion caused by tooth–tooth contact explains the greater definition of wear features seen in real *Bradypus* and *Choloepus* teeth in comparison to features produced by tooth–tooth wear on the models. Even using tooth–tooth contact only, however, only the harder outer dentin wore unevenly. In some places irregular pits were formed, while in others an area remained more resistant to wear for longer than expected (the expectation was that the model would generate essentially even wear on the artificial teeth). This fact was not explained by the materials used in the models, as the plaster used for the outer hard layer of dentin was homogeneous. However, it was observed that small flakes of plaster rubbed off during the occlusion trials, and these may have contributed to the gouging effect seen, as the flakes frequently adhered to the occluding surfaces subsequent to their separation from the tooth surface itself. In this manner, the plaster flakes mimicked to a small degree the effect of a bolus of food remaining on the occlusal surface for more than one masticatory stroke. However, this effect was always more limited in mass of material than would have been true for a partially masticated food bolus, and bulky or extremely fibrous food would have greatly increased this effect on the teeth.

The presence of a softer center material in the models predisposed the harder outer plaster to crumble slightly more rapidly toward the center of the tooth, thus explaining the pits often seen in the center of juvenile tooth shells, prior to breakthrough into the softer core of dentin. Once the breakthrough has occurred, the inner aspect of the hard outer dentin shells tends to wear slightly more easily than other areas, perhaps because it has less rigid support than the middle or leading edges of the hard outer dentin.

The order of the formation of wear features, (i.e., basins first, ridges parallel to the direction of the masticatory stroke second, and pointed, sharp-edged cusps and anterior and posterior wear facets last) is dictated in both sloths by the relatively small occlusal contact area in juvenile sloths. As wear on the teeth continues, the small conical tips disappear, and the occlusal surface approaches the maximum dimension of the tooth, that of the root. In some cases, the surface exposed to wear is even greater than the maximum diameter of the root, as the wear surface occurs at a slant relative to the angle of growth of the tooth. Wearing away of the small conical tips

alone results in a larger surface area for surface contact, but because the teeth also become increasingly more obliquely oriented in their alveoli by the differential growth (greater anteriorly) of the mandibles and maxillae, the surface area exposed for contact becomes ever more extensive in larger and older sloths. Nevertheless, the occlusal surface area in the largest sloths is typically somewhat less than in many herbivores and much less than in some hypsodont mammals, such as elephants (Maglio 1973), which have almost continuous occlusal surfaces between teeth.

REFERENCES

Bown, T.M., & Kraus, M.J. 1979. Origin of the tribosphenic molar and metatherian and eutherian dental formulae. In *Mesozoic Mammals*, ed. J.A. Lillegraven, Z. Kielan-Jaworowska, & W.A. Clemens, pp. 172–181. Los Angeles: University of California Press.

Cassiliano, M.L., & Clemens, W.A. 1979. Symmetrodonta. In *Mesozoic Mammals*, ed. J.A. Lillegraven, Z. Kielan-Jaworowska, & W.A. Clemens, pp. 150–161. Los Angeles: University of California Press.

Clemens, W.A. 1979. Marsupialia. In *Mesozoic Mammals*, ed. J.A. Lillegraven, Z. Kielan-Jaworowska, & W.A. Clemens, pp. 192–220. Los Angeles: University of California Press.

Costa, R.L., & Greaves, W.S. 1981. Experimentally produced tooth wear facets and the direction of jaw motion. *Journal of Paleontology* 55, 635–638.

Crompton, A.W., & Hiiemae, K.M. 1970. Molar occlusion and mandibular movements during occlusion in the American opossum, *Didelphis marsupialis* L. *Zoological Journal of the Linnean Society* 49, 21–47.

Crompton, A.W., & Jenkins, F.A., Jr. 1978. Mesozoic mammals. In *Evolution of African Mammals*, ed. V.J. Maglio & H.B.S. Cooke, pp. 46–55. Cambridge: Harvard University Press.

Crompton, A.W., & Jenkins, F.A., Jr. 1979. Origin of mammals. In *Mesozoic Mammals*, ed. J.A. Lillegraven, Z. Kielan-Jaworowska, & W.A. Clemens, pp. 59–73. Los Angeles: University of California Press.

Every, R.G., & Kuhne, W.S. 1971. Bimodal wear of mammalian teeth. In *Early Mammals*, ed. D.M. Kermack & K.A. Kermack, pp. 23–38. London: Academic Press.

Gingerich, P. 1971. Dental function in the Paleocene primate *Plesiadapis*. In *Prosimian Anatomy, Biochemistry and Evolution*, ed. R.D. Martin, G.A. Doyle, & A.C. Walker, pp. 531–541. London: Duckworth & Company.

Gordon, K.D. 1982. A study of microwear on chimpanzee molars: implications for dental microwear analysis. *American Journal of Physical Anthropology* 59, 195–215.

Greaves, W.S. 1973. The inference of jaw motion from tooth wear facets. *Journal of Paleontology* 47, 1000–1001.

Hiiemae, K.M. 1978. Mammalian mastication: a review of the activity of the jaw muscles and the movements they produce in chewing. In *Development, Function and Evolution of Teeth*, ed. P.M. Butler & K.A. Joysey, pp. 359–398. London: Academic Press.

Hiiemae, K.M., & Ardran, G.M. 1968. A cineradiographic

study of feeding in *Rattus norvegicus*. *Journal of Zoology, London* 154, 139–154.

Hiiemae, K.M., & Crompton, A.W. 1971. A cinefluorographic study of feeding in the American opossum *Didelphis marsupialis*. In *Dental Morphology and Evolution*, ed. A.A. Dahlberg, pp. 299–334. Chicago: University of Chicago Press.

Illiger, C. 1811. *Prodromus systematis mammalium et avium additis terminus zoographicus utriudque classis*. Berlin: C. Salfeld.

Kay, R.F., & Hiiemae, K.M. 1974. Jaw movement and tooth use in Recent and fossil primates. *American Journal of Physical Anthropology* 40, 227–256.

Kielan-Jaworowska, Z., Bown, T.M., & Lillegraven, J.A. 1979. Eutheria. In *Mesozoic Mammals*, ed. J.A. Lillegraven, Z. Kielan-Jaworowska, & W.A. Clemens, pp. 221–258. Los Angeles: University of California Press.

Kielan-Jaworowska, Z., Eaton, J.G., & Bown, T.M. 1979. Theria of metatherian–eutherian grade. In *Mesozoic Mammals*, ed. J.A. Lillegraven, Z. Kielan-Jaworowska, & W.A. Clemens, pp. 182–191. Los Angeles: University of California Press.

Kraus, M.J. 1979. Eupantotheria. In *Mesozoic Mammals*, ed. J.A. Lillegraven, Z. Kielan-Jaworowska, & W.A. Clemens, pp. 162–171. Los Angeles: University of California Press.

Krause, D.W. 1982. Jaw movement, dental function and diet in the Paleocene multituberculate *Ptilodus*. *Paleobiology* 8, 265–281.

Kron, D.G. 1979. Docodonta. In *Mesozoic Mammals*, ed. J.A. Lillegraven, Z. Kielan-Jaworowska, & W.A. Clemens, pp. 91–98. Los Angeles: University of California Press.

Maglio, V.J. 1973. Evolution of mastication in the Elephantidae. *Evolution* 26, 638–658.

Martin, L.D. 1980. Functional morphology and the evolution of cats. *Transactions, Nebraska Academy of Science* 8, 141–154.

Naples, V.L. 1981. "Cusp" development in sloth "cuspless" teeth. *Journal of Dental Research* 60A, 433.

Naples, V.L. 1982. Cranial osteology and function in the tree sloths, *Bradypus* and *Choloepus*. *American Museum Novitates* 2739, 1–41.

Naples, V.L. 1985. Form and function of the masticatory musculature in the tree sloths, *Bradypus* and *Choloepus*. *Journal of Morphology* 183, 25–50.

Naples, V.L. 1986. The morphology and function of the hyoid region in the tree sloths, *Bradypus* and *Choloepus*. *Journal of Mammalogy* 67, 712–724.

Naples, V.L. 1987. Reconstruction of cranial morphology and analysis of function in the Pleistocene ground sloth *Nothrotheriops shastensis* (Mammalia, Megatheriidae). *Contributions in Science* 389, 1–21.

Naples, V.L. 1989. The feeding mechanism in the ground sloth, *Glossotherium*. *Contributions in Science* 415, 1–23.

Naples, V.L. 1990. Morphological changes in the facial region and a model of dental growth and wear pattern development in *Nothrotheriops shastensis*. *Journal of Vertebrate Paleontology* 10, 372–389.

Parker, W.K. 1885. On the structure and development of the skull in the Mammalia. Part II. Edentata. *Philosophical Transactions of the Royal Society of London* 1885, 1–119.

Rensberger, J.M. 1973. An occlusion model for mastication and dental wear in herbivorous mammals. *Journal of Paleontology* 47, 515–528.

Rensberger, J.M. 1975. Function in the cheek tooth evolution of some hyposdont geomyoid rodents. *Journal of Paleontology* 49, 10–22.

Rensberger, J.M., & Koenigswald, W. von 1980. Functional and phylogenetic interpretation of enamel microstructure in rhinoceroses. *Paleobiology* 6, 477–495.

Romer, A.S. 1966. *Vertebrate Paleontology*. Chicago: University of Chicago Press.

Ryan, A.S. 1979. Wear striation direction on primate teeth: a scanning electron microscope examination. *American Journal of Physical Anthropology* 50, 155–168.

Scott, W.B. 1937. *A History of the Land Mammals of the Western Hemisphere*. London: Macmillan.

Sicher, H. 1944. Masticatory apparatus of the sloths. *Fieldiana: Zoology* 29, 161–168.

Thewissen, J.G.M., & Domning, D.P. 1992. The role of phenacodontids in the origin of the modern orders of ungulate mammals. *Journal of Vertebrate Paleontology* 12, 494–504.

Walker, A.C. 1978. Prosimian primates. In *Evolution of African Mammals*, ed. V.J. Maglio & H.B.S. Cooke, pp. 90–99. Los Angeles: University of California Press.

Wible, J.R. 1991. Origin of the Mammalia: the craniodental evidence reexamined. *Journal of Vertebrate Paleontology* 11, 1–28.

Winge, H. 1941. *The Interrelationships of the Mammalian Genera*. Copenhagen: C.A. Reitzel Forlag.

9

Determination of stresses in mammalian dental enamel and their relevance to the interpretation of feeding behaviors in extinct taxa

JOHN M. RENSBERGER

ABSTRACT

Inferences about feeding behaviors in fossil mammals have traditionally utilized associations of tooth shape and recently patterns of microwear with diets in modern mammals. These associations exist because of different ways that food (or associated materials) and teeth interact during food processing, but the mechanics of these relationships are still imperfectly understood. Integral to the mechanics of chewing is the transfer of stress to the food at levels needed to fracture the items, and, therefore, the strength of the dental structures is a limiting factor in this mechanism. The occlusal shapes that achieve higher occlusal stresses increase the likelihood of dental fracture, as does increase in body size of the mammal, and these relationships lead to selection for stronger dental materials. Enamel has been responsive to selection for increased strength, and the derived enamel microstructures in Cenozoic mammals provide information about the stresses that the teeth are adapted to withstand. Differences in both direction and relative magnitude of stress are reflected in the evolutionary responses of the microstructure. The evolution of the enamel has been sensitive to different directions of stresses in different teeth, in different regions of a tooth, and to differences through the enamel thickness. Although absolute values for the magnitudes of stresses are not as yet known, differences in relative magnitude of stresses that can be withstood by a given microstructure can be seen in different regions of individual cusps in the early perissodactyl *Hyracotherium*. In the spotted hyaena *Crocuta crocuta*, which apparently develops exceptionally high occlusal stresses, a highly derived microstructure occurs at the position in the canine where the stresses are highest. The microstructures in the upper canine of *Crocuta* and in the lower canine tusk of the astrapothere *Parastrapotherium* suggest behavioral uses of these teeth that are not evident from other sources.

Two independent sources of information about stresses are useful in confirming the relationships between stress and microstructure: Finite element modeling enables the calculation of stresses under postulated loading conditions, and enamel cracks provide empirical confirmation of the directions of premortem stresses.

INTRODUCTION

Cenozoic mammals have acquired a diverse array of dental structures. Inferences about the dietary behaviors of fossil taxa have been made largely on the basis of associations between diet and dental morphology in living taxa. In recent years, the morphologic foundation for such inferences has been extended to include the microscopic patterns of abrasion on the surfaces of dental materials, especially the enamel (see Teaford 1988 for a review of the literature).

Both the shapes of teeth and the microwear patterns are related to diet through the mechanical (and sometimes chemical) interactions between teeth and food objects. Dental shapes are in part related to dietary behaviors because no single pair of surfaces can be uniformly efficient at the different tasks performed by teeth: subdividing a wide variety of food materials, grasping variously shaped objects, or killing the many different prey animals mammals feed upon. Microwear is related to diet to the extent that foods of different physical or chemical properties wear or corrode dental materials in different ways or different degrees. Differences in occlusal mechanics, which are dependent on dietary specialization, affect patterns of microwear (Gordon 1982; Rensberger 1986; Stern, Crompton, & Skobe 1989; Teaford & Byrd 1989; Maas 1991). These two aspects of teeth, shape and microwear, have a complex interrelation to one another and to the mechanics of occlusion.

A general model relating all of these aspects to dietary behavior might be built upon the basic mechanical relationships between the dental structures and the objects they contact, and there have been formulations of some elements of such a model (Rensberger 1973; Lucas 1982; Van Valkenburgh & Ruff 1987).

Mechanical processing and occlusal stress

The mechanical subdivision of food particles is generally important in mammals because a high rate of nutrient intake and short digestive time are necessary to sustain endothermy. Mechanical processing is particularly critical in herbivores because the cellulose in plant cell walls resists digestion, and the cells must be fractured to expose their nutrients (Janis & Fortelius 1988). In both ruminant and nonruminant digestive systems, the dentition subdivides fibers to a uniformly small size that is characteristic of each taxon (Rensberger 1973; Laub 1992). The stresses between opposing dental surfaces apparently exceed the rupture strength of the plant fibers in every chewing stroke. This high degree of efficiency is indicated by (1) the small particle size that clusters around a modal value for each taxon that has been studied – *Equus* and *Bos* (Rensberger 1973), Mammutidae and Elephantidae (Laub 1992) – and (2) and the small number of chewing strokes to produce this size (Rensberger 1973).

The projections of cusps, lophs, and enamel edges reduce the total area of tooth–food–tooth contact during mastication, and thereby concentrate stresses at the sites of contact (Rensberger 1973, 1975; Lucas 1982; Fortelius 1985). The stresses due to forces generated by the masticatory muscles are sufficiently magnified at the points of contact to subdivide the food. The penalty paid is increased attrition through wear or fracture at these points, and the requirements of dental durability therefore limit the tooth shapes and the degree of optimization for chewing or biting efficiency, as has been shown, for example, in carnivore canines (Van Valkenburgh & Ruff 1987; fig. 6 in Pfretzschner 1988).

Occlusal stress and dental strength

A major change in tooth form and mechanics occurred in the transition between late Cretaceous therians and early Tertiary ungulates. In the late Cretaceous insectivores such as *Procerberus* and *Zalambdalestes,* the cusps have steep sides and wear striae, indicating that the power stroke in chewing was at a high angle. Shearing occurred as the jaws were closing, and the occlusal edges adjacent to the nearly vertical surfaces passed one another in a translating movement. Edges were critical because they concentrated the stress that is transferred to the food material.

Later, in the ungulates, the occlusal surfaces again contain edges, and the shearing of food materials occurs as the edges are translated past one another, but the shape of the mechanism is different. In both cases the mechanism involves translation and edges that concentrate stress at the tooth–food interface.

The main distinction is that in advanced ungulates the translation occurs in a plane that is more nearly horizontal and the shapes of the crowns are modified to allow this change in direction (Butler 1972, 1973; Rensberger, Forstén, & Fortelius 1984).

In the transition between insectivores and early Tertiary ungulates, the changes include a reduction and loss of intercusp crests, a reduction of vertical relief, and a reduction in the slope of the sides of the cusps (Figure 9.1a,b; Rensberger 1986, 1988; Janis & Fortelius 1988). For example, in the insectivores *Procerberus* and *Zalambdalestes,* the anterior and posterior sides of the paracone form angles ranging from about 26° to 47°, but in early Tertiary ungulates this increases to a range of about 50° to 70°. In many later ungulates in which shearing blades were reestablished, the movement became almost horizontal and the number of edges multiplied, so that only enough vertical relief is present as required to guide the movement (Figure 9.1c). In these later taxa attrition rates are such that higher relief cannot be maintained.

Correlating with the early changes were increases in size (Figure 9.1d), which in the early Paleocene represents one of the highest sustained evolutionary rates for mammals (Sloan 1987). The long dominance of thin stress-concentrating mechanisms in Mesozoic therians, the reduction of these structures in early ungulates in correlation with the increase in body size, and the eventual reestablishment of stress concentrating edges in occlusal surfaces of lower relief in ungulates of the later Cenozoic suggest that a size-related constraint prevented the early ungulates from retaining edge-dominated teeth. Dental strength seems to have been this limiting factor.

Because the maximum forces that can be exerted by the masticatory muscles depend largely on their cross-sectional areas, the maximum chewing force in the much larger early ungulates was greatly increased compared to that in the insectivores. Occlusal areas of the cheek teeth tend to scale isometrically with body size in mammals (Fortelius 1985). Therefore the relationship between occlusal force and occlusal area should remain roughly the same with increase in body size, so that an increase in body size would not in itself be expected to change average occlusal stresses. However, because the chewing forces in the early herbivores were much larger than in the insectivores, greatly increased dental stress would be developed when stress concentration occurred, as, for example, when a small, hard object was trapped between opposing teeth, or during certain variations in occlusal movements. The decrease in the cusp aspect ratio of height to width that occurred in early ungulates reduces the stresses and allows the cusps to tolerate larger muscular forces before fracture.

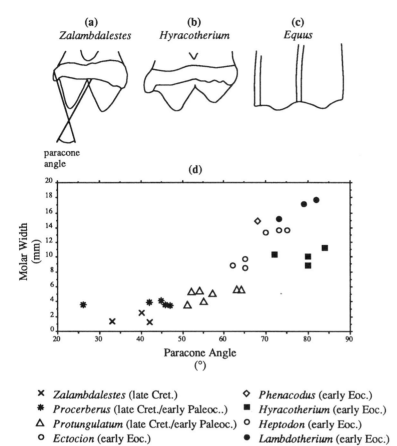

Figure 9.1. Change in cusp shape in early Cenozoic mammals, as shown by differences in the paracone angle. Buccal views of °an upper molar in (a) a late Cretaceous insectivore (*Zalambdalestes*), (b) a primitive Eocene perissodactyl (*Hyracotherium*), and (c) a late Cenozoic perissodactyl (*Equus*). Broader cusps in (a) than (b) resist larger vertical occlusal forces. In (c) chewing mechanism is changed to accommodate horizontal shear, which does not require high vertical relief. (d) Change in paracone angle with increase in size (tooth width) from insectivores to early ungulates.

The change to blunter occlusal structures would, however, have resulted in reduced stress concentration at the tooth–food interface for most food materials (excepting extremely stiff food materials, such as hard seeds, which have the effect of concentrating stress somewhat independent of the shape of the surface [Lucas 1982]). The occlusal structures in the early ungulates would have been inefficient in the mastication of fibrous plant materials, in contrast to structures in modern browsing and grazing ungulates, suggesting that such resources could not be exploited to the extent they were by the later ungulates.

If the diversity of materials that could be masticated efficiently was limited by the early ungulate cusps, such a constraint must have increased the selective value of any innovation in dental materials that would increase resistance to fracture and allow greater occlusal stress concentration. Stronger materials would allow thinner, sharper structures that would generate higher occlusal stresses needed to subdivide tougher structural tissues. Similar advantages would accrue to carnivorous mammals requiring sharp cusps or carnassial blades because carnivory includes the special hazard of teeth encountering bones (Van Valkenburgh & Ruff 1987; Van Valkenburgh 1988). As we shall see below, in a section on

the relationship between stresses and microstructure, there is good evidence that ungulates and carnivores did develop stronger enamels.

Sensitivity of enamel to stress

Enamel is the most resistant to wear of the dental materials, due to its high mineral content – 92 percent by volume compared with 69 percent for dentin (Waters 1980). Because wear blunts sharp structures, enamel is critical in the maintenance of cusp tips and thin edges, and this must be the reason why enamel or enameloid is so ubiquitous in vertebrate teeth. On the other hand, enamel is more brittle than dentin, which means it tends to fracture under less deformation or has a lower work of fracture. The work of fracture in human enamel in its weakest direction has been measured as 13 Jm^{-2} compared to 270 Jm^{-2} for dentin in its weakest direction, a twentyfold difference (Rasmussen et al. 1976). The ability of occlusal structures to concentrate stress on food items depends on their sharpness, but as stresses increase with sharpness, so does the likelihood of the edge fracturing. Maintaining sharp edges is, therefore, limited by the brittleness of enamel.

STRESS AND FRACTURE

Behavior of brittle materials

Many engineering materials undergo plastic deformation before fracturing. For example, in most metals under stress, adjacent layers of atoms slide over one another in a process called dislocation. If a ductile metal rod is subjected to axial (longitudinal) tension, it will deform first by dislocations aligned about 45° with the axial direction, which indicate that shearing is occurring. As this deformation proceeds, the rod will become thinner ("neck") in the region of the dislocations, and finally the stress concentration in the thin part causes it to fracture on a ragged surface normal to the axial direction. On the other hand, if a rod composed of a brittle solid that responds elastically to stress up to failure is loaded axially in tension, it fractures in a plane normal to the axis without any prior necking. Therefore, in contrast to the effects of resolved shear stress, which is the criterion for plastic deformation, it is the maximum tensile stress that is the criterion for fracture in brittle solids (Kingery, Bowen, & Uhlmann 1976). Brittle materials tend to be stronger in compression than in tension, because compression forces atoms together whereas tension pulls them apart. Additionally, it is believed that there are numerous microscopic cracks in brittle materials, and the effect of tension is to expand these cracks, whereas compression has the opposite effect.

Fracturing in brittle solids occurs through crack propagation, which requires a critical stress at the crack tip. For a completely brittle solid in which crack propagation occurs elastically without plastic flow, the required stress for a crack to propagate is:

$$\sigma \cong \sqrt{\left[\frac{2E\gamma}{\pi c} \left(\frac{\rho}{3a} \right) \right]}$$

where σ is the stress at the crack tip, E is Young's modulus or stiffness of the material, γ is the surface energy or work required to create the new crack surface, c is the crack length (one-half the length of an internal crack), ρ is the radius of curvature at the crack tip, and a is the equilibrium spacing between atomic planes in the absence of applied stress (Griffith 1921; Tetelman & McEvily 1967).

The magnitude of the stress at the crack tip depends on the length of the crack. For a tensile stress σ induced in a thin elastic plate containing an elliptical crack in which the major axis of length $2c$ is perpendicular to σ, the maximum tensile stress occurs at the tip of the crack and is (Tetelman & McEvily 1967):

$$\sigma_{max} = \sigma[1 + 2\sqrt{(c/\rho)}]$$
$$\sigma_{max} \cong 2\sqrt{(c/\rho)} \qquad \text{(for } c \gg \rho)$$

The quantity $2\sqrt{(c/\rho)}$ is the stress concentration at the crack tip. When ρ is much smaller than c, the stress concentration factor becomes large. In practice, at least in brittle materials, ρ may have a value approximating atomic distances and remains constant as c increases (Gordon 1976).

The effects of the stress concentration factor associated with cracks on the distribution of stresses in a solid may be visualized by considering the stresses in a plate of uniform thickness (Figure 9.2a) anchored at one edge (the left in this case) and subjected to a tensile load on the opposite edge. In the absence of cracks, the stresses are distributed fairly uniformly, as in Figure 9.2a. However, when a notch or crack is introduced under identical loading conditions (Figure 9.2b), a large increase in stress appears at the crack tip because the forces on the lower right quadrant of the plate must be balanced by opposing forces in the left side. The trajectories of these forces can no longer run directly between the right and left sides, owing to the absence of continuity across the crack, but must run around the crack tip (Figure 9.2c). Thus the forces from the lower right quadrant converge on the small number of atomic bonds at the crack tip. If the crack were lengthened vertically, the stress at the crack tip would be increased (by the stress concentration factor, above) because a still greater part of the load would converge at the crack tip.

As noted above, crack propagation requires a critical amount of energy at the tip. Some energy is absorbed in forming the additional crack surface, and if this is greater than the energy at the crack tip, the crack will not grow. However, the amount of energy required to form new surface remains constant for any unit increase in crack length (provided the relief of the crack surface remains constant). Therefore, as the crack becomes incrementally longer, at some length the energy gained at its tip will exceed that amount required to form an increment of new surface, and without increase in the applied forces the crack runs away (at a velocity equal to the rate of sound transmission in the material) and the material fractures (Gordon 1976).

Resistance to brittle fracture

When a cracked plate is loaded in tension (Figure 9.3), stresses are induced parallel to the nominal stress (Figure 9.3a; σ), but also in other directions. The plate contracts in the width and thickness directions because of the Poisson effect (when a body elongates under stress it contracts laterally). This contraction is strongest near the tip of the crack (Figure 9.3b), where the stress σ_x in the length direction is at maximum (Figure 9.3c) due to the stress concentration factor caused by the crack. Because

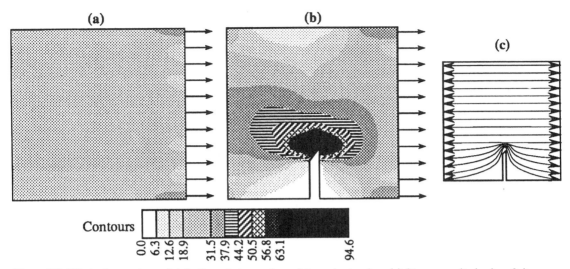

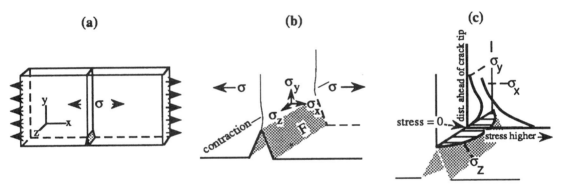

Figure 9.2. Effect of a crack on distribution of stresses in a plate under tension. (a) Stress magnitudes in a finite-element model of plate anchored on left and loaded in tension along right edge. (b) Same as (a), but with crack. (c) How nominal stress trajectories are affected by the crack. In (a) all trajectories would be parallel.

Figure 9.3. Stresses near a crack tip in a plate under tension. (a) Load and direction of nominal stresses. (b) Tensile stresses generated near crack tip, acting in x, y, and z directions. F is one face of the crack. (c) Graphs (*heavy lines*) of differences in magnitudes of these stresses along each axis at varying distances from crack tip and from center of plate in thickness direction.

the faces F of the crack are unstressed, they tend to retain their dimensions and therefore resist the contraction, causing transverse tensile stresses (Figure 9.3c; σ_z, σ_y) in the thickness and width directions of the plate (Parker 1957; Tetelman & McEvily 1967). The transverse stress σ_y is at a maximum some distance ahead of the notch (Figure 9.3c).

These transverse stresses can help prevent the crack from spreading in materials that are not homogeneous, especially when the microstructure is highly oriented. If the material contains elongated grain or fiber boundaries aligned parallel to the nominal stress (Figure 9.3b; σ) and perpendicular to the plane of an advancing crack, and the boundaries of the grains or fibers are weaker than the internal cohesion of the material, the transverse stress σ_y will tend to cause the material to split along the grain or fiber

boundaries ahead of the advancing crack (Figure 9.4a). When the crack reaches the split area (Figure 9.4b), the tip is blunted (Cook & Gordon 1964) and the stress concentration factor, which varies with the tip radius, is reduced.

The effect of tip blunting is similar to that of yielding (plastic flow), which occurs in the region of high stress at the crack tip in most metals and increases the stress required for fracture. The required stress increases because the material at the crack tip deforms nonelastically, enlarging the crack tip and thus decreasing the stress concentration factor. The total energy requirement for propagation becomes the sum of the work to form the new surfaces plus the work to produce the nonelastic deformation (Tetelman & McEvily 1967).

For the crack-resisting properties of fibers to be

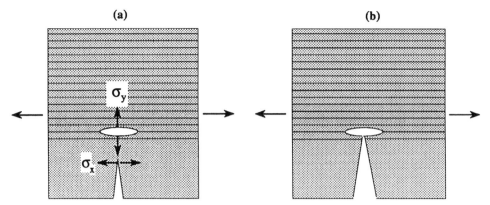

Figure 9.4. Crack-stopping effect of boundaries of fibers (prisms) perpendicular to plane of advancing crack. (a) Material splits ahead of crack in region of high σ_y (Fig. 9.3). (b) Crack tip is blunted as it reaches the void, reducing stress σ_x perpendicular to crack plane.

optimally effective in preventing an unstable crack from developing, the fiber direction must be parallel to the nominal tensile stress direction, and the intervals between the fibers must be small so that the splitting occurs repeatedly as the crack advances. These considerations of the properties of materials under load can be applied to the evolution of mammalian enamel.

Crack resistance in enamel

Enamel is formed of hydroxyapatite crystals by a wall of ameloblasts (mobile enamel-depositing cells) that advance in an outward direction from the enamel–dentin junction. Although these crystals tend to be outwardly directed on average, they are not entirely parallel but tend to radiate somewhat from centers. In most enamels of Cenozoic mammals, these discontinuities in crystallite direction are related to the shapes and positions of the ameloblasts. The differential crystallite directions cause the paths followed by the ameloblasts to be visible in thin sections of enamel examined with light microscopy owing to the anisotropy (directional differences) of light transmission inherent in the crystals. These pathways appear as parallel, rodlike structures called prisms or rods. These discontinuities also respond differentially to fracture, so that fracture surfaces reveal the prism structure. Each prism extends through the entire thickness of the enamel; consequently its long axis is very long compared to its thickness, which is usually close to 5 μm.

Hydroxyapatite, like apatite in general, fractures more easily parallel than perpendicular to the C axes of the crystals. Because the average crystallite direction tends to be parallel to the long axes of prisms, and probably also because of small amounts of weak organic deposits around crystallites and prisms, enamel fractures more easily parallel than perpendicular to the prism direction and cracks tend to follow prism boundaries. For example, in ground sections cracks are typically observed passing around prisms, and only rarely transecting them (Boyde 1976; Rasmussen et al. 1976).

The fact that prismatic enamel typically fractures along prism boundaries suggests that the prisms behave mechanically as fibers and that enamel would be weakest under tensile stresses acting perpendicular to the prism direction. In enamel in which the prisms are relatively straight and parallel, tensile stress normal to the prism direction and sufficient to initiate a crack between a pair of prisms could extend through the entire enamel thickness along the weak prism interfaces and effect fracture without transecting a prism.

An apparent selective response to this limitation, and one accompanying the early Cenozoic emergence of eutherian herbivores, was the appearance of a new type of tooth enamel that was more resistant to crack propagation. This new structure first appears in a few arctocyonid condylarths in the early Paleocene (Koenigswald, Rensberger, & Pfretzshner 1987). In these forms, as in many later mammals, the prisms in a given small region of enamel are bounded on either side by differently oriented, or decussating, prisms (Figure 9.5). The boundaries between adjacent sets of decussating prisms will be referred to below as *decussation planes*. These planes are frequently relatively flat over short distances, and the planes bounding adjacent sets of prisms tend to be parallel. This tabular structural arrangement reduces the number of directions (Figure 9.6a) in which a crack can propagate without transecting prisms to only planes parallel to the decussation plane (Figure 9.6b).

Prism decussation helps limit the crack length in several ways: (1) by forcing local deviation of the crack plane from the optimal direction, (2) by

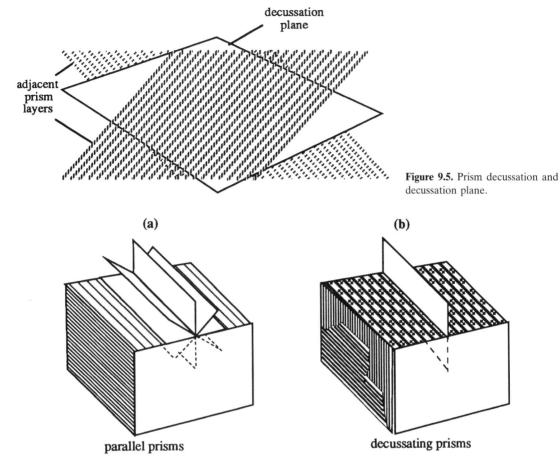

Figure 9.5. Prism decussation and decussation plane.

(a)

(b)

parallel prisms

decussating prisms

Figure 9.6. Directions in which fracture planes may run without transecting prisms in (a) enamel lacking decussation and (b) enamel with prism decussation (vertical in this illustration).

making it pass through prism diameters, (3) by enlarging the radius of the crack tip, and (4) by increasing the relief of the crack surface and therefore the required energy for crack extension.

DETERMINATION OF DENTAL STRESSES

Measurement of principal stresses

In this section I discuss possible ways of determining dental stresses in the teeth of mammals. These possibilities include: in vivo strain-gauge techniques for extant forms, inference from microstructure, mathematical modeling, and observation of crack directions in enamel.

When forces are applied to a structure, its dimensions change. The amount of deformation divided by the original dimension is termed the strain. Strain gauges (miniature transducers whose electrical resistance changes in proportion to a change in length) have frequently been used to measure strains in bone

(Lanyon 1976; Hylander 1979; Biewener et al. 1983; Swartz 1991). Using a rosette gauge, from which the deformation in three known directions is assessed, directions and magnitudes of the principal strains may be calculated (Dally & Riley 1991). The principal stresses can be calculated using moduli of elasticity, constants that describe the anisotropic magnitudes of the strain response of the material to stress (Carter 1978).

There are a number of practical difficulties in using this method for determining the principal stresses in teeth during occlusal activity. The strain gauge indicates deformation only at the position of its application. Because dental surfaces, especially of mammalian cheek teeth, are more complex than surfaces of equivalent areas in bones, a greater number and density of gauges would be needed to define a stress pattern in a tooth. A tooth in a small mammal may have numerous cusps or crests in an area that is as small or smaller than the size of the smallest gauge. At the enamel–dentin junction there is an abrupt transition in physical properties, causing changes in stresses through the thickness of the enamel.

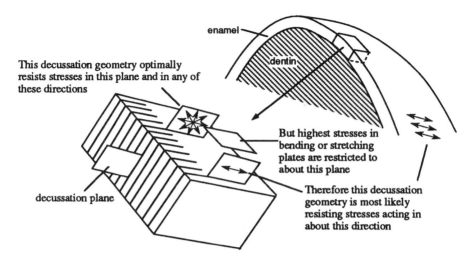

Figure 9.7. How prism decussation gives an indication of the directions of the stresses, assuming that selection has optimized the decussation plane for resisting crack propagation. Decussation is nonspecific in its resistance to any tensile stress direction within the decussation plane. However, distribution of stresses in a plate under axial tension or bending stresses (see Fig. 9.8) narrows possible directions of maximum stress to approximately the single direction shown.

Resolving these differences would require embedding gauges within the enamel, introducing a flaw in the tooth, and thereby altering the magnitudes and directions of stresses near the gauge. Furthermore, the durability of gauges mounted on the surface of the tooth is a problem because occlusal surfaces are adapted to crush or shred materials contained between them. This method is clearly inapplicable for fossils, and is hampered by major technical difficulties for the teeth of most extant mammals.

Inference of stresses from enamel microstructure

If the prism decussation planes are maintained by selection to optimally resist the maximum tensile stresses that occur in the teeth, the microstructural information may indicate the directions in which the stresses were acting in the taxon.

Decussation optimally resists tensile stresses acting parallel to the decussation plane. This information alone does not precisely identify the stresses that are being resisted by the enamel microstructure because the structure is equally effective in resisting stresses acting in any direction within the decussation plane (Figure 9.7). However, the possible stresses that a given structure has become adapted to resist can be narrowed further by considering how the shape of enamel constrains its behavior.

When a cusp is loaded, the enamel does not behave as if it were merely a continuation of the body of the cusp. Because enamel is stiffer than dentin, it transmits most of the load (Yettram, Wright, &

Pickard 1976). This, and the fact that enamel is always thin in one dimension and broad in the other two, makes it behave like a plate. When a plate is subjected to tensile stress, the major stress acts within the plane of the plate, parallel to its surfaces. When this restriction is applied to the multiple possible stress directions within the decussation plane of Figure 9.7, it leaves only the stress acting within the plane of the enamel plate as the stress that the structure was selected to resist. On the other hand, when an enamel plate is subjected to a compressive load parallel to the plane of the plate, because enamel is seldom truly flat, its response to compressive force is usually bending. Bending also is the response when the load is perpendicular to the plane of the plate. In a bending plate, the maximum stresses occur in the region where the principal stresses are parallel to the surface of the plate, even when the load is perpendicular to the surface (Figure 9.8). For that reason, the maximum tensile stress (as well as the maximum compressive stress) will occur nearest the outer or inner surfaces and parallel to the plane of the enamel.

Therefore, information about the attitude of the decussation plane frequently leads to a prediction that the direction of the maximum tensile stresses is within a plane normal to the decussation plane and parallel to the plane of the enamel (Figure 9.7). However, where the enamel is formed in strongly bent shapes, the stresses within the enamel become more three-dimensional in their differentiation, and the selective response of the enamel and its decussation planes may be more complex. Furthermore,

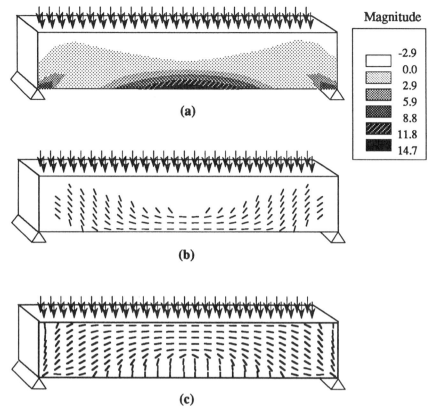

Figure 9.8. Magnitudes and directions of major stresses through the thickness of a plate supported at its ends and uniformly loaded (*arrows*). (a) Magnitudes of major stress. (b) Directions of maximum tensile stresses, represented by short lines. (c) Directions of maximum compressive stresses.

if the loading directions vary to a high degree, flat planes of decussation would not offer maximum resistance to all states of stress.

Mathematical modeling

Enamel microstructure has taken a great diversity of form through the Cenozoic, departing from or elaborating beyond simple tabular decussation in various groups and in different ways, and making the functional interpretation of the microstructural geometry less obvious in some cases than in others. Furthermore, selection for abrasion resistance may have influenced enamel prism direction because prisms are anisotropic in abrasive wear (Rensberger & Koenigswald 1980; Boyde & Fortelius 1986). Prism direction may therefore be influenced by the demands of resisting both stress and abrasion. For these reasons, it is useful to have independent methods of verifying predictions about the directions of stresses and hypotheses about the functional morphology of enamel microstructure.

Mathematical modeling of stresses in dental structures under specified loading conditions offers a means of independently determining the stresses

induced by chewing or other uses of teeth. Stress problems for engineering beams and for long bones can be solved using the fairly straightforward equations of beam theory (Thomason, this volume). Dental structures, however, are seldom regular enough to allow such an easy solution. An alternative is to consider the tooth as consisting of an interacting aggregate of simply shaped units or elements. This allows the equations describing the behavior of each of these elements to be combined in a larger set of interdependent equations describing the behavior of the structure as a whole. This method, known as finite-element analysis (FEA), allows the stresses in any part of an indefinitely complex structure to be calculated (limited mainly by the computing resources available).

The required input data for a finite element model, in addition to the sizes and positions of the elements, include some properties of the material: Young's modulus of elasticity (E, the tensile or compressive stress required to produce unit strain in the material), shear modulus for the material (G, shear stress per shear strain), and Poisson's ratio for the material (v, ratio of longitudinal strain to lateral strain). Vincent (1982) and Currey (1984) discuss these

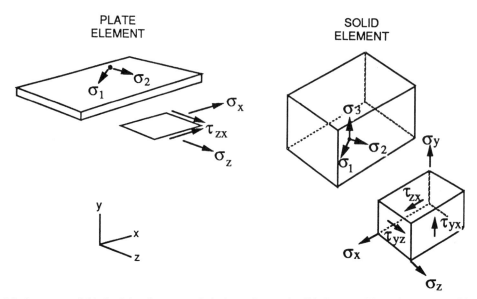

Figure 9.9. Stresses available in finite-element analysis from plate and solid elements. Plate elements provide stresses along axes of a coordinate system within the plane of the plate, σ_x and σ_z, which can be resolved to the major (maximum) and minor (minimum) stresses, σ_1 and σ_2, and the shear stress τ_{zx}, within that plane. Solid elements provide stresses σ_x, σ_y, and σ_z, along the axes of a global coordinate system, and these can be resolved to principal stresses σ_1, σ_2, and σ_3. Solid elements also provide shear stresses τ within planes yx, zx, and yz and the maximum shear stress. Elements of a variety of shapes and types other than those illustrated are also available to facilitate defining the model.

constants with reference to biologic materials and problems involved in determining their values.

A variety of element types are available in FEA software. Plate elements have thicknesses that are a small fraction of the length or width and allow solution of the major and minor principal stresses and maximum shear stress within the plane of the element (Figure 9.9). Plate elements may be used, for example, for the calculation of stresses occurring within a cross-section of a tooth. Yettram et al. (1976) calculated the principal stresses within a vertical section through a human molar at about 42 positions in the dentin and 200 in the enamel. Pfretzschner (1992) calculated the stresses in a vertical section of a generalized ungulate molar. Plate elements may also be used for analyses of stresses in a shell (a curved structure of small thickness). Because enamel is stiffer than dentin, it behaves under load like a shell. Rensberger (1992) calculated the stress directions and relative magnitudes in a shell representing the protoloph enamel of a rhinocerotoid upper molar.

Solid finite elements are fully three-dimensional and may be as thick as wide or long. Furthermore, whereas plate elements may only be joined edge to edge, solid elements may be joined along any face. Solid elements allow the calculation of the three principal stresses σ_1, σ_2 and σ_3; the normal stresses σ_x, σ_y, and σ_z; and the shear stresses τ_{xy}, τ_{yz} and τ_{zx} (Figure 9.9) and therefore allow three-dimensional

evaluation of the state of stress in each element. A variety of shapes are possible – tetrahedral, pentahedral, and hexahedral or "brick"-shaped (Zienkiewicz & Taylor 1989), etc. – which facilitate modeling objects of different shapes.

A practical problem in implementing irregular three-dimensional models is the difficulty in specifying the Cartesian coordinates of the nodes (corners) of the elements. FEA software may include the ability to generate nodes, but usually this facility is restricted in its usefulness in defining irregularly shaped objects. Separate programs are often used to facilitate the design of three-dimensional objects.

The volume of data comprising the calculated stresses is very large, for it usually includes the principal or normal stresses and shear stresses at the centers and corners of the elements, which may number several thousand. This information can be handled most easily using software for postprocessing the results. Postprocessing is useful in filtering the output for selected data, generating three-dimensional stress lines, and rotating the model together with its stresses in real time.

Inference of stresses from empirical data

The stresses in a solid depend upon its shape, the elastic behavior of its materials, and how it is loaded. It is not easy to precisely provide these requirements and values in a model. The directions in which teeth

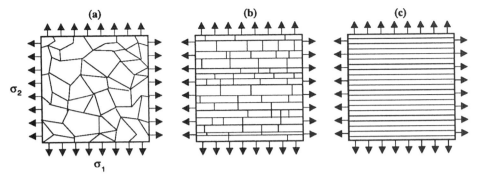

Figure 9.10. Effect of differences in relative magnitude of within-plane stresses acting on a plate, as indicated by cracks in a brittle coating. All loads are tensile. Loadings: (a) σ_1 equals σ_2; (b) σ_1 is greater than σ_2; (c) σ_2 is zero. (After Dally & Riley 1991 with permission of McGraw-Hill Inc.)

are loaded are sometimes uncertain. Measurements of the elastic moduli of enamel and dentin have not been entirely satisfactory, because different workers have arrived at different values, in part because of anisotropy in elasticity of the materials and practical problems in preparation of the materials for testing. This has been particularly true of the Young's modulus for enamel (Waters 1980). It is useful, therefore, to have one or more independent sources of information about the state of stress that can corroborate values determined from models.

Cracks provide an important source of information about the stresses in enamel. Unlike bone, enamel is never repaired after cracking or fracture. The attitudes of crack or fracture planes provide a source of information about the directions of stresses that exceeded the strength of the enamel. A difficulty in utilizing cracks or fractures as indications of the stresses resulting from chewing or other activities is that one must distinguish the premortem from the postmortem cracks, because enamel is subject to cracking both during life and after death. This is obviously a consideration when examining fossil specimens, which may have been subjected to trampling shortly after death and also to stresses by being buried in load-bearing strata for millions of years after death. Even in the case of Recent animals collected by museum personnel, one cannot be sure that damage has not been incurred after death during the process of trapping and preparing specimens. Furthermore, changes in moisture and temperature apparently subject dental materials to stresses that result in the cracking of teeth, judging from the frequency in museum specimens of cracked teeth that could not have survived chewing activities during life.

However, premortem cracks in tooth enamel are distinguishable from postmortem cracks provided that they existed some time before death and in positions sufficiently near the occlusal surface to have received abrasion from food materials (Rensberger

1987). Abrasion rapidly smoothes the sharp edges of cracks or fractures, probably because the sharpest structures concentrate stresses during abrasion and are the most rapidly worn away. The mirrorlike rounded edge of an abraded crack can be recognized under low magnification (10× to 20×) light microscopy as a bright line of reflected light. The edge of a postmortem crack, in contrast, is rough and thin and reflects no light, so that one sees only the black line of the crack itself. Many premortem cracks in enamel are closed and would be invisible except for the specular reflection from the polished edges. Postmortem cracks in teeth of modern mammals are usually much more conspicuous than the premortem cracks, and appear as dark gaps that are visible to the unaided eye. In fossils the postmortem cracks are often filled with rock matrix and are also readily visible. Detecting subtle premortem cracks is aided by slowly rotating the tooth to cause the reflected incident illumination to scan across the surface. The premortem crack will appear as a thin bright line near but slightly outside the broader band of specular reflection on the surface. It can be distinguished from a surface scratch by rotating the specimen so that the plane of the crack beneath the surface reflects light back toward the eye.

Crack patterns in brittle materials reflect the directions of the stresses. Experiments show, for example, that when a solid coated with a brittle material is loaded in various ways, distinctive patterns of cracks differentiate the stress conditions that arise (Dally & Riley 1991). When a plate is loaded under tension by equal stresses at 90° to one another, a random crack pattern is developed (Figure 9.10a). When the plate is loaded by tensile stresses of different magnitudes in directions differing by 90°, crack systems normal to each of those directions appear, but those normal to the major stress dominate (Figure 9.10b). When the plate is loaded by tensile stress in only one direction, the cracks are all perpendicular to the direction of the stress (Figure 9.10c).

The question arises whether the anisotropy of the microstructure gives a false indication of the direction of the stresses that produced a crack or fracture. It is likely that if stresses are equal and are acting at 90° to one another, or if the major stress direction occurs both normal and parallel to the decussation planes with equal frequency, there would be a strong bias against cracks normal to the decussation planes. However, because decussation seldom extends through the entire enamel thickness, and the region in which it is absent is usually the outer enamel, the visible cracks that occur in the outer enamel will lack that bias. Also, if the major stresses are acting in different directions in a taxon, a microstructure that is anisotropic in its resistance to fracture would have little selective value, and a more complex structure that is more isotropic would tend to be acquired; an isotropic microstructure would diminish the bias in crack direction, and the frequencies of the crack directions would reflect the true stress direction. The empirical data discussed below support these expectations.

Fractures passing through a decussation zone are influenced, at some scale, by the microstructure, depending on the form of the crack-resisting geometry. Therefore the analysis of cracks as indications of stress patterns is most useful when undertaken in conjunction with a study of the microstructure.

RELATIONSHIPS OF STRESSES AND MICROSTRUCTURE

Early ungulates

The directions of striae and the attitudes of occlusal facets in insectivorous mammals of the late Cretaceous and earliest Paleocene indicate that they chewed with dominantly vertical movements of the mandibular dentition, forming an angle with the vertical plane of about 23° to 35° in the palaeoryctids (Butler 1972; Rensberger 1986). In the earliest ungulate, *Protungulatum,* the angle becomes 40° to 55° and in the phenacodontids about 57°, a direction still dominated by the vertical component (Rensberger 1986).

When cusps with circular or elliptical shapes are loaded by vertical or near vertical forces, shell theory and finite element and physical models indicate that the dominant direction of the resulting major stress in the enamel is horizontal and tensile (Pfretzschner 1988; Rensberger 1992).

The earliest occurrences of prism decussation are in some of the early Paleocene arctocyonids (Koenigswald et al. 1987). The decussation planes in these taxa appear to be more or less perpendicular to vertical sections, as in many Eocene and later ungulates, and this is the direction that would optimally resist the horizontal tensile stresses predicted by the models.

Rhinocerotoids and astrapotheres

The most prominent deviations of the decussation plane from the primitive condition that has been observed in ungulates is the vertical attitude that characterizes the rhinocerotoids and astrapotheres (Rensberger & Koenigswald 1980; Fortelius 1985; Boyde & Fortelius 1986).

The geometry and mechanics of rhinocerotoid cheek teeth are highly modified from the primitive ungulate condition. The cusps of primitive ungulates have essentially been replaced by straight-walled lophs, and the chewing direction has a strong horizontal component. When a finite element model of the enamel in narrow straight lophs, like those in the rhinocerotoids, is loaded by forces with a strong transverse component, the major stress tends to be vertical where its magnitudes are highest (Rensberger 1992).

The discrimination of premortem and postmortem cracks in the cheek teeth of the perissodactyls *Mesohippus* and *Diceros* confirms this 90° difference in the direction of the major stress predicted by finite element modeling (Rensberger 1992). The enamel in the less derived *Mesohippus* still retains the horizontal decussation planes of the primitive ungulates, and the premortem cracks visible in the outer enamel are overwhelmingly vertical, the direction that is resisted best by the horizontal decussation planes. In contrast, the modal direction of the premortem cracks is horizontal in *Diceros,* whereas the decussation planes are vertical. However, the premortem cracks in *Diceros* and other rhinocerotoids are slightly bent in a lenticular shape, in close correspondence to the shape predicted by the finite element model.

The extinct South American ungulate order Astrapotheria is also characterized by vertical decussation in the molars (Fortelius 1985). The resemblance of the microstructure to that in the rhinocerotoids is remarkable, considering that the derived nature of the structure was independently acquired in the two groups (Rensberger & Pfretzschner 1992). In both groups the premortem cracks are horizontal and have the lenticular shape described above. It seems highly probable that the similarity of the microstructure in the cheek teeth of the two groups resulted from selective responses to similar stresses, because the chewing motion, defined by facet and chewing *stria* orientation, and cheek tooth shapes (especially the flat lophs) are similar.

The resemblances of the prism decussation pattern

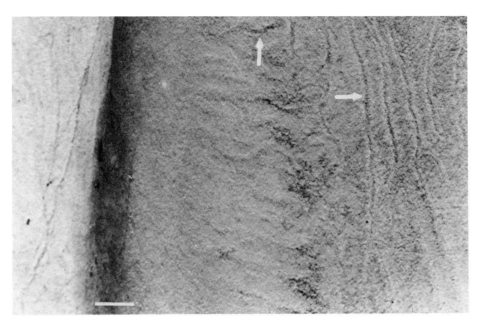

Figure 9.11. Scanning electron micrograph showing horizontal decussation adjacent to vertical decussation in Oligocene rhinocerotoid *Subhyracodon.* A tangential section in which the outermost enamel was removed by air abrasion. Vertical decussation occurs in flat part of the enamel (right side of micrograph), grading into horizontal decussation (middle of micrograph) where enamel bends around the paracone. Arrows identify zones of horizontally and vertically decussating prisms. Occlusal direction toward the top; metacone toward right. Scale (white bar) = 200 μm.

in the astrapotheres and rhinocerotoids include more than the shared presence of vertical decussation. The astrapotheres have two types of enamel structure through the thickness of the enamel, one with vertical decussation planes in the inner half or so of the enamel thickness, and another with horizontal decussation planes in the outer enamel (Rensberger & Pfretzschner 1992). Rhinocerotoids have a similar condition (Figures 9.11 & 9.12), although in earlier studies the horizontal decussation was not observed (Rensberger & Koenigswald 1980; Boyde & Fortelius 1986). The degree of development of horizontal decussation varies with position in the tooth. It is best developed where the enamel bends prominently (Figures 9.11 & 9.12), for example, around the paracone, protocone, and hypocone.

I generated solid element FE models of the protocone where the enamel bends most strongly, and of the flattest part of the ectoloph across the metacone and metastylar regions (Figure 9.13; center). The respective chewing directions for these regions were established on the basis of facet orientation and stria direction. Forces derived from these chewing directions were applied to each model. The results (Figure 9.13) show that the major stresses in the outer enamel of the protocone are tensile and nearly horizontal, whereas the major stresses in the flat part of the ectoloph are nearly vertical

throughout. I have been unable to find premortem cracks in the convex areas of the protocone, hypocone, or paracone that would provide empirical evidence of the stresses there. Premortem cracks do occur in the flat regions of the protoloph and metaloph and are dominantly horizontal, consistent with the vertical stresses through the thickness of the enamel predicted by the model.

In astrapotheres, horizontal decussation, as viewed by light microscopy from the outer surface, occurs in the outer enamel of the paracone but does not appear to be developed in the flat parts of the ectoloph. The protoloph and metaloph are not as uniformly elongate and flattened as in rhinocerotoids, and, in those areas of one specimen examined by Pfretzschner (personal communication), horizontal decussation may be more extensive than in rhinocerotoids.

Anterior teeth in mammals are subject to more diverse uses and loading patterns than cheek teeth, and the enamel exhibits considerable diversity, often differing from that in the cheek teeth in the same taxon (Koenigswald 1988). The decussation planes in the enamel of the canine tusks of astrapotheres are more complex than in the molars. The decussation planes in the Oligocene *Parastrapotherium* contain wavelike bends that make it resistant to fracture in any plane (Rensberger &

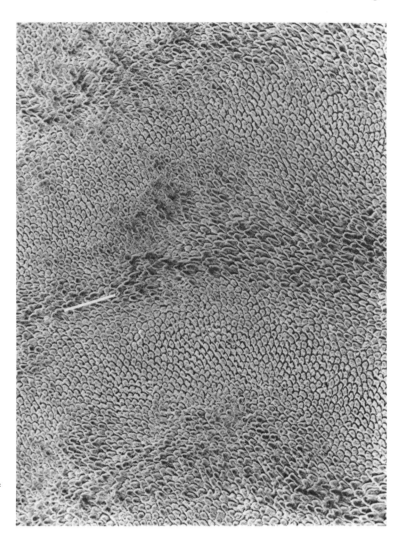

Figure 9.12. Enlargement of region of horizontally decussating prisms in paracone enamel of *Subhyracodon*. Scale = 50 μm and is aligned parallel to the decussation plane in that region.

Pfretzschner 1992). The enamel of a canine in *Parastrapotherium* contains numerous premortem cracks running in at least two divergent planes, indicating different loading directions (Rensberger & Pfretzschner 1992).

The fact that the highest stresses in a bending plate occur near the surface (Figure 9.8) predicts that fractures would be initiated at either the outer or inner surfaces of the enamel, depending on the direction in which the plate is bent. Cracks like these, initiated on both inner and outer surfaces, occur in the tusk enamel of *Parastrapotherium*. A finite element model shows that such cracks will be inclined in their course through the enamel thickness; inclinations corresponding to loads near the tip of the tusk were observed in the astrapothere enamel, implying that the tusk was engaging objects while being swung to either side (Rensberger & Pfretzschner 1992).

Differences in stress magnitude and microstructure

The examples described above indicate a correspondence between prism decussation and the *direction* of the major stress. It is clear that the fracture of different food materials requires different *magnitudes* of stress. For example, the fracture strengths of flax and hemp fibers under tensile stress differ by an order of magnitude (Gordon 1976). If different food materials require different occlusal stress magnitudes for their fracture, differences may be reflected in the degree or pattern of prism decussation that has evolved to resist those stresses.

INCREASES IN STRESS MAGNITUDE IN EARLY PALEOCENE THERIANS. A nominal relationship of prism decussation to occlusal stress magnitude seems

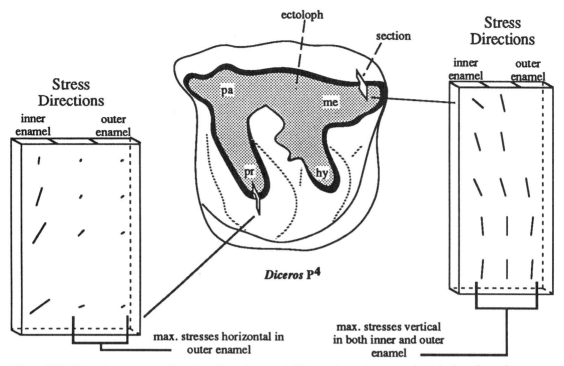

Figure 9.13. Finite-element stress directions through enamel thickness in protocone and ectoloph regions of a rhinocerotoid upper cheek tooth. Enamel near metacone was loaded with dominantly horizontal forces normal to outer surface; enamel at protocone was loaded with dominantly horizontal forces normal to enamel surface where it projects above dentinal platform.

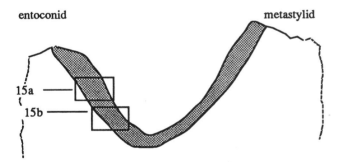

Figure 9.14. Vertical anteroposterior section through a lower molar talonid of the primitive perissodactyl *Hyracotherium*. Positions of images of Figure 9.15a,b are indicated.

to exist. Koenigswald et al. (1987) found that the presence or absence of decussation in a taxon is related to body size, which suggests a relationship to stress magnitude. They also noted that the angular divergence of decussating prisms and the extensiveness of the decussation in the early Eocene *Arctocyon* are greater than in some early Paleocene arctocyonids, but whether the occlusal stresses in the early Eocene form were higher is not clear.

DIFFERENTIAL STRESS MAGNITUDE IN PRIMITIVE EOCENE UNGULATES. Decussation has varied degrees of development from the tips to the bases of the cusps in the molars of the early Eocene perissodactyl *Hyracotherium* (Figures 9.14 & 9.15). In the anterior enamel of the entoconid, decussation

is moderately well developed in the middle part of the vertical extent of the cusp (Figure 9.15a), but diminishes toward the talonid basin (Figure 9.15b) and in the region nearest the tip. If this differential pattern represents a response to different stress magnitudes, it predicts that the highest tensile stresses occur in the middle regions of the vertical section.

The stress environment can be reconstructed in a finite element model of the cusp. The stress analysis models a basically ellipsoidal cusp that is worn at the tip. The base departs from the ellipsoidal shape as it becomes concave to join the floor of the talonid basin. The analysis only requires the stress environment on one side of the cusp. However, for economy of construction, the model is symmetrical in

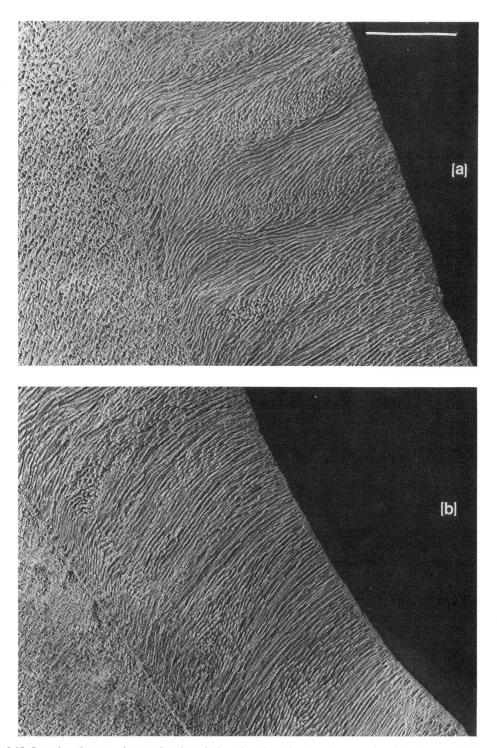

Figure 9.15. Scanning electron micrographs of vertical section through enamel on anterior side of entoconid in *Hyracotherium.* (a) Higher on the wall of the talonid basin, showing marked decussation. (b) Deeper in the basin the decussation is reduced and at the bottom disappears altogether. Scale = 100 μm.

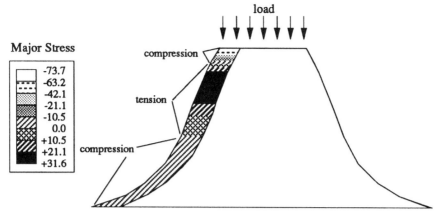

Figure 9.16. Section through a three-dimensional finite-element model of a modified ellipsoidal cusp, showing differentiation of the stress magnitude in the radial section of enamel. Major stress is horizontal where indicated as tensile.

having the concave base extend entirely around the structure.

Chewing in primitive ungulates involved a dominantly vertical motion (Rensberger 1986). As the motion acquired an increasingly horizontal component in later forms, the crest and enamel edge alignments lose the primitive circular or random patterns characteristic of vertical chewing motion and become aligned perpendicular to the direction of the transverse component. The pattern in *Hyracotherium* (fig. 3, Rensberger et al. 1984) is still rather random compared to that in later ungulates and indicates retention of a mainly vertical chewing motion.

When the model is loaded uniformly at the occlusal surface by vertical forces, the stresses in the enamel shell have their highest positive (tensile) values in the upper middle region of the cusp, with the values decreasing in either direction and becoming negative (compressive) at the top and bottom (Figure 9.16). The pattern of decreases in tensile stresses in the lower and upper parts of the model is consistent with the diminished decussation in the corresponding regions of the talonid cusps. Compressive stresses exerted on the cusp at the occlusal surface result in circumferential tensile stresses at a lower position, and these stresses diminish as the enamel reverses curvature near the talonid basin. The degree of decussation corresponds to these differences in magnitude of the stresses.

HIGH, VARYING STRESSES. Structures subjected to stresses that vary in direction are not well strengthened by tabular decussation, which offers a minimal resistance to fracture in the plane of decussation. The anterior teeth are, more often than cheek teeth, subjected to loads of varying direction. Small angular differences in the lateral excursion of the mandible result in differences in positions of opposing teeth

at occlusion that are greatest at the anterior end of the jaws. Also, anterior teeth are often used for grasping, pulling, or twisting, which impose diverse loading directions related to head and postcranial body movements. The resistance of objects grasped by the anterior teeth is derived from the mass of the object or resistance of the substrate to which it is attached. The resistance of the small food fragments that are processed by the posterior teeth is derived from the opposing occlusal surfaces, the motion of which is constrained by the consistency of the chewing motion.

Tusks are especially subject to different loading directions. Astrapotheres are characterized by elongate canine tusks, and these have a more complex enamel microstructure than the molars. Furthermore, because the canines have a large aspect ratio of length to width, the enamel is subject to higher bending stresses than the molars, confirmed by a larger number of premortem cracks. The premortem cracks in the canine of the Oligocene *Parastrapotherium* occur in at least two modally divergent planes, indicating that multiple loading directions are involved (Rensberger & Pfretzschner 1992). As in the molars, prisms tend to decussate vertically (in a plane parallel to the longitudinal axis of the tooth), but the decussation planes bend in a wavelike pattern so that there are characteristics of both vertical and horizontal decussation. The structure is further modified by differences in the attitude of the decussation plane through the thickness of the enamel. This microstructure offers the type of resistance to crack propagation that is characteristic of decussation, but is as resistant to longitudinal radial cracks (in planes parallel to the long axis of the tooth) as to transverse cracks (perpendicular to the long axis of the tooth).

The hyenas may attain the highest dental stresses

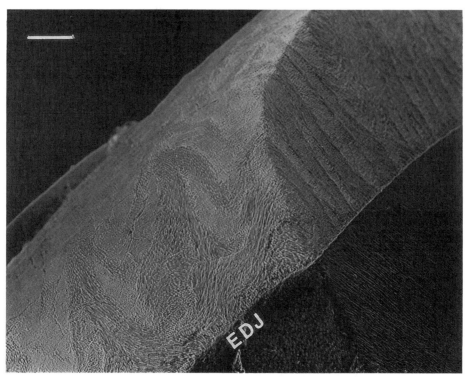

Figure 9.17. Scanning electron micrograph of folded decussation planes in anterolabial canine enamel of *Crocuta crocuta*. Oblique occlusal view, outer surface of enamel at upper left, enamel–dentin junction (EDJ) labeled at occlusal surface. Light area in center and lower left is naturally worn surface of the canine tip that has been lightly etched with acid to reveal structure. Slightly darker area at the upper right of center is a radial (vertical) section through the enamel. Scale = 100 μm.

among the extant Carnivora. Savage (1955 and personal communication) experimentally measured the compressive load required to fracture flakes from a horse humerus. He found that, to artificially produce flakes of sizes equivalent to those splintered from a horse humerus by a spotted hyena (*Crocuta crocuta*), stress on the order of 70 MNm^{-2} was required. A brown bear fed the same bones, although it tried, was able to remove only a few minute flakes. In a survey of modern felids, canids, and hyaenids, the bone-eating taxa had the highest incidence of dental fracture and *C. crocuta*, which is probably the greatest consumer of bone among living taxa (Van Valkenburgh, Teaford, & Walker 1990), ranked highest of all (Van Valkenburgh 1988).

The highest frequency of dental fracture observed in *Crocuta* occurs in the canines (Van Valkenburgh 1988), apparently because, even though most of the bone crushing is accomplished by the postcanine teeth, the powerful jaw musculature results in exertion of exceptional force in the anterior teeth as well. Therefore, one might expect to find a high degree of microstructural modification of the canine enamel in *Crocuta*.

The canines in *Crocuta* are subjected to different loading directions, judging from the diverse directions of striae on the teeth. The decussation planes in the enamel of *Crocuta* are bent in wavelike undulations, which provides resistance to tensile stresses acting in any plane, from transverse to vertical. The amplitude of the folded decussation planes varies in different positions on the canine. In general, the amplitude increases toward the tip of the tooth. Figure 9.17 shows some of the most deeply folded decussation planes, which occur on the anterior and anterolabial sides of the tooth near the tip. In this region the sides of the folds are almost vertical over much of the fold amplitude. As the amplitude of folding diminishes from the tip toward the base of the crown, the average direction of the decussation planes becomes more horizontal.

A finite element analysis of the stresses in the canine of *Crocuta* indicates a relationship between the regional differences in the stresses and the intensity of folding of the decussation planes. When the model is loaded at the tip by a distributed force directed perpendicular to the wear facet (Figure 9.18), the load induces major stresses that are tensile over most of the enamel and greatest on the anterolabial side of the tooth near the tip (Figure

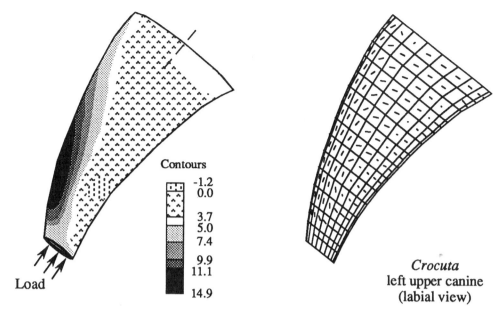

Figure 9.18. Major stress magnitudes and directions in upper canine of *Crocuta*, loaded at tip in the direction parallel to central axis of tooth near base of crown, as calculated in a three-dimensional finite-element model of enamel and dentin. Labial view, anterior toward left.

9.18a). The stresses diminish toward the posterior side of the tooth and toward the base of the crown, corresponding regionally to the reduction in amplitude of the folded decussation planes.

The stress direction in the anterolabial region is vertical, whereas the direction becomes increasingly horizontal toward the posterior side of the tooth (Figure 9.18b). The vertical direction of the tensile stress on the anterolabial surface is confirmed by horizontal premortem cracks that frequently occur there in *Crocuta* and are longest in the area where the modeled stresses are highest.

The loading direction perpendicular to the wear facet at the tip is approximately parallel to the central axis at the base of the crown. When the loading orientation is made parallel to the central axis near the tip, the distribution of the stresses changes markedly (Figure 9.19). If the force is parallel to the axis at the tip, stress is reduced at the tip, and the region of greatest stress magnitude moves to near the base of the tooth and to the posterolingual side (Figure 9.19a). The axis of the canine in *Crocuta* is curved three-dimensionally, in both anteroposterior and mediolateral planes, which makes the position of the maximum strain asymmetric with respect to the anteroposterior plane. This relationship suggests that the angle of the tooth with respect to the loading direction is controlled so that the load is parallel to the base, where the largest bending moment could occur. This, however, places most of the strain near the tip and on the anterolabial side, and this seems to be the source of selection for reinforcement of

the enamel microstructure against the vertical stresses in that area. Although the occurrence of diversely directed abrasive striae on the canine and the presence of folded decussation planes throughout the enamel indicate movements of hard objects in many different directions while in contact with the canine, the position of the most derived microstructure indicates that the greatest deformation of the tooth occurs during orthal biting when a hard object is caught between the opposing canines.

In *Felis concolor* (cougar), the upper canine is similar in size and shape to that in *Crocuta*, but is a little less robust. The loading directions are as diverse in *F. concolor* as in *Crocuta*, judging from the directions of striae in the same position on the tooth. However, while the decussation planes in *F. concolor* are also folded, the amplitude of folding varies relatively little in the different regions of the tooth and lacks the specialization seen in *Crocuta*. The amplitude of folding on the anterolabial side of the canine near the tip is similar to that in other parts of the crown and most closely resembles the minimal folding that occurs near the base of the crown in *Crocuta*.

The pattern of stress directions in the canine of *F. concolor* must be similar to that in *Crocuta*, because the shapes of the canines, the cross-sectional properties of the jaws (Biknevicius & Ruff 1992), and the movements of the jaws during biting are similar in the two taxa. The intensified regional folding of the decussation planes in *Crocuta* and its absence in *Felis* therefore appear to be related to the greater

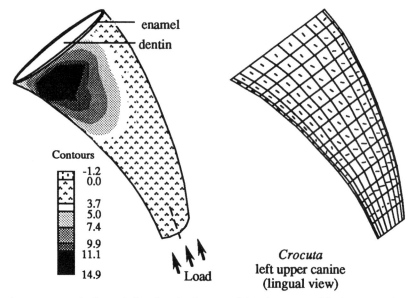

Figure 9.19. Major stress magnitudes and directions in the same finite-element model of upper canine of *Crocuta* shown in Figure 9.18, but loaded parallel to central axis of tooth near tip. There is no stress concentration on the anterior and labial sides. Anterior toward right.

magnitude of the stresses in *Crocuta*. When an axially straight, symmetrically ellipsoidal, oval, or cylindrical shape is loaded parallel to its long axis by a compressive force, circumferential tensile stresses are produced near the outer surface. But when the axis is slightly concave, the compressive load results also in a bending moment and vertical tensile stresses on the convex side of the axis. The greater load in *Crocuta* increases the vertical stresses on the anterolabial side of the tooth without increasing the circumferential stresses, and this apparently led to selection for greater resistance to vertical stresses in that area. The details of the enamel microstructure in *Crocuta* are even more complex when the differences through the thickness of the enamel are considered, and these are being described in a separate paper (Rensberger, in preparation).

CONCLUSIONS

The strengths of dental materials limit the types of food materials that can be efficiently processed. With an improved understanding of how the gross shapes of occlusal structures are dictated by trade offs between the strengths of the materials and the efficiency of the design for food capture and processing, we should be able to more closely approach mechanical models that will allow us to separate the effects of the various factors influencing the evolution of the masticatory apparatus. Such models, together with data from microwear and jaw mechanics, would help us understand how the numerous and

complex differences in the dentitions of fossil mammals are related to differences in dietary behaviors.

The relationship of enamel microstructures to dental stresses can be interpreted by considering how brittle materials fracture and how structurally complex materials (composites) inhibit fracturing. Hypotheses developed by examination of microstructures can be tested by empirical data from premortem cracks in tooth enamel and from models, including finite element stress analyses, using wear features and jaw and tooth morphology to infer loading directions. The results obtained so far indicate a responsiveness in the evolution of enamel microstructure to changes in dental stresses. Although absolute magnitudes of dental stresses cannot yet be inferred from the microstructure, evidences of differences in relative stress magnitudes and in stress directions have been found, and these have been consistent with empirical data from cracks and/or theoretical models.

ACKNOWLEDGMENTS

This study benefited from discussions about feeding behaviors and jaw and dental mechanics with Alan Boyde, Laurence Frank, Sue Herring, Fred Grine, William Hylander, Mary Maas, Hans Ulrich Pfretzschner, and Blaire Van Valkenburgh. Elizabeth de Wet kindly provided *Crocuta* teeth from the archaezoology collections of the Transvaal Museum, Pretoria, South Africa, where the specimens examined in this study and another in preparation will be permanently preserved. Laurence Frank provided access to the *Crocuta* colony in Berkeley, California, for the author to observe feeding behaviors. I am grateful to

William Clemens, J. Howard Hutchison, and Donald E. Savage of the University of California Museum of Paleontology, and to John Flynn and William Turnbull of the Field Museum of Natural History for providing access to fossil ungulates in those collections.

REFERENCES

Biknevicius, A.R., & Ruff, C.B. 1992. The structure of the mandibular corpus and its relationship to feeding behaviors in extant carnivorans. *Journal of Zoology, London* 228, 479–507.

Biewener, A.A., Thomason, J., Goodship, A.E., & Lanyon, L.E. 1983. Bone stress in the horse forelimb during locomotion at different gaits: A comparison of two experimental techniques. *Journal of Biomechanics* 16, 565–576.

Boyde, A. 1976. Enamel structure and cavity margins. *Operative Dentistry* 1, 13–28.

Boyde, A., & Fortelius, M. 1986. Development, structure and function of rhinoceros enamel. *Zoological Journal of the Linnean Society* 87, 181–214.

Butler, P.M. 1972. Some functional aspects of molar evolution. *Evolution* 26, 474–483.

Butler, P.M. 1973. Molar wear facets of early Tertiary North American Primates. In *IVth Symposium, International Congress of Primatology* 3, 1–27.

Carter, D.R. 1978. Anisotropic analysis of strain rosette information from cortical bone. *Journal of Biomechanics* 11, 199–202.

Cook, J., & Gordon, J.E. 1964. A mechanism for the control of cracks in brittle systems. *Proceedings of the Royal Society* A282, 508–520.

Currey, J.D. 1984. *The Mechanical Adaptations of Bones.* Princeton University Press.

Dally, J.W., & Riley, W.F. 1991. *Experimental Stress Analysis.* 3rd ed. New York: McGraw-Hill.

Fortelius, M. 1985. Ungulate cheek teeth: developmental, functional, and evolutionary interrelations. *Acta Zoologica Fennica* 180, 1–76.

Gordon, J.E. 1976. *The New Science of Strong Materials.* 2nd ed. Princeton University Press.

Gordon, K.D. 1982. A study of microwear on chimpanzee molars: implications for dental microwear analysis. *American Journal of Physical Anthropology* 59, 195–215.

Griffith, A.A. 1921. The phenomena of rupture and flow in solids. *Philosophical Transactions of the Royal Society* A221, 163.

Hylander, W.L. 1979. Mandibular function in *Galago crassicaudatus* and *Macaca fascicularis.* An in vivo approach to stress analysis of the mandible. *Journal of Morphology* 159, 253–296.

Janis, C.M., & Fortelius, M. 1988. On the means whereby mammals achieve increased functional durability of their dentitions, with special reference to limiting factors. *Biological Reviews* 63, 197–230.

Kingery, W.D., Bowen, H.K., & Uhlmann, D.R. 1976. *Introduction to Ceramics.* New York: John Wiley & Sons.

Koenigswald, W. von 1988. Enamel modification in enlarged front teeth among mammals and the various possible reinforcements of the enamel. In *Teeth Revisited: Proceedings of the VIIth International Symposium on*

Dental Morphology, ed. D.E. Russell, J.-P. Santoro, & D. Sigogneau-Russell. *Mémoires du Muséum national d'Histoire naturelle, Paris (series C)* 53, 147–167.

Koenigswald, W. von, Rensberger, J.M., & Pfretzschner, H.U. 1987. Changes in the tooth enamel of early Paleocene mammals allowing increased diet diversity. *Nature* 328, 150–152.

Lanyon, L.E. 1976. The measurement of bone strain in vivo. *Acta Orthopaedica Belgica* 42, 98–108.

Laub, R.S. 1992. The masticatory apparatus of the American mastodon *(Mammut americanum). Journal of Vertebrate Paleontology* 12, 38A–39A.

Lucas, P.W. 1982. Basic principals of tooth design. In *Teeth, Form, Function and Evolution,* ed. B. Kurtén, pp. 154–162. New York: Columbia University. Press.

Maas, M.C. 1991. Enamel structure and microwear: an experimental study of the response of enamel to shearing force. *American Journal of Physical Anthropology* 85, 31–49.

Parker, E.R. 1957. *Brittle Behavior of Engineering Structures.* New York: John Wiley & Sons.

Pfretzschner, H.U. 1988. Structural reinforcement and crack propagation in enamel. In *Teeth Revisited: Proceedings of the VIIth International Symposium on Dental Morphology,* ed. D.E. Russell, J.-P. Santoro, & D. Sigogneau-Russell. *Mémoires du Muséum national d'Histoire naturelle, Paris (series C)* 53, 133–143.

Pfretzschner, H.U. 1992. Enamel microstructure and hypsodonty in large mammals. In *Structure, Function and Evolution of Teeth,* ed. P. Smith & E. Tchernov, pp. 147–162. London & Tel Aviv: Freund Publishing House.

Rasmussen, S.T., Patchin, R.E., Scott, D.B., & Heuer, A.H. 1976. Fracture properties of human enamel and dentin. *Journal of Dental Research* 55, 154–164.

Rensberger, J.M. 1973. An occlusion model for mastication and dental wear in herbivorous mammals. *Journal of Paleontology* 47, 515–528.

Rensberger, J.M. 1975. Function in the cheek tooth evolution of some hypsodont geomyoid rodents. *Journal of Paleontology* 49, 10–22.

Rensberger, J.M. 1986. Early chewing mechanisms in mammalian herbivores. *Paleobiology* 12, 474–494.

Rensberger, J.M. 1987. Cracks in fossil enamels resulting from premortem *vs.* postmortem events. *Scanning Microscopy* 1, 631–645.

Rensberger, J.M. 1988. The transition from insectivory to herbivory in mammalian teeth. In *Teeth Revisited: Proceedings of the VIIth International Symposium on Dental Morphology,* ed. D.E. Russell, J.-P. Santoro, & D. Sigogneau-Russell. *Mémoires du Muséum national d'Histoire naturelle, Paris (series C)* 53, 351–365.

Rensberger, J.M. 1992. Relationship of chewing stress and enamel microstructure in rhinocerotoid cheek teeth. In *Structure, Function and Evolution of Teeth,* ed. P. Smith & E. Tchernov, pp. 163–183. London & Tel Aviv: Freund Publishing House.

Rensberger, J.M., Forstén, A., & Fortelius, M. 1984. Functional evolution of the cheek tooth pattern and chewing direction in Tertiary horses. *Paleobiology* 10, 439–452.

Rensberger, J.M., & Koenigswald, W. von. 1980. Functional and phylogenetic interpretation of enamel microstructure in rhinoceroses. *Palaeobiology* 6, 477–495.

Rensberger, J.M., & Pfretzshner, H.U. 1992. Enamel

structure in astrapotheres and its functional implications. *Scanning Microscopy* 6, 495–510.

Savage, R.J.G. 1955. Giant deer from Lough Beg. *The Irish Naturalists' Journal* 11, 1–6.

Sloan, R.E. 1987. Paleocene and latest Cretaceous mammal ages, biozones, magnetozones, rates of sedimentation, and evolution. In *The Cretaceous-Tertiary boundary in the San Juan and Raton Basins, New Mexico and Colorado*, ed. J.E. Fassett et al. *Special Papers of the Geological Society of America* 209, 165–200.

Stern, D., Crompton, A.W., & Skobe, Z. 1989. Enamel ultrastructure and masticatory function in molars of the American opossum, *Didelphis virginiana. Zoological Journal of the Linnean Society* 95, 311–334.

Swartz, S.M. 1991. Strain analysis as a tool for functional morphology. *American Zoologist* 31, 655–669.

Teaford, M.F. 1988. A review of dental microwear and diet in modern mammals. *Scanning Microscopy* 2, 1149–1166.

Teaford, M.F., & Byrd, K.E. 1989. Differences in tooth wear as an indicator of changes in jaw movement in the guinea pig *Cavia porcellus. Archives of Oral Biology* 34, 929–936.

Tetelman, A.S., & McEvily, A.J. 1967. *Fracture of Structural Materials*. New York: John Wiley & Sons.

Van Valkenburgh, B. 1988. Incidence of tooth breakage among large predatory mammals. *American Naturalist* 131, 291–302.

Van Valkenburgh, B., & Ruff, C.B. 1987. Canine tooth strength and killing behavior in large carnivores. *Journal of Zoology, London* 212, 379–397.

Van Valkenburgh, B., Teaford, M.F., & Walker, A. 1990. Molar microwear and diet in large carnivores: inferences concerning diet in the sabretooth cat, *Smilodon fatalis. Journal of Zoology, London* 222, 319–340.

Vincent, J.F.V. 1982. *Structural Biomaterials*. London: Macmillan Press Ltd.

Waters, N.E. 1980. Some mechanical and physical properties of teeth. In *The Mechanical Properties of Biological Materials*, ed. J.F.V. Vincent & J.D. Currey, pp. 95–135. Cambridge University Press.

Yettram, A.L., Wright, K.W.J., & Pickard, H.M. 1976. Finite element stress analysis of the crowns of normal and restored teeth. *Journal of Dental Research* 55, 1004–1011.

Zienkiewicz, O.C., & Taylor, R.L. 1989. *The Finite Element Method*, 2 vols. London: McGraw-Hill.

10

The structural consequences of skull flattening in crocodilians

ARTHUR B. BUSBEY

ABSTRACT

Crocodilian evolution has been marked by the progressive evolution of a highly akinetic, flattened skull and the formation of a bony secondary palate. Whereas the rostrum of crocodilian precursors was lightly constructed with vertical to subvertical margins (oreinirostral), the rostrum and rostral anchor areas in crocodilians became suturally well integrated, thickened, and flattened, subtubular (platyrostral) to tubular in cross section. Rostral flattening had profound effects on a variety of mechanical parameters related to cross-sectional geometry that are interpreted herein as adaptive measures that strengthened the skull, reflecting different feeding behaviors. Secondarily, in at least two and probably more crocodilian clades, rostral morphology converged on that of the oreinirostral condition, probably reflecting changes in feeding habit.

INTRODUCTION

Crocodilians are extant archosaurs that have undergone numerous adaptive radiations best reflected in a wide variety of head shapes (Langston 1973; Meyer 1984). Although widely regarded as unchanged "living fossils" (but see Meyer 1984), crocodilians have attained a variety of cranial morphologies, not all of which are reflected in extant taxa, and the living species provide a valuable tool for interpreting the anatomical and behavioral limits of the group through time. The most obvious differences in head shape, within crocodilians and between them and other archosaurs, are related to rostral shape (Figure 10.1). The rostrum is that portion of the skull, anterior to the orbits, that is primarily concerned with trophic functions and whose structural anchor consists of most of the dermal and palatal bones in the vicinity of the orbits.

Early in crocodilian evolution, rostral features appeared that readily enable us to distinguish croco-

dilians from other archosaurs (Langston 1973; Benton & Clark 1988; Walker 1990; Sereno & Wild 1992). Many of these features reflect changes in the functional morphology of the head that are probably related to feeding (Busbey 1977). The first significant papers on the functional aspects of cranial and rostral anatomy of crocodilians are those by Iordansky (1964, 1973) and Langston (1973). These seminal papers contain abundant observations on functional cranial characters and speculation about the ultimate origin of the peculiar cranial anatomy of crocodilians.

Iordansky (1964, 1973) commented on various functional aspects of cranial structure within the extant crocodiles. Langston (1973) speculated on the functional significance of the main structural features of the crocodilian rostrum during crocodilian evolution (see section below on crocodilian phylogeny and rostral structural change) and provided numerous examples of convergent evolution of rostral morphologies. Most importantly, Langston suggested that certain features of the crocodilian rostrum are adapted to withstand axial torsion and bending stress.

Busbey (1977), in a preliminary mechanical study, concurred with Langston's suggestions that the crocodilian secondary palate may have initially formed to strengthen the skull rather than to provide a physical separation between the nasal and oral cavities. Pioneering engineering studies on the mechanical role of the secondary palate in mammals (Thomason & Russell 1986) support an initial strengthening function either in conjunction with or independently of any respiratory function.

In this study I will focus on several functional aspects of the crocodilian rostrum. First, there will be a brief investigation of the structural consequences of rostral shape-change in crocodilians, based on engineering beam theory. Second, the change in beam characteristics along the rostra of several taxa will

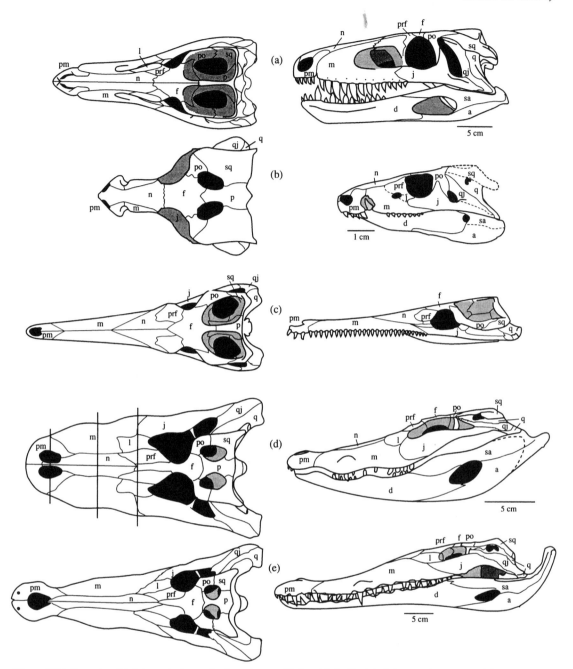

Figure 10.1. Dorsal and left-lateral views of selected crocodiloformid skulls. (a) *Sphenosuchus acutus* (after Walker 1990), (b) *Protosuchus richardsoni* (after Crompton & Smith 1982 and Busbey & Gow 1984), (c) *Metriorhynchus superciliosum* (after Steel 1973), (d) *Alligator mississippiensis* (after Iordansky 1973), (e) *Crocodylus acutus* (after Iordansky 1973). Skull (d) shows the approximate levels at which cross sections were measured (see text). Abbreviations: a, angular; d, dentary; f, frontal; j, jugal; l, lacrimal; m, maxilla; n, nasal; p, parietal; pm, premaxilla; po, post orbital; prf, prefrontal; q, quadrate; qj, quadratojugal; sa, surangular; sq, squamosal.

be discussed. Third, the historical pattern of changes in rostral shape will be considered.

CROCODILIAN ROSTRAL SHAPE

Crocodilians have been informally grouped into two categories based on rostral shape, *longirostrine* and *brevirostrine*. Longirostrine crocodilians (Figure 10.1c) have elongated, tubular rostra and brevirostrine crocodilians are the remaining forms (Figure 10.1d,e). These terms, however, are inadequate to describe the spectrum of shape differences. I propose additional terms to categorize the variation in crocodilian rostral shape. These terms do not represent discrete morphometric subdivisions but rather broad shape categories which are useful in the subsequent functional analysis.

Two primary terms describe rostral shape in lateral profile: *Platyrostral* means flat-snouted, and *oreinirostral* ("hill-like snout") describes forms having a deeper rostrum with a convex upper margin. Platyrostral may be subdivided into *tubular* (dorsoventral and lateromedial diameters are subequal), *broad* (lateromedial diameter more than twice the dorsoventral one), and *narrow* (lateromedial diameter between 1.2 and 1.9 times the dorsoventral one).

The ratio of rostral-to-skull length may also be used to define the relative length of the rostrum to the entire skull. Those taxa with a rostral length greater than 70 percent of basal skull length may be said to have *long* rostra, those between 70 percent and 55 percent have *normal* rostra and those less than 55 percent have *short* rostra (see Figure 10.2).

Some of the categories are shown in Figure 10.1. From the top: *Sphenosuchus* and *Protosuchus* are short oreinirostral forms; *Metriorhynchus* is long tubular platyrostral; *Alligator mississippiensis* is broad, short-to-normal platyrostral, and *Crocodylus acutus* is a long, narrow platyrostral form.

These categories are useful descriptors of some aspects of crocodilian cranial shape, but certainly do not define the entire shape. Because the categories are broad, they should not be viewed as taxonomic characters, and there is clear taxonomic overlap between taxa with similar head shapes. Both points are illustrated in Figure 10.2 in which the axes are ratios of cranial width and lengths for 127 extant and extinct crocodilian taxa. The abscissa represents the ratio of rostral length (anterior edge of orbits to tip of premaxilla) to basal skull length (basioccipital condyle to tip of premaxilla). The ordinate is the ratio of rostral width at the external nares to the width at the postorbital bars. It partially expresses rostral taper and relative rostral broadness (and is not included in the shape categories above). The

degree of overlap of taxa show that these terms are only broadly descriptive. Genera are almost entirely unsegregated, showing that rostral shape is not strongly correlated with phylogeny. Furthermore, there is considerable ontogenetic shape change within a species, shown by the gray polygon that encompasses a growth series of *A. mississippiensis* (data from Dodson 1974).

CROCODILIAN PHYLOGENY AND ROSTRAL STRUCTURAL CHANGE

It is now generally accepted that a study of the functional aspects of any group is useful only within the context of a reasonable phylogenetic model of the group (see Weishampel or Witmer, this volume). Although the intent of this chapter is to focus on the mechanical consequences of particular rostral shapes, a brief description of crocodilian phylogeny is required. An extended discussion of the crocodilian phylogeny is beyond the scope of this chapter, but an understanding of the limitations and problems with current phylogenetic models is necessary and must preface a functional morphological study.

Figure 10.3 is a schematic cladogram, based on the "preferred" cladograms of Benton and Clark (1988) and the suggested sphenosuchid relationships of Sereno and Wild (1992) superimposed on time. For the most part the Benton–Clark cladogram is composed to minimize convergence between clades. It is a reasonable working approximation of relationships, although the position of some taxa is problematic. As the functional relationships between many cranial characters are not yet understood, the degree to which some characters may reoccur convergently is not well known.

For example, the complex of characters associated with oreinirostral morphology is not currently well understood. This may have resulted in the branching times of several Late Cretaceous/Early Tertiary oreinirostral genera (e.g., *Notosuchus, Baurusuchus, Libycosuchus, Araripesuchus,* and *Sebecus*) being pushed back into the Middle Jurassic. Their intermediate position on the cladogram, between the Metasuchia and Neosuchia, requires a host of precursors to have survived from the Jurassic to the Late Cretaceous and Tertiary without leaving a fossil record. This presumed clade is largely based on the presence of a subvertical maxilla and a relatively vertical postorbital bar. The latter is the morphological consequence of the former and should be expected in any oreinirostral crocodilian (and it certainly occurs in other oreinirostral archosaurs). In the general list above these characters probably represent independent, secondary reversion to an oreinirostral form rather than retention of primitive rostral morphology.

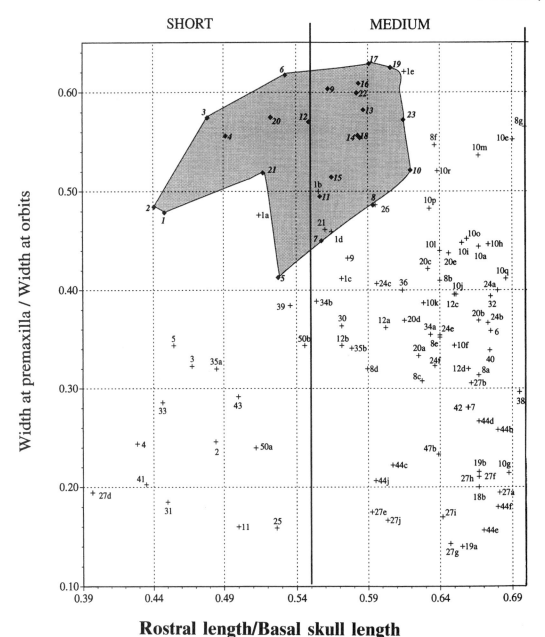

Figure 10.2. Scatter chart of rostral width and length ratios (described in text) for 127 extinct and extant crocodilian taxa. The gray area is the inclusive polygon for an ontogenetic series of 21 specimens of *Alligator mississippiensis* (from Dodson 1975) to indicate the ontogenetic shape variation in a single species. 1a, *Alligator mcgrewi*; 1b, *A. olseni*; 1c, *A. prenasalis*; 1d, *A. sinensis*; 1e, *A. visheri*; 2, *Alligatorellus pusillus*; 3, *Alligatorium meyeri*; 4, *Allognathosuchus haupti*; 5, *Araripesuchus gomesii*; 6, *Asiatosuchus germanicus*; 7, *Brachyuranochampsa eversolei*; 8a, *Caiman crocodilus apaporiensis*; 8b, *C. c. crocodilus*; 8c, *C. c. fuscus*; 8d, *C. c. yacare*; 8e, *C. latirostris*; 8f, *C. l. chacoensis*; 8g, *C. neivensis*; 9, *Ceratosuchus burdoshi*; 10a, *Crocodylus affinis*; 10b, *C. acer*; 10c, *C. actus*; 10d, *C. cataphractus*;

10e, *C. checchiai*; 10f, *C. intermedius*; 10g, *C. johnsoni*; 10h, *C. megarhinus*; 10i, *C. moreletii*; 10j, *C. niloticus*; 10k, *C. novae-guineae*; 10l, *C. palaeindicus*; 10m, *C. palustris palustris*; 10n, *C. porosus*; 10o, *C. rhombifer*; 10p, *C. robustus*; 10q, *C. siamensis*; 10r, *C. sivalensis*; 11, *Dakosaurus maximus*; 12a, *Diplocynodon darwini*; 12b, *D. gervaisi*; 12c, *D. hantoniensis*; 12d, *D. ratelii*; 13, *Dollosuchus dixoni*; 14, *Dyrosaurus phosphaticus*; 15, *Eosuchus lerichei*; 16, *Euthecodon nitriae*; 17, *Gavialis lewisi*; 18a, *Gavialosuchus americanus*; 18b, *G. eggenburgensis*; 19a, *Geosaurus gracilis*; 19b, *G. suevicus*; 20a, *Goniopholis felix*; 20b, *G. gilmorei*; 20c, *G. lucasii*; 20d, *G. simus*; 20e, *G. stovalli*; 21, *Hispanochampsa mulleri*; 22, *Ikanogavialis gameroi*; 23, *Kentisuchus (Croc.) spenceri*; 24a, *Leidyosuchus acutidentatus*; 24b, *L.*

LONG

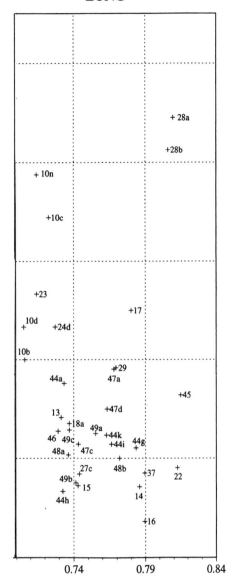

Vertical rostral edges arose independently in other crocodilian clades whose ancestors were platyrostral, including the eusuchians *Planocrania* (Li 1976, 1984), *Pristichampsus* (Langston 1975) and *Quincana* (Molnar 1977), and to a lesser degree in modern paleosuchid caimans. If Buffetaut (1986) and Chiappe (1988) are correct, then at least some oreinirostral taxa may have had platyrostral ancestors.

The most significant rostral changes within the Crocodylomorpha, from a functional perspective, were identified and listed by Langston (1973) and include:

1. Depression of the rostrum (i.e., increasing platyrostry).
2. Progressive development of a bony secondary palate, which proceeded in a well known sequence from protosuchian crocodilians, through "mesosuchians", into eusuchians.
3. Extreme development of scarf joints in the rostrum anterior to the orbits, best developed in platyrostral neosuchians (Figure 10.4).

After consideration of the cladogram and appropriate taxa, several additional changes should be appended to Langston's list:

4. Widening of the entire skull, starting posteriorly with the quadrates in protosuchians and eventually proceeding anteriorly into the rostrum in nonmarine forms.
5. Increase in adult size, and therefore in adult skull size.

The Sphenosuchia (Figures 10.1a & 10.3) had a kinetic (Walker 1990) oreinirostral skull with vertical maxillary margins, a vertical suspensorium that is primitive for the Archosauriformes (sensu Sereno 1991), and showed few tendencies toward platyrostry (though Walker 1990 noted some posterior inclination of the quadrates and slight development of palatal shelves in *Sphenosuchus*). Other archosaurs, such as *Proterochampsa* (Sill 1967) and the phytosaurs, attained a platyrostral skull independently of the Crocodyliformes, but retained the narrow rostrum and vertical suspensorium.

Protosuchians – including *Orthosuchus* and *Gobiosuchus* which were excluded from the Protosuchia by Benton and Clark (1988) – were relatively small as adults and retained an oreinirostral rostrum, but showed widening of the skull posteriorly to form

canadensis; 24c, *L. gilmorei*; 24d, *L. riggsi*; 24e, *L. sternbergi*; 24f, *L. wilsoni*; 25, *Libycosuchus brevirostris*; 26, *Melanosuchus niger*; 27a, *Metriorhynchus acutus*; 27b, *M. hastifer*; 27c, *M. blainvillei*; 27d, *M. brachyrhynchus*; 27e, *M. cultridens*; 27f, *M. laeve*; 27g, *M. moreli*; 27h, *M. palpebrosus*; 27i, *M. superciliosum*; 27j, *M. temporalis*; 28a, *Mourasuchus amazonensis*; 28b, *M. atopus*; 29, *Mycterosuchus nasutus*; 30, *Navajosuchus novomexicanus*; 31, *Notosuchus terrestris*; 32, *Orthogenysuchus olseni*; 33, *Orthosuchus stormbergi*; 34a, *Osteolaemus tetraspis osborni*; 34b, *O. t. tetraspis*; 35a, *Paleosuchus palpebrosus*; 35b, *P. trigonatus*; 36, *Paralligator sungaricus*; 37, *Pelagosaurus typus*; 38, *Pristichampsus rollinati*; 39, *Procaimanoidea utahensis*; 40, *Prodiplocynodon langi*; 41, *Protosuchus ricardsoni*; 42, *Sebecus icaeorhinus*; 43, *Sphenosuchus acutus*; 44a, *S. bollensis*; 44b, *S. brevidens*; 44c, *S. durobrivensis*; 44d, *S. edwardsi*; 44e, *S. intermedius*; 44f, *S. larteti*; 44g,

S. latifrons; 44h, *S. leedsi*; 44i, *S. megistorhynchus*; 44j, *S. obtusidens*; 44k, *S. telesauroides*; 45, *Stomatosuchus inermis*; 46, *Teleorhinus robustus*; 47a, *Teleosaurus cadomensis*; 47b, *T. calvadosi*; 47c, *T. deslongchampsi*; 47d, *T. gladius*; 48a, *Thoracosaurus macrorhynchus*; 48b, *Thoracosaurus scanicus*; 49a, *Tomistoma lusitanica*; 49b, *T. machikanense*; 49c, *T. schlegelii*; 50a, *Uruguaysuchus aznaresi*; 50b, *U. terrai*.

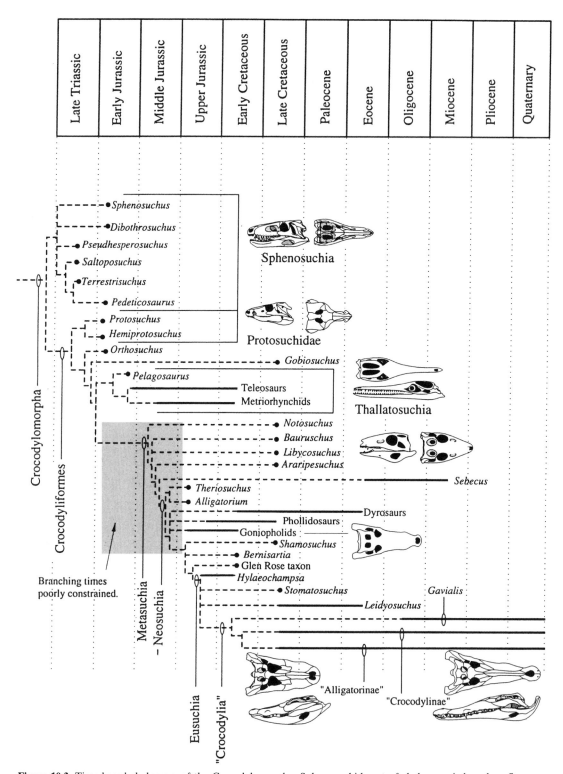

Figure 10.3. Time-based cladogram of the Crocodylomorpha. Sphenosuchid part of cladogram is based on Sereno and Wild (1992); remainder is largely based on cladograms in Benton and Clark (1988).

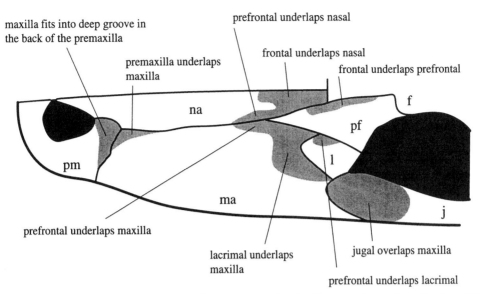

maxilla fits into deep groove in
the back of the premaxilla

premaxilla underlaps
maxilla

prefrontal underlaps nasal

frontal underlaps nasal

frontal underlaps prefrontal

na

pm

ma

f

pf

l

j

prefrontal underlaps maxilla

lacrimal underlaps
maxilla

jugal overlaps maxilla

prefrontal underlaps lacrimal

Figure 10.4. Dorsal view of left half of skull of *Alligator mississippiensis* with areas of sutural overlap (scarf joints) indicated in gray stippling. Abbreviations as in Figure 10.1.

the first true crocodilian skull deck. This posterior widening is due to posterolateral "migration" of the quadrate condyles and concomitant lateral spreading of the squamosals and postorbitals (Figure 10.1b). *Orthosuchus stormbergi* (Nash 1975) shows some tendency toward platyrostry (although the type specimen is slightly dorsoventrally crushed, somewhat exaggerating the degree of platyrostry). In comparison with sphenosuchians, the protosuchians have reduced antorbital fenestrae (potentially weak areas in the maxillary wall), concomitantly thickened bone in the maxillary wall, and variably developed palatal shelves. Protosuchians also have a fused basicranialaxis resulting in an akinetic skull.

Although there is some ventrolateral expansion of the pterygoid flanges in protosuchians, they do not appear to have provided much medial support function for the mandibles as occurs in more platyrostral taxa. *Sphenosuchus* had relatively large lateral pterygoid flanges (Walker 1990), but as the pterygoids were unfused to the braincase they could not have acted as medial mandibular braces as do the pterygoids of more derived crocodyliformids.

The marine thallatosuchians (sensu Benton & Clark 1988; Figure 10.3) retained a narrow snout, wherein the rostrum became greatly depressed anterior to the orbits and elongated into a tubular platyrostral shape (synonymous with the longirostrine condition) appropriate for piscivory (Figure 10.1c). These were also the first crocodyliformids to attain large size as adults.

In thallatosuchians the pterygoid extends back under the braincase and dorsally around part of the basisphenoid, forming a dorsoventrally oriented pillar and giving a vertical appearance to the back

of the skull (Busbey & Gow 1984). Structurally this pillar firmly unites the dermal bones of the skull (upper plate) and the palatal region (lower plate) posteriorly, eliminating any tendency for these two plates to move relative to one another. Within this group is the earliest evidence of extensive suturing between the opisthotic and other laterally oriented braincase elements (Langston 1973) and the lower and upper plates, finalizing the form of the braincase pillar.

The first appearance of narrow and broad platyrostral taxa occurs in the large Jurassic goniopholid metasuchians and in the diminutive taxa *Theriosuchus* and *Alligatorium* (Figure 10.3). These taxa also have a more extensive secondary palate (Langston 1973) and show the extensive overlapping scarf joints that anchor the rostrum to the postorbital portion of the skull. In these forms a diminutive antorbital fenestra is variably present. Goniopholids are also the first taxa to possess greatly ventrolaterally expanded pterygoid flanges for stabilizing the mandibles posteriorly. This pterygoid stabilization, so ubiquitous in all more derived crocodilians, probably serves (1) to resist medial traction of the mandibles produced by the more medial orientation of the jaw musculature (Busbey 1977, 1982) and (2) to resist any mediolateral movements of the mandibles during active use of the jaws.

Interestingly, some of the oreinirostral metasuchian forms and eusuchian forms share several characteristics of the rostrum, aside from the elevated rostrum itself, that hearken back to oreinirostral ancestors. These include reduced rostral scarf joints and much thinner bone (see cross sections of *Sebecus iceaorhinus* in Figures 10.7 & 10.8).

The rostral morphologies discussed above, then, reoccur in various taxa at subsequent times in crocodilian history as the result of convergent evolution (Langston 1973).

Bones of the secondary palate

One of the most significant and well-known structural changes of the crocodilian skull was the development of an extensive bony secondary palate. Langston (1973) described the progressive roofing of the oral chamber and its separation from the nasal cavities, following the grade approach first used by Huxley (1875) when he defined the Protosuchia, Mesosuchia (now considered paraphyletic), and Eusuchia. These groups were largely defined on the position of the internal nares and, therefore, the relative completeness of the secondary palate. Though the cladogram of Figure 10.3 no longer contains the Mesosuchia as a clade, the progressive formation of the secondary palate, as Langston described, did occur.

The functional separation of the nasal and oral cavities was accompanied by many other transformations in the rostrum itself and, presumably, in the cartilaginous nasal conchae and the anterior portions of the anterior pterygoid muscle that partially arise from the posterior wall of the posterior nasal conchae and surrounding regions (Iordansky 1964; Schumacher 1973; Busbey 1986). These other changes are not considered here.

The eusuchian crocodilian secondary palate is composed, from front to back, of medial projections of the premaxillae, maxillae, palatines, and pterygoids. In addition, nasal diverticulae may pass laterally into other pneumatic spaces of the skull. The narial passageway enters through the premaxilla, along the dorsum of the premaxillary and maxillary shelves, into a partial enclosure in the palatines and eventually into a tunnel formed from the posterior portions of the palatines and pterygoids. It exits at the internal nares at the far posterior edge of the pterygoids.

The efficient submerged respiration of modern crocodilians is possible because of the close proximity of the internal nares to the glottis and the ability to seal off the pharynx with an enormous gular fold that arcs across the pterygoids just anterior to the internal choanae. It is difficult to imagine that the glottis was in such close, functional contact with the internal nares in protosuchians because the glottis would have been so far forward that there would have been little space for the tongue. The close relationship between the choanae and pterygoids suggest that no gular fold was present in protosuchians. The unfused secondary palate was well developed long before the posterior edge of the choanae had

migrated far enough posteriorly to approach the glottis. Unless the long anterior nasal vacuities seen in mesosuchians were completed ventrally in a tough connective tissue, I think it improbable that a functionally significant gular fold was present in a primitive "mesosuchian" such as *Eutreturanosuchus*. An open palate might also have caused problems for animals with the feeding habits of modern crocodilians (Langston 1973). In any case, the respiratory advantage must have been slight during the early stages of palate formation.

I will return to a discussion of the possible strengthening role of the incipient palatal shelves and of the entire palate in later sections.

Scarf joints in the crocodilian rostrum

The aim of describing the structural changes in the crocodilian rostrum is to set the scene for the following mechanical analysis. In the context of rostral mechanics, the nature of the sutures between the bones is important, particularly as many form robust overlapping scarf joints rather than simple butt joints. (See Hildebrand 1974 for illustrations of these joint types.)

Scarf joints are well developed in *A. mississippiensis*, for example (Figure 10.4). The largest scarf joint occurs between the maxilla and jugal, oriented so as to prevent medial and longitudinal turning. Scarf joints are also present between the maxilla and lacrimal, lacrimal and prefrontal, prefrontal and frontal, prefrontal and nasal, and frontal and nasal. The configuration is complex. The lacrimal underlies the maxilla anteriorly, its superior surface fitting into a groove on the ventral side of the maxilla to form a ventral lacrimal process. A supralacrimal process of the prefrontal extends forward into a groove floored by the lacrimal and roofed by the maxilla, in effect forming the reverse of the ventral prefrontal process that Bolt (1974) noted in some reptiles and amphibians. The sublacrimal process of the prefrontal protrudes anterior to the ventral lacrimal process and inserts into a shallow groove in the roof of the maxilla to produce yet another groove that is floored posteriorly by the ventral lacrimal process, floored anteriorly by the supralacrimal process of the prefrontal, and roofed by the maxilla. A posterior submaxillary process of the nasals extends into this groove. Thus a triple overlap is effected, supporting the rostrum against lateral torsion and longitudinal bending. An anterior frontal process extends forward beneath the nasals and laterally under the prefrontal pillar. This effectively transfers stress from the nasals into the temporal arcade. The nasals and maxillae are anchored anteriorly by inserting together as a "tongue" into a "groove" formed by the premaxilla.

MECHANICAL CONSEQUENCES OF PLATYROSTRY

I have discussed the secondary palate and scarf joints as mechanical mechanisms within the crocodilian skull. Undoubtedly mechanical factors have had a role in cranial design in crocodilians. I now want to consider the mechanical consequences of evolving the platyrostral form, and to consider features of the rostrum that seem to be consequent to rostral flattening.

I will consider tubular platyrostral forms only briefly in this discussion, as that form appears to be a specialization for piscivory. Tubular platyrostral (longirostrine) taxa are also hyperdentate to varying degrees. The teeth are uniformly thin and elongate and seem to be adapted primarily for puncturing and detaining relatively docile prey in water. This tubular rostral form probably first appeared in the problematical *Eopneumatosuchus* (taxonomic status unclear, Crompton & Smith 1982) and was present in a variety of taxa. A tubular hyperdentate rostrum is extremely efficient for piscivores because (1) the rostrum has approximately the same cross section in all potential directions of movement and thus minimizes drag in all directions, and (2) the large number of teeth increases the chances of catching prey and facilitates transport towards the pharynx. As this morphology is apparently an adaptation for piscivory it will not be considered further in this chapter, and I will concentrate on the mechanics of nontubular platyrostral and oreinirostral morphologies.

The platyrostral and oreinirostral forms under consideration may all be assumed to have been carnivorous and, therefore, to have had feeding behaviors broadly similar to those in extant reptilian carnivores. Under ideal circumstances we would have data on specific feeding behaviors for extant crocodilians and on the forces acting on the rostrum during its use. We would then compare known force patterns with the mechanical properties of the rostrum in extant forms, and use this comparison as a basis for inferring functional relationships for the fossil forms. As is usual in this type of study, we are constrained to work with an incomplete set of data. First, there are no extant oreinirostral crocodilians and, second, we only have behavioral observations for a subset of extant platyrostral forms, with no data on the forces associated with those behaviors.

My approach is to first consider the types of loading that known feeding behaviors in extant reptilian carnivores may exert on the rostrum. My examples are *A. mississippiensis* and the Komodo dragon, *Varanus komodoensis*, whose rostral morphology approximates the oreinirostral condition. Then I introduce the concepts of beam theory that will be used to assess the mechanical properties of selected

rostra. I use these methods on some modeled shapes to illustrate the general consequences of shape change on structural strength, and I finally quantify the strength of sections through some real rostra.

Feeding behavior and rostral loading

Several behaviors would seem to have the potential to exert the greatest stresses exerted on the rostrum of modern (and, by inference, extinct) platyrostral crocodilians. These behaviors include: (1) *biting down on prey* in the mouth; (2) *rolling* (Cott 1961; Busbey 1977, 1982), which is invoked during prey dismemberment, prey destabilization, and while attacking other conspecific crocodilians (Meyer 1984); and (3) *pitching or yawing of the head* with large prey in the mouth (McIlhenny 1935; Pooley & Gans 1976) as part of the prey acquisition and manipulation behavior of inertial feeding (Gans 1969; Busbey 1986, 1989).

Simple biting (mandibular adduction) should produce vertical loading on the rostral margins (Figure 10.5). Vertical loading components produced simultaneously on both jaw margins produce compression in a vertical plane and shear stress in an inclined plane (the side of the rostrum). These components should also cause torque around sutures (with torque increasing medially) and dorsoventral bending of the snout. Unilateral biting causes medially directed forces and increases axial torque (Bolt 1974). In *V. komodoensis* (see the excellent summary in Auffenberg 1981), large prey are taken by a series of rapid bites where, subsequent to the bite, the head is pulled back, slicing out the skin and large pieces of either viscera or muscle. For these larger prey there appears to be no behavior that causes other than compressive loading along the rostral margins (see "rolling" below). If similar feeding behaviors were present in oreinirostral crocodilians, then these vertical compressive forces are expected to be the dominant forces acting on the rostrum of oreinirostral taxa.

Rolling is used both in the water and on land and as a mechanism for both destabilizing and dismembering prey. During rolling prey is tightly held in the mouth while the entire body is rotated, sometimes at very high rates. Rolling should cause (1) increased axial torsion about the rostral centroid and (2) medial deflection of loading vectors along the leading upper jaw margin and ventromedial deflection of loading on the leading edge of the lower jaw (Figure 10.5). Medial deflection of the loading due to rolling should cause greater compression in a more oblique plane (the sides of the snout) and shear in a more vertical plane.

Rostral bending (produced by flexing or pitching of the head with large struggling prey) should pro-

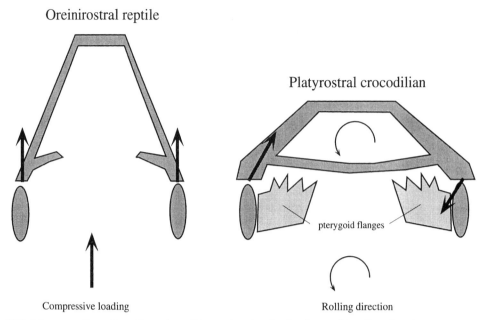

Figure 10.5. Diagrammatic sections through reptilian rostra showing loading and bending directions.

duce large vertical components along the entire margin of the jaw that are not localized as are those generated during biting, due to bending of the rostrum in both the sagittal and transverse planes. The stress is concentrated at the posterior end of the rostrum (Langston 1973; Bolt 1974) though bending occurs throughout the length of the snout. Mediolateral snout movements produce laterally directed forces, causing compression in the side of the face in the direction of movement and tension on the opposite side and mediolateral bending. Anteroposterior compression of the snout parallel to the midline would occur during frontal attacks on large prey.

Rolling is not exhibited by large terrestrial reptilian carnivores, such as *V. komodoensis* (Auffenberg 1981), which have a more oreinirostral skull. *Varanus* attacks its prey by biting down swiftly and retracting its head, which has the effect of slashing then slicing its prey. *V. komodoensis* has a rostral cross section not unlike that of many carnivorous dinosaurs and oreinirostral crocodilians. The largest loads induced in the rostrum of such a "conventional" feeding behavior should only be the result of biting (Figure 10.5), causing bilateral compression along the rostrum sides and greater dorsoventral (vertical plane) bending moments.

I suggest that the rostrum of a platyrostral crocodilian should function under a different set of mechanical constraints than the rostrum of *V. komodoensis*. It is reasonable to hypothesize that crocodilians with oreinirostral skulls fed in a manner more consistent with terrestrial carnivorous reptiles (slash and slice) than with modern platyrostral crocodilians (rolling).

Rostral shape may largely be a function of the interaction of several interdependent factors, including (1) adaptation of rostral shape for feeding specializations, (2) mechanical adaptation for minimizing the stress produced in loading, and (3) canalized developmental/anatomical constraints. With regard to point 2, I think there are primarily two activities responsible for such loads: (1) the feeding behavior used to take the largest prey in the diet (Busbey 1977), and (2) agonistic behavior (Meyer 1984). In effect, the rostrum is used similarly in both of these activities, so that the mechanical consequences of such behaviors are similar.

We must consider the following questions: (1) What are the mechanical consequences of platyrostry, and (2) what rostral features seem to represent an evolutionary response to these consequences?

BEAM THEORY MODELING OF SKULLS

Frazzetta (1968) employed a simple beam mechanical model to study the effects of the fenestration of the vertebrate skull in terms of section area and bending stress. Thomason and Russell (1986) used beam theory, but with more sophisticated techniques, in their study of the rostrum of the opossum *Didelphis* and the mechanical significance of the mammalian secondary palate. The beam model may be employed to investigate changes in the relative bending stress resulting from uniform rostral flattening or narrowing, as occurred during crocodilian evolution.

I will treat the crocodilian rostrum as a beam, using simplified beam section models, to investigate the mechanical consequences of rostral flattening and widening. The models are strictly heuristic and only approximate the actual cross-sectional anatomy of crocodilians.

A strict application of the beam model may not be entirely appropriate for the crocodilian snout for several reasons; for example, bone is both viscoelastic and anisotropic (Wainright et al. 1976; Currey 1984). Such a model is, however, a useful tool for understanding the mechanical implications of changes in rostral cross section. Additionally, extensive sutures may also affect stress distribution, wherein each bone would be treated as an individual beam with greater stress concentrated around the sutures. To simplify this study, sutures are assumed to have at least the same strength as bone and therefore, at least in the mechanical portion of the study, are considered to be inconsequential. Visual inspection of modern crocodilian skulls (mostly in museum collections and zoos) that suffered in vivo skull damage due to bullet wounds or clubs, invariably shows that fractures cross sutures and do not run along them. This suggests that, at least in living crocodilians, the sutures are not zones of rostral weakness.

Several mechanical parameters of the rostrum are dependent upon the distribution of mass in the rostral cross section. In a rostrum the distribution of mass (neglecting the effects of soft tissues such as nasal cartilages) is affected by bone thickness, the width and height of the rostrum, and by the presence or absence of additional features such as a secondary palate or bony internal struts, plates, and spicules. The mechanical parameters considered herein are the vertical (I_{xx}) and horizontal (I_{yy}) second moments of area of the section with respect to x- and y-axes through the section centroid. Relative increases or decreases in bending stress can be investigated through the relationship of bending stress to the moments of inertia.

Stress (σ) is related to the second moment of area by the basic beam theory formula:

$$\sigma = My/I$$

where M is the bending moment, y is the distance from the centroid (neutral axis) of the section to the section of mass under consideration, and I is the second moment of area. The ratio I/y is called the section modulus (Z) and reflects the distribution of mass around the neutral axis. To assess the effect of shape change on stresses, M was held constant and *relative stress* was calculated as the inverse of the section modulus. Relative stress, therefore, varies inversely with I and directly with y.

For any given rostral section, both the vertical and horizontal second moments can be determined with respect to the section centroid. The definitional equations for these parameters are presented in engineering and biomechanics texts (see Wainright et al. 1976) and can be calculated for any arbitrary plane section (see fig. 15.3c in Thomason, this volume). To model changes in rostrum cross section, a computer program and spreadsheet template were written to permit calculation of the parameters for arbitrary plane sections.

Geometrical models to show the effects of shape change

Simple models were calculated, starting with an inverted U-shaped section five units long on a side with a thickness of 0.5 units (representing a rostrum without a secondary palate). Section moments were calculated for a series of mock rostral sections where snout narrowness was increased and for sequences where the section was flattened (see insets in Figure 10.6). The same series of sections were calculated for a section with a "secondary palate" 0.25 units thick.

Model results are presented in Figure 10.6 in terms of vertical and horizontal moments of inertia (upper graphs) and relative stress (lower graphs). They imply the following:

1. A relative decrease in rostral depth (progressive platyrostry) results in exponential increases in dorsoventral bending stress but has little effect on mediolateral bending stress concentration, the results for σ_y even overlap (Figure 10.6d).
2. Narrowing the rostrum dorsally, while maintaining a wide palate, results in relative increases in stress concentration both mediolaterally and dorsoventrally. However, the increases due to narrowing are negligible compared to those that result from ongoing platyrostry.
3. In all cases the presence of a "secondary palate" (and probably any other bony plates or supports) results in relatively lower bending stress levels; the greatest differences are seen between progressively platyrostral sections.

The results for the models have several implications for the change in shape of the crocodilian rostrum. One of the most important mechanical consequences of increasing platyrostry was a concomitant increase in vertical bending stress (Figure 10.6b). As the relative bending rostral bending stress increased from oreinirostral protosuchians to platyrostral neosuchians, the likelihood of bending failure at the preorbital rostral anchor would increase.

The sides of the crocodilian rostrum became oriented at successively higher angles to the midsagittal plane with increasing platyrostry (Figure 10.4); that is, they became oriented more normal to

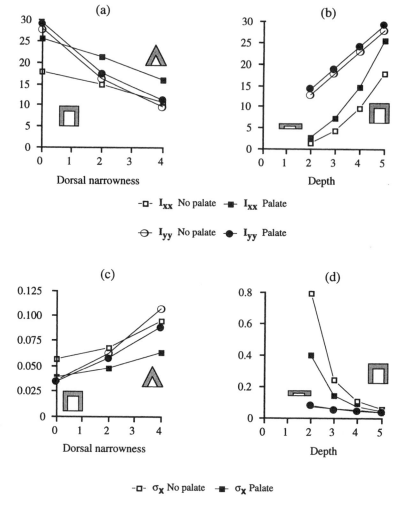

Figure 10.6. Second moments of area, (a) and (b), and relative bending stress, (c) and (d), for models of crocodilian rostra calculated from plane section equations (see text for details). I_{xx} and σ_x refer to dorsoventral moments and bending stress, and I_{yy} and σ_y refer to mediolateral moments and bending stress. Small insets represent the relative cross section shape, regardless of actual section thickness or presence or absence of the "secondary palate."

dorsoventral bending forces. Increasing rostral margin obliquity decreases the cross-sectional area of the snout, further increasing the relative bending stress (Figure 10.6) and orienting the side of the rostrum to increase the shear component of vertical loading. Because the shear resistance of bone is less than half its compressive strength (Hildebrand 1974), this would have the effect of reducing the structural integrity of the rostrum. Rostral broadening (acquisition of the narrow platyrostral condition, followed by the broad one) should also increase the torque on the center of the rostrum because it increases the moment of the jaw margins with respect to the section centroid.

The model suggests that relative dorsoventral bending stress levels increased greatly with increasing platyrostry. The addition of additional mass to the section, by way of thickening the bone and adding additional structures such as the secondary palate (even an incomplete one), significantly contributes

to the ability of the section to withstand vertical bending and axial torsion. Thomason and Russell (1986) showed that torsional strength and stiffness are also greatly increased by the presence of the secondary palate in *Didelphis*.

Beam theory analysis of crocodilian rostra

In addition to the purely theoretical portion of the study described above, sections of the rostra of eight different crocodilian taxa (seven extant and one extinct) were used (Figures 10.7, & 10.8). Exterior outlines were measured at three levels in each skull (see the three dark lines in Figure 10.1d). These are sections at the maxilla–premaxilla suture, at the fifth maxillary tooth, and at the level of the fronto–nasal suture. Functionally, these sections provide an indication of the distribution of mass at the point where the anterior biting segment (premaxilla) is anchored to the posterior biting segment (maxilla), at the

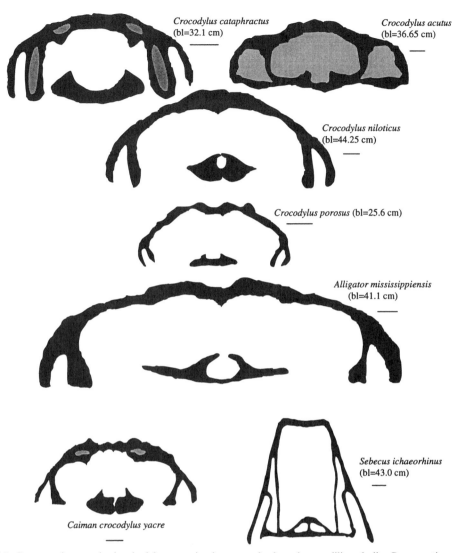

Figure 10.7. Cross sections at the level of frontoparietal suture of selected crocodilian skulls. Cross sections are slightly diagrammatic. Gray areas in cross sections are reconstructed based on inspection of skulls and are largely problematic. Cross section for *S. icaeorhinus* is based on a plaster reconstruction and inspection of original material. Abbreviation: bl, body length. Bar scales = 1 cm.

position usually associated with the largest (fifth) maxillary tooth, and at a position where the rostrum is anchored to the skull.

Outer skull boundaries were made with an adjustable engineer's contour. Internal boundaries were estimated, where possible, by visual inspection of the interrostral space through the external nares or from the orbits. As such, the internal measurements should be regarded as somewhat crude approximations of the actual sections. However, since the purpose of this portion of the study is to produce relative comparisons, these internal approximations should not invalidate the results.

Input of sectional boundaries for calculation was accomplished in two ways: (1) the program for cal-culating moments was modified to allow digitizer input, allowing direct digitizing of field notes; and (2) a macro was written for the National Institutes of Health Image application to permit interactive digitizing and calculation of section parameters from 8-bit grayscale images that were scanned in with a flatbed scanner or drawn by hand.

These real skulls provide some idea of the anteroposterior distribution of rostral mass and therefore the ability of the skull to resist bending stress. Moments calculated are from the raw skull measurements and do not account for size differences among specimens, which are to some extent reflected in the relative vertical positions of the data on the result charts. Snout length allometry changes

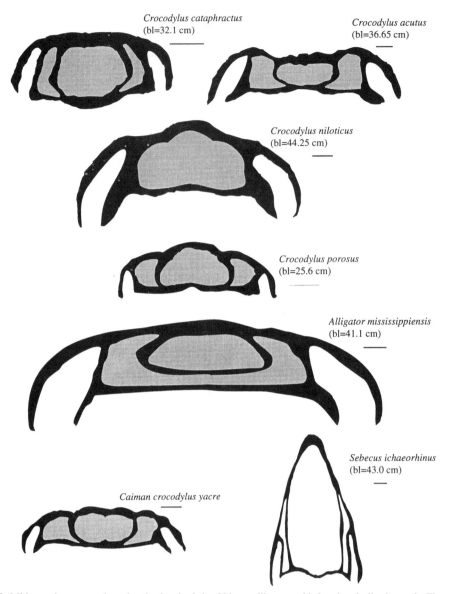

Figure 10.8. Midrostral cross sections (at the level of the fifth maxillary tooth) for the skulls shown in Figure 10.7, and the same comments apply. Abbreviations as in Figure 10.7. Bar scales = 1 cm.

relatively little during ontogeny in some platyrostral forms. Dodson (1974) reported a significantly positive coefficient of allometry (1.1; Bartlett's β) of rostral length with respect to basal skull length in *Alligator*.

Erickson (1976) published reconstructions of a growth series of *Leidyosuchus riggsi* from which I calculated reduced major axis allometric coefficients for two measures of rostral length. The allometric coefficient from the anterior tip of palatal fenestra to the tip of snout compared to basal skull length is 0.988, and for the anterior edge of orbits to tip of snout compared to basal skull length it is 0.960; neither is significantly different from 1.0 at the small

sample size (n = 5). Rostral width in *Alligator* is only slightly positive, especially toward the anterior tip of the snout (Dodson 1974), and in *L. riggsi* there is no significant departure from isometry in rostral width during growth.

The data from actual specimens are shown in Figures 10.9 and 10.10. The individual lines (specimens) are uncorrected for differences in size, so the absolute position along the abscissa (Figure 10.9) is not as important as are the shapes of the lines themselves. The profiles for *Sebecus* (oreinirostral), *Crocodylus niloticus* (narrow platyrostral), and the large *A. mississippiensis* (broad platyrostral) may be compared, since these skulls have about the same

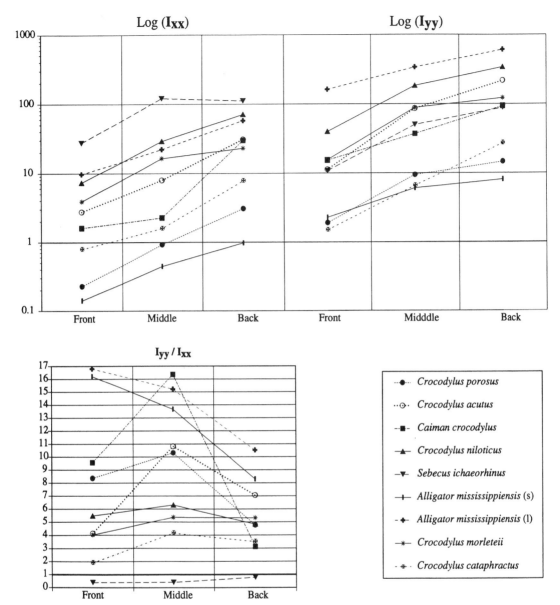

Figure 10.9. (*Upper panel*) log of the second moments of area (I_{xx} and I_{yy}) with respect to vertical axis through the section centroid for selected crocodilians. (*Lower left panel*) ratio of second moments of area through the section centroid for selected crocodilians. Large values represent greater difference between moments in the horizontal plane compared to moments in the vertical plane. Value $I_{yy}/I_{xx} = 1$ represents a cross section with two axes of symmetry, in this case one that is perfectly square.

basal length. The skulls of *C. cataphractus* and *C. acutus* have similar basal lengths and contrast a relatively tubular platyrostral taxon with a relatively narrow platyrostral form. In addition, data from small (15.4 cm basal length) and large (44.1 cm basal length) skulls of *A. mississippiensis* may be compared. Figure 10.10 isolates the comparable curves mentioned above.

Although the data set is small and the results too different to generalize about all crocodilians, several conclusions can be drawn that are either

obvious or might be testable with larger sample sizes (see Figure 10.10):

1. For a given basal skull length, taxa with a more tubular platyrostral morphology tend to have smaller section moments and therefore, for a comparable load, should experience greater bending stress.

2. Although there is variation among taxa, there are two basic patterns to the increase of inertial moments along the rostrum. Some have a nearly

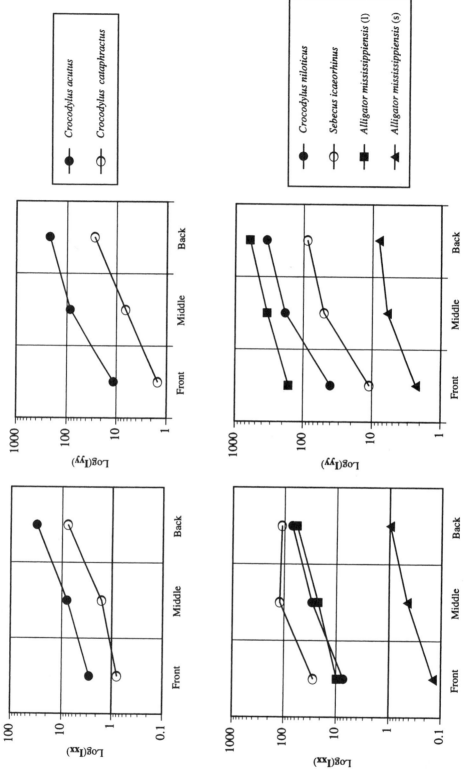

Figure 10.10. Comparisons of second moments of areas for selected crocodilians of approximately equal basal skull length (see text).

log-linear increase in moment (which in the actual animal represents an exponentially increasing ability to withstand bending stress along the rostrum in an anteroposterior direction), whereas most others seem to have a greater increase in section moment anterior to the largest maxillary teeth than from those teeth to the rostral anchor (which may indicate greater susceptibility to bending stress failure in the anterior portion of the rostrum).

3. An oreinirostral skull is much less susceptible to dorsoventral bending stress, for the same size, then is a more platyrostral skull. A platyrostral skull is much less susceptible to mediolateral bending stress, for the same size, than an oreinirostral skull. Therefore, a platyrostral skull is less prone to failure when experiencing loads out of a dorsoventral plane than is an oreinirostral skull.

4. There is a fairly constant log-linear increase in moment from the tip of the snout to the area of the orbits during the growth of *A. mississippiensis*, although there is an overall increase in the vertical and horizontal section moments as a function of size (change in the vertical positions of the curves).

FUNCTIONAL IMPLICATIONS OF PLATYROSTRY

We are now in a position to integrate the mechanical implications of shape in crocodilian rostra with topics introduced previously, including feeding behavior, the secondary palate, and scarf joints.

The feeding apparatus and feeding behavior

As the suspensorium flattened in protosuchians and aquatic "mesosuchians," the jaw adducting muscles would have been aligned more medially, reducing the vertical force component of adduction. This change suggests that effective power for slicing or slashing prey may have been decreased (both activities may have been compensated for by the buoyant effect of water). With this power loss and accentuation of medial and horizontal force components (indicated by the development of the lateral pterygoid flanges), the vertical forces acting along the edges of the rostrum decreased.

That feeding habits were changing as these modifications were acquired is indicated by (1) the modification of a primitive shearing dentition into conical, peglike teeth better adapted for puncturing and holding; (2) the broadening of the snout; (3) the development of the secondary palate; and (4) the posterior deepening of the mandible (Langston 1973). Posterior mandibular deepening is seen in animals whose mandibles are adapted for holding with little shearing (Hildebrand 1974).

Other modifications of feeding behavior are reflected in rostral shape. If increasing platyrostry has the potential for decreasing the mechanical integrity of the snout, then what are the advantages of such a shape? The flattened skulls of some labyrinthodont amphibians and chelonians have been shown to increase efficiency for "gape-and-suck" feeding (Stewart 1976; Taylor 1987). No modern crocodilians, however, exhibit such feeding behavior. Although it is possible that increasing platyrostry merely reflected an enhanced ability to remain cryptic near the water surface, I think the change in shape was due to more complicated factors.

Some indication of the answer may be gained by looking at the feeding habits of modern crocodilians. Gans (1969) noted that inertial feeders frequently possess rostra that have much smaller cross sections than the orbital and postorbital portions of their respective skulls. This reduction in cross section is especially important for those animals that move their heads in a viscous medium, since the smaller cross section minimizes initial drag and allows for faster angular velocities during lateral head movement. As previously stated, the swift side movement of the head is a useful technique in crocodilian feeding, and it is usually used with prey smaller than the predator.

Rolling and the loads it produces have been previously mentioned. A rostrum with low sides is more effective in withstanding compression and axial torsion during rolling than would an oreinirostral snout. A low rostrum might also be effective, as Bolt (1974) suggested, in minimizing the moment of the sutures relative to mediolateral forces.

The secondary palate and respiration

The initial structural impetus for the development of the secondary palate, as suggested by Langston (1973), is consistent with the results above in light of my suggestion (Busbey 1977) that it was the evolution of a rolling feeding behavior that provided the structural requirement for modifications of the rostrum and addition of the secondary palate (making the secondary palate of crocodilians an exaptation sensu Gould & Vrba 1982). Therefore, the correct paradigm for viewing the overall structure of the crocodilian rostrum is one related to expected changes in loading from a terrestrial reptilian carnivore feeding mode to that seen in crocodilians. Moreover, it suggests that it is inappropriate to view the secondary palate as merely a device that evolved to separate breathing and feeding functions.

Increasing platyrostry accompanies the structural

modifications associated with the lateral loadings and increased axial torque resulting from the evolution of rolling feeding behavior, and this would have the effect of decreasing both the vertical and horizontal moments of area of the rostral cross section. The structural results of a decrease in the moments would be decreasing resistance to axial torque and increased stress concentration in the "remaining" rostral cross section. The formation of a secondary palate adds mass back into the rostral section and helps to offset sectional mass losses due to increasing platyrostry.

Secondarily the bony palate acts as a structural tie across the bottom of the rostrum to support the sides of the face during the inequal distribution of forces that must accompany rolling. Collapse between the side of the rostrum and the palate is prevented by the thick internal buttressing between the lateral and palatal maxillary processes and the braincase "pillar" that stabilizes the skull posteriorly.

The significance of scarf joints at the rostral anchor

As discussed by Bolt (1974), the arrangement of sutures must be considered in an analysis of cranial functional morphology. Bolt summarized the functions of two sutural types considered as end members with respect to design for resisting compression, shear, and torsion. *Butt* joints tend to form in areas under little torsion, although frequently a complex interdigitating suture forms where torsion tends to occur. Bolt called such sutures *longitudinal*, because they are usually oriented parallel to the midline, and noted that anterior to the orbits they resist the compression produced along the marginal tooth row. The scarf joints in the crocodilian rostrum are oriented so as to be effective both parallel and perpendicular to the midline.

In crocodilians, longitudinal sutures appear well medial to the tooth row (Figure 10.2). The only butt joints occurring near the orbits are in the sutures between the lacrimal and jugal, jugal and prefrontal, and prefrontal and frontal just where they enter the orbit. These butt joints did not evolve because of force exerted on the side of the face by the dentition (a view that might be taken if only cross sections were used) but probably because of the stresses that concentrate around the orbits. The distribution of butt joints around the orbits indicates that the major stress directions are parallel to the orbit edges and are probably compressive. Butt joints also occur in the rostrum between opposing nasals and between opposing maxillae (the suture between the maxilla and nasal is slightly oblique with a straight dorsal edge that, in effect, is functionally a butt joint but is morphologically a tongue-and-groove joint). In the nasals, the primary force trajectories are probably

parallel to the midline, indicating that an antero-posterior shear component might exist during bending.

The force components caused by vertical loading near the largest maxillary teeth should tend to push the nasals together, and therefore the nasals have two major structural functions. The most important function is to transmit compressive forces back along the dorsal midline to the rostral anchor (thus helping to prevent upward bending) and the second function is to act as the keystone for the rostral arch, preventing the face from splitting along the midline when vertical components are produced along the tooth row. At the outer margins, the stress is parallel to the midline. This conclusion reinforces the bending-stress interpretation for the function of the palate.

Secondarily oreinirostral crocodilians

If a crocodilian fed like *Varanus*, that is, without the rolling feeding behavior, increased vertical compression at the jaw margins would be accompanied by increased vertical adductor forces, decreased axial torsion, and decreased nonvertical bending moments. Such changes would affect the rostrum and the preorbital bones that anchor the rostrum to the face. The elevation of the preorbital and orbital regions of the skull and/or narrowing of the palatal region into an oreinirostral shape should correlate with an increase in the magnitude of the vertical loading components and a decrease in axial torsion and mediolateral rostral loading; presumably this occurred in secondarily oreinirostral crocodilians such as *Pristichampsus* or *Planocrania*.

Langston (1975) noted that the palate in *Pristichampsus vorax* is vaulted rather than being slightly convex as it is in platyrostral crocodilians. Vaulting places the palate closer to the neutral axis of the rostrum and therefore decreases the usefulness of the palate in resisting torsion and dorsoventral bending. It does, however, form a sort of internal "flying buttress" that would help dissipate largely vertically directed forces along the tooth rows.

In *Pristichampsus* the oreinirostry increases posteriorly, whereas in *Sebecus* (Colbert 1946; Gasparini 1972; Busbey 1986), and *Baurusuchus* (Price 1955) both the rostral and postrostral portions of the skull are elevated. Although the secondary palate is well developed in the relatively small oreinirostral Cretaceous taxa and Sebecosuchia (sensu Buffetaut 1982), it is arched, relatively thin, and does not exhibit the internal buttressing present in platyrostral crocodilians. Although the basic *bauplan* of the platyrostral crocodilian skull is present in pristichampsines, the structural modifications represent a return to more conventional terrestrial reptilian carnivore feeding.

In *Pristichampsus* the bone in the sides of the snout and palate is relatively thinner than the bone in comparable positions in an *A. mississippiensis* or *C. niloticus* of similar size. Although the rostra of the available specimens of *P. vorax* cannot be disarticulated, the scarf joints are smaller and the area of the jugomaxillary suture is much reduced.

In summary, increasing platyrostry in early crocodilians was accompanied by structural modifications to facilitate new demands on rostral mechanics. The development of the secondary palate is possibly the most significant crocodilian adaptation for strengthening the rostrum. Analogous, though not homologous, adaptations are seen in the rostra of phytosaurs, turtles, and mammals (Thomason & Russell 1986; though Bramble 1977 suggested that the initial impetus for the development of the mammalian secondary palate was the development of a strong, agile tongue).

ACKNOWLEDGMENTS

I am indebted to Wann Langston for his keen insights on crocodilians. This research was partially funded by Texas Christian University Research Foundation Grant 5-3802.

REFERENCES

Auffenberg, W. 1981. *The Behavioral Ecology of the Komodo Monitor.* Gainesville: University of Florida.

Benton, M.J., & Clark, J.M. 1988. Archosaur phylogeny and the relationships of the Crocodylia. In *The Phylogeny and Classification of Tetrapods*, Volume 1: *Amphibians, Reptiles, Birds*, ed. M.J. Benton, *Systematics Association Special Volume* 35A, 295–338. Oxford: Clarendon Press.

Bolt, J.R. 1974. Evolution and functional interpretation of some suture patterns in Paleozoic labyrinthodont amphibians and other lower tetrapods. *Journal of Paleontology* 48, 434–458.

Bramble, D.M. 1977. Functional integration and the origin of the mammalian masticatory-auditory complex. *Journal of Paleontology, Supplement to No. 2*, 51, 4.

Buffetaut, E. 1982. Radiation évolutive, paléoécologie et biogéographie des crocodiliens mésosuchiens. *Mémoires de la Société Géologique de France. Nouvelle Série* 142, 1–88.

Buffetaut, E. 1986. Un Mésosuchien ziphodonte dans l'Eocene supérior de la Liviniere (Hérault, France). *Geobios* 19, 101–108.

Busbey, A.B., III 1977. Functional morphology of the head of *Pristichampsus vorax* (Crocodilia, Eusuchia) from the Eocene of North America. Unpublished M.A. Thesis, University of Texas at Austin.

Busbey, A.B., III 1982. Form and function of the jaw musculature of *Alligator mississippiensis*. Unpublished Ph.D. thesis, University of Chicago.

Busbey, A.B., III 1986. New material of *Sebecus* cf. *huilensis* (Crocodilia: Sebecosuchidae) from the Miocene La Venta Formation of Colombia. *Journal of Vertebrate Paleontology* 6, 20–27.

Busbey, A.B., III 1989. Form and function of the jaw apparatus of *Alligator mississippiensis*. *Journal of Morphology* 202, 99–127.

Busbey, A.B., III & Gow, C. 1984. A new protosuchian crocodile from the Upper Triassic Elliot Formation of South Africa. *Palaeontologia Africana* 25, 127–149.

Chiappe, L.M. 1988. A new trematochampsid crocodile from the Early Cretaceous of north-western Patagonia, Argentina and its palaeobiogeographical and phylogenetic implications. *Cretaceous Research* 9, 379–389.

Colbert, E.H. 1946. *Sebecus*, representative of a peculiar suborder of fossil Crocodilia from Patagonia. *Bulletin of the American Museum of Natural History* 87, 217–270.

Cott, H.B. 1961. Scientific results of an inquiry into the ecology and economic status of the Nile crocodile (*Crocodylus niloticus*) in Uganda and Northern Rhodesia. *Transactions of the Zoological Society of London* 29, 211–356.

Crompton, A.W., & Smith, K.K. 1982. A new genus and species of crocodilian from the Kayenta Formation (Late Triassic?) of northern Arizona. In *Aspects of Vertebrate History*, ed. L. Jacobs, pp. 193–217. Flagstaff: Flagstaff Museum of Northern Arizona.

Currey, J.D. 1984. *The Mechanical Adaptations of Bones.* Princeton: Princeton University Press.

Dodson, P. 1974. Functional and ecological significance of relative growth in *Alligator*. *Journal of Zoology* 175, 315–355.

Erickson, B.R. 1976. Osteology of the early eusuchian crocodile *Leidyosuchus formidabilis*, sp. nov. *Monograph Vol. 2: Paleontology*, The Science Museum of Minnesota, pp. 1–61.

Frazzetta, T.H. 1968. Adaptive problems and possibilities in the temporal fenestration of tetrapod skulls. *Journal of Morphology* 125, 145–158.

Gans, C. 1969. Comments on inertial feeding. *Copeia* 1969, 855–857.

Gasparini, Z.B. 1972. Los Sebecosuchia (Crocodilia) del territori Argentino. Consideraciones sobre so "status" taxonomico. *Ameghiniana* 9, 23–34.

Gould, S.J., & Vrba, E.S. 1982. Exaptation – a missing term in the science of form. *Paleobiology* 8, 4–15.

Hildebrand, M. 1974. *Analysis of vertebrate structure.* New York: John Wiley & Sons.

Huxley, T.H. 1875. On *Stagonolepis robertsoni*, and the evolution of the Crocodilia. *Quarterly Journal of the London Geological Society* 31, 423–438.

Iordansky, N.N. 1964. The jaw muscles of crocodiles and some relating structures of the crocodilian skull. *Anatomischer Anzeiger* 115, 256–280.

Iordansky, N.N. 1973. The skull of the Crocodilia. In *Biology of the Reptilia*, Part 4(D), ed. C. Gans, pp. 256–284. New York: Academic Press.

Langston, W., Jr. 1973. The crocodilian skull in historical perspective. In *Biology of the Reptilia*, Part 4(D), ed. C. Gans, pp. 263–289. New York: Academic Press.

Langston, W., Jr. 1975. Ziphodont crocodiles: *Pristichampsus vorax* (Troxell), new comb., from the Eocene of North America. *Fieldiana* 33, 291–314.

Li, J.L. 1976. [Fossils of Sebecosuchia discovered from

Nanxiong, Guangdong.] *Vertebrata PalAsiatica* (In Chinese) 14, 169–173.

Li, J.L. 1984. [A species of *Planocrania* from Hengdong, Hunan.] *Vertebrata PalAsiatica* (In Chinese with English summary) 22, 123–134.

McIlhenny, E.A. 1935. *The Alligator's Life History*. Boston: Christopher Publishing House.

Meyer, E.R. 1984. Crocodiles as living fossils. In *Living Fossils*, ed. S.M. Stanley & N. Eldridge, pp. 105–131. New York: Springer Verlag.

Molnar, R.E. 1977. Crocodile with laterally compressed snout: first find in Australia. *Science* 197, 62–64.

Nash, D. 1975. The morphology and relationships of a crocodilian, *Orthosuchus stormbergi*, from the upper Triassic of Lesotho. *Annals of the South African Museum* 67, 227–329.

Pooley, A.C., & Gans, C. 1976. The Nile crocodile. *Scientific American* 234, 114–124.

Price, L.I. 1955. Novos crocodilídeos dos Arenitos da Série Baurú, Cretáceo do Estado de Minas Gerais. *Anais da Academia Brasileira de Ciências* 27, 487–498.

Schumacher, G.H. 1973. The head muscles and hyolaryngeal skeleton of turtles and crocodilians. In *Biology of the Reptilia*, Part 4(D), ed. C. Gans, pp. 101–199. New York: Academic Press.

Sereno, P.C. 1991. Basal archosaurs: phylogenetic relationships and functional implications. *Journal of Vertebrate Paleontology* 11(Supplement to part 4), 1–51.

Sereno, P.C., & Wild, R. 1992. *Procompsognathus:* theropod, "thecodont" or both? *Journal of Vertebrate Paleontology* 12, 435–458.

Sill, W.D. 1967. *Proterochampsa barrionuevoi* and the early evolution of the Crocodilia. *Bulletin of the Museum of Comparative Zoology, Harvard University*, 135, 415–446.

Steel, R. 1973. Crocodylia. In *Encyclopedia of Paleoherpetology,* Part 16, ed. O. Kuhn, p. 116. Stuttgart: Gustav Fischer Verlag.

Stewart, J.B. 1976. A functional morphological study of gape and suck feeding in some aquatic vertebrates. Unpublished M.A. thesis, University of Texas at Austin.

Taylor, M.A. 1987. How tetrapods feed in water: a functional analysis by paradigm. *Zoological Journal of the Linnean Society* 91, 171–195.

Thomason, J.J., & Russell, A.P. 1986. Mechanical factors in the evolution of the mammalian secondary palate: a theoretical analysis. *Journal of Morphology* 189, 199–213.

Wainright, S.A., Biggs, W.D., Currey, J.D., & Gosline, J.M. 1976. *Mechanical Design in Organisms*. London: Renshaw.

Walker, A.D. 1990. A revision of *Sphenosuchus acutus* Haughton, a crocodylomorph reptile from the Elliot Formation (late Triassic or early Jurassic) of South Africa. *Philosophical Transactions of the Royal Society (Biological Sciences)* 330, 1–120.

11

Graphical analysis of dermal skull roof patterns

KEITH S. THOMSON

ABSTRACT

The pattern of dermal bones in the skull roof is one of the most useful of taxonomic "characters" for vertebrates, particularly the lower vertebrates where the large number of individual elements creates a greater quantity of potential variation and, therefore, information. These patterns also have potential significance as indicators of the underlying cranial mechanics. A simple method of graphical analysis is presented and tested, with the tentative conclusion that such methods can be used to look for the orientation of major stresses acting within the dermal skull due to the action of the jaw and respiratory musculatures.

INTRODUCTION

It has been known since the beginning of formal vertebrate paleontology that the pattern of the bones of the dermal skull roof contains a great deal of taxonomically useful information. In most cases, the general patterns will allow one to identify an organism to major group. The details will often identify the animal to genus and even species (as far as species can be recognized in paleontology).

Most paleontologists have been content to use dermal skull bone patterns for taxonomic purposes rather than to attempt to read any more from them. When it comes to matters of cranial function, for example, paleontologists, leaning heavily on extrapolations from living models, have turned to the jaw apparatus, in great part perhaps because of the importance of tracing out the course of jaw–ear evolution (see, for example, Crompton, this volume). But, if the dermal skull roof has so much taxonomic information, it must surely be possible also to explore other areas, and the most obvious of these is function.

In this paper I propose the use of simple methods of graphical analysis to give a new, integrative view of dermal bone patterns. By the word *pattern* I mean an arrangement of dermal bones comprising a consistently recognizable unit, such as the skull table or cheek plate, so that both the number and shape of bones is repeated from taxon to taxon. The definition of the unit, by convention, is topographic – cheek or skull table, for example – but at the same time this is usually functional also. I will take my examples principally from the lower vertebrates because they tend to have more elements within the dermal units and show a much wider diversity of structure than higher vertebrates. Thus, there is more information to deal with. The dermal skull roof of lower vertebrates differs from the complex cranial vault of mammals or modern reptiles in that very little of the jaw and gill musculature attaches directly to it. Because of the kinetism of the skull in most forms, the marginal dermal bones receive a large portion of the force of the bite. The cheek unit is usually involved in respiratory as well as feeding movements.

THE "CAUSES" OF DERMAL BONE PATTERNS

At least since Aristotle, science has been the study of cause. As within any morphological feature or set of features, the pattern of the bones in the skull roof is obviously a reflection of a number of different causal factors. In particular we may expect that phylogenetic, developmental, material, functional, and holistic factors are prominent in "causing" a particular pattern. As discussed in the next four subsections on influencing factors, any morphological feature is a function of its history, of the way in which it is made, of the material of which it is made, of the ways it works, and of the properties of other intersecting systems. Any given head bone may be part of, or affected by, several mechanical systems (feeding, respiration, posture) and bone as a tissue has both mechanical and physiological roles in the organism. It is not easy to tease out all the different factors, but it is worth asking whether causation of dermal bone patterns has a strong functional component.

Our present understanding of the cause and significance of these patterns is, almost by default, that they are produced by a combination of "packing" of discrete elements within a unit, modified by relative growth of the individual elements, and under the overall control of morphogenetic systems, particularly as influenced by the lateral line system. Because of our preoccupation with homology, we tend to concentrate on individual bones, and we assume that the pattern itself is determined first by the fact that a given number of bones is a given and, second, that the basic shape of each of these bones is predetermined. We assume that the existence of "bones" is real.

However, as Goodwin and Trainor (1983) have argued for the vertebrate limb, bones themselves may not be the "real" entities. Individual bones may not have discrete causation in the sense that they are specifically coded for genetically. Instead bones, their number, shape, and pattern of arrangement in the skull, are simply part of a much broader nexus of causes and controls. The phenotypic plasticity of bones, especially as revealed by experiments with mechanical loading, adds an additional element for analysis.

Phylogenetic factors

Historical constraint, of course, always seems to be a major factor in the control of dermal bone pattern. This is obvious from the fact that important commonalities are observed in clades where the individual taxa differ widely in size, shape, and function. For example, in certain lineages of vertebrates symmetrically shaped bones, located in the median line, are rare. In tetrapods, there is a median bone in the posterior skull table in only one taxon: *Ichthyostega*. In the Arthrodira, on the other hand, median elements are the rule.

However, phylogenetic constraint is really only the sum of all other constraints integrated over time, because what is inherited is a genetic/developmental pattern-generating mechanism and a set of functions, rather than structure itself. One difficulty with an exclusively phylogenetic approach to paleontology is that it tends to "ossify" the notion that "bones" are real entities that are transmitted through genealogical sequences, as if there were genes for parietals, nasals, and frontals. In order to understand the mechanisms and causes of evolutionary change, however, it is necessary to look beyond the bones themselves to the factors that cause and control their manifestation in the skull.

Developmental factors

Developmental processes actually create the phenotype, so it is the *process* that is inherited, not the bone. For example, any given bone in the skull roof of a fish forms from a complex primordium. The fact that the "same" bone forms in related fishes means that those fishes have inherited the same pattern-forming mechanism and also the controlling developmental conditions that produce a similar result. The capacity to produce different manifestations of the "same" bone, or a quite different element (say two bones instead of one), will be a product of the extent to which the pattern-forming mechanism can be changed genetically and epigenetically. But that capacity to change will not be infinite nor indeed will all conceivable variants be practically realizable. For example, it may be impossible for the mechanism to produce a pair of bones in the place of one (or even what is simpler, apparently, one bone instead of two). This constraint is a property of the pattern-generating process and may in part explain evolutionary "trends" (Thomson 1991).

Holistic factors

Perhaps the most difficult sets of factors to analyze in any evolutionary, morphogenetic, or functional system are those that we can call holistic – that is, having to do with the integrated properties of (in this case) the skull as a whole rather than the separate attributes of its various parts. For example, an especially important morphogenetic constraint in the generation and proliferation of dermal bone patterns in lower vertebrates is the connection between bone primordia and lateral line neuromast primordia (discussion and review in Thomson 1987, 1993). The direction of causality still is not fully understood – which causes the patterning of the other? – but the connection is often very strongly conserved throughout lineages. The position of a bone primordium on the skull surface reflects not only the migration of cells to the primordium and the growth of cells within the primordium, but also the interaction of the primordium with lateral-line neuromasts that have themselves migrated to that particular site. The final shape of the bone is therefore strongly determined by the functional requirements of the lateral-line sensors it encloses. These in turn will reflect such matters as the overall shape of the head and the lines of streamlined flow of water over the head. In short, the manifestation of any part of the head is strongly conditioned by the function, development, and history of the head as a whole. This makes analysis more difficult than at first would appear.

Functional factors

In this chapter I am particularly concerned with looking for functional factors in the control of dermal bone patterns. Can we find functional explanations (causalities) of the differences or similarities of pattern among taxa? The positions and shapes of elements in dermal bone patterns must in part reflect the mechanical functions of those elements. These functions are obviously multiple. The bones exist as protective structure for the underlying structures – braincase, jaw muscles, palate, etc. They serve as points of anchor for the dentition (marginal and palatal). But, most importantly, the bony elements serve to receive and transmit from place to place forces generated in the skull during feeding and respiration. They act both as buttresses and as levers.

Traditionally, the most useful aspect of dermal bones to paleontologists has been the fact that all the bones within a given skull are of different and consistent shapes. This information is vital to the interpretation of homology, and paleontologists have always attempted to explain differences in the patterning of bone arrangement in the skull in terms of evolutionary sequences. However, we need to look further at the shapes of the dermal bones in vertebrates to try to see what the particular mechanical attributes of the dermal bone patterns might be.

It has been known for at least a hundred years (Wolff 1892; Roux 1895; Carter 1987; Carter, Wong, & Orr 1991) that the internal architecture and external shapes of individual bones are adapted for particular functional regimes and can be experimentally modified by changes in loading (see Thomason, this volume). There has been less work on the patterning of groups of bones, although Wong and Carter (1988) used computer modeling to match different stress regimes to structural patterning in the human sternum.

In the case of fossil forms, we cannot experiment with dermal bone patterns to find their function although there are some curiosities concerning the shapes of bones in vertebrates that we would like to explain. They are almost always quite regular in shape, for example. Triangular bones rarely exist. There are few odd shapes like "keyholes." There are no tori: Any aperture in the dermal skull is almost always enclosed within a pair of flanking bones.

GRAPHICAL ANALYSIS OF DERMAL BONE PATTERNS

The number, size, shape, and relative position of individual dermal elements creates the dermal bone pattern. Although an isolated bone has a shape that we define by its margins, the *patterns* that we see when we draw a skull or partial skull are, in fact, a drawing of the sutures. What follows is an attempt to find a new way of looking at dermal bone patterns, by concentrating on the position of bones and the orientations of the sutures, rather than the shapes of individual bones.

In order to do this I will use a very simple graphical technique, applied as an example first to the skull roof of a common Paleozoic amphibian (*Colosteus*, as described by Romer 1947).

STEP 1. The simplest approach (Figure 11.1a) is to take an outline of a dermal bone pattern and to draw a series of lines perpendicular to the sutures at the midpoint of the segment of that suture between two bones. Although there seems to be very little system to the shapes of the dermal bones of the skull (and therefore their sutural connections), these short perpendicular line segments turn out to be highly oriented. They reveal a different kind of order in the skull pattern. We can call these lines *trans-suture lines*.

STEP 2. A very simple refinement of this method is to join up the perpendicular lines to form a *trans-suture web* (Figure 11.1b). It is immediately apparent that, with only very minor "artistic license," these webs form a series of very graceful curves and straight lines. They iron out all the irregularities of the bones and sutures and integrate dermal bone shape across the whole skull. (The problems of graphically representing a three-dimensional object in two dimensions makes it impossible to continue the analysis accurately around the sides of the skull within a single image.)

STEP 3. As a further elaboration, the lines of the web can be drawn to a width proportional to the length of the sutures across which they pass (Figure 11.1c). This allows us to represent graphically a sense of both the net orientations of sutures and their relative dimensions.

Obviously the method is extremely simplistic. However, even with this very simple analysis some generalizations appear. For example, the shapes of many bones seem to be such that the trans-suture lines come to meet at one or more points *within* the body of the bone (Figure 11.2a), usually near its center. This is a function of the fact that many dermal bones are roughly rhomboidal in shape. However, it is perfectly possible to imagine shapes of bones in which this would not happen (Figure 11.2b). Such shapes are common, for example, in marginal elements. An example of this is seen in the cheek of an osteolepiform (Figure 11.3) with its long, low, triangular maxilla. These trans-sutural lines and webs

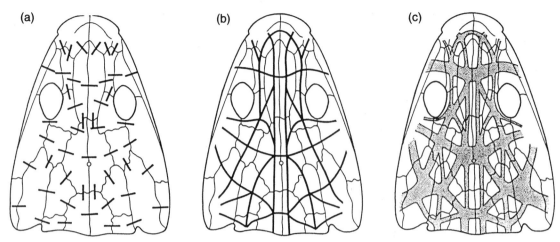

Figure 11.1. Simple graphical analyses of suture patterns in skull roof of Paleozoic amphibian *Colosteus*. (a) Short *trans-suture* lines drawn perpendicular to sutures indicate orientations of stresses that sutures resist. (b) Lines in (a) are linked together to form a *trans-sutural web*, showing the web of probable lines of stress resistance acting in the skull. (c) Lines in (b) are drawn to a width corresponding to length of sutures through which they pass, thus indicating relative strengths of the sutures. Note that in all cases it is not possible to represent accurately the lines acting at the curved edges of the skull.

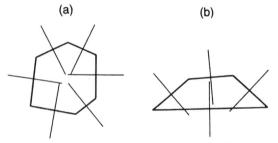

Figure 11.2. Two representative bone shapes, with lines drawn perpendicular to the sutural edges. (a) An irregular rhomboid, typical of the shape of bones in most skulls. Note that the lines meet near the center of the bone. (b) A hypothetical bone drawn to show the extreme shape that is necessary for the lines to fail to meet within the body of the bone.

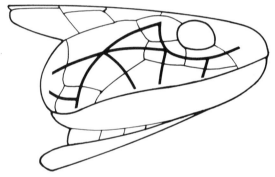

Figure 11.3. Cheek of an osteolepiform osteichthyan showing the orientation of trans-suture lines.

are, therefore, not identical with webs drawn simply to connect the centers of all bones.

Sensory lines usually pass through the centers of growth of dermal bones, and therefore there is some overlap between the arrangement of the trans-sutural webs and the pattern of the lateral-line system over the skull. However, because many of the major dermal bones of the skull of lower vertebrates lack a lateral-line component, the congruence is not complete. Furthermore, the geometric center of many bones is not the center of growth. Therefore the trans-sutural lines and webs are not a one-to-one map of morphogenesis.

Eventually it will be necessary to modify these simple graphic representations according to the nature of the suture – compressive, tensile, etc. In most cases, sutures are a complex mixture of butting, overlapping, and interlocking structures (see, for example, Figure 11.5a). Scarf joints, of the sort seen in the mammalian zygomatic arch, are rare in lower vertebrates. Were we able to identify the mechanical attributes of these different suture types, an added layer of refinement of functional analysis would be available. Even with its obvious limitations, this simple method provides a new level of description that begs interpretation.

INTERPRETATION OF THE GRAPHICAL ANALYSIS

What is the significance of these trans-sutural lines and webs? Obviously they are strongly determined by morphogenetic factors, because they are, after all, a simplified graphical integration of where the bones are and how their sutures are oriented. Can they tell us anything else about why bones have particular shapes and fall in particular positions? That is, does the pattern of packing of dermal bones in the skull represent anything more than an epiphenomenon of morphogenesis and, for example, the arrangement of the lateral lines? One way to answer this question is to look for congruence between these lines and other identifiable factors acting in the skull.

I propose as a hypothesis to be tested that in the patterning of dermal bone arrangements *the nature and arrangement of the bones, and therefore their sutures, are in part functionally determined.* This hypothesis is based on a number of assumptions concerning sutures, as follows.

1. Sutures exist in the dermal skull primarily because of the need for growth. When growth stops, the sutures often become fused. The skull grows from a number of discrete centers, as a historical constraining fact. Growth occurs at the margins of the discrete elements, and therefore the marginal contacts among elements are kept open as sutures. An alternative to having sutures among the bones would be to delay the appearance of the dermal skull until full size is reached, then it could be produced as a solid structure (as in some agnathan fishes; White 1958; Moy-Thomas & Miles 1971). However, in most vertebrates not only is growth continuous throughout life but, in any case, the dermal skeleton is functional from the earliest age.
2. Sutures may add positive attributes to the skull such as flexibility and shock absorption. They may also form potential lines of weakness in the skull, especially in the immature skull, and the points of union of multiple sutures may represent potential centers of weakness compared with a continuous solid skull table. Again this has not been studied in lower vertebrates but see Jaslow (1990) for a review of the mechanical properties of mammalian sutures.
3. The dermal skull exists not only as a protective structure but also serves to receive and transmit stresses and strains due either directly or indirectly to the action of the jaw, respiratory, and postural musculature. These stresses and strains may be expected to be highly oriented because they are created by the action of specific muscle groups. Buckland-Wright (1978) has shown this to be true in the cat skull. The quantity of force involved is not known.
4. If the preceding assumptions are correct, then it follows that dermal bone patterns (the shapes of the bones and the nature and orientation of the

sutures) can be analyzed rationally with respect to these lines of action and reaction.
5. The orientation, length, and precise nature of the sutures may each have functional significance (Moss 1957; Jaslow 1990).
6. Each suture is potentially subject to one or more of at least four types of insult: compression, tension, torsion and shearing – possibly in combination. Sutures subject to compressive stress may be simply butting contacts; those subject to tension are likely to be extended by interdigitation; and those subject to torsion and shear are likely to be overlapping.
7. In general, the property of any suture to resist disrupting forces will be a function of (a) the size (principally length) of the suture, (b) its composition, and (c) its orientation with respect to the force.
8. Other things being equal, a skull unit consisting of fewer bones is stronger than one consisting of many. Hence there is a universal tendency in lineages toward reduction of number of elements by loss or fusion.

In summary, the hypothesis that the nature and arrangement of the sutures has functional significance can be expressed as follows: If the dermal bones of the skull roof and cheek serve to receive and transfer various forces developed in the skull due to muscular action, the shapes of the bones (centers of strength) and the orientation of sutures (potential zones of flexibility and/or weakness, even of reinforcement) should reflect the orientation of these forces. The change of a dermal bone pattern through evolution would then reflect a complex feedback between the processes of morphogenesis and natural selection of adaptive changes in cranial function.

For fossil lower vertebrates, in the absence of direct experimentation, we may test the hypothesis by means of comparisons among different taxa, using graphical analysis to look for congruence between sutural patterns and known functional attributes, particularly the orientation of stress transfer within the dermal skull.

APPLICATIONS OF THE GRAPHICAL ANALYSIS

The osteichthyan cheek

Among the more unusually shaped dermal bones in the lower vertebrates are the arc-shaped elements of the cheek region of very early actinopterygians (Figure 11.4). In one of the best-worked examples of evolutionary functional morphology in fossil fishes, Schaeffer and Rosen (1961) related changes in skull

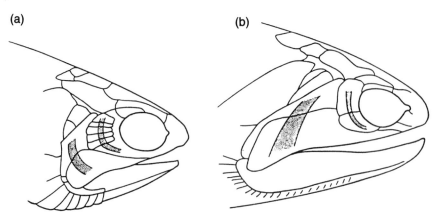

Figure 11.4. Views of cheek in two actinopterygian fishes, (a) the "palaeonscid" *Pteronisculus* and (b) the "sub-holostean" *Boreosomus*, showing the general correspondence between the orientation of the sutures between main cheek elements and orientation of jaw adductor musculature. (After Schaeffer & Rosen 1961.)

morphology and jaw function in the transitions among what were then called "chondrosteans," "holosteans," and "teleosts." They demonstrated changes in the suspensorium and palate that produce a major shift in the orientation of the principal adductor musculature of the jaw. In osteichthyans the jaw muscles do not attach directly to the dermal cheek elements. The dermal bones of the cheek act to transfer stress produced by movement of the whole suspensorium relative to the neurocranium and skull table during feeding and respiration, and to receive the force of the bite when the lower jaw is brought up against the maxilla. We would predict, therefore, that dermal bone shape and the orientation of sutures would correlate with the angle of the suspensorium and the orientation of the adductor masses.

Because there are so few bones in the dermal cheek region, drawing trans-suture lines is very easy (Figure 11.4) and, as predicted, they are completely congruent with the orientation of the jaw musculature and suspensorium. The shape of the cheek elements thus appears to be functionally determined.

Placoderm versus osteichthyan dermal bone patterns

One of the great puzzles of fish phylogeny is the relative phylogenetic position of the Placodermi. Can they be related in any way to other gnathostomes? The problem is that their dermal skulls are arranged in a totally different pattern from other vertebrates, and it has never been possible satisfactorily to find homologies between the bones of the placoderm skull and those of any other vertebrate (see Gross 1962; discussion in Thomson 1993).

As shown in Figure 11.5, when one makes reference to obvious neurocranial landmarks such as the otic capsules, pineal complex, and branchial complex, only the middle and anterior part of the placoderm skull table is equivalent to the skull of an osteichthyan (Thomson 1993). In a "typical" placoderm such as *Coccosteus* (Miles & Westoll 1968), the back of the skull consists of a huge median nuchal (Figure 11.5; N) and paired paranuchals (Figure 11.5; PN). Anterior to this, in the space equivalent to that covered by the osteichthyan "parietals" (postparietals in sarcopterygians) are the paired, eccentrically butterfly-shaped centrals (Figure 11.5; C) covering the otic region. In comparison, the skull table of an early actinopterygian osteichthyan such as *Moythomasia* (Jarvik 1980) is dominated by two pairs of almost square elements (usually termed frontals and parietals). The more posterior pair covers the otic region. Behind this there are various extrascapular elements but these do not cover any part of the neurocranium.

Drawing trans-suture lines in a placoderm is difficult because the sutures tend to be very complex with considerable overlap (Figure 11.5a). In practice one has to average out a general orientation of each suture, and this opens up the possibility of error. However, comparing trans-suture lines webs in a placoderm and an osteichthyan gives us a new way to compare their skulls. As Figure 11.6 shows, the trans-sutural web in the placoderm skull is mostly oriented at a diagonal to the anteroposterior axis, together with transverse lines around the junction of cheek and table. In the osteichthyan (Figure 11.7) the trans-suture web is almost exclusively longitudinal and transverse. In the placoderm very little of the web is oriented transversely, whereas in the actinopterygian the transverse lines are greater in sum than the anteroposterior components. A particularly interesting feature of both skulls is that so many of the long trans-suture lines are rectilinear, even though bone shape is far more complex in placoderms than in osteichthyans.

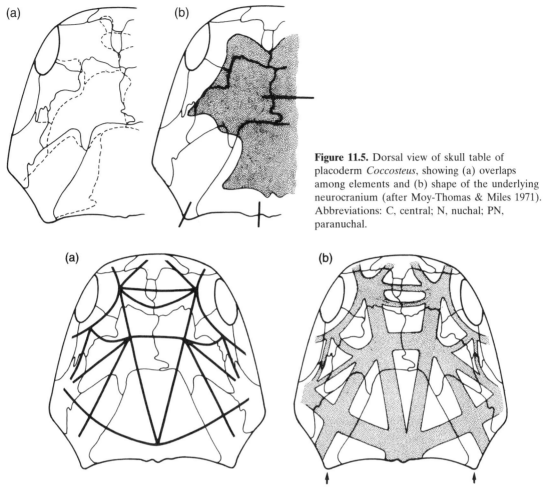

Figure 11.5. Dorsal view of skull table of placoderm *Coccosteus*, showing (a) overlaps among elements and (b) shape of the underlying neurocranium (after Moy-Thomas & Miles 1971). Abbreviations: C, central; N, nuchal; PN, paranuchal.

Figure 11.6. Skull table of *Coccosteus*, showing two different versions of trans-sutural webs (compare Fig. 11.5). Position of craniothoracic joint shown with an arrow.

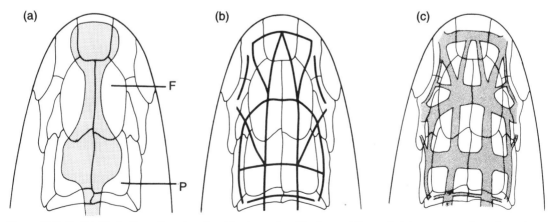

Figure 11.7. Skull table in the actinopterygian *Moythomasia*, with shape of the underlying neurocranium shown in (a), trans-sutural lines shown in (b), and trans-sutural webs shown in (c). Compare with Figure 11.6. Abbreviations: F, frontal; P, parietal.

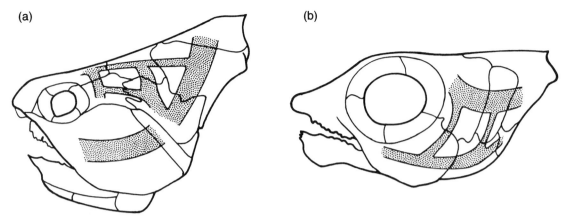

Figure 11.8. Comparison of the trans-suture lines in the cheek of two placoderms. (a) *Coccosteus* and (b) *Pholidosteus.*

The differences in orientation of the sutures in the two types of skull correlate strongly with the different arrangement of their cranioaxial joints. In placoderms there is both the regular median cranio-occipital joint and the paired lateral craniothoracic joint (between the dermal head and shoulder shields, marked by arrows on Figure 11.6b). In the case of the osteichthyan skull, it seems that the stresses and strains related to movement of the head relative to the trunk were principally oriented anteroposteriorly, and those related to movement of the cheek on the skull table were transverse. In the placoderm, the presence of the lateral craniothoracic joints prob-ably added a strong oblique component to the sys-tem of stress transfer in the dermal skull (Miles 1967, 1969). If this interpretation is correct, the large median nuchal bone was the *keystone* of the whole skull. Furthermore, the fact that the suspensorium lies on a plane anterior to the rear margin of the skull table may have added yet another oblique stress component. This might correlate with the presence of a major arc-shaped trans-sutural band across the back of the skull. Therefore, in the placoderm and the osteichthyan one can propose correlations be-tween the orientation of the trans-sutural webs and reconstructed patterns of stresses transmitted through the dermal skull from articulations between head and cheek, and head and trunk.

As a further note, in the placoderm skull most of the sutures are convoluted and serpentine. This cre-ates, for example, the extraordinarily shaped central bones. It may indicate that there is a considerable element of shear stress within the skull. This is con-firmed by the great extent of overlapping rather than simple butting sutures in the skull (Figure 11.5a) and possibly correlates again with the combination of stresses acting in the skull due to the unusual head-trunk connections.

In summary, the two types of skull appear to be functionally distinct and, if function has as strong a determining role as I propose, it is not surprising that no homologies can be found between the shapes and arrangements of the dermal bones of placoderms and osteichthyans. It is impossible to reconstruct any phylogenetic intermediates between placoderms and osteichthyans because graded structural or functional intermediates between the basic skull conditions (with and without the craniothoracic joint) are pre-sumably impossible.

Cheek region in placoderms

That there is considerable variation in function within placoderms is shown in a comparison of the cheek region in two well-known genera, *Coccosteus* and *Pholidosteus* (Figure 11.8). The bone shapes and trans-sutural lines are quite differently aligned, prob-ably in connection with different mechanisms of ventilating the branchial chamber – by lateral exten-sion of the cheek in *Coccosteus* and by raising and lowering the head in *Pholidosteus.*

The dipnoan skull roof

As is well known, the skull roof of the earlier dipnoans is totally different from that of other fishes. Instead of consisting of a relatively small number of dermal bones of different shape according to the region of the head, it is essentially a mosaic, albeit one in which the posterior elements are larger than the more anterior ones. A great deal of effort has been spent in trying to homologize these bones with those of other fishes (reviews in White 1965; Thomson & Campbell 1971; Campbell & Barwick 1987, 1990). A trans-sutural web for a Devonian dipnoan such as *Dipterus* is truly a web, with virtu-ally no overall structure (Figure 11.9). This case

Figure 11.9. Trans-suture web analysis of left side of skull of Devonian dipnoan fish *Dipterus*.

then makes a good test of our assumptions and methodology.

If the underlying cranial structures of the dipnoan head were essentially the same as those of other fishes, then the thesis that patterning of dermal bone shape and sutures in the skull has functional significance related to the transfer of stresses would be falsified. However, the Dipnoi are unique among osteichthyan fishes for their extreme holostyly (fusion of neurocranium and palatal complex into one unit). In direct contrast to other fishes, the cheek is not movable relative to the head, the bite is not borne on a marginal dentition, and in almost every case maxillae are lacking (review in Miles 1977). Instead, the palate is fused as one solid unit with the neurocranium and there is a crushing palatal dentition. The dermal skull roof is fused to, or closely applied to, the massive neurocranium. In dipnoans, therefore, it is very likely that the dermal skull roof does not act to transfer the same sorts of forces as in fishes with kinetic skulls. Specifically, it does not receive the force of the bite. Almost all that function has passed to the neurocranial–palatal complex. I conclude that the lack of orientation in the arrangement of the dermal elements probably reflects this functional condition. The conclusion is further strengthened by the fact that the rostral region of the skull in many early fishes (especially sarcopterygians), where the dermal bones are fused to the underlying nasal capsule, also shows a dipnoanlike mosaic of small elements (see fig. 14 in Thomson 1988).

Cheek and mandible in osteichthyan fishes

When one draws a trans-sutural web for the cheek of lobe-finned fishes, the unique status of the dipnoans is again immediately obvious (Figure 11.10a). The postorbital cheek region is a very minor part of the skull. In the osteolepiforms (Figure 11.3) and porolepiforms (Figure 11.10b) the cheek is a large unit with a very strong vertical component to the orientation of the sutures, quite in contrast to the early actinopterygians (Figure 11.4).

As previously noted, one of the characteristic features of the actinopterygian fishes, especially the earlier forms with a complete dermal cheek series, is the orientation of the bones in the cheek region. This changes through the history of the Actinopterygii (see Figure 11.4), but all actinopterygians have a fundamentally different pattern from those of the lobe-finned fishes. These differences are only in part attributable to differences in the orientation of the main adductor muscle masses of the jaw. The nondipnoan sarcopterygians are characterized by the intracranial joint and thus have a very different pattern of branchial ventilation (analogous in some ways to that of placoderms) and jaw mechanics. The posteriorly directed hyomandibular and posterior palate orient the jaw-closing muscles almost directly downward. The lungfish represent yet a third functional system, in which the cheek region is not a major transmitter of forces between the skull and mandible.

The dermal skull table in rhachitomes and embolomeres

A well-known marker for these two groups of temnospondyl amphibians is the pattern of sutural contact in the posterior skull table. In rhachitomes (Figure 11.11b), a sutural contact between the postparietal and supratemporal excludes the tabular from meeting the parietal. In embolomeres (Figure 11.11a), the opposite condition applies: The tabular is sutured to the parietal, excluding the postparietal from the supratemporal. For years this has been considered an interesting, taxonomically useful, but not particularly illuminating quirk, probably due to the fact that the tabular horn, on its lateral aspect, defines the otic notch and is thus connected with hearing rather than general cranial mechanics (Romer 1947; compare Lombard & Bolt 1988; Clack 1989).

Graphical analysis by means of trans-sutural webs (Figure 11.11) suggests that this difference is, in fact, related to major differences in the way the cheek attaches to the skull table and the posterior skull table attaches to the trunk (see also fig. 15 in Thomson 1988). In the embolomeres the tabular,

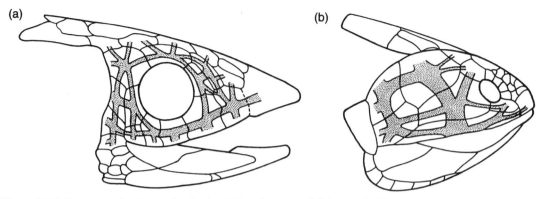

Figure 11.10. Trans-sutural webs on the cheek of (a) a dipnoan and (b) a porolepiform.

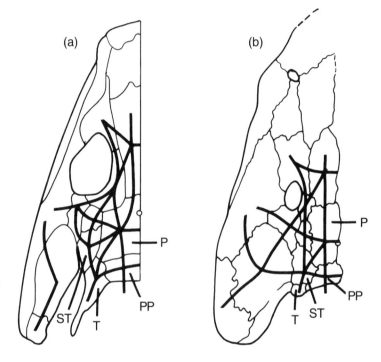

Figure 11.11. Trans-sutural webs on the posterior skulls of (a) an "embolomere" and (b) a "rhachitome" (compare Figure 11.10). Abbreviations: P, parietal; PP, postparietal; ST, supratemporal; T, tabular.

with its elongate horn, was probably a major focus of stresses acting between skull table and trunk which were, in general, anteroposteriorly directed. The skull table–cheek connection is kinetic and presumably was particularly weak postorbitally. In rhachitomes, where the cheek is broadly and firmly sutured to the skull table, the stresses are more strongly transverse, and the tabular is only a minor element in the table–trunk connection.

The posterior skull table in *Ichthyostega*

An intriguing feature of the posterior skull table in the Devonian amphibian *Ichthyostega* is the pres-

ence of a large median postparietal bone where all other amphibians and osteolepiform fishes have paired elements (see descriptions in Jarvik 1980). The question arises whether this reflects some major difference in skull function, parallel perhaps to the conditions in placoderms, the only other case whether there is a large median element in the rear of the dermal skull roof. Graphical analysis of the suture lines (Figure 11.12) suggests that this is not the case. Apart from differences due to the relative breadths of the skulls, the overall pattern of the trans-suture web in the rest of the skull of *Ichthyostega* is entirely consistent with that in the other Devonian amphibian *Acanthostega*, which has a paired postparietal like all other amphibians (Jarvik 1980).

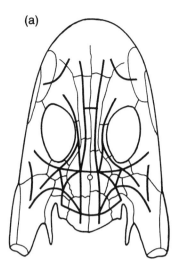
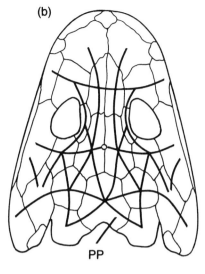

Figure 11.12. Comparison of trans-sutural webs on skulls of (a) *Acanthostega* and (b) *Ichthyostega.* Abbreviation: PP, postparietal.

CONCLUDING REMARKS

The immediate aims of this exercise have been (1) to propose a method of graphical analyses that can be used to derive a simple integration of dermal bone size and shape within the skull of lower vertebrates, and (2) to test whether, using these techniques, correlations can be found between dermal bone patterns and functional attributes of the skull. In other words, to ask: Is dermal bone shape functionally determined, at least in part, and can we therefore use analysis of dermal bone patterns to assist in decoding the cranial function of fossil vertebrates? I suggest that the results show that a great deal of functional information is encoded within the patterns of arrangement of the dermal bones of the skull. The shapes of dermal bones and the patterns of their arrangement are probably strongly functionally determined.

The analytical techniques described here are extremely simple (and simple-minded), far out of proportion to the subtle complexities of function of even the simplest skull. To progress beyond this stage, far more sophisticated analyses will be needed. The use of computers to model skull surfaces in three dimensions immediately comes to mind.

A third, and ultimately far more important, aim of this chapter is to open the door to a change in our way of thinking about dermal bones – from a concentration on the individual bones (their shape and thus their homologies) to a point of view in which all the bones in the dermal skull form a single evolutionary, morphogenetic, and functional unit, and even to the view that the bones themselves are only the epiphenomenon of deeper factors involved in the "causes" of skull morphology. In this new rhetoric, it is not the individual bones themselves that are real but the factors that create the sutures.

ACKNOWLEDGMENTS

I am grateful to Peter Dodson, and an anonymous reviewer for valuable advice in the preparation of this manuscript. The figures were kindly prepared by Linda Price Thomson.

REFERENCES

Buckland-Wright, J.C. 1978. Bone structure and the patterns of force transmission in the cat skull (*Felis catus*). *Journal of Morphology* 155, 325–362.

Campbell, K.S.W., & Barwick, R.E. 1987. Paleozoic lungfishes – a review. *Journal of Morphology*, Supplement 1: 93–131.

Campbell, K.S.W., & Barwick, R.E. 1990. Paleozoic dipnoan phylogeny: functional complexes and evolution without parsimony. *Paleobiology* 16, 143–169.

Carter, D.R. 1987. Mechanical loading history and skeletal biology. *Journal of Biomechanics* 20, 1095–1109.

Carter, D.R., Wong, M., & Orr, T.E. 1991. Musculoskeletal ontogeny, phylogeny, and functional adaptation. *Journal of Biomechanics* 24, Supplement 1, 3–16.

Clack, J.A. 1989. Discovery of the earliest-known tetrapod stapes. *Nature* 342, 425–427.

Goodwin, B.C., & Trainor, L.E.H. 1983. The ontogeny and phylogeny of the pentadactyl limb. In *Development and Evolution*, ed. B.C. Goodwin, N. Holder, & C.C. Wylie. Cambridge University Press.

Gross, W. 1962. Peut-on homologuer les os des arthrodires et des téléostomes? *Colloques International du Centre de la Recherche Scientifique* 104, 67–74.

Jarvik, E. 1980. *Basic Structure and Development of Vertebrates*. London: Academic Press.

Jaslow, C.R. 1990. Mechanical properties of cranial sutures. *Journal of Biomechanics* 23, 313–321.

Lombard, R.E., & Bolt, J.R. 1988. Evolution of the stapes in Paleozoic tetrapods. In *The Evolution of the Amphibian Auditory System*, ed. B. Fritsch. New York: Wiley.

Miles, R.S. 1967. The cervical joint and some aspects of the origin of the Placodermi. *Colloques Internationaux du Centre National de la Recherche Scientifique* 163, 49–71.

Miles, R.S. 1969. Features of placoderm diversification and the evolution of the arthrodire feeding mechanism. *Transactions of the Royal Society of Edinburgh* 68, 124–170.

Miles, R.S. 1977. Dipnoan (lungfish) skulls and the relationships of the group: a study based on new species from the Devonian of Australia. *Zoological Journal of the Linnean Society of London* 61, 1–328.

Miles, R.S., & Westoll, T.S. 1968. The placoderm fish *Coccosteus cuspidatus* Miller *ex* Agassiz from the Middle Old Red Sandstone of Scotland. Part I, Descriptive morphology. *Transactions of the Royal Society of Edinburgh* 67, 373–476.

Moss, M.L. 1957. Experimental alteration of sutural area morphology. *Anatomical Record* 127, 569–590.

Moy-Thomas, J.A., & Miles, R.S. 1971. *Palaeozoic Fishes.* Philadelphia: Saunders.

Romer, A.S. 1947. Review of the Labyrinthodontia. *Bulletin of the Museum of Comparative Zoology, Harvard* 99, 1–368.

Roux, W. 1895. *Gesammete Abhandlungen uber die Entwicklungsmechanik der Organismen. I. Funktionelle Anpassung.* Leipzig: Enmgelmann.

Schaeffer, B., & Rosen, D.E. 1961. Major adaptive levels in the evolution of the actinopterygian feeding mechanism. *American Zoologist* 1, 187–204.

Thomson, K.S. 1987. Speculations concerning the role of the neural crest in the morphogenesis and evolution of the vertebrate skeleton. In *The Neural Crest,* ed. P.F.A. Maderson. New York: Wiley.

Thomson, K.S. 1988. *Morphogenesis and Evolution.* New York: Oxford University Press.

Thomson, K.S. 1991. Parallelism and convergence in the horse limb: the external-internal dichotomy. In *New Perspectives on Evolution,* ed. L. Warren & H. Kaprowski. New York: Wiley-Liss.

Thomson, K.S. 1993. Segmentation, the adult skull, and the problem of homology. In *The Vertebrate Skull,* ed. J. Hanken & B.K. Hall. Chicago: University of Chicago Press.

Thomson, K.S., & Campbell, K.S.W. 1971. The structure and relationships of the primitive Devonian lungfish *Dipnorhynchus sussmilchi* (Etheridge). *Bulletin of the Peabody Museum of Natural History, Yale University* 38, 1–109.

White, E.I. 1958. Original environment of the craniates. In *Studies on Fossil Vertebrates,* ed. T.S. Westoll. London: Athlone Press.

White, E.I. 1965. The head of *Dipterus valenciennesi* Sedgwick and Murchison. *Bulletin of the British Museum (Natural History), Geology* 11, 1–45.

Wolff, J. 1892. *Die Transformationen der Knochen.* Berlin: Hirschwald.

Wong, M., & Carter, D.R. 1988. Mechanical stress and morphogenetic endochondral ossification of the sternum. *Journal of Bone and Joint Surgery* 70A, 992–1000.

12

The forelimb of *Torosaurus* and an analysis of the posture and gait of ceratopsian dinosaurs

ROLF E. JOHNSON and JOHN H. OSTROM

ABSTRACT

The discovery and subsequent mounting of a well-preserved forelimb and shoulder referrable to *Torosaurus* (Ornithischia: Ceratopsia) provide important new evidence pertaining to forelimb posture and possible locomotory movements in ceratopsian dinosaurs. Posture and movement are topics currently under considerable debate within the broader context of the reinterpretation of dinosaurian biomechanics. The complete scapulo-coracoid, humerus, radius, and ulna are uncrushed and provide data on morphology, arthrology, and limb proportions that allow reconstruction of posture and gait for *Torosaurus* and most other ceratopsians that have a similar osteology. The manus and wrist are missing here, but are known from other ceratopsian taxa. All limb elements are massive and of graviportal proportions.

We describe a technique for flexibly mounting a high-fidelity cast to test hypotheses of limb configuration and kinematics. Combining osteological descriptions with results of manipulating the mounted cast, we conclude that, in the normal standing posture, the humerus was inclined posteroventrally and somewhat laterally and that the elbow was flexed to about 45° from the vertical. The large size of the deltopectoral crest is consistent with this bent-arm interpretation: The massive pectoral musculature inferred to insert on this crest would have been necessary to maintain the inclination of the humerus. The implications for the animal's gait – that the forelimbs moved in a somewhat sprawling manner – are supported by trackways ascribed to ceratopsians. These conclusions contradict Bakker (1975), who proposed, on energetic grounds, an upright stance and parasagittal forelimb movements during locomotion. Manipulations of the mounted limb preclude such a position and movements. The importance of including the primary paleontological database – osteological features in the case of vertebrates – in the development of any functional hypotheses is strongly reaffirmed.

INTRODUCTION

When we watch the often graceful ways in which terrestrial vertebrates move, their apparently effortless locomotion belies the complexity of the underlying anatomy and muscular activity. The analysis of vertebrate locomotion is a challenging and diverse field of study (Hildebrand 1989). For extinct organisims with no extant representatives, the challenge of biomechanical interpretation is all the greater. Limb kinematics must be inferred from osteology and ichnology (e.g., trackways) with no direct means to test conclusions. We just cannot go to a zoo and watch, film, or otherwise dissect the actual postures, gaits, and behaviors of dinosaurs, for example. Perhaps one of the most prolific areas of scientific research is now occurring within the context of the rather broad reexamination of dinosaur biology and paleoecology. The specific case study herein deals with the forelimb biomechanics of the ceratopsian dinosaur *Torosaurus*. Neontological comparison, which is so important in studies of the functional morphology of extinct forms, is difficult in the case of dinosaurs because there are no living close relatives of similar form.

Ceratopsians were a diverse group of small (*Protoceratops*) to large (*Triceratops*) dinosaurs that radiated during the late Cretaceous Period. Even the most casual observer of dinosaurs knows that over the last two decades, graphic representations of these animals have changed dramatically. (Many popular restorations are now the work of dinosaur artists, based on the inferences of paleontologists.) No longer depicted as slow moving, lumbering beasts, the ceratopsians, along with the other dinosaurs, are now commonly restored in full gallop, exhibiting a rather racehorselike demeanor (Bakker 1986; Paul 1988). We are interested in reevaluating the evidence for these new restorations.

A central controversy is ceratopsian forelimb posture: Were the forelimbs held in an upright mammalianlike stance (Figure 12.1a) with limb excursion in a parasagittal plane, or were the elbows bent and the fore feet wide apart (Figure 12.1b) giving a

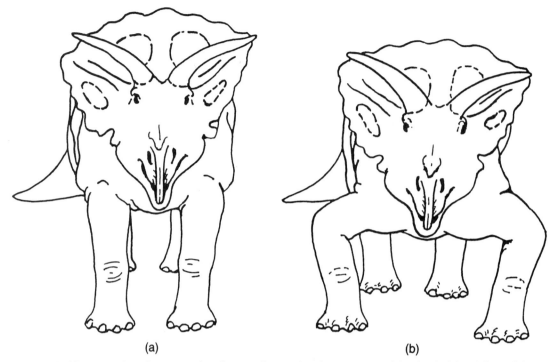

Figure 12.1. (a) Restoration of a ceratopsian dinosaur, in anterior view, as a cursorial animal with a fully upright posture. (b) Restoration with the traditional, bent-elbow, sprawling forelimb posture advocated in this paper.

sprawled stance and limb excursion analogous to that of extant reptiles?

The collection, and subsequent mounting for display at the Milwaukee Public Museum (MPM), of the most complete skeleton of the massive ceratopsian dinosaur *Torosaurus* (Figure 12.2) has both provided data for postural restorations of ceratopsians and presented compelling implications for assertions of galloping ceratopsians. *Torosaurus* was known from less than ten specimens, all of which were skulls or skull fragments, prior to the discovery of the Milwaukee specimen, which includes the skull and associated postcranial elements of one individual.

Reconstructing the largest dinosaurs is a difficult task. (*Reconstruction* means attempting to assemble the skeleton in an anatomically correct manner, whereas *restorations* are artistic representations of how the live animal might have appeared in its normal habitat.) They have no extant relatives of the same size, and many of them were much larger than any living terrestrial vertebrate (the largest of which are mammals). Consequently, all functional inferences about dinosaurs must be obtained from available data, in this case preserved osteological elements. Previous reconstructions of the forelimb posture in ceratopsian had largely been based on composite material. The quality and completeness of this new *Torosaurus* specimen allows us a previously unavailable opportunity to reexamine the most

recent reconstructions of the forelimb posture in this animal and in other ceratopsians with a similar limb morphology.

After reviewing the history of the debate about ceratopsian forelimb posture, including a look at the earliest reconstructions of ceratopsians, we will critically examine the two competing interpretations (upright versus sprawling) in light of evidence from the recently completed reconstruction of the Milwaukee *Torosaurus*. Along with a description of the osteology of the forelimb of the new *Torosaurus,* we will briefly describe a new way of physically testing the competing ideas and ranges of possibility for forelimb orientation. The broader application of this technique to postural reconstruction is also outlined.

A HISTORY OF CERATOPSIAN RECONSTRUCTION

The family Ceratopsidae was established in 1888 by O.C. Marsh based entirely on skull fragments and some postcranial elements that he referred to as a new species *Ceratops montanus*. By 1895, only seven years after the initial descriptions, the Ceratopsidae included at least 15 species, and in that year the first illustration of a fully reconstructed ceratopsian skeleton was published in a poster collage entitled "Restorations of Extinct Animals; by O.C. Marsh –

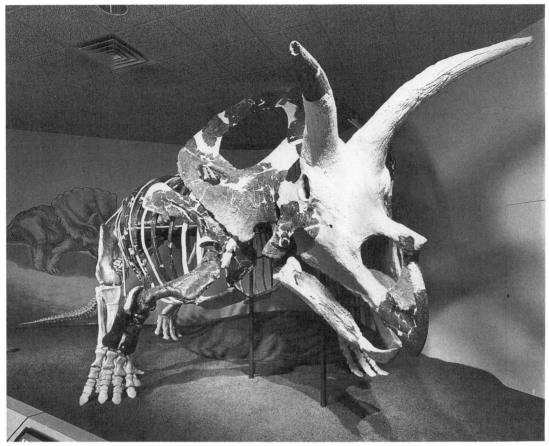

Figure 12.2. Photograph of completed mount of MPM VP6841, *Torosaurus* cf. *latus*, showing the proportion of preserved material (dark) versus reconstruction (white plaster). Forelimbs are placed at the inferred extremes of their range of motion, implying an active gait while retaining a sprawling position.

Original Specimens in Museum of Yale University." This poster has more than a dozen skeletal reconstructions of Mesozoic and Cenozoic creatures, including one of *Triceratops prorsus* in lateral view that appears to show the forelimbs in an upright position directly below the shoulder sockets. The poster is labeled "Plate I" and carries the date 1895, and presumably was intended, but did not appear, as the first plate of Marsh's (1896) classic monograph "The Dinosaurs of North America" published the next year as part of the Sixteenth Annual Report of the U.S. Geological Survey. This collage did appear in Marsh's paper on "Fossil Vertebrates" in volume VIII of *Johnson's Universal Cyclopedia* of 1895, but here it bears no plate number. If this image is indeed the very first attempt to illustrate the full skeletal posture of a ceratopsian, it is of historical as well as precedential significance by placing Marsh's stamp of approval on an upright ceratopsian stance. The *Triceratops* reconstruction appears as plate LXXI in Johnson's (1895) publication and was republished as figure 125 in "The Ceratopsia"

(Hatcher, Marsh & Lull 1907). It was preceded in that paper by a photograph (fig. 124) that depicts a posterior view of the standing skeletal mount of *Triceratops* just completed a few months earlier at the U. S. National Museum (USNM). In this view, the forelimbs are clearly positioned in a bent-elbow sprawl. This USNM skeletal mount was the work of C.W. Gilmore, which he reported and described in 1905. In his brief report, Gilmore (1905) referred to the "turtle-like flexure of the anterior extremities" and stated (p. 434):

The position of the forelimbs in the present mount appears rather remarkable for an animal of such robust proportions, but a study of the articulating surfaces of several parts *precludes* [our emphasis] an upright mammalian type of limb, as was represented by Marsh in the original restoration.

Lateral and three-quarter views of the same USNM #4842 skeletal mount are presented as Plate XLIX A and B in Hatcher, Marsh, and Lull's "Ceratopsia" volume, further documenting Gilmore's conviction that ceratopsian forelimbs could not have been placed in erect, parasagittal orientation.

More than 20 years later, Sternberg (1927) came to a similar conclusion as he described the morphology and position of the head of the humerus in *Chasmosaurus belli*, Lambe. Sternberg noted (p. 69) that the head of the humerus:

was on the external side of the proximal end of the bone, instead of on the end as is the case with most dinosaurs. Consequently, the only way the limb could be posed, so the head of the humerus fitted into the glenoid cavity was to place the humerus at almost right angles to the perpendicular. This made the animal very low in front and extremely bow-legged. The humerus placed in this position made a very much better articulation with the ulna and radius than could be gained otherwise.

Tait and Brown (1928) also noted that the posterior limbs in ceratopsians were "long and vertically disposed" but in contrast "the strong, stout humeri are oriented not vertically, as are the femora, but horizontally" with the animal's weight borne through the distal and laterally carried ends of the projecting humeri. Tait and Brown equated this bizarre posture with unusual feeding habits in ceratopsians. Their hypothesis, including discussions of agonistic behaviors and ceratopsian display structures, was recently amplified by Farlow and Dodson (1975).

In 1933, H.F. Osborn, in his description of the mounted skeleton of *Triceratops elatus* also made the observation that the forelimbs are set out widely from the body (p. 6):

After repeated attempts to pose the forelimb with the humerus subvertical and the elbow more or less pointed backward (as in mammalian quadrupeds), it appeared that neither of these poses could be worked out without disjointing the articulations in a quite impossible manner. Nothing short of a horizontal humerus and a completely everted elbow would permit the proper articulation of the facets and place the chief muscle attachments in proper and mechanically possible relations.

Russell (1935), in his review of ceratopsian musculature, observed this same characteristic: "The most striking feature of the ceratopsian limbs is the marked difference in pose and mechanical arrangement between the anterior and posterior pair" in which the forelimbs are "typically reptilian, projecting out from the body" and requiring considerable muscular effort to support the thoracic region. Farlow and Dodson (1975) alluded to the powerful deltopectoral crest of the ceratopsian humerus which provided expanded leverage for the insertion of the pectoralis musculature, noting that "a large deltopectoral crest is necessary . . . for animals with a sprawling or semi-erect posture."

Erickson (1966) mounted a composite skeleton of *Triceratops* at the Science Museum of Minnesota

(SMM). His observations of the forelimb are in agreement with the authors mentioned above, and his description of the SMM mount includes a section on the rationale and biomechanics involved in deciding to position the forelimbs in a bent-elbowed sprawl.

It is clear that the earliest workers on ceratopsians both noted and analyzed the postural dichotomy between hindlimbs and forelimbs. All of them, with the possible exception of Marsh, were convinced through their morphologic and arthrologic studies that the most biomechanically reasonable reconstruction of the forelimb of ceratopsian dinosaurs demanded that the humerus be placed in a near horizontal plane, with the aspect of the front end of the animal being that of a reptilelike sprawl.

Bakker (1968) proposed that ceratopsians had upright forelimb posture and considerable cursorial capability, essentially reviving the earlier (implied) interpretations of Marsh (1895) and contradicting the consensus of intervening workers. His thesis maintained that the horned dinosaurs not only had fully upright four-legged posture, but that even the largest ceratopsians (*Triceratops, Torosaurus, Chasmosaurus*), approaching 10 metric tons in live weight, were capable of galloping at speeds of up to 45 kilometers per hour. A major point in his argument was the seemingly contradictory limb arrangements of museum ceratopsian skeletal mounts, with upright hindlimbs and sprawling forelimbs, giving these animals an awkward appearance. In addition, arguments for upright forelimb posture were placed in the context of activity and mobility for implied endothermic dinosaurian physiology.

Bakker (1986) continued to argue for upright forelimb posture and rapid cursorial velocities in ceratopsians, and most recently Alexander (1989, p. 59) has suggested that "*Triceratops* may have been more athletic [than elephants], and may have galloped like a rhinoceros." His conclusion is based on analysis of bone strength indices and endothermic physiology, both of which may be inferred with caution from bone structure, but are indirect ways in which to infer stance and gait. This fully upright stance has now influenced both restorations and reconstructions of ceratopsian dinosaurs.

There apparently is no quarrel about ceratopsian hind-limb posture – it was upright. The head of the femur, where known, is well offset from the femoral shaft, and the acetabulum is directed laterally and slightly downward. Moreover, the distal femoral condyles for the tibia and fibula (epipodials) are positioned in a transverse axis essentially parallel to the plane of projection of the femoral head. The hindlimb epipodials consequently appear to have flexed and extended in the same near vertical plane through which the femur rotated, and that could

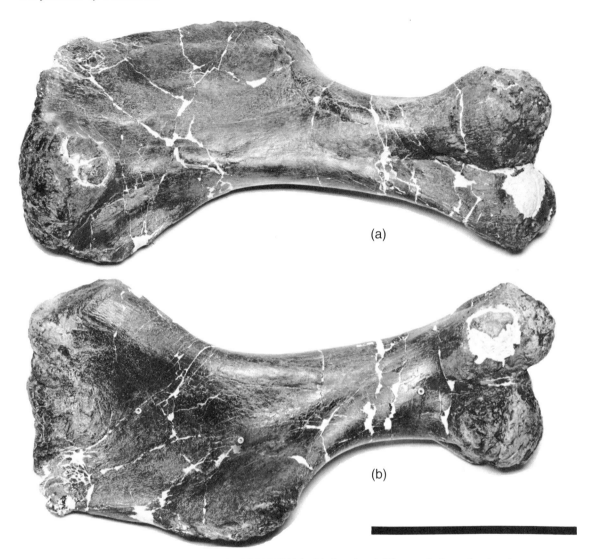

Figure 12.3. Photographs of right humerus of MPM VP6841 in (a) dorsal and (b) ventral views, showing the excellent degree of preservation. Scale = 30 cm.

only have been in a near-parasagittal plane. The ceratopsian hindlimb was upright and vertical.

OSTEOLOGICAL EVIDENCE FOR RECONSTRUCTING THE FORELIMB POSITION IN *TOROSAURUS*

In 1981, a collecting team from the Milwaukee Public Museum discovered the remains of the ceratopsian dinosaur *Torosaurus* in the Hell Creek Badlands of northeastern Montana (MPM Locality #3291). Along with a major part of the skull, numerous postcranial elements, including an uncrushed right forelimb (lacking only the carpus and manus; Figure 12.2), and both right and left scapulo-coracoids, were collected from a fine grained siltstone deposited in a

fluvial floodplain facies. Preservation of all recovered elements is excellent. There is little or no distortion, and muscle scars and external morphologic characters are exceptionally well preserved (Figures 12.3–12.6). The specimen is catalogued as MPM VP6841.

We focus here on the forelimb for several reasons. First, and most important, it is in the fore quarters where the contradictory evidence for ceratopsian postural reconstructions is found. Second, it is this complete forelimb set of elements preserved in this new specimen of *Torosaurus* that provides the best noncomposite evidence contrary to arguments about the parasagittal placement of the forelimb. We use osteology as our primary data in an effort to decrease the amount of speculative inference about posture and function.

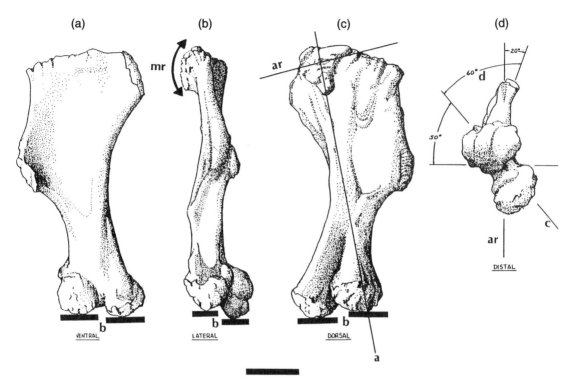

Figure 12.4. Depictions of right humerus of MPM VP6841 in (a) ventral, (b) lateral, (c) dorsal, and (d) distal views. Marked features are discussed in the text: a, plane of rotation about shoulder; ar, axis of rotation at shoulder; b, offset of distal condyles; c, axis of rotation at elbow; d, approximate angle between axes of rotation at shoulder and elbow (ar and c); mr, directions of rotation of humeral head in lateral view. Scale = 15 cm.

Today, and in marked contrast to many popular restorations of ceratopsians, the majority of scholarly opinion seems to favor a bent-limb "push-up" arrangement with the humerus projecting out to the side. This sprawling orientation is consistent with and reinforced by the morphology of forelimb elements of MPM VP6841. We will document the relevant features of each bone separately below. The essence of our argument is that the nature of the shoulder, elbow, and wrist articulations preclude an upright stance and parasagittal motion of the limb.

Flexible mount of the forelimb

Prior to mounting the Milwaukee specimen, and as an important and rather graphic part of our testing of the two competing models for ceratopsian forelimb reconstruction, we utilized a simple mechanical model of the forelimb of *Torosaurus* (Figure 12.7). This technique has potential application for similar functional and postural studies on other taxa.

High-fidelity casts of each of the forelimb elements and the scapulo-coracoid were produced from RTV silicone rubber molds. These casts were then mounted in a large wooden gantry (Figure 12.7). The use of a universal jointing apparatus, itself capable of being positioned wherever desired on the articulating surfaces of the limb elements, allowed us to place the limb in any pose that we might want to examine. Elastic straps representing the major muscles of the forelimb were attached to their presumed origins and insertions. These were included as part of a more in-depth study of the locomotory mechanics in *Torosaurus* (Ostrom & Johnson, in press) but did not restrict the range of joint motion we wished to analyze for the present study. We began with the forelimb in a parasagittal posture and moved it incrementally to a sprawling posture. At each position we analyzed the extent of articulation of all joints (and inferred possible lines of force for the major muscle groups).

This forelimb model was presented to the members of the Society of Vertebrate Paleontology at the 1990 meetings in Lawrence, Kansas. Society members were able to engage in lively debate and explore their own ideas by freely manipulating the limb. This limb model was of great importance in defining both the range of possibilities, impossibilities, and biomechanical constraints for the posture of *Torosaurus* and ceratopsians that share a similar limb morphology. Manipulation of the model confirmed the conclusions based on detailed examination of the isolated bones: The forelimb stance was not upright.

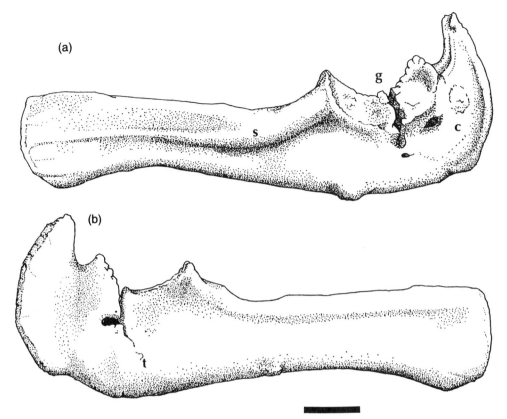

Figure 12.5. Depictions of scapulo-coracoid of MPM VP6841 in (a) medial, and (b) lateral view. The upper border of the figures would face posteroventrally during life (see Fig. 12.7b), and see text for discussion of orientation of glenoid in life. Abbreviations: c, coracoid; g, glenoid; s, scapula; t, suture between scapula and coracoid. Scale = 15 cm.

The use of this type of physical model has potential for many functional studies of vertebrate anatomy. In addition, it allows the researcher to experiment with cast replicas of the bones without putting the original fossil material in jeopardy of damage as a result of the rigorous physical manipulation of these elements that is desirable in studies of this sort.

The forelimb osteology of *Torosaurus* cf. *latus* (MPM VP6841)

THE SCAPULO-CORACOID. The scapula is 100 cm long and quite broad with subparallel blade margins (Figure 12.5). We may suppose that the scapulo-coracoid supported the attachment sites for the trapezius and rhomboideus muscle analogs, and was capable of some small degree of rotation (less than 10°) against the thorax. Based on observations made by Sternberg (1951) regarding the probable orientation of the pectoral girdle in an exceptionally well-preserved specimen of *Leptoceratops gracilis*, the scapula most probably lay against the lateral surface of the ante-

rior thoracic ribs and was close to parallel to the longitudinal axis of the thoracic vertebral column. The scapulo-coracoid was free of the axial skeleton and was slung by muscle and tendon from the rib cage and was therefore free to move against the thorax. The coracoid most probably made sutural cartilagenous contact with the sternal segment as in living saurians. Preserved scars on the anterior-dorsal margin of the coracoid of MPM VP6841 support this assessment. Considering the mass of the animal and the immense cranial burden borne entirely by the forelimbs, it is quite unlikely that the orientation of the glenoid ever deviated by much from a lateroventral plane. In this position, we can also be confident that the glenoid faced downward, with a significant anterolateral component to its long axis, and only slightly backward (see Figure 12.7b). The broad blade of the scapula is oriented up and backward, in this position, at an angle of approximately 30°–45° to the thoracic spinal column, with the coracoid below and anterior to the glenoid.

In mounting the Milwaukee specimen, numerous orientations of the scapulo-coracoid were tried, based on the physical constraints imposed by the restored

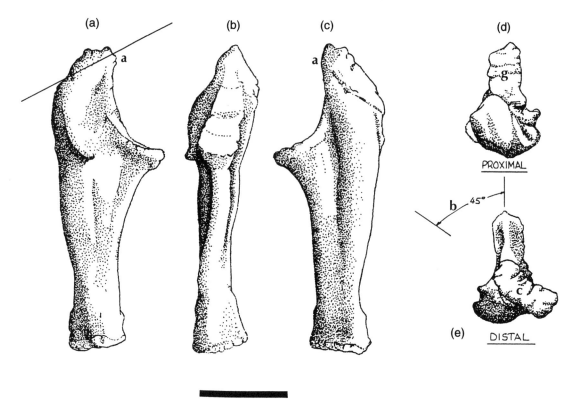

Figure 12.6. Depictions of ulna of MPM VP6841 in (a) anterior, (b) medial, (c) posterior, (d) proximal and (e) distal views. Abbreviations: a, olecranon process; c, surface for articulation with carpus; g, trochlear surface for articulation with humerus. Scale = 15 cm.

sternal plates, rib cage, and vertebral column. The final mounted position and orientation of the pectoral girdle was derived from this rigorous testing of multiple orientations (Figure 12.2). It is important to point out that the biomechanically feasible range for the orientation of the scapulo-coracoid is not critical for changing the interpretation that follows here. The different orientations of this element tried in our tests yielded the same results regarding forelimb posture.

THE HUMERUS. The humerus (propodial) has robust proportions (Figures 12.3 & 12.4) and is the longest segment of the forelimb. It is 24 percent longer than the associated ulna and 45 percent longer than the associated radius; these proportions are not characteristic of or expected in a cursorial animal (Coombs 1978). Its stocky design speaks more for a solid undercarriage base than it does for enhancement of stride, that is, stability of support rather than velocity of motion. Even more important is the enormous size of the deltopectoral crest. This feature is very long, extending more than 38 cm (50 percent of the humeral length), and projects approximately 18 cm from the humeral shaft. It is by far the most conspicuous feature of the humerus and casts doubt on

any possibility of an upright or near-vertical posture of the forelimb. For example, when positioned in a vertical orientation (the adducted position), the deltopectoral crest projects anterolaterally. In that position, it would have caused excessive torque and subsequent leverage for adduction by the pectoralis muscle attaching to it. In fact, the humerus, in a fully upright position with such an enormous deltopectoral crest lever, would have been twisted by contraction of the pectoralis, quite the opposite of the desired protraction movement for the upright propodial (Figure 12.8). Instead, the leverage of the large deltopectoral crest "dictates" a near-horizontal transverse attitude of the humerus. In this position the immense size of the deltopectoral crest makes good mechanical sense. With the humerus projecting sideways from the glenoid, the deltopectoral crest now is directed forward and ventrally. The pectoralis muscle most likely extended from this crest in a medial direction, exerting the powerful adducting force necessary for supporting the body and head, and supplementing the retraction of the humerus during locomotion.

The head of the humerus cannot be precisely delineated because of the absent cartilagenous articular cap of unknown dimensions. However, the

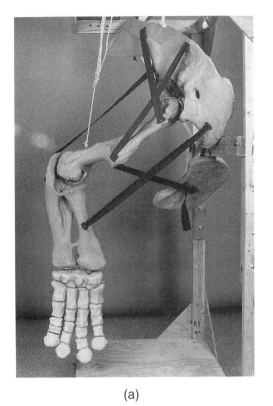

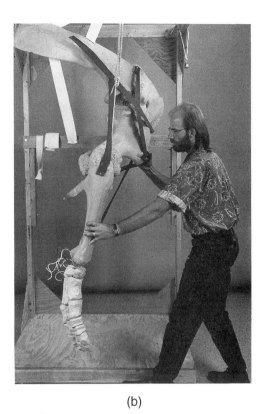

(a)   (b)

Figure 12.7. Photographs of apparatus used to test range of possibilities of forelimb motion and posture in *Torosaurus.* (a) Anterior view with the limb in a bent-elbow posture. Note congruence of the surfaces forming the elbow joint. (b) Senior author shown manipulating the model into a test position. Elastic bands represent the major muscles inferred to have attached between shoulder and limb, and allow for graphic visualization of the lines of force of relevant muscles in any position of limb (see Fig. 12.8). Carpus and manus are reconstructions based on composites of *Triceratops.*

well-preserved osseous portion indicates a large head that projected at 20°–30° to the shaft axis. The head also extends onto the dorsal or external side of the proximal end of this element, forming a subtriangular articular surface (Figure 12.4c). The concavity of the glenoid on the scapulo-coracoid (Figure 12.5) indicates a subspherical articular prominence that might have been from 14 to 18 cm across in the proximal-distal dimension and 20 to 24 cm across in the anteroposterior dimension. The position and relationship of the head of the humerus relative to the deltopectoral crest further documents the requirements for a near-horizontal orientation of the humerus, in part because of the purely physical constraints for obtaining articulation, without obstruction, with the glenoid cavity.

The distal end of the humerus is marked by two distinct articular condyles, the capitulum and the trochlea, for articulation with the radius and ulna, respectively. These condyles are not placed one behind the other, as in typical mammalian cursors. Rather, they are positioned nearly side by side. However, a transverse axis through these two con-

dyles is twisted 60°–70° to the plane of the humeral head, 40° to the axis of rotation at the shoulder (Figure 12.4d) and 120° to the plane of the delto-pectoral crest. That being the case, the plane of elbow extension and flexion could not possibly have coincided with any parasagittal plane. Accordingly, the epipodials (radius and ulna) must have flexed in a transverse plane. The condyles are also offset from each other, with the ulnar condyle 8 cm proximally to the radial condyle (Figure 12.4a–c: b). This disparity in the internal relationship between these articular surfaces is also a noncursorial feature (W.P. Coombs, personal communication).

THE RADIUS AND ULNA. The epipodials are also massive, but are shorter than the humerus. As preserved, the maximum epipodial length is slightly more than 80 percent that of the humerus. Coupled with those proportions is the relative size of the olecranon (Figure 12.6). That feature, though prominent and robust, seems short for an animal of such massive weight in the front quarters. That may only be an impression generated by the hypothesized action of

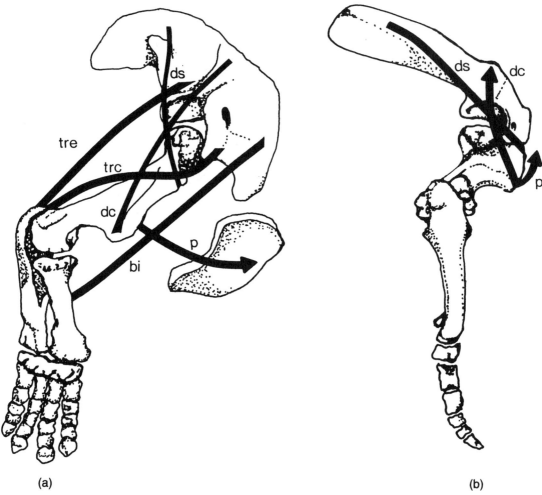

(a)   **(b)**

Figure 12.8. Reconstruction of the lines of action of principal muscles of shoulder and forelimb in (a) anterior and (b) lateral views. Only major forelimb muscles have been reconstructed, based on analogous muscle groups in extant archosaurs. Forelimb is depicted in bent-elbow stance we advocate here based on osteological considerations. Scapulo-coracoid is foreshortened in (a) as is the humerus in (b). Note that the action of the pectoralis, p, on the massive deltopectoral crest is primarily to adduct the humerus, but also to rotate it about the long axis (thereby causing epipodials and manus to swing in a parasagittal plane). Axial rotation of humerus is largely precluded in mammalian cursors. This interpretation of the muscles supports the inference of a sprawling stance, but is not necessary for making that inference. Abbreviations: bi, biceps; dc, deltoideus clavicularis; ds, dorsalis scapulae; p, pectoralis; trc, triceps coraco-scapularis; tre, triceps scapularis.

the epipodial segment in which the olecranon could not function as a lever powering ulnar extension as it does in the parasagittal excursion in true mammalian cursors with relatively long olecranons.

Considerable effort went into testing the range of possible articular arrangements between the ulna and radius of the MPM VP6841, both as the skeletal elements comprising the epipodial pair and in their relationship to the humerus. Unlike the horse or rhinoceros (representing cursorial and mediportal mammals, respectively), there is not the strong keel-and-groove arrangement that dictates one and only one relationship between these elements. However,

there is a best fit that can be attained in articulating these two elements together as a unit, in which the external morphology of both the proximal and distal ends of the ulna and radius nest one against the other. In addition, the surface presented for articulation with the manus demands that the distal ends of these two elements be articulated with their long axes in the same parabolic plane. The resulting articular relationship between these epipodial elements supports the stance proposed here.

The bent-arm posture of ceratopsians would have the nearly horizontal, transverse humerus acting in part like the axis of a wheel. In this analogy, the epipodials correspond to one of the wheel spokes. The reduced ceratopsian olecranon probably provided the leverage required to position the epipodial support to the forefoot, rather than contributing to the forearm extension during the stride. The olecranon also is oriented back toward the sigmoid and articular face of the proximal end of the ulna (Figure 12.6). This orientation, and the resulting line of force, is in a plane fully 180° from that found in the cursorial horse or rhinoceros, where its function in propelling the epipodial in the sagittal plane is obvious.

As noted above, the carpus and manus are not represented in this new *Torosaurus* specimen. Consequently, we can only speculate upon the actual relationships between the distal extremities of the radius and ulna and the carpus and manus. The probable arrangement has the radius and ulna sharing almost co-equal support, with the carpus and manus elements arranged in transverse rows, as restored in two specimens referable to *Triceratops* (American Museum of Natural History AMNH #3350 and USNM #4842).

DISCUSSION

The above description of the osteological morphology and positional modeling for the forelimb of *Torosaurus* support the inference that the forelimb was held in a sprawling posture with the humerus being held in a subhorizontal plane and with principal motion occurring in a transverse plane. Although the range of motion for the forelimb excludes a parasagittal positioning of this element, there is a need to continue to define the range of allowable motion at the various joints and the subsequent limb excursion pathways during various gaits.

All of the inferences that we document are derived from direct observations of the preserved osteology and from the manipulations of the physical model of flexibly mounted forelimb element casts. Although no inference of muscle reconstruction or directions of the lines of force is required to arrive at our conclusions, reconstructions of the major muscle groups and their lines of action are a logical next step in the process of refining our postural model for ceratopsians, and we have included inferences on the action of the pectoralis muscle, in particular, in our arguments; the large deltopectoral crest implies a large and powerful pectoralis musculature. This implication is consistent with that of vertebrates exhibiting a sprawling posture. Inferential data, such as muscle size and lines of force, should be factored into any complete biomechanical analysis of these animals, but our arguments for *Torosaurus* rest with the primary observation of osteological features.

In criticisms of Gilmore's (1915, 1919, 1920) and Lull's (1933) papers for their bent elbow restorations, Bakker (1986) argued as follows:

the biomechanics and soft-tissue anatomy at the shoulder presents, I believe, an unambigous case for rhino-like fore quarters in the big dinosaurian quadrupeds. Extant crocodilians have a semi-erect forelimb that moves very much like the classic Gilmorean restoration of the horned dinosaur forelimb, . . . If dinosaurs had elbows bowed out like those of crocodilians, then the shoulder anatomy [of *Triceratops*] should agree with the crocodilian condition. It does not.

It does not because *Triceratops* is not a crocodile, it does not share the limb osteology of a crocodile, nor did it, presumably, live or behave like a crocodile. Bakker continues: "*Triceratops* has a glenoid that is a true socket" and in effect is more like that of a rhino, and "All stegosaurs, horned dinosaurs, ankylosaurs and sauropods have shoulder joint anatomy similar to that of *Triceratops.* Forelimb function in all must have been fully erect." No data are presented for either of these claims and the anatomical evidence does not support this statement: (1) The shoulder sockets of all of the dinosaurs listed are not exactly alike, and (2) the degree of upright or parasagittal forelimb posture cannot be determined by the shoulder anatomy alone. The degree and kind of mobility may be determined approximately, but design and particular shape of the humerus and the more distal segments of the limb are critical factors as well.

This is exactly our point: The entire anatomy of the forelimb elements is required to assess the normal forelimb posture and its exact excursions. Proximally, the shape of the head of the humerus and the configuration of the glenoid permit a wide degree of humeral protraction and retraction, together with some unknown latitude of transverse abduction-adduction. In addition, the distal condyles of the elbow are aligned in an axis that is twisted at 60°–70° to the plane of the humeral head. This is in startling contrast to, for example, the distal condyles of the humerus in the horse and rhinoceros, in which the limbs are parasagittal.

We conclude that the ceratopsian humerus could not possibly have been positioned in a vertical, parasagittal plane. The reason is because the radius and ulna could not have articulated with those distal condyles and still been within a vertical parasagittal plane as well. Moreover, the distal extremities of the epipodials must have formed a transverse axis hinge for the carpals in order for the manus to align approximately with the intended path of progression, and because of rotation along the axis of the epipodials, the alignment of the infracarpal axis of rotation can be shifted considerably. This in itself is a noncursorial feature.

A number of ceratopsian skeletons recently placed on display in major natural history museums have been mounted with the forelimbs in an upright, mammallike stance. For example, skeletal mounts of ceratopsians at the Royal Tyrrell Museum of Paleontology (Drumheller, Alberta, Canada), with the forelimbs placed in an upright posture, have elbows and shoulders that are disarticulated in order to allow this parasagittal, upright posture. These reconstructions are now acknowledged to be in need of remounting with the forelimbs placed in a sprawl (P.J. Currie, personal communication)

Trackway data supplied by Martin Lockley (1986) were also factored into our study. Although these ichnofossils have been used as one of the lines of evidence for upright forelimb posture in ceratopsians, these data, if anything, are rather inconclusive. The tracks Lockley tentatively assigned to *Triceratops* show the manus placed somewhat lateral to the pes, which would in fact support our contention that the forelimbs were held laterally relative to the hindlimbs.

The above primitive posture proposed here might simply have been retention of this mode of support, perhaps having a basis reflective on ceratopsians being secondarily quadrupedal, derived from the supposed bipedal ancestor *Psittacosaurus,* but it could just as well have been a special adaptation in response to the ceratopsian condition. Consider the ceratopsian state: (1) a very large and massive head, (2) a diverse array of cranial horns and spikes, and (3) the almost certainty that these features were used in physical combat. MPM VP6841 has an exceptionally well-preserved indication of paleopathology in the form of a healed puncture wound and companion "stress fracture" on the anterolateral surface of the left squamosal. Whether the horns were used in interspecific or intraspecific aggressive behavior, the selective advantage of a broad-based forelimb undercarriage compared to a narrow base of an upright parasagittal forelimb stance is obvious – enhanced aggressive stability.

As pointed out by Walter (1986) in her study of kannemeyeriid dicynodonts, the divergent arrangement between the forelimbs and hindlimbs in those animals might in fact represent, or strongly indicate, a dichotomy in function. This dichotomy of function most probably applies to ceratopsians as well. The posture and locomotion of *Torosaurus* and other members of the Ceratopsia must have been quite unlike that of any animal we can directly observe today.

Implications for ceratopsian gait

The following brief observations are discussed more completely elsewhere (Ostrom and Johnson, in press). The ability to gallop has been claimed for ceratopsian dinosaurs (Bakker 1968, 1971, 1986, 1987; Paul 1988, 1991), not so much on the basis of compelling anatomical evidence, but rather on assertions that such energetic locomotion was consistent with the suggestion that all dinosaurs were endothermic with a high metabolic rate (see papers in Thomas & Olson 1980 for discussion).

Coombs (1978) reviewed in detail the anatomical adaptations and physical constraints that are to be expected in animals that exhibit cursorial modes of locomotion. A list of morphologic and biomechanical attributes for extant graviportal, mediportal, subcursorial and cursorial levels of running ability allowed Coombs to hypothesize and evaluate the running potential for various groups of dinosaurs. Special recognition was paid to noting the importance of osteological features in any assessment of an animal's cursorial ability.

Parasagittal limb placement is energetically efficient in mammalian cursors, but *Torosaurus* is neither a cursor nor a mammal. *Torosaurus* was a very large quadruped 10 m long or more, weighing in the vicinity of 5–8 metric tons. It had a large and massive head that might have reached and possibly exceeded one-third of its total body length. Such an enormous head must have placed an unusual burden on the forelimbs, and has to be factored into any speculation about ceratopsian locomotion, or even head or shoulder movement. The much more common *Triceratops*, with a smaller but still proportionately large head, featured nearly identical forelimb elements of nearly the same proportions.

There are references to a "hare-like" bounding gait observed in some crocodilian taxa (Zug 1974). This galloping gait is observed infrequently in only a few species and in individuals of small (less than 2 m) total body length. Although of interest to us in the context of this study, we have concentrated on the mammalian gallop, due in large part to the direct reference of mammalian cursors as locomotor analogs to ceratopsians.

Although certainly not capable of galloping, our results indicate that *Torosaurus* was undoubtedly able to move, when so motivated, at a pace well beyond a lumbering amble. Nor does a sprawling forelimb posture neccesarily negate a possible endothermic physiology for ceratopsian dinosaurs. We would also like to note that the Milwaukee specimen was ultimately mounted with the animal's hind- and forelimbs mounted at the extreme end phases of the step cycle. In this way, visitors have the opportunity to view the proposed limits of forelimb posture and orientation that are deemed biomechanically and arthrologically defensible within the animal's inferred step cycle . The final aspect of MPM VP6841 is, we believe, that of a dynamic and energetic animal (Fig. 12.2).

CONCLUSIONS

Forelimb posture in *Torosaurus,* and by extension other ceratopsian dinosaurs with analogous limb morphology, was a sprawl similar to that seen in extant reptiles. Differences in anatomy are readily apparent in the arthrology and osteology of these archosaurs, yet similar arrangements and biomechanical constraints indicate that the aspect and subsequently the posture of the forelimbs was similar.

Although some range of motion is possible at each joint, their structure precludes a parasagittal position and subsequent ability to achieve a cursorial gait. The morphology of the shoulder joint and the resulting kinematics of the forelimb about the possible axes of rotation at the shoulder are exceedingly complex. This motion is not the simple hingelike motion at the shoulder that is a key assertion made by advocates of an upright stance.

Functional reconstructions of dinosaurs are a challenge in the absence of good extant comparators. The empirical evidence available from osteological studies is, however, adequate in determinations of limb posture and range of motion. In this case, important functional information for testing the competing hypotheses of the stance of *Torosaurus* and other ceratopsian dinosaurs was directly available from osteological observations.

Subsequent studies need to expand on both the functional and ecological implications of postural reconstructions for ceratopsian dinosaurs. For example, there is a need to examine the implications of energetics studies and stress indices of limb elements as they relate to postural and locomotion studies. In addition, velocity ratios for the forelimb of *Torosaurus* and other members of the suborder should also be quantified.

Studies of herding behavior and possible long-distance seasonal migrations of ceratopsian and other dinosaurs are also an important part of the locomotion equation. Our understanding of the biology of ceratopsian dinosaurs is still evolving, and it is our hope that recent discoveries of new material and ongoing reexamination of research collections will add considerably to our expanding body of knowledge regarding the functional morphology of these dinosaurs.

Studies of functional morphology can be fraught with ambiguity, where absolute conclusions regarding form and function of extinct organisms are educated guesses at best. How we choose appropriate models or analogies for our functional studies should be among the first questions asked, especially when we are not able to draw on the data derived from neontological works. Studying the basic osteological features of fossil taxa is an important traditional method for gaining an understanding of functional

morphology, and to ignore the evidence available through this line of study can lead to inadequate conclusions.

REFERENCES

Alexander, R.M. 1989. *Dynamics of Dinosaurs and Other Extinct Giants.* New York: Columbia University Press.

Bakker, R.T. 1968. The superiority of dinosaurs. *Discovery, Yale Peabody Museum* 3, 11–23.

Bakker, R.T. 1971. Dinosaur physiology and the origin of mammals. *Evolution* 25, 636–658.

Bakker, R.T. 1975. Dinosaur renaissance. *Scientific American* 232, 58–78.

Bakker, R.T. 1986. *The Dinosaur Heresies.* New York: Wm. Morrow & Co.

Bakker, R.T. 1987. The return of the dancing dinosaurs. In *Dinosaurs Past and Present,* vol. I, S.J. Czerkas & E.C. Olson, pp. 38–69. Seattle: University of Washington Press.

Coombs, W.P., Jr. 1978. Theoretical aspects of cursorial adaptations in dinosaurs. *Quarterly Review of Biology* 53, 393–418.

Erickson, B.R. 1966. Mounted skeleton of *Triceratops prorsus* in the Science Museum. *Scientific Publication of the Science Museum* 1, 1–16.

Farlow, J.O., & Dodson, P. 1975. The behavioral significance of frill and horn morphology in ceratopsian dinosaurs. *Evolution* 29, 353–361.

Gilmore, C.W. 1905. The mounted skeleton of *Triceratops prorsus. Proceedings of the U.S. National Museum* 29, 433–435.

Gilmore, C.W. 1915. Osteology of *Thescelosaurus,* an ornithopodous dinosaur form the Lance Formation of Wyoming. *Proceedings of the U.S. National Museum* 49, 511–616.

Gilmore, C.W. 1919. A new restoration of *Triceratops,* with notes on the osteology of the genus. *Proceedings of the U.S. National Museum* 55, 97–112.

Gilmore, C.W. 1920. Osteology of the carnivorous Dinosauria in the United States National Museum, with special reference to the genera *Antrodemus* (*Allosaurus*) and *Ceratosaurus. Bulletin of the U.S. National Museum* 110, 1–154.

Hatcher, J.B., Marsh, O.C., & Lull, R.S. 1907. The Ceratopsia. *Monographs of the United States Geological Survey* 19.

Hildebrand, M. 1989. Vertebrate locomotion: an introduction. *BioScience* 39, 764–765.

Johnson, A.J. 1895. *Johnson's Universal Cyclopedia,* vol. VIII, New York: A.J. Johnson.

Lockley, M. 1986. *A Guide to Dinosaur Tracksites of the Colorado Plateau and American Southwest.* Produced in conjunction with The First International Symposium on Dinosaur Tracks and Traces, Albuquerque, New Mexico. Denver: University of Colorado Geology Department.

Lull, R.S. 1933. Revision of the Ceratopsia, or horned dinosaurs. *Memoir of the Peabody Museum, Yale* 3, 1–135.

Marsh, O.C. 1895. Fossil vertebrates. In *Johnson's Universal Cyclopedia 1895,* vol. VIII. New York: A.J. Johnson.

Marsh, O.C. 1896. The dinosaurs of North America.

Sixteenth Annual Report of the U.S. Geological Survey, vol. 2, pp. 133–244.

Osborn, H.F. 1933. Mounted skeleton of *Triceratops elatus. American Museum Novitates* 654, 1–24.

Ostrom, J.H., & Johnson, R.E. In press. Why ceratopsian dinosaurs could never have galloped! *Paleobiology.*

Paul, G.S. 1988. *Predatory Dinosaurs of the World.* New York: Simon & Schuster.

Paul, G.S. 1991. Giant horned dinosaurs did have fully erect forelimbs. *Journal of Vertebrate Paleontology* 11, 50A.

Russell, L.S. 1935. Musculature and function in the Ceratopsia. *Bulletin of the National Museum of Canada* 77, 39–48.

Sternberg, C.M. 1927. Horned dinosaur group in the National Museum of Canada. *The Canadian Field-Naturalist* 4, 67–73.

Sternberg, C.M. 1951. Complete skeleton of *Leptoceratops gracilis* Brown from the Upper Edmonton Member on Red Deer River, Alberta. *Annual Report of the National Museum of Canada, Bulletin No. 123.*

Tait, J., & Brown, B. 1928. How the Ceratopsia carried and used their head. *Transactions of the Royal Society of Canada* 3, 13–23.

Thomas, R.D.K., & Olson, E.C., ed. 1980. A cold look at the warm-blooded dinosaurs. *American Association for the Advancement of Science, Selected Symposium* 28, 1–514.

Walter, L.R. 1986. The limb posture of kannemeyeriid dicynodonts: functional and ecological considerations. In *The Beginning of the Age of Dinosaurs,* ed. K. Padian, pp. 89-97. Cambridge University Press.

Zug, G.R. 1974. Crocodilian galloping: an unique gait for reptiles. *Copeia* 2, 550–552.

13

Functional evolution of the hindlimb and tail from basal theropods to birds

STEPHEN M. GATESY

ABSTRACT

Reconstructions of bipedal locomotion in extinct theropods have been based primarily on information from crocodilians and birds, their closest living relatives. Much of the conflict between opposing hypotheses of bipedalism appears to be due to different phylogenies and categories of locomotor analysis. I present a new hypothesis of theropod locomotor evolution based on a methodology combining functional and morphological data. I propose that not all theropod limbs operated the same way; there was a shift from the primitive saurian limb retraction mechanism to a novel "avian" mechanism during theropod evolution. An integral aspect of this transition was modification of the tail. In basal theropods the tail was a substantial part of the body serving as a base for the caudofemoral musculature. More derived theropods evolved a reduced tail that became decoupled from its primitive linkage to the hindlimb. In birds the tail serves an entirely different function as a highly specialized structure for controlling the tail feathers during flight. The terrestrial locomotion of birds – flying, short-tailed theropods – is similar to, but distinctly different from, walking and running in basal members of the Theropoda.

INTRODUCTION

Birds are often characterized as vertebrates with a dual locomotor system: the wings for flying and the legs for walking, running, hopping, or swimming (Pennycuick 1986; Duncker 1989; Butler 1991). Not surprisingly, most studies of avian origins have focused on the evolution of flight (e.g., contributions in Hecht et al. 1985). Although attention to the pectoral appendage is warranted, a bird's locomotor system is more than simply a pair of wings. The hindlimbs and the tail are also integral constituents, but little consideration has been given to functional changes in these posterior components during the origin of birds.

Osteological data, including those of the leg and tail, provide strong support for an ancestry of birds within theropod dinosaurs, as depicted in the phylogeny used in this chapter (Figure 13.1). Convergence has been invoked to explain these similarities (e.g., Heilmann 1926; Tarsitano & Hecht 1980), but a theropod ancestry of birds has gained wide acceptance (see Witmer 1991 for a review of the debate over avian origins). Character distributions establish that many aspects of modern birds actually appear in the theropod lineage prior to the origin of the avian subclade (Ostrom 1975, 1976; Gauthier & Padian 1985; Gauthier 1986). The hindlimb of *Archaeopteryx* and other birds retains numerous primitive features, implying that it underwent less dramatic modification than the forelimb during theropod evolution. This impression may account for the dearth of functional consideration of the hindlimb during the terrestrial–aerial transition. A significant reduction in tail length is often noted, but the implications of this change have been given little recognition.

In this chapter I outline modifications in hindlimb and tail function that may have taken place between basal theropods and modern birds. I begin by presenting two conflicting reconstructions of bipedal locomotion in extinct theropods. I then describe an experimental approach using locomotor studies of living archosaurs (crocodilians and birds) used to generate a third hypothesis of theropod locomotor history. Functional morphological and paleontological data are integrated to assess the significance of anatomical changes in the hindlimb and tail throughout theropod evolution. Studies of this type typically ask "What can birds tell us about other theropods?" Rephrasing this question as "What do other theropods tell us about birds?" compels us to view avian form and function in a new light. The reciprocal use of neontological and paleontological inquiry should lead to a better understanding of major locomotor transitions such as the origin of avian flight.

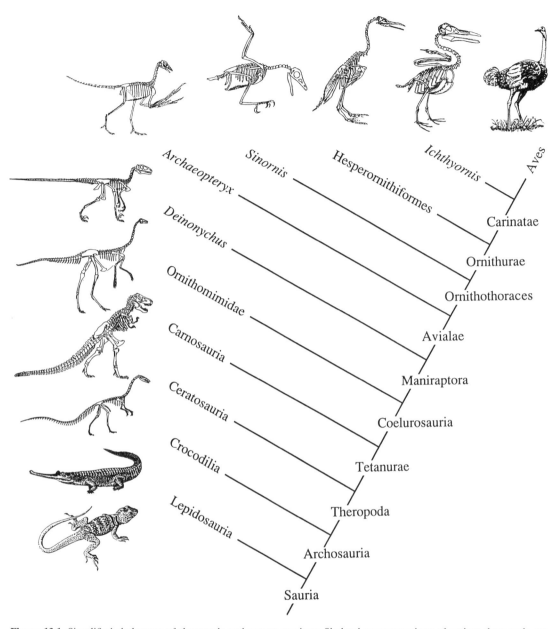

Figure 13.1. Simplified cladogram of theropods and extant saurians. Skeletal reconstructions of extinct theropods are *Coelophysis, Tyrannosaurus, Struthiomimus, Deinonychus, Archaeopteryx, Sinornis, Hesperornis* and *Ichthyornis* (from Carroll 1988 except *Sinornis,* from Sereno & Rao 1992; living saurians from Carr 1963 and Stebbins 1966). Cladogram combines results of Gauthier (1986), Cracraft (1986), Sereno & Rao (1992) and Chiappe (in press).

PHYLOGENY AND CONFLICTING RECONSTRUCTIONS OF THEROPOD LOCOMOTION

Paleontologists focusing on reconstructions of theropod locomotion fall into two main camps, which differ in their choice of phylogeny and methodology. Tarsitano (1981, 1983) analyzed theropod locomotion using a traditional (noncladistic) hypothesis of relationship (Figure 13.2c; Tarsitano & Hecht 1980). Dinosaurs, birds, crocodilians, and pterosaurs were considered independent lineages from a basal archosaur stock called "thecodonts" (now considered paraphyletic). With no living descendents of theropods to study, Tarsitano based his reconstruction on what I call the "ancestor" model. In this methodology, similarity is hypothesized up the phylogeny to include more distal branches (Figure 13.2c).

Tarsitano (1983) focused on pelvic anatomy, where he identified similarities among theropods,

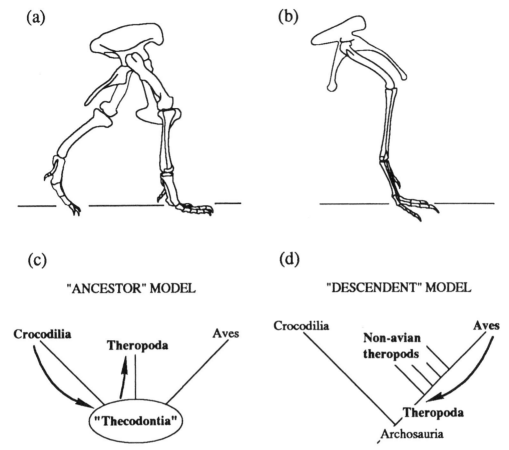

Figure 13.2. Reconstructions of hindlimb position during locomotion in extinct theropods based on different methodologies. (a) *Tyrannosaurus* limbs showing a large arc of femoral retraction during the propulsive phase. Reconstruction was made with an "ancestor" model (c) using data from "thecodonts" and living crocodilians (modified from Tarsitano 1983). (b) *Coelophysis* limb position at maximum femoral retraction showing much less movement than in (a) (after Padian & Olsen 1989). Reconstruction was made with the "descendent" model (d), which stresses the birdlike nature of theropod bipedalism. Arrows indicate the direction of inference between groups emphasized in each analysis.

crocodilians, and "thecodonts" that were not shared by birds. He proposed that important aspects of locomotion in the ancestral group (Thecodontia) were retained in some descendent lineages (Theropoda and Crocodilia), but not all (Aves). Thus, the myology and function of living crocodilians was used to reconstruct "thecodonts," which were then employed to reconstruct extinct theropods, such as *Tyrannosaurus* (Figure 13.2a).

Other workers have reached very different conclusions using a cladistic phylogeny of archosaurs (Figure 13.2d), which supports Ostrom's hypothesis of a theropod ancestry for birds (e.g., Ostrom 1975). If such a relationship is accepted, modern birds are highly informative about bipedalism in their extinct theropod forbears. Birds have been used in a method of paleontological reconstruction that I call the "descendent" model, in which information from a living group is attributed to the extinct members of a more

inclusive clade. In the descendent model, similarity is hypothesized down the cladogram to include more basal branches (Figure 13.2d).

Using this approach, Padian and Olsen (1989) postulated that bipedalism in living members of the clade Aves is representative of bipedalism in the larger clade Theropoda (Figure 13.2d). Close phylogenetic and anatomical affinity have been used to support a hypothesis in which all theropods (avian and nonavian) are essentially identical in their manner of bipedal progression (Figure 13.2c; Padian 1986; Paul 1988; Padian & Olsen 1989; Campbell & Marcus 1991). This was outlined most explicitly in a study comparing the skeleton and footprints of modern rheas to those of Mesozoic theropods such as *Coelophysis* (Padian & Olsen 1989). These workers proposed that among theropods the "basic patterns of stance and gait have not changed since the late Triassic" (p. 235). They concluded: "on the basis of comparative anatomy and kinematics of theropods, living and extinct, that there were no substantial differences in hindlimb function, despite sweeping changes in the pelvis, tail, and mode of life that occurred during the evolution of birds" (p. 239).

Before presenting a different methodology and hypothesis of theropod locomotion, it may be helpful to specify different categories for which locomotor similarity can be analyzed and to clarify terminology.

ASSESSING LOCOMOTOR SIMILARITY

Categories of locomotor analysis

The six categories described below are among those that have been considered by paleontologists studying theropod locomotion. These are by no means exhaustive, but help in the evaluation of competing hypotheses by making differences between interpretations clearer. More general, hierarchical levels of analysis have been proposed for the study of complex morphological systems in other vertebrates (Lauder 1991).

CATEGORY A: SKELETAL ANATOMY. This is the traditional approach for most paleontological research. Data are collected directly from skeletal material and consist of qualitative and/or quantitative descriptors of bone shape, articular geometry, muscle scars, internal structure, etc. This information can be assembled to assess skeletal proportions and ontogenetic or phylogenetic scaling.

CATEGORY B: LIMB POSTURE. Although the term "posture" has been used in many different ways, herein it designates the degree to which the limb is abducted. Limb abduction is the smallest angle between the femur or humerus and the sagittal plane of the body. In extant forms it can be quantified; angles or ranges of angles express the change in limb posture during the stride cycle. However, the true abduction angle often has to be calculated since it is typically distorted in dorsal or lateral views of the limb. In extinct tetrapods it must be estimated from osteological evidence (category A) and from the relative width of fossil trackways. Postural grades such as "sprawling," "semi-erect," and "erect" have been proposed to categorize the degree of limb abduction in different tetrapods, but the reality and utility of such grade systems have been questioned (Jenkins 1971; Gatesy 1991a; Sereno 1991).

CATEGORY C: LIMB-SEGMENT ORIENTATION. This type of analysis describes the position of individual limb elements during standing or at a single time in the stride cycle. Positions of segments can be quantified individually as deviations from specified anatomical planes. More often, orientations are expressed as joint angles between segments. For most animals limb movements are highly three dimensional, meaning that angles measured directly from dorsal or lateral films do not accurately represent true joint angles. To be meaningful, angles should be calculated in the plane of the segments forming the joint. Orientations and angles such as these may be quantified precisely in living forms. In fossil tetrapods the geometry of articular surfaces has been used to infer orientations, but these orientations often the subject of much contention.

CATEGORY D: LIMB KINEMATICS. An even more detailed type of locomotor analysis is the change in limb segment orientation through time (kinematics). These data reveal how the limb elements are actually moved during locomotion to propel and support the animal. Ideally, descriptions in this category include kinematics at different speeds so that the locomotor repertoire of the animal can be assessed. Such data have been quantified for very few living archosaurs (*Columba*, Cracraft 1971; *Gallus*, Manion 1984; *Alligator*, Gatesy 1991a; *Numida* and *Colinus*, Gatesy, unpublished data), and are clearly unavailable for extinct forms.

CATEGORY E: MUSCULAR ANATOMY. Information of this kind can be acquired in living animals by methods such as gross dissection and histology. In extinct forms muscular anatomy is reconstructed using skeletal anatomy and comparative data. For many muscles the information necessary to estimate attachments and size is simply not preserved. Studies of living birds (McGowan 1979, 1982, 1986) and mammalian carnivores (Bryant & Seymour 1990) have revealed the speculative nature of any detailed reconstruction. One compromise is to restrict reconstruction "to only those muscles that leave obvious and reasonably interpretable scars or are judged to be reasonably unequivocal from previous study and established phylogenies" (Bryant & Seymour 1990, p. 116). This caution is rarely heeded by paleontologists. Complete restorations of extinct theropod hind limb musculature are intriguing illustrations, but have questionable scientific significance.

CATEGORY F: NEUROMUSCULAR CONTROL. This type of analysis addresses the control of limb muscles by the nervous system during locomotion. It is typically studied by electromyography (EMG), in which electrodes are used to record electrical changes associated with muscle activation. What muscles are generating the forces that control and mediate limb movement? When and with what intensity are they active? How do various muscles interact and contribute to limb movement? Are some muscles less important to locomotor behavior than others? Motor pattern data of this type are only available for the hindlimbs of a limited number of archosaurs: *Gallus* chicks (Jacobson & Hollyday 1982; Bekoff et al. 1987), *Branta* (Weinstein, Anderson, & Steeves 1983), and *Alligator* and *Numida* (Gatesy 1989). Few paleontologists deal directly with this subject (see Giffin, this volume), but neuromuscular control is often implied from hypotheses in other categories of analysis. It is an important component because it helps link the anatomical information (categories A and E) with limb kinematics (category D). It is also the most difficult to reconstruct in anything more than the most general terms (see Lauder, this volume).

Correlation among categories

Although the locomotor categories outlined above are interrelated, the correlation among some categories may not be absolute. For example, similar postures are shared by organisms with diverse morphologies and kinematics. Humans and birds both have a highly adducted limb posture (category B), but the orientations of the femur, tibia, and foot (category C) differ dramatically and undergo very different movements (category D) during walking and running (Gatesy & Biewener 1991). Reconstructions of locomotion in extinct animals should explicitly state what is known and what is being hypothesized about each category.

In addition, all aspects of opposing reconstructions may not be in conflict. This appears to be the case for the two hypotheses presented above. Padian and Olsen (1989) proposed that theropods are birdlike in their skeletal morphology, posture, limb position, and limb kinematics (categories A–D,) but they did not directly address musculature (category E) or neuromuscular control (category F). These authors contrasted their reconstruction with Tarsitano's, which they viewed as based on a crocodilian paradigm. Tarsitano (1983) did invoke a crocodilian model, but this was used to reconstruct pelvic musculature and to identify what he called the primary "bone-muscle complex" of extinct theropods. Thus, Tarsitano was using crocodilians to reconstruct myology (category E) and something akin to neuromuscular control (category F). Padian and Olsen assumed that Tarsitano's acceptance of a crocodilian model in these two categories signified his acceptance of a crocodilian model in all cat-egories. Contrary to Padian and Olsen's assertion, Tarsitano never invoked a crocodilian limb posture (category B) for theropods; both camps actually agree that the limbs were highly adducted.

These authors are in disagreement about limb-segment orientations (category C) and kinematics (category D) in extinct theropods. Padian and Olsen proposed birdlike limb orientation and movement, but Tarsitano's (1983) theropod differed from either the avian or crocodilian model. His *Tyrannosaurus* (Figure 13.2a) is standing and moving unlike any living archosaur. This chimera has an avianlike posture, a crocodilianlike pelvic musculature, and a unique pattern of limb orientation and kinematics. Such a reconstruction stands and moves in a way not seen in any living animal.

Questions to be addressed

How can these conflicting reconstructions be evaluated? Is either hypothesis adequate? Like most paleontologists, I accept the outcome of a cladistic analysis in which birds are flying theropods (Figure 13.1). However, the reconstruction of all theropods standing, walking, and running like modern birds is not an inevitable outcome of this phylogeny. If Padian and Olsen (1989) are correct, it must be presumed that the origin of flight had no significant effect on the theropod hindlimb. Is it possible to evolve the dual locomotor system of birds (aerial and terrestrial) from the single system of basal theropods (terrestrial) without affecting the primitive morphology, function, and behavior (Figure 13.3)? Does Tarsitano's "ancestor" model provide crucial evidence about primitive archosaur features overlooked by the "descendent" model? How well are the living archosaurs used to reconstruct extinct theropods really understood? In the following section I will introduce data from recent functional studies in an attempt to clarify the reconstruction of bipedal locomotion in extinct theropods. I try to be explicit in my use of phylogenetic information and the locomotor categories for which data are informative.

AN EXPERIMENTAL APPROACH TO ARCHOSAUR FUNCTION

I have used an experimental approach to analyze changes in hindlimb and tail function during archosaur evolution (Gatesy 1989, 1990, 1991a, 1991b, in press; Gatesy & Dial in press). I follow a methodology presented by functional morphologists (Lauder 1986; Schaefer & Lauder 1986), but incorporate data from extinct taxa. This consists of four steps: (1) the adoption of a cladistic phylogeny and selection of living taxa; (2) the analysis of locomotor categories A–F of living forms; (3) the assessment of differences between living taxa and their functional significance; and (4) the tracing of osteological correlates of important muscular subsystems onto

Basal theropods: fully terrestrial   Avian theropods: aerial and terrestrial

Figure 13.3. Living birds combine aerial and terrestrial locomotion, whereas nonavialian theropods had only a single locomotor behavior. Could the morphology and functional mechanisms of basal theropod bipedalism survive relatively unchanged in birds, or did the evolution of flight significantly alter the primitive terrestrial system? (Figures modified from McFarland et al. 1985.)

the cladogram to elucidate function in extinct taxa. In this way it is possible to combine information from multiple sources to reconstruct a clade's locomotor history.

The recent appearance of cladistic analyses of diapsid phylogeny (Benton 1985; Gauthier 1984, 1986; Gauthier & Padian 1985; Cracraft 1986; Benton & Clark 1988; Sereno 1991) has greatly enhanced the potential for evolutionary study in this group. I follow the phylogeny of Gauthier (1986), which unites birds and crocodilians in the clade Archosauria (Figure 13.1). Archosauria and Lepidosauria (squamates and *Sphenodon*) constitute the extant members of the clade Sauria. "Birds," synonymous with Avialae unless further specified, are considered a subclade of theropod dinosaurs. Recent discoveries and phylogenetic analyses (e.g., Sereno & Rao 1992; Sanz & Bonaparte 1992; Chiappe in press) are helping to resolve the early evolution of birds, but this area is currently in a state of flux.

For living archosaurs I studied the American alligator (*Alligator mississippiensis*) and the helmeted guineafowl (*Numida meleagris*). Hindlimb posture (category B), segment orientations (category C), and kinematics (category D) were quantified from high speed X-ray films. Simultaneously, activity patterns of hip and thigh muscles were recorded with surgically implanted fine-wire EMG electrodes (category F). Using these techniques I could document movement of the hindlimb skeleton and correlate this with muscle activity. These data were compared to limb

kinematics (Snyder 1954; Brinkman 1981) and muscle activity (K.K. Smith & F.A. Jenkins, Jr., unpublished data) in members of the lepidosaur outgroup.

Kinematic and caudal differences between birds and crocodilians

An analysis of hindlimb kinematics and motor pattern in the guineafowl (16 hip and thigh muscles) and alligator (14 hip and thigh muscles) distinguished major differences in locomotion in these living archosaurs (Gatesy 1989). One result of this study was a confirmation of the importance of the tail and caudofemoral musculature during terrestrial locomotion in the alligator, but not in the guineafowl (Gatesy 1990). The caudofemoralis longus (CFL) is a large muscle in crocodilians that originates from the tail base and inserts on the femur (Figure 13.4b). In birds the CFL is either a small muscle, as in the guineafowl (Figure 13.5b), or is completely absent (George & Berger 1966). Although it is an oversimplification to restrict attention to only one muscle of the limb, the difference in the morphology, activity pattern, and functional significance of the CFL between crocodilians and birds is particularly striking. Data for each locomotor category are summarized in Table 13.1.

HINDLIMB KINEMATICS IN THE ALLIGATOR. During walking, alligators retract the hindlimb by extending the hip and knee joints (Figure 13.4a,c; Gatesy 1990,

ALLIGATOR

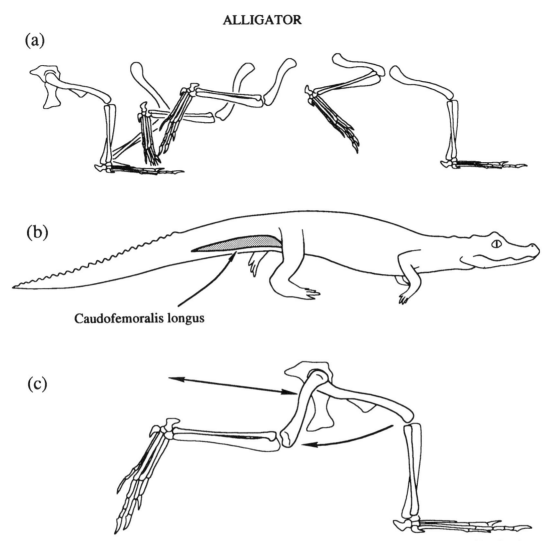

Figure 13.4. Terrestrial locomotion in the alligator, *Alligator*. (a) Hind limb kinematics during one stride of a slow walk. (b) Caudofemoralis longus (CFL) in the alligator is a large, cylindrical muscle originating from tail base and inserting on fourth trochanter of femur. (c) Position of hindlimb at beginning and end of propulsive phase. The majority of foot displacement is produced by the large arc of femoral retraction (*arrow*) powered by caudofemoralis longus (*double-headed arrow*).

1991a). The 60°–80° arc of femoral retraction (hip extension) is primarily powered by the CFL, which shows high amplitude EMG activity at even the slowest speed walks. Other potential synergists, such as the hamstrings, do not show significant activity during walking. These data verify that crocodilians employ a caudofemoral retraction system to move the entire hindlimb about the hip joint, as was predicted from their morphology (e.g., by Tarsitano 1981). The situation in *Alligator* is likely to be the primitive archosaurian condition, since a caudofemoral retraction system appears to be employed by *Sphenodon* and most limbed lizards (Snyder 1954, 1962; Gatesy 1990; Russell & Bauer 1992).

HINDLIMB KINEMATICS IN THE GUINEAFOWL. Compared to other living saurians, birds are highly unusual in the morphology of their tail and caudofemoral musculature. It must be stressed, however, that guineafowl also differ from lepidosaurs and crocodilians in limb posture, position, kinematics, and many muscle motor patterns (Gatesy 1989, 1990, in press). During walking, the femur undergoes a very small propulsive arc of only 5°–15°. The limb is not retracted significantly by hip extension, as in crocodilians and lepidosaurs, but by knee flexion of 50°–70°. Coactivation of the hamstring and quadriceps musculature controls this movement; the CFL is inactive. As guineafowl reach faster speeds, the kinematic

Table 13.1 *Summary of anatomical, cineradiographic, and electromyographic analyses of hindlimb function in the alligator and guineafowl during treadmill locomotion*

Locomotor category	Alligator	Guineafowl
A: Skeletal anatomy		
Caudal vertebrae number	ca. 40	5 + pygostyle
CFL insertion scar	Fourth trochanter	None
B: Limb posture (femoral abduction)		
Range of angles during a stride	20°–50°	12°–15°
C: Limb segment orientation		
Stance	Quadrupedal	Bipedal
Foot position	Plantigrade	Digitigrade
D: Limb kinematics (propulsion)		
Hip extension arc	60°–80° (walk)	5°–15° (walk), 20°–40° (run)
Knee extension arc	30°–40° (walk)	0°–5° (walk), 5°–25° (run)
Knee flexion arc	None (walk)	50°–70° (walk and run)
E: Muscular anatomy		
Caudofemoralis longus	Well developed	Minute
F: Neuromuscular control		
1° propulsive mechanism of hind limb	CFL retraction of femur	Hamstrings/quadriceps' control of knee

pattern is modified slightly. Knee flexion is still the primary means of foot displacement, but this becomes augmented by a period of hip-and-knee extension late in the propulsive phase prior to foot liftoff (Figure 13.5a,c). The femur does not pass vertical at even the fastest runs recorded. CFL activity may assist other, more massive, muscles in retracting the femur during running, but its force contribution must be almost negligible.

"AVIAN" BIPEDAL LOCOMOTION. I suggest that the pattern of limb use and muscle activity in the guineafowl can be used to exemplify "avian" bipedal locomotion. Unlike alligators, guineafowl do not use a caudofemoral retraction system to power the hindlimb. This can be inferred from the minute size of the CFL, and is firmly established by limb kinematics and muscle activity. There is no reason to suspect that other modern birds differ from the guineafowl in this regard (see "Potential methodological drawbacks" section below). In birds a knee-based limb mechanism takes the place of the hip-based mechanism utilized by crocodilians and lepidosaurs (Tarsitano 1983; Gatesy 1990). If a caudofemoral retraction system is primitive for the clade Archosauria, when was this means of propulsion replaced by knee flexion in the lineage leading to birds? Is loss of the primitive caudofemoral system restricted to birds, or is it representative of a more inclusive group of archosaurs? To answer this question, CFL size must be estimated in extinct taxa.

Caudofemoral musculature in fossil theropods

Fortunately, the CFL of archosaurs has a distinctive insertion scar on the femur, the fourth trochanter (Figure 13.6b). The fourth trochanter is only lost in maniraptoran theropods and pterosaurs. A fourth trochanter is a primitive character for archosaurs (Gauthier 1984) and its association with the caudofemoral musculature is well accepted (e.g., Romer 1956). The CFL can also be assessed from the size of its potential site of origin at the base of the tail. In theropods other than birds there is a demarcation between the proximal and distal parts of the tail known as the transition point (Figure 13.6a; Russell 1972; Gauthier 1986). The transition point denotes the shift from more typical proximal vertebrae to distal caudals with no transverse processes, more elongate prezygapophyses, and "boat-shaped" chevrons (Gauthier 1986). These changes make the posterior part of the tail quite narrow and probably form a maximum limit of CFL origin in fossil theropods. In modern crocodilians and birds, the loss of caudal transverse processes correlates relatively well with the distal-most extent of the CFL's origin. By tracing modifications of CFL origin and insertion on the cladogram (Figure 13.6c), it is possible to follow the history of this muscle through theropod evolution with some certainty.

An ample site of origin (large tail base) and pronounced insertion scar (fourth trochanter) are primitive archosaurian characters retained in the clades

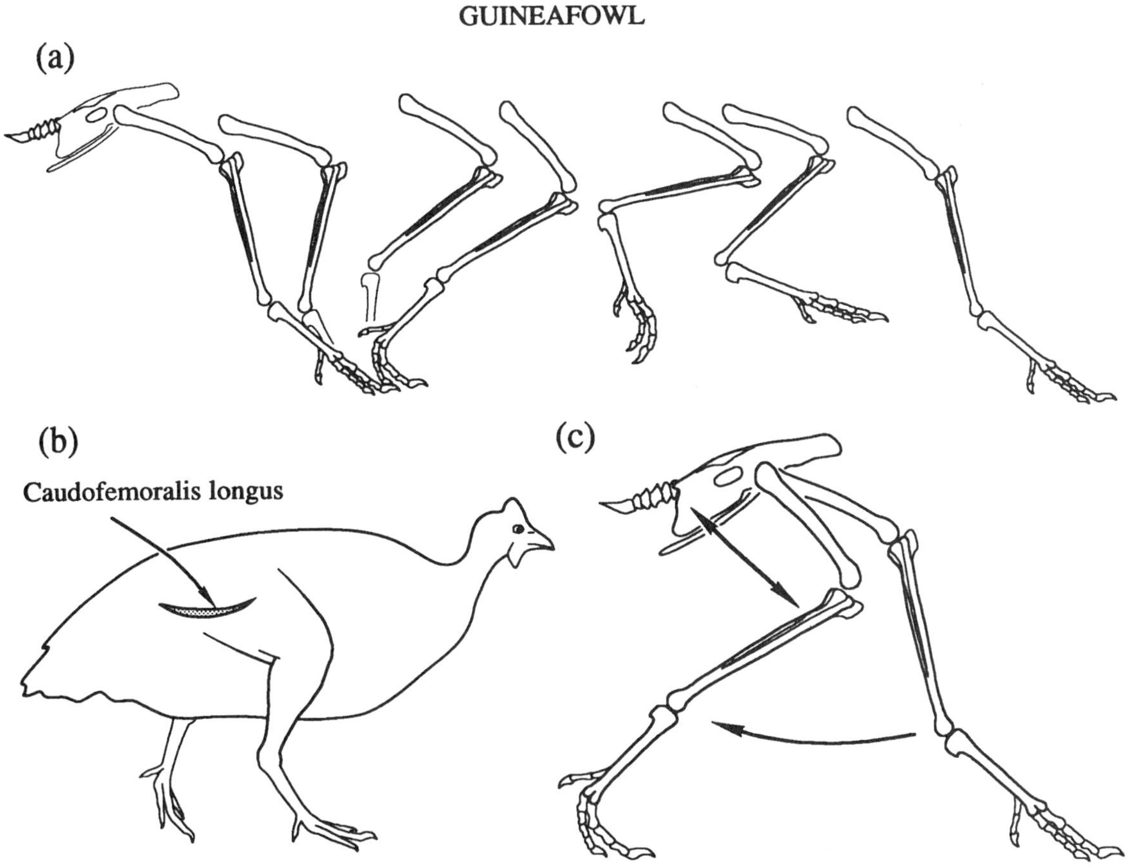

GUINEAFOWL

Figure 13.5. Terrestrial locomotion in guineafowl, *Numida.* (a) Hindlimb movements during one stride at a slow run (1.0 m/s). Note small arc of femoral retraction during propulsive phase. (b) Caudofemoralis longus (CFL) in guineafowl is a thin, straplike muscle that lacks the broad origin and prominent insertion of crocodilians and lepidosaurs. (c) Hindlimb position at beginning and end of propulsive phase at a slow run. Knee flexion (*single-headed arrow*) produced by hamstrings (*double-headed arrow*) accounts for majority of foot displacement.

Dinosauria and Theropoda (Gauthier 1986). In theropods ancestrally, the tail is a long, presumably well-muscled structure composed of a large number of caudal vertebrae (ca. 40–50). The femur has a marked fourth trochanter for CFL insertion. In basal theropods, such as ceratosaurs, the transition point is located about halfway down the tail (Gauthier 1986). Changes in this primitive condition occur at different nodes along the cladogram (Figure 13.6c; Gauthier 1986; Cracraft 1986; Sanz, Bonaparte, & Lacasa 1988; Sanz & Bonaparte 1992; Sereno & Rao 1992). The transition point's position is sequentially more more proximal in Tetanurae, Maniraptora, and Avialae (Figures 13.6c & 13.7). The number of caudal vertebrae bearing transverse processes reflects this incremental shift: at least 17 in ceratosaurs, at least 14 in carnosaurs, 12–15 in ornithomimids, 11–12 in *Deinonychus,* and eight or less in *Archaeopteryx* and other birds (Gatesy 1990). The fourth trochanter undergoes a similar pattern of reduction. The caudofemoral insertion scar is more

weakly developed in coelurosaurs than in more basal theropods and is completely absent in advanced maniraptorans (dromaeosaurs and birds). In the Ornithothoraces the terminal caudal vertebrae are consolidated into a pygostyle (Figures 13.1 & 13.6c).

Trends in tail and caudofemoral evolution

The character distributions summarized above document a dramatic transformation in tail size and CFL indicators during theropod evolution. Although the reduction in tail among *Archaeopteryx* and all other birds is cited in most biology texts, diminution of the tail has occurred at nodes both above and below *Archaeopteryx* on the cladogram. When total tail length (ilium to tail tip) is compared relative to hindlimb length (femur + tibia + metatarsus), the magnitude of this decline can be appreciated (Figure 13.7). Tails of fossil theropods are rarely completely preserved, but those of basal theropods such as ceratosaurs and carnosaurs are typically about

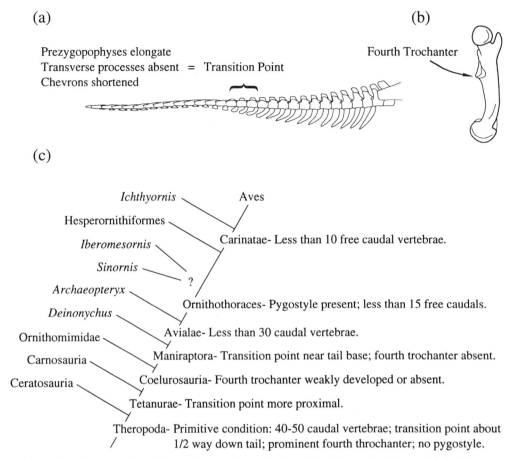

Figure 13.6. Skeletal indicators of caudofemoral musculature. (a) Caudal skeleton of ornithomimid *Struthiomimus* illustrating transition point between proximal and distal portions of tail. Note loss of transverse processes, elongation of prezygapophyses and change in chevron shape near transition point (after Russell 1972). (b) Medial view of left femur of *Tyrannosaurus* showing fourth trochanter, the insertion site of caudofemoral musculature (after Osborn 1916). (c) Distribution of characters relating to caudofemoral musculature through theropod evolution (cladogram from Gauthier 1986; Sereno & Rao 1992; Sanz & Bonaparte 1992; Chiappe, in press; data from Gauthier 1986 and Cracraft 1986).

twice as long as the hind limb. This is in sharp contrast to modern birds, in which the tail is only about 15 to 20 percent of hindlimb length. Extant birds have tails that are approximately ten times shorter than those of basal theropods, relative to hindlimb length. Between these extremes are theropods with intermediate relative tail lengths. However, this single measure does not account for the narrowing of the tail or the proximal shift in transition point in ornithomimids, *Deinonychus*, and *Archaeopteryx*. As the tail is shortened and decreased in diameter, the potential origination site of the CFL is also reduced. This can be seen in the relative length of the tail base (Figure 13.7), as shown by the extent of caudal transverse processes. A gradual decrease in relative tail-base length is apparent in Coelurosauria more closely related to modern birds. Concordantly, the

insertion scar of the CFL on the femur progressively decreases in size between basal theropods and coelurosaurs, leading to its eventual loss in maniraptorans.

AN ALTERNATIVE HYPOTHESIS OF THEROPOD LOCOMOTOR EVOLUTION

Loss of the CFL retraction system during theropod evolution

If the relationship between musculature (category E), motor control (category F), and kinematics (category D) found in modern lizards, crocodilians, and birds holds for extinct saurians, a functional caudofemoral retraction system can be reconstructed

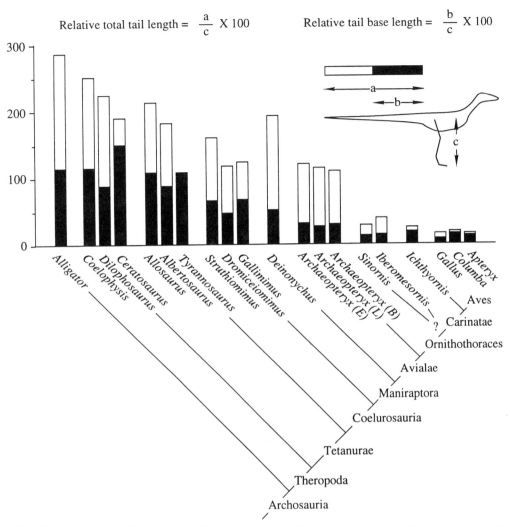

Relative total tail length = $\frac{a}{c}$ X 100   Relative tail base length = $\frac{b}{c}$ X 100

Figure 13.7. Plot of relative tail length and relative tail base length through theropod evolution. A ratio of total tail length and tail base (caudals with transverse processes) length to hind limb length (femur + tibia + metatarsal III) yielded a percentage for each genus. Note dramatic reduction in relative tail size and proximal migration of transverse processes in theropods more closely related to birds.

as the primitive mechanism of hindlimb movement in archosaurs, dinosaurs, and theropods. Archosaurs appear to have simply retained the plesiomorphic saurian limb mechanism, which was then passed on to basal theropods. A retraction mechanism linking the tail and femur is used by lepidosaurs and crocodilians despite differences in their limb posture. If caudofemoral musculature is accepted as the primary limb-retraction mechanism in basal theropods, it is extremely difficult to reconstruct these forms as birdlike, since it is clear that modern birds do not use this system to power their hindlimbs.

Character modifications along the cladogram reveal that the ancestral theropod retraction mechanism was gradually phased out. The size and significance of the CFL appear to have been reduced as the tail and hind limb changed in the theropod lineage. I propose that during theropod evolution there was a shift from a primitive limb mechanism using the CFL to a novel mechanism using knee flexion to propel the body forward with each stride (Figure 13.8). Because of this transition, it is difficult to generalize about bipedalism in theropods; theropods did not all operate in exactly the same way.

Basal forms such as ceratosaurs and carnosaurs appear to have inherited the caudofemoral retraction system relatively unchanged from primitive dinosaurs. I suggest that forms such as ornithomimids used caudofemoral retraction, but that an incipient knee-based system may also have been present. I

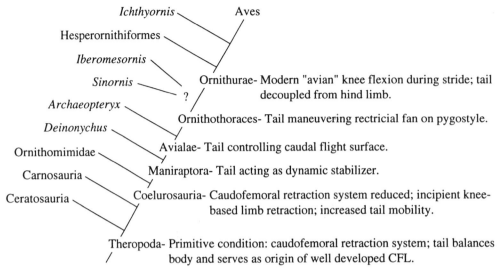

Figure 13.8. Proposed distribution of changes in hindlimb and tail function through through theropod evolution (cladogram from Gauthier 1986; Sereno & Rao 1992; Sanz & Bonaparte 1992; Chiappe in press).

hypothesize further enhancement of the knee-based system as the tail and CFL were progressively negated, eventually leading to what I now call the "avian" condition. Although it is termed "avian," this condition probably reached its present form in the Ornithurae, where pelvic structure (prominent antitrochanter, loss of pubic and ischiadic symphyses, etc.) suggests a femoral orientation similar to that of modern birds. Extant birds show little evidence of the primitive CFL retraction system of basal theropods because it has been gradually replaced by an alternative mechanism during the evolution of this clade (Figure 13.8).

Changes in tail function

The adoption of a novel limb mechanism in derived theropods is linked to tail reduction in at least two ways. As noted above, diminution of the tail affected the CFL retraction system. At the same time, the dramatic size decrease of a major body part changed the way that theropods balanced. Ancestrally, the theropod tail most likely functioned as a cantilever beam, balancing the front of the body about the acetabulum (Newman 1970; Tarsitano 1981, 1983; Alexander 1985). Since the body's center of mass was near the hip joint, the feet could be placed directly below the pelvis. As the tail diminished, so did its usefulness as a cantilever. This reduction in body mass behind the hip joint appears to have resulted in a shift in femoral orientation. In modern birds the body's center of mass is located well in front of the acetabulum (Manion 1984). Most birds maintain the femur in a subhorizontal position to

bring the feet anteriorly under the center of mass (Storer 1971; Alexander 1983; Manion 1984). With a reduced CFL and a femur constrained by its role in balancing the body, femoral retraction was no longer tenable as the primary means of foot movement, particularly at low speeds. In this new orientation, flexion of the relatively stationary knee joint proved to be an adequate solution for foot movement. Thus, I consider the adoption of an "avian" limb segment orientation and kinematics as a solution to locomotion, rather than as an arrestor mechanism during landing in early flying birds (Walker 1977).

Why did the theropod tail become smaller? As noted above, tail reduction occurred over several nodes, rather than at a single node, and began prior to the origin of flight (Figure 13.8). Therefore, a simple aerodynamic explanation is not adequate. I suggest that tail reduction in coelurosaurs and maniraptorans may indicate a different role for the tail in these forms. A reduced body mass and more mobile tail may have increased the agility and speed of ornithomimids. Ostrom (1969) postulated that the highly specialized tail in *Deinonychus* could have been used as a dynamic stabilizer during acrobatic maneuvers. The tail's ability to stabilize would likely be compromised if a caudofemoral retraction system was necessary to power the raptorial hind limb in predators such as *Deinonychus*. A slimmer, more mobile tail would thus go hand in hand with a reduced CFL. If true, this would represent a shift in function from basal theropods, where mobility of the long, heavy tail was probably much less. As the tail took on a new role in coelurosaurs, the hindlimb may have been evolving a more "avian" retraction mechanism. This system would be powered by muscles originating from the pelvis rather than the tail base, thus freeing the caudal appendage.

In Avialae the role of the tail appears to change once again (Figure 13.8). With the acquisition of flight the tail's primary task became control of the main tail feathers (rectrices). The rapid reduction in tail length in the Ornithothoraces was produced by a reduction in caudal number and the fusion of distal vertebrae into the pygostyle. This modified the frond-shaped tail of *Archaeopteryx* into the fan-shaped tail of other birds. The origin of the pygostyle is probably related to refining caudal maneuverability and the ability to fan the rectrices (Raikow 1985; Baumel 1988; Sereno & Rao 1992; Gatesy & Dial in press). The transformation of a relatively simple saurian tail into the highly specialized rectricial control system of modern flying birds is rarely appreciated. Interestingly, the maniraptoran tail was in some ways preadapted for flight since it had already evolved into a more mobile appendage that was almost freed of its primitive connection to the hindlimb. Anatomical, functional, and neuromuscular evidence of decoupling of the tail from the hindlimb can be found in modern birds (Gatesy & Dial in press). This decoupling of the ancient caudofemoral linkage may have facilitated the novel allegiance of the tail with the forelimbs during flight.

POTENTIAL METHODOLOGICAL DRAWBACKS AND WEAKNESSES OF THE ANALYSIS

Conclusions are entirely dependent on the phylogeny

In the absence of a phylogeny it is impossible to make any meaningful evolutionary statement. Attempts to trace character evolution are contingent on a well-supported phylogenetic analysis. As the first major work on fossil and living saurians, Gauthier's cladogram (1986) has become a starting point for analyses of theropod phylogeny. However, there is little doubt that it will be modified in the future as taxa are discovered, described, and reanalyzed. As alternative phylogenies appear, the character distributions can be assessed to evaluate if my hypothesis of locomotor evolution is weakened or strengthened.

Living archosaurs may not adequately represent their group

This issue has received little attention in paleontological circles. Since so few birds, crocodilians, and lepidosaurs have been well studied, it is possible that the chosen species are not representative of each group. The likelihood of significant variation among crocodilians is probably much lower than among birds, which is a larger clade with much greater morphological and locomotor diversity. The potential hazard of misrepresentation also differs between locomotor categories. Much more is known about osteological (category A) and myological (category E) diversity within groups, for example, than limb kinematics (category D) and neuromuscular control (category F). The hindlimb musculature of crocodilians is remarkably conservative (Tarsitano 1981). Therefore, the choice of *Alligator* may be appropriate to represent crocodilians, but does bipedal locomotion in the guineafowl exemplify "avian" locomotion?

It is impossible to deny that many birds (particularly those specialized for aerial or aquatic lifestyles) deviate from this so-called "typical" locomotion in one or more categories. What is the primitive condition for Aves? Unfortunately, the phylogeny of the main branches of modern birds is still poorly resolved. A common solution is to study locomotion in ground-dwelling or flightless birds that are thought to resemble nonavian theropods most closely (although derived from flying ancestors). Forms traditionally studied include the chickenlike galliformes (chickens, pheasants, grouse, turkeys, quail, and guineafowl) and the flightless ratites (ostriches, rheas, emus, cassowaries, and kiwis). I chose the helmeted guineafowl (*Numida meleagris*) as a ground-dwelling bird that was of appropriate size for the experimental analysis, easily available, and conducive to training.

Some confidence about the use of the guineafowl as the model avian biped comes from a study of locomotion in eight species of galliformes and ratites spanning a 2000-fold range of body mass (Gatesy & Biewener 1991). We found that all birds (including the guineafowl) employ the same basic pattern of limb movement and gait parameters, despite the difference in size between a quail and an ostrich. Hindlimb motor patterns are known from very few birds other than the guineafowl, yet these data are all qualitatively similar. EMG recordings from the CFL of pigeons (Gatesy & Dial in press) are similar to those of guineafowl, confirming that the lack of a caudofemoral retraction system is not restricted to galliformes. The generalization of locomotor data from individual or, at best, several species can be justified if the limits of such a procedure are kept in mind. Additional data will either support the generalization or require that it be reformulated. As the understanding of modern forms improves, so can the resolution with which avian locomotor history is evaluated.

Similar trackways indicate similar kinematics in all theropods

This premise was used by Padian and Olsen (1989) to "test" for similarity between Triassic and modern theropods. Although this interesting approach has merit, I question the resolution at which details of limb-segment orientation, kinematics, muscular anatomy, and neuromuscular control (categories C–F) can be addressed. Footprints may not be equally informative about all locomotor categories. Trackways have helped confirm that birds retained the obligatory, digitigrade bipedalism, and highly adducted limb posture of their theropod ancestors. But, would similar tracks require complete kinematic similarity? It is quite possible that such bipeds could make almost identical footprints even if they differed in several locomotor categories. Trackways may not provide enough details to discriminate between hip-based and knee-based limb-retraction mechanisms, but additional research may prove fruitful.

Assessment of limb function is too subjective

The interpretation of a "primary" limb mechanism and a muscle's relative importance is subjective. However, such hypotheses should be based on as much data as possible. Prior to the collection of functional data from living forms, reconstructions were made solely from comparative anatomy and a general impression of limb use in living archosaurs. Documenting limb movement and muscle activity provides new data to this old topic. Interpretations of these data can change as new information is acquired, or other workers can make their own judgements.

Obviously, by stressing the reduction of the CFL during theropod evolution, other parts of the limb are not being treated adequately. The CFL, however, represents modification of more than just a single muscle. As the primary limb retraction mechanism, it holds a central place in saurian hindlimb function. In addition, the CFL is linked to a reduction in tail size. The dramatic diminution of a major appendage of the body has significance well beyond that of a single muscle. Finally, the CFL is one of the few muscles in the archosaur hindlimb with an unambiguous skeletal correlate, the fourth trochanter. Other muscles clearly change (Gatesy in press), but the attachments and homology of these muscles are more controversial. Restricting discussion to those aspects of the locomotor system for which good evidence exists is preferable to making uninformed hypotheses about the entire limb that cannot be substantiated. The resolution of any functional analysis is distressingly low for any extinct animal, but this limitation must be accepted (see Lauder, this volume).

SUMMARY AND CONCLUSIONS

Every organism is a mosaic of primitive and derived features. The task facing a functional paleomorphologist is the interpretation of this mosaic. Basal theropods retained many plesiomorphic features, but they also acquired novelties that set them apart from other archosaurs. Similarly, birds retained many characteristics of their ancestors, but also evolved enough to be recognized as a distinct clade of theropods.

Judging the relative functional significance of plesiomorphic and apomorphic traits in the theropod locomotor system is by no means simple. However, there is no a priori reason to restrict the source of information to either birds or crocodilians. Padian and Olsen (1989, p. 239) assert that "as the phylogenetic level becomes more specific, the functional inference becomes more powerful," but the condition in birds may differ dramatically from that in other theropods, despite their close relationship. Without information from outgroups, there is no justification for attributing avian characteristics to all theropods (Bryant & Russell 1992; see Weishampel or Witmer, this volume). The phylogenetic position of crocodilians makes them critical to understanding the primitive organization of archosaurs. Crocodilians allow us to evaluate the function of characters that have been so dramatically altered during theropod history that they are no longer recognizable in modern birds.

Based on my functional interpretation of character distributions relating to the tail and caudofemoral musculature, I propose a gradual transition between two different mechanisms of hindlimb movement during theropod history. Basal theropods appear to have inherited the primitive saurian limb-retraction mechanism using caudofemoral musculature. As the tail was reduced in coelurosaurs and maniraptorans, so was the CFL. This affected both limb retraction and overall balance of the body, resulting in the adoption of a novel, knee-based limb mechanism. Further reduction of the tail to enhance rectricial control precluded the use of the primitive limb system, resulting in the condition found in birds today (Fig. 13.3).

Among living tetrapods, birds remain as the best single model for locomotion in extinct theropods. However, birds must be viewed as a distinct clade of short-tailed theropods specialized for flight. Functional and morphological modifications of the long-tailed, fully terrestrial body plan of basal theropods must be recognized when reconstructing theropod locomotor evolution.

ACKNOWLEDGMENTS

I thank A.W. Crompton and F.A. Jenkins, Jr. for their support during this project. Discussions about locomotion and evolution with K.P. Dial, K. Padian, K. Schwenk, P. Sereno, L. Chiappe, D. Brinkman, and J. Gatesy were extremely helpful, as were the comments of K.P. Dial, J.J. Thomason, and an anonymous reviewer on earlier versions of this chapter. T. Joanen and L. McNease of the Rockefeller Wildlife Refuge provided alligators for my work, portions of which were supported by NIH training grant 5-T32-GM07117-05, Sigma Xi, the Frank M. Chapman Fund of the American Museum of Natural History, and a NSF Postdoctoral Research Fellowship IBN-9203275.

REFERENCES

Alexander, R.M. 1983. Allometry of the leg bones of moas (*Dinornithes*) and other birds. *Journal of Zoology, London* 200, 215–231.

Alexander, R.M. 1985. Mechanics of posture and gait of some large dinosaurs. *Zoological Journal of the Linnean Society* 83, 1–25.

Baumel, J.J. 1988. Functional morphology of the tail apparatus of the pigeon (*Columba livia*). *Advances in Anatomy, Embryology and Cell Biology* 110, 1–115.

Bekoff, A., Nusbaum, M.P., Sabichi, A.L., & Clifford, M. 1987. Neural control of limb coordination. I. Comparison of hatching and walking motor output patterns in normal and deafferented chicks. *Journal of Neuroscience* 7, 2320–2330.

Benton, M.J. 1985. Classification and phylogeny of the diapsid reptiles. *Zoological Journal of the Linnean Society* 84, 97–164.

Benton, M.J., & Clark, J.M. 1988. Archosaur phylogeny and the relationships of the Crocodylia. In *The Phylogeny and Classification of the Tetrapods*, vol. 1, ed. M.J. Benton, pp. 295–338. Oxford: Clarendon Press.

Brinkman, D. 1981. The hind limb step cycle of *Iguana* and primitive reptiles. *Journal of Zoology, London* 181, 91–103.

Bryant, H.N., & Russell, A.P. 1992. The role of phylogenetic analysis in the inference of unpreserved attributes of extinct taxa. *Philosophical Transactions of the Royal Society of London B* 337, 405–418.

Bryant, H.N., & Seymour, K.L. 1990. Observations and comments on the reliability of muscle reconstruction in fossil vertebrates. *Journal of Morphology* 206, 109–117.

Butler, P.J. 1991. Exercise in birds. *Journal of Experimental Biology* 160, 233–262.

Campbell, K.E., & Marcus, L. 1991. The relationship of hindlimb bone dimensions to body weight in birds. *Natural History Museum of Los Angeles County Science Series* 36, 395–412.

Carr, A. 1963. *The Reptiles*. New York: Time-Life.

Carroll, R.L. 1988. *Vertebrate Paleontology and Evolution*. New York: W.H. Freeman and Company.

Chiappe, L.M. In press. The phylogenetic position of the Cretaceous birds of Argentina. In *Third Symposium of the Society of Avian Paleontology and Evolution*, ed. D.S. Peters. *Senckenburghiana*.

Cracraft, J. 1971. The functional morphology of the hind limb of the domestic pigeon, *Columba livia*. *Bulletin of the American Museum of Natural History* 144, 173–268.

Cracraft, J. 1986. The origin and early diversification of birds. *Paleobiology* 12, 383–399.

Duncker, H.-R. 1989. Structural and functional integration across the reptile-bird transition: locomotor and respiratory systems. In *Complex Organismal Functions: Integration and Evolution in Vertebrates*, eds. D.B., Wake & G. Roth, pp. 147–169. Chichester: John Wiley & Sons.

Gatesy, S.M. 1989. Archosaur neuromuscular and locomotor evolution. Unpublished Ph.D. thesis, Harvard University.

Gatesy, S.M. 1990. Caudofemoral musculature and the evolution of theropod locomotion. *Paleobiology* 16, 170–186.

Gatesy, S.M. 1991a. Hind limb movements of the American alligator (*Alligator mississippiensis*) and postural grades. *Journal of Zoology, London* 224, 577–588.

Gatesy, S.M. 1991b. Hind limb scaling in birds and other theropods: implications for terrestrial locomotion. *Journal of Morphology* 209, 83–96.

Gatesy, S.M. In press. Neuromuscular diversity in archosaur deep dorsal thigh muscles. *Brain, Behavior and Evolution*.

Gatesy, S.M., & Biewener, A.A. 1991. Bipedal locomotion: effects of size, speed and limb posture in birds and humans. *Journal of Zoology, London* 224, 127–147.

Gatesy, S.M., & Dial, K.P. In press. Tail muscle activity patterns in walking and flying pigeons (*Columba livia*). *Journal of Experimental Biology*.

Gauthier, J. 1984. A cladistic analysis of the higher systematic categories of the Diapsida. Unpublished Ph.D. thesis, University of California at Berkeley.

Gauthier, J. 1986. Saurischian monophyly and the origin of birds. *Memoirs of the California Academy of Sciences* 8, 1–55.

Gauthier, J., & Padian, K. 1985. Phylogenetic, functional, and aerodynamic analyses of the origin of birds and their flight. In *The Beginnings of Birds: Proceedings of the International Archaeopteryx Conference, Eichstätt 1984*, eds. M.K. Hecht, J.H. Ostrom, G. Viohl, & P. Wellnhofer, pp. 185–197. Eichstätt: Freunde des Jura-Museums Eichstätt.

George, J.C., & Berger, A.J. 1966. *Avian Myology*. New York: Academic Press.

Hecht, M.K., Ostrom, J.H., Viohl, G., & Wellnhofer, P. 1985. *The Beginnings of Birds: Proceedings of the International Archaeopteryx Conference, Eichstätt 1984*. Eichstätt: Freunde des Jura-Museums Eichstätt.

Heilmann, G. 1926. *The Origin of Birds*. London: Witherby.

Jacobson, R.D., & Hollyday, M. 1982. A behavioral and electromyographic study of walking in the chick. *Journal of Neurophysiology* 48, 238–256.

Jenkins, F.A. Jr. 1971. Limb posture and locomotion in the Virginia opossum (*Didelphis marsupialis*) and in other non-cursorial mammals. *Journal of Zoology, London* 165, 303–315.

Lauder, G.V. 1986. Homology, analogy, and the evolution of behavior. In *Evolution of Animal Behavior*, ed. M.H. Nitecki & J.A. Kitchell, pp. 9–40. New York: Oxford University Press.

Lauder, G.V. 1991. Biomechanics and evolution: integrating physical and historical biology in the study of complex systems. In *Biomechanics in Evolution*, ed. J.M.V. Rayner & R.J. Wootton, *Society for Experimental Biology Seminar Series 36*, pp. 1–19. Cambridge University Press.

Manion, B.L. 1984. The effects of size and growth on the hindlimb locomotion in the chicken. Unpublished Ph.D. thesis, University of Illinois at Chicago.

McFarland, W.N. Pough, F.H. Cade, T.J., & Heiser, J.B. 1985. *Vertebrate Life*, 2nd ed. New York: Macmillan.

McGowan, C. 1979. The hind limb musculature of the brown kiwi, *Apteryx australis mantelli. Journal of Morphology* 160, 33–74.

McGowan, C. 1982. The wing musculature of the brown kiwi *Apteryx australis mantelli* and its bearings on ratite affinities. *Journal of Zoology, London* 197, 173–219.

McGowan, C. 1986. The wing musculature of the weka (*Gallirallus australis*), a flightless rail endemic to New Zealand. *Journal of Zoology, London* 210, 305–346.

Newman, B.H. 1970. Stance and gait in the flesh-eating dinosaur *Tyrannosaurus. Biological Journal of the Linnean Society* 2, 119–123.

Osborn, H.F. 1916. Skeletal adaptations of *Ornitholestes, Struthiomimus, Tyrannosaurus. Bulletin of the American Museum of Natural History* 35, 733–771.

Ostrom, J.H. 1969. Osteology of *Deinonychus antirrhopus*, an unusual theropod from the Lower Cretaceous of Montana. *Bulletin of the Yale Peabody Museum of Natural History* 30, 1–165.

Ostrom, J.H. 1975. The origin of birds. *Annual Review of Earth and Planetary Sciences* 3, 55–77.

Ostrom, J.H. 1976. *Archaeopteryx* and the origin of birds. *Biological Journal of the Linnean Society* 8, 91–182.

Padian, K. 1986. On the type material of *Coelophysis* Cope (Saurischia: Theropoda), and a new specimen from the Petrified Forest of Arizona (late Triassic: Chinle Formation). In *The Beginning of the Age of Dinosaurs*, ed. K. Padian, pp. 45–60. Cambridge University Press.

Padian, K., & Olsen, P.E. 1989. Ratite footprints and the stance and gait of Mesozoic theropods. In *Dinosaur Tracks and Traces*, ed. D.D. Gillette & M.G. Lockley, pp. 231–241. Cambridge University Press.

Paul, G.S. 1988. *Predatory Dinosaurs of the World.* New York: Simon & Schuster.

Pennycuick, C.J. 1986. Mechanical constraints on the evolution of flight. In *The Origin of Birds and the Evolution of Flight*, ed. K. Padian, pp. 83–98. San Francisco: California Academy of Science.

Raikow, R.J. 1985. Locomotor system. In *Form and Function in Birds*, vol. 3, ed. A.S. King & J. McLelland, pp. 57–147. London: Academic Press.

Romer, A.S. 1956. *Osteology of the Reptiles.* Chicago: University of Chicago Press.

Russell, A.P., & Bauer, A.M. 1992. The m. caudifemoralis longus and its relationship to caudal autotomy and locomotion in lizards (Reptilia: Sauria). *Journal of Zoology, London* 227, 127–143.

Russell, D.A. 1972. Ostrich dinosaurs from the late Cretaceous of western Canada. *Canadian Journal of Earth Sciences* 9, 375–402.

Sanz, J.L., & Bonaparte, J.F. 1992. A new order of birds (Class Aves) from the Lower Cretaceous of Spain. *Natural History Museum of Los Angeles County, Science Series* 36, 39–49.

Sanz, J.L., Bonaparte, J.F., & Lacasa, A. 1988. Unusual Early Cretaceous birds from Spain. *Nature* 331, 433–435.

Schaefer, S.A., & Lauder, G.V. 1986. Historical transformation of functional design: evolutionary morphology of feeding mechanisms in loricarioid catfishes. *Systematic Zoology* 35, 489–508.

Sereno, P.C. 1991. Basal archosaurs: phylogenetic relationships and functional implications. *Journal of Vertebrate Paleontology* 11 (supplement), 1–53.

Sereno, P.C., & Rao, C. 1992. Early evolution of avian flight and perching: new evidence from the Lower Cretaceous of China. *Science* 255, 845–848.

Snyder, R.C. 1954. The anatomy and function of the pelvic girdle and hindlimb in lizard locomotion. *American Journal of Anatomy* 95, 1–36.

Snyder, R.C. 1962. Adaptations for bipedal locomotion of lizards. *American Zoologist* 2, 191–203.

Stebbins, R.C. 1966. *A Field Guide to Western Reptiles and Amphibians.* Boston: Houghton Mifflin.

Storer, R.W. 1971. Adaptive radiation in birds. In *Avian Biology*, vol. 1, ed. D.S. Farner & J.R. King, pp. 149–188. New York: Academic Press.

Tarsitano, S. 1981. Pelvic and hindlimb musculature in archosaurian reptiles. Unpublished Ph.D. thesis, City University of New York.

Tarsitano, S. 1983. Stance and gait in theropod dinosaurs. *Acta Paleontologica Polonica* 28, 251–264.

Tarsitano, S., & Hecht, M. 1980. A reconsideration of the reptilian relationships of *Archaeopteryx. Zoological Journal of the Linnean Society of London* 69, 149–182.

Walker, A.D. 1977. Evolution of the pelvis in birds and dinosaurs. In *Problems in Vertebrate Evolution*, ed. S.M. Andrews, R.S. Miles, & A.D. Walker, pp. 319–357. London: Academic Press.

Weinstein, G.N. Anderson, C., & Steeves, J.D. 1983. Functional characterization of limb muscles involved in locomotion in the Canada goose, *Branta canadensis. Canadian Journal of Zoology* 62, 1596–1604.

Witmer, L.M. 1991. Perspectives on avian origins. In *Origins of the Higher Groups of Tetrapods*, ed. H.P., Schultze & L. Trueb, pp. 427–466, Ithaca: Comstock.

14

Functional interpretation of spinal anatomy in living and fossil amniotes

EMILY B. GIFFIN

ABSTRACT

The conservatism, complexity, and direct role in motor control of the nervous system make it an ideal subject for functional study. Support of the central nervous system by the skeleton allows reconstruction of the gross neural anatomy of extinct taxa, and cerebral endocasts have long been used to infer intelligence, behavior, and metabolic regime in fossil amniotes. Gross spinal anatomy of each amniote class is distinctive, and the pattern within each class varies with the extent and pattern of postcranial innervation of somatic musculature. Documentation of the relationship of gross spinal anatomy and function in living taxa allows prediction of postural style and locomotor mode in related fossil taxa. Application of this methodology allows prediction of "improved" posture in most dinosaur taxa, and of hindlimb-dominated swimming in the desmatophocid carnivore *Allodesmus*.

THE NERVOUS SYSTEM AND VERTEBRATE PALEONTOLOGY

The choice of the nervous system as a subject of study for a vertebrate paleontologist may seem counter-intuitive. Nervous tissue is never preserved in the fossil record, and the complexities of its organization are cellular. Nevertheless, the potential of the nervous system to reveal aspects of the history of vertebrates has been exploited for more than a century by a series of paleontologists. Among the traits that make the nervous system so amenable to study are the following:

1. The central nervous system is evolutionarily conservative, largely a by-product of its complexity. Silurian cephalaspidomorphs display cranial neuroanatomy remarkably similar to that of living vertebrates (Stensiö 1963), and the postcranial pattern of segmental innervation almost certainly has still more ancient origins. As noted by Thomson (1988), structural complexity exerts a conservative influence over change, constraining possible pathways. Conservation of the basic outline of ancestral patterns aids in the recognition of homologous characters.

2. The central nervous system is the commnication and coordination system of the body, and is a key to an animal's ability to perceive and respond to its environment. The particular sensory modalities and motor responses utilized by a given animal are reflected in the adaptive variations on otherwise highly constrained neural patterns. As a result, there is potentially a very high information content in these structures.

3. The cellular organization of the central nervous system is reflected in its gross anatomy. This gross anatomy is in turn reflected by the surrounding skeleton, which intimately supports and protects the delicate neural tissues. Foramina in the skeleton document the course of peripheral nerves, and the central nervous system is encased by the braincase and vertebrae.

In 1975, Leonard Radinsky suggested that the neuroanatomy of viverrid carnivores could be used to address three different topics. These can be summarized as (1) phylogenetic relationships of living taxa and their extinct relatives, (2) evolution of the nervous system, and (3) functional characteristics of extinct taxa. Each of these topics has been explored by a number of neuroanatomists during the past century.

Neuroanatomy is a rich source of traits for phylogenetic analysis because of its complexity and conservatism. The basic architecture of the vertebrate central nervous system is recognizable in some of the oldest vertebrates known. Variations in this ground plan have been used as characters in phylogenetic reconstructions of many vertebrate groups, including agnathans (Stensiö 1963), reptiles (Ulinski

1983, 1986), and mammals (Switzer, Johnson, & Kirsch 1980; Kirsch & Johnson 1983).

Evolution of the nervous system has been accessible to paleontologists since the availability of internal molds of the cranial cavity (usually called endocranial casts) allowed inference of cerebral structure of extinct taxa. Marsh (1874) used observations of endocranial casts to establish several "laws" of brain evolution, some of which have been questioned by later workers (Edinger 1962). The negative allometry of the brain relative to body size noticed by Marsh has been treated exhaustively by Jerison (1971, 1973). Jerison (1973) also established the *encephalization quotient* (the ratio of actual to expected brain size for an animal of a given body size), documented variations in brain size by vertebrate class, and summarized trends in vertebrate brain evolution.

Functional morphology of paleontological specimens relies on the relationship between observed structure and function in living taxa. Convergent structure (homoplasy) between living and fossil species is interpreted as the product of adaptive natural selection and as an indicator of functional similarity. It is accordingly important to recognize other (nonadaptive, nonfunctional) causes of homoplasy and to attempt to eliminate them from our analyses. Potential nonadaptive causes of homoplasy have been discussed by a variety of authors (see especially Gould & Lewontin 1979; Wake 1991), and include constraints of tissue mechanics, pleiotropic or allometric effects, nonheritable phenotypic variation, and restricted genetic flexibility.

Despite the existence of nonadaptive causes of homoplasy, functional analysis of fossil taxa is not only desirable but possible, at least within general limits. This is because structure very often *does* reflect function. Accuracy of prediction of function in extinct taxa may be improved, although not ensured, by a variety of methodological precautions. Such precautions have been practiced by neuroanatomists more frequently over the past 20 years than earlier in the century. They include:

1. Choice of characters from conservative body systems, in which homology and homoplasy are more likely to be differentiated.
2. Limitation of analysis to taxonomically restricted groups that share a similar "essential structure" or *bauplan* of the system under consideration.
3. Documentation of soft- and hard-tissue relationships in living taxa.
4. Documentation of the range of variation of both structure and function in living members of the chosen taxon.

The use of gross neuroanatomy to infer function relies primarily on variations of size and structure of the entire brain or of brain subunits evident from fossil endocasts. Throughout her distinguished career, Edinger (1926, 1948, 1966) described the gross anatomy of a wide range of endocasts, often making functional predictions. In other studies, she chose a particular brain component (for example, the pituitary gland), and examined its expression across taxa (Edinger 1942). Roth, Roth, and Hotton (1986) adopted a similar approach to the study of the parietal foramen, parietal eye, and pineal body. They correlated loss of the parietal foramen in the history of therapsids with a decreased need for a "photothermal monitor" as seasonality of the environment decreased. Hopson (1977) used brain size relative to body size as a predictor of behavioral complexity and thermal regime in dinosaur subgroups. His comprehensive documentation of reptilian endocranial casts established the range of cerebral anatomy throughout living and extinct members of the class. Radinsky (1968b) used the patterns of cerebral sensory projections mapped by neurophysiologists in living carnivores as a basis for predicting relative size of sensory fields in fossil taxa. He was able to predict the division of extinct otters into facial-dominant and hand-dominant functional groups by comparing the sulcal patterns of otter endocasts with those of the cerebra of living species of known functional group (Figure 14.1). This and other works of Radinsky (1968a, 1969, 1971, 1973) are particularly elegant and powerful because he restricted analysis to a single family or subfamily of mammals, documented variation in living genera, and used the cellular basis of gross anatomy to support functional interpretations.

CELLULAR AND GROSS SPINAL ANATOMY OF AMNIOTES

Gross structure of the spinal cord as inferred from the neural canal has rarely been assumed to be a rich or reliable source of information about the function and life-style of fossil amniotes. Additionally, only a limited amount of information is available about the gross spinal cord anatomy of living taxa. However, the traits that make the nervous system accessible to functional interpretation are as true for the spinal cord as for the brain. The general structural pattern of the spinal cord and its range of variation in living amniotes will be presented in some detail to facilitate functional interpretations suggested below.

The basic outline of spinal anatomy has been documented by many workers over the past century, and most notably and comprehensively summarized by Ariens Kappers, Huber, and Crosby more than 50 years ago (1936). In its most basic

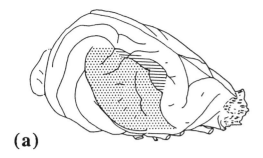

(a)

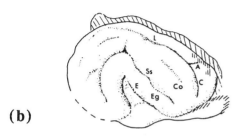

(b)

Figure 14.1. (a) Inferred primary somatic sensory cortical projections areas of head (*stippled*) and forelimb (*horizontal lines*) in *Lutra canadensis.* (b) Cranial endocast of *Potamotherium* with sulci and gyri identified. Both figures ×0.75. Abbreviations: A, ansate sulcus; C, coronal sulcus; Co, coronal gyrus; E, anterior ectosylvian sulcus; Eg, anterior ectosylvian gyrus; L, lateral sulcus; Ss, suprasylvian sulcus. (Figures reprinted by permission of Wiley-Liss, a division of John Wiley & Sons, Inc., from L. Radinsky in *Journal of Comparative Neurology*, Copyright © 1968.)

organization, the spinal cord of vertebrates is segmental, with a one-to-one relationship between somatic segments and the spinal nerves, which carry both sensory and motor fibers. Details of cellular anatomy recorded for reptiles (Cruce 1979), birds (Leonard & Cohen 1975), and mammals (Rexed 1954) display a remarkable consistency across vertebrate classes. Cell bodies of sensory neurons are located external to the cord in dorsal ganglia. The centrally located spinal gray matter contains cell bodies of interneurons and motor neurons. It can be subdivided into ten laminae (Figure 14.2) on the basis of cell shape, staining pattern, and connections. Interneurons are concentrated in the intermediate gray, and motoneurons in the ventral horn. The gray is surrounded by fibers of the dorsal, lateral, and ventral funiculi. These fibers act as connections among cells of different sides, of different segmental levels, and with long ascending and descending fibers.

This pattern is supplemented in spinal segments that supply the limbs by an increased number of incoming and outgoing fibers (white matter) and also by increases in number of cell bodies (gray matter),

particularly those of motor neurons (lamina IX) and interneurons (lamina IV). Kusuma, ten Donkelaar, and Nieuwenhuys (1979) documented the increase in spinal cord cross section caused by increases in white and gray matter at limb levels (Figure 14.3a,b). The dominant contribution to this increase is made by fibers rather than cell bodies, and is particularly marked at anterior regions of the cord. Variation in the size of the increase between taxa clearly reflects both differences in trunk and limb territory and innervation pattern (motor unit size). For example, Kusuma et al. (1979) noted the extremely small thoracic cord of turtles, relating it to the lack of trunk musculature (Figure 14.3b). They also documented the lack of limb enlargements in limbless reptilian genera. Ariens Kappers et al. (1936) noted the large increase in cord cross section in segments supplying the hand of *Homo sapiens*, relating this to fine motor control.

Gross anatomy of the spinal cord is also affected by numbers of long ascending and descending fibers which connect cells of the spinal cord to those of the brainstem. The gradual anterior (cranial) increase in the number of these fibers seen in vertebrates with extensive connections between the brain and spinal cord augments cord diameter and is termed *frontal accumulation.*

Ariens Kappers et al. (1936), Nieuwenhuys (1964), and Petras (1976), among others, have documented variations of basic spinal organization unique to the amniote classes. Some of these variations are accessible to gross observation, and can be recognized in measurements of spinal cord cross section by segment. Particularly evident are differences in the relationship between vertebral and spinal segments seen in reptiles and birds, and in mammals.

The reptilian spinal cord (Figure 14.4) extends the entire length of the vertebral column, despite considerable variation in number of vertebrae (exceeding 300 in some snake genera). As a result, one spinal segment lies within the neural canal of each vertebra, a condition here termed *isosegmental.* Increases in cross section of the cord at brachial and lumbosacral levels (Figure 14.5a; *Tupinambis*) reflect the presence of additional motoneurons, interneurons, and sensory fibers innervating the limbs. The cervical cord is of approximately the same cross section as the interlimb cord. The small size of the cervical cord reflects the relative paucity of long ascending and descending fibers and restricted cranial control of locomotion in reptiles.

The isosegmental relationship of cord segment and vertebra, and restricted cranial control of locomotion, are also typical of birds (Figure 14.5a; *Columba*). The very short avian interlimb area is recognizable in the close placement of limb enlargements, and the tail is very short. Limb enlargements are larger

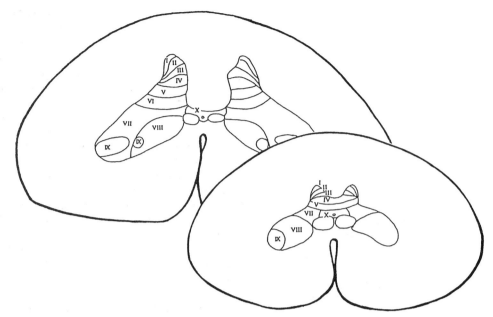

Figure 14.2. Thoracic (*front*) and lumbar (*rear*) level cross sections of spinal cord of tegu lizard, *Tupinambis*, showing laminae (*Roman numerals*) of spinal gray matter. (Redrawn from Cruce 1979.)

relative to the interlimb cord than in reptiles, reflecting the innervation of the large wings and bipedal legs, as well as the relatively immobile torso.

Mammals, like birds, are known for the intimate relationship between neural and skeletal material. However, another peculiarity of mammalian spinal development significantly alters cord–column relationships. Mammals lack the isosegmental relationship of cord and canal seen in other amniotes (Figure 14.4), a product of at least two different phenomena. First, the mammalian cord stops growth earlier in development than the neural canal, and is somewhat shorter than the canal in most genera, terminating at slightly but progressively more anterior vertebral positions as ontogeny progresses. Second, and more dramatically, somatic segments with small muscle territory and/or sparse innervation pattern are small not only in cross section but also in length, further shortening the cord. Spinal nerves travel posterior to the filum terminale within the neural canal as a *cauda equina*, passing through the segmentally appropriate intervertebral space.

The lack of alignment of cord and canal can be recognized in area measurements of the canal (not the cord), because enlargements occur in vertebrae anterior to the position predicted by their spinal segmentation. Most noticeably, the enlargement for the brachial plexus occurs within the posterior cervical vertebral canals in almost all mammalian taxa, and the enlargement for the lumbosacral plexus occurs within lumbar or even thoracic vertebrae when segments with sparse innervation exist.

The mammalian cord (Figure 14.5a; *Didelphis*) is also clearly differentiated from those of reptiles and birds by a large cervical cord cross section, which reflects the presence of extensive long ascending and descending tracts between the brainstem and body. Additionally, the tail of mammals is innervated by homologs of only the first two caudal segments of reptiles, and loses its segmental organization. The extremely rapid postlumbosacral decrease in cord size reflects this innervation pattern. Both of these spinal characteristics are clearly reflected by neural canal measurements. They allow osteological methods of discriminating between reptilian and mammalian neuroanatomy, as demonstrated by canal measurements of a reptile (*Alligator*) and a mammal (*Gorilla*) of similar body size with the same number of presacral vertebrae (Figure 14.5b).

HOW GOOD AN INDICATOR OF GROSS SPINAL CORD ANATOMY IS THE NEURAL CANAL?

Size relationships of the brain and the surrounding braincase have been estimated for a variety of living and fossil taxa. Avian and mammalian brains fill the braincase nearly completely, as demonstrated by details of external brain anatomy recorded by the braincase. Contact of braincase and brain is much less intimate in reptiles. Reptilian brains are typically estimated to fill approximately half of the braincase (Jerison 1973; Hopson 1979).

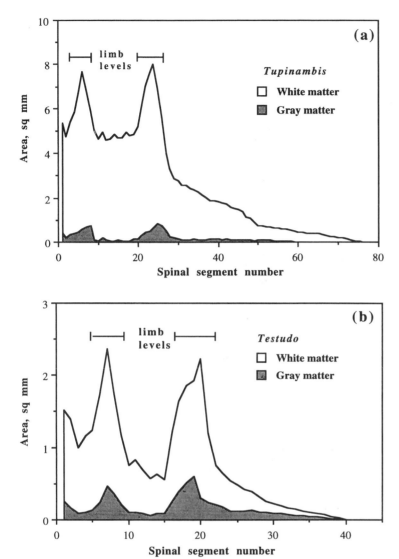

Figure 14.3. Cross-sectional areas of gray and white matter by segment of (a) *Tupinambis* and (b) *Testudo.* (Data from Kusuma et al. 1979.)

The size relationship of spinal cord and neural canal is reasonably hard to measure, as removal of the cord for measurement destroys the canal. Dissection of fresh tissue is not particularly useful, as the cord is so soft that its shape is compromised. These problems can be overcome by sectioning specimens that are fixed, either by freezing or in formalin, depending on their size. These methods and some of the resulting data describing cord-canal relationships are detailed below. Cord size and canal size have been highly correlated in all taxa examined, and canal size has therefore been used to predict cord size in a wide range of vertebrates for which soft tissue was not available or not examined.

The vertebral column and surrounding musculature of a 6-foot (juvenile) specimen of *Alligator mississippiensis* were frozen and then sectioned by segment with a band saw. Scaled photographs of the frozen sections were measured using the image analysis software IMAGE on a Macintosh computer, generating the data in Figure 14.6. The cord filled an average of 43 percent of the canal, with a correlation between cord and canal areas of 0.86. Despite relatively low "fill percentage," the high correlation between cord and canal dimensions allows the inference of variations in cord size from canal measurements alone. Several lizard specimens *(Crotaphytus, Draco, Iguana)* were preserved in formalin, decalcified (Gooding and Stewart's fluid), dissected, sectioned, videorecorded, and measured using the same methods as for larger specimens. Average percentage of canal filled by cord in these smaller taxa ranged from 66 percent to 71 percent, and correlation between cord and canal areas always exceeded 0.95.

Estimates of cord size from canal measurements in birds is complicated by the presence of the

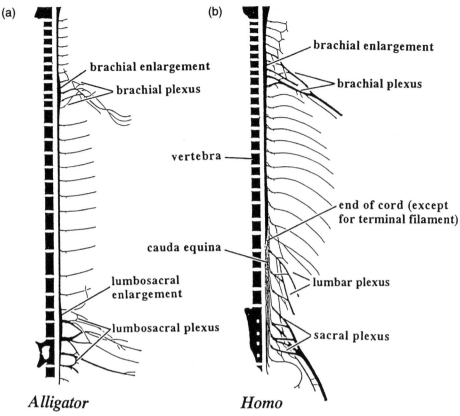

Figure 14.4. Relationship of vertebrae to spinal segments in (a) a reptile and (b) a mammal. Reptilian vertebrae and spinal segments have a one-to-one, isosegmental relationship. Mammalian spinal cord typically terminates at lumbar vertebral levels, and spinal nerves travel caudally as a *cauda equina* within neural canal to segmentally determined points of passage.

glycogen body. This uniquely avian structure is a carbohydrate storehouse of uncertain function (Imhof 1905; De Gennaro 1982), and is housed between the dorsal horns of the gray matter in the lumbosacral portion of the cord. As a result, canal cross section is greatly increased in this area. Relative size of the glycogen body varies between taxa, and seems to be correlated more closely with body size ($r = 0.79$) than with limb use ($r = 0.28$) (Giffin 1991). Sections of a fixed, decalcified adult pigeon (*Columba*) yielded the data in Figure 14.6, in which the combined cord and glycogen body filled an average of 63 percent of the canal, with a correlation of 0.95. Again, the high correlation suggests that cord size can be predicted reliably from canal size except in lumbosacral segments.

DOCUMENTATION OF NEURAL CANAL–LOCOMOTOR RELATIONSHIPS IN LIVING AMNIOTES

Variations within each of the phylogenetically constrained patterns of neural canal anatomy presented above reflect the territory and innervation pattern of the segmental musculature in different taxa. Examples of the range of variation within each pattern are presented separately, and will serve as standards against which to compare the data from fossil taxa.

Reptiles

Variation in locomotor style is restricted in living reptiles, but comparison of size of limb enlargements can distinguish between taxa with strictly terrestrial, semiaquatic, and arboreal lifestyles. Most living reptiles are quadrupedal sprawlers, with limbs that extend laterally from the body. The girdles are relatively immobile, and the brachial girdle is braced by the clavicle. Both lateral undulation and limb excursion contribute to step length, meaning that both torso and limbs are intimately involved in locomotion (Snyder 1962; Jenkins & Goslow 1983; Peterson 1984). The neural canal trace of *Varanus* (Figure 14.7) is typical of that of terrestrial reptiles and looks very similar to traces constructed from cord data. The nearly equal peaks representing forelimb and hindlimb enlargements of the canal exceed the size

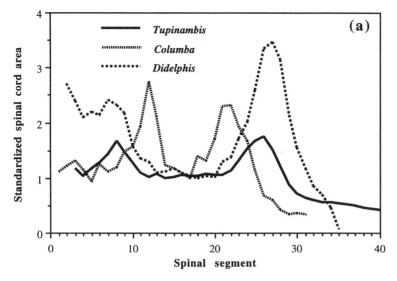

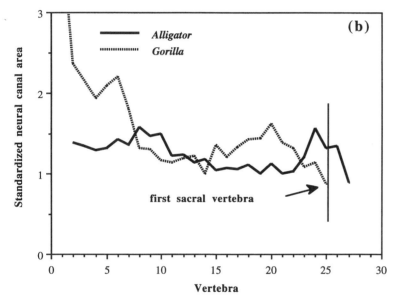

Figure 14.5. Comparisons of spinal cord and neural canal area by segment in different amniote classes. Values for each specimen have been standardized by division by lowest value between limb levels. X axis represents segments, but not necessarily metameric number. (a) Spinal cord area in reptile *Tupinambis* (data from Kusuma et al. 1979), bird *Columba* (data are original), and mammal *Didelphis* (data from Voris 1928). Difference in distance between forelimb and hindlimb peaks reflects difference in number of interlimb segments. (b) Neural canal area in a reptile (*Alligator*) and mammal (*Gorilla*) of the same presacral segment number. Note large size of cervical canal in *Gorilla*, as well as anterior position of its limb enlargements relative to those of *Alligator*. The atlas measurement of *Alligator* specimen is missing.

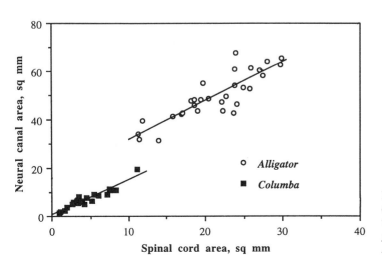

Figure 14.6. Relationship of spinal cord area and neural canal area in two living vertebrates. For *Columba*, $y = 0.61 + 1.46x$; $r = 0.95$; for *Alligator*, $y = 15.33 + 1.62x$; $r = 0.86$.

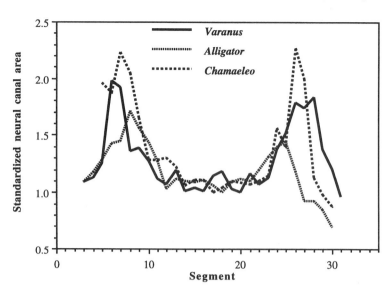

Figure 14.7. Neural canal area by segment for three reptile genera of different locomotor mode. *Varanus* is terrestrial; *Alligator* is semiaquatic; and *Chamaeleo* is arboreal. Values of each specimen have been standardized by division with the least interlimb neural canal area. X axis represents segments, but not necessarily metameric number; difference in distance between forelimb and hindlimb peaks reflects difference in number of interlimb segments.

of the interlimb canals by a factor of about 2. This value may be used as a standard against which to compare values for reptiles that have different locomotor modes.

No living reptile is truly aquatic, but a variety of genera display semiaquatic life-styles. Seymour (1982) described the usual form of swimming in marine iguanas and crocodilians as sculling, in which the limbs are pressed close to the body and the posterior region of the trunk and the tail undulate laterally. This increased dependence on movement of the torso, and decreased emphasis on limbs, leads to the expectation of a smaller differential in extent of innervation between limb and interlimb areas of the cord when compared to terrestrial genera. This prediction is met (Figure 14.7; *Alligator*), and is statistically significant for a grouping of semiaquatic reptiles (Giffin 1990).

Variation in the opposite direction is evident in reptiles with arboreal locomotion, whose movement is characterized by a decrease in lateral undulation and an increase in brachial excursion (Peterson 1984). The change in locomotion is accompanied by a change in posture, which is more nearly "upright," with the legs supporting the body from beneath. These differences in limb use are reflected by a relative increase in cord size at limb levels compared to interlimb levels (Giffin 1990), as demonstrated by *Chamaeleo* (Figure 14.7).

Birds

Variations in the relative size and use of both forelimbs and hindlimbs of birds can be correlated with neural canal anatomy. Relative muscle territory of the wing increases from flightless birds to weak fliers and again to strong fliers. This progressive increase

is reflected in the brachial enlargement of the neural canal relative to that of interlimb segments (Giffin 1990). Lumbosacral neural canal measurements are complicated mechanically by fusion of the synsacrum and interpretively by their enlargement for the glycogen body. Consequently, a relatively small number have been measured. Nevertheless, avian cursors (e.g., the ostrich *Struthio*) have relatively much larger lumbosacral enlargements than do "walkers" (e.g., the pigeon *Columba*) or "waddlers" (e.g., the duck *Anas*). If glycogen body size does correlate well with body size over a wider range of taxa than presently surveyed, such data may be interpreted as a reflection of size of limb territory.

Mammals

Step length in most mammalian genera is largely a product of limb excursion, not of lateral undulation. Most tailed terrestrial mammals show a neural canal trace with two well-defined limb enlargements, and with spinal segments located within vertebrae of similar segmental position. Variations from typical terrestrial locomotion exhibited by several aquatic carnivore lineages are reflected in distinctive patterns of postcranial innervation (Giffin 1992). Otters are semiaquatic quadrupedal paddlers (Tarasoff et al. 1972), and display a trace that varies only subtly from the generalized mammalian pattern described above (Figure 14.8; *Lutra*). Sea lions (otariids) use the large forelimbs as primary force generators (English 1976) in a bilateral stroke. Hindlimbs typically trail behind the body during swimming, although they may be rotated forward to enable quadrupedal locomotion on land. The sea lion neural trace is radically foreshortened (Figure 14.8; *Callorhinus*), reflecting the very short length of the spinal segments

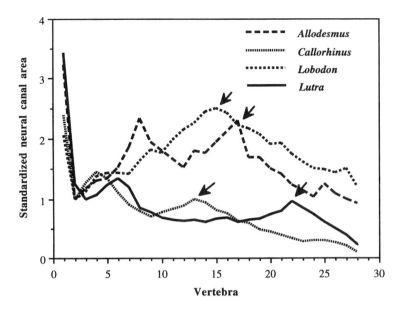

Figure 14.8. Neural canal cross section by segment in a lutrine (*Lutra*), an otariid (*Callorhinus*), a phocid (*Lobodon*), and a desmatophocid (*Allodesmus kernensis*, Los Angeles County Museum 1557/4320). Values of each specimen have been standardized by division by the least cervical neural canal area. Arrows indicate maximal areas of hindlimb enlargement for each genus. Values for vertebrae 10, 11, and 16 are missing in *Allodesmus*.

that supply the reduced thoracic and hindlimb musculature. The lumbosacral enlargement occurs within the neural canals of the midthoracic vertebrae and is much smaller than the brachial enlargement, reflecting the relative territory of the two limbs. Seals (phocids) use alternating hindlimb strokes to generate power, while the small forelimbs play a steering role (King 1983). During swimming the torso swings forcefully from side to side, and the thoracic musculature is massive (Howell 1928). The seal neural trace (Figure 14.8; *Lobodon*) reflects not only the modest neural supply to the small forelimbs, but the massive innervation of torso and hindlimbs as well.

Because a large number of aquatic carnivore specimens have been measured, a variety of statistical analyses are possible. Neural canals of 58 specimens separable into three groups based on swimming mode (quadrupedal paddlers, forelimb swimmers, hindlimb paddlers) were standardized for body size and subjected to analysis of variance (ANOVA) and principal components analysis (PCA). The repeated measures ANOVA test indicated that the means of the three categories are significantly different from each other in neural canal anatomy ($p < 0.001$). The first two factors extracted from the PCA accounted for 85.1 percent of the variance of the data, and are presented in an unrotated orthogonal plot in Figure 14.9a. The distinctive clustering of individuals of each swimming mode is a visual confirmation of the results of the ANOVA test. The first axis separates specimens by extent of torso innervation, and the second axis reflects height of limb maxima. Note that otariids are more uniform in neural anatomy than either phocids or lutrines. Phocids can be separated into overlapping subgroups (Giffin 1992) on the basis of torso and hindlimb innervation. The

innervation of the large hindlimbs used as primary propulsive organs separates the sea otter *Enhydra* from other lutrines.

PUTTING THE METHOD TO WORK: LOCOMOTOR PATTERN PREDICTION IN FOSSIL TAXA

The demonstration of a link between the cellular anatomy of spinal cord, gross structure of the spinal cord and neural canal, and locomotor function in a variety of extant vertebrates allows several kinds of questions to be asked about extinct taxa and about their evolution. As in most paleontological studies, incomplete osteological material and ignorance of actual soft-hard tissue relationships in extinct taxa introduce some level of speculation into the answers proposed. Additionally, care must be taken not to place extinct taxa into functional categories rigidly defined by living taxa. The range of function in extinct taxa may far exceed, and almost certainly is quite different from, that of living species. Brief analyses from several different fossil groups are presented below.

Predictions of limb use and posture

The ratio of limb-to-torso innervation is a reliable indicator of posture and locomotor mode in living reptiles. A comparison of limb and torso innervation can therefore be used to predict relative importance of limb excursion and undulation in extinct quadrupedal taxa. The relative size of limb-level neural canals in dinosaurs demonstrates a range very similar to that of extant taxa, and suggests that the

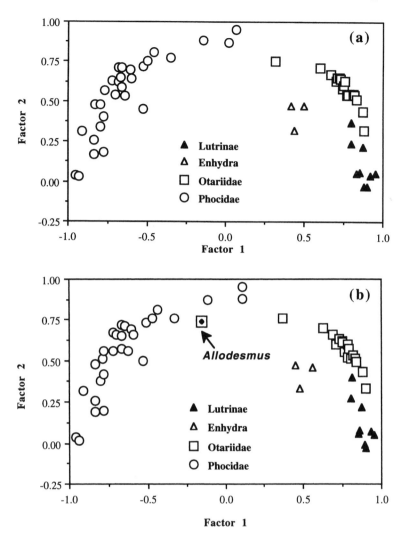

Figure 14.9. Scattergrams of first two principal components of variation in neural canal anatomy. (a) 58 specimens of living marine mammals (lutrines, otariids, and phocids). (b) The same 58 living specimens and desmatophocid *Allodesmus kernensis* (Los Angeles County Museum 1557/4320). Values for vertebrae 10, 11, and 16 of all specimens have been removed from data matrix analyzed.

great majority of dinosaurs walked with an upright posture, with a minimum of torso undulation (Giffin 1990). *Stegosaurus* is an exception to this pattern, with enlargements in the size range of living sprawlers, and *Apatosaurus* falls into an intermediate range.

Forelimbs of bipedal dinosaurs display a wide range of innervation. The neural canals housing spinal segments supplying the forelimbs of *Deinonychus*, *Allosaurus*, and *Coelurus* indicate a neural supply relatively much larger than that supplying locomotory limbs of either living or extinct taxa. The small size of these limbs indicates that a dense pattern of innervation, and not limb territory, is implicated, and supports an interpretation of manipulative ability predicted on the basis of osteology (Ostrom 1969).

Measurement of the sacral neural canals of living reptiles and birds served as comparisons for interpretation of the massive endosacral enlargements

present in stegosaur and sauropod dinosaurs (Giffin 1991). Although sometimes popularly believed to house a "second brain," neural supply to enlarged hindlimbs and extensive innervation of an agile tail have been more reasonably proposed by a series of paleontologists over the last half century (Janensch 1939; Steel 1970; Galton 1990). However, the ratio of endosacral canal to interlimb canal far exceeds that necessary to supply even the most massive and densely innervated hindlimb, and segmental location of the enlargements does not correspond with that predicted for tail musculature. An alternative suggested by the phylogenetic relationship of birds and dinosaurs (Gauthier 1984) is that some, but not all, dinosaurs possessed a glycogen body. This suggestion was first made more than a century ago (Krause 1881), but confirmation will have to await further determination of glycogen body use in living birds and the availability of more dinosaur endosacra.

The segmental pattern of postcranial neural supply

that very robustly predicts swimming mode in living mammalian carnivores can also be applied to fossil taxa. Predictions can be based on either a comparison of neural traces or by principal components analysis. The Miocene desmatophocid carnivore *Allodesmus* is an ideal genus for such analysis, because nearly complete vertebral columns exist. Mitchell (1966) and Barnes (1972) have used osteological evidence to suggest that *Allodesmus* was a forelimb swimmer. The neural trace of *Allodesmus* (Figure 14.8) indicates massive enlargements of spinal segments supplying the hindlimbs, suggesting that they were the dominant swimming limbs. There is also significant interlimb innervation, although not across the entire torso. The principal components analysis of the aquatic taxa analyzed above (Figure 14.9a) was rerun including the *Allodesmus* specimen. Vertebral measurements missing from the fossil specimen (vertebrae 10, 11, 16) were eliminated from the matrix for all specimens. These two alterations of the matrix produce a slightly different PCA scattergram (Figure 14.9b), but different swimming modes are still clearly grouped. *Allodesmus* is clustered among members of the hindlimb-dominated mode of locomotion that possess the most restricted torso innervation. This conclusion contrasts with that proposed by Mitchell (1966) and Barnes (1972).

Characterization of locomotor pattern and phylogenetic pattern transitions

The neural patterns typical of reptiles, birds, and mammals have undergone differentiation within various subgroups with changes in the musculature they innervate. Observation of neural structure in fossil taxa may help us determine not only the speed, but the course of locomotor transitions in the fossil record. The transition from terrestrial to aquatic life-style among phocids between the Miocene and the Recent offers the chance to "catch" extensive changes in neural pattern in progress.

Two phocids (the Miocene phocine *Leptophoca lenis* and the Early Pliocene monachine *Acrophoca longirostris*) were measured and analyzed functionally and statistically (Giffin 1992). Both appear to be nearly fully "phocid" in postcranial neuroanatomy, with small forelimb and large hindlimb enlargements. However, each has less extensive torso innervation than related living genera. Such early establishment of distinctively phocid neuroanatomy suggests that the transition from locomotion typical of terrestrial carnivores to that of seals was very rapid. The lack of extensive torso innervation in the fossil forms implies that reliance on the hindlimbs preceded extensive involvement of the torso in the development of this swimming style. A similarly rapid transition in locomotion appears to be true for early whales, which show neuroanatomy remarkably similar to that of living taxa despite the presence of structurally reduced hindlimbs (Giffin, unpublished results).

Reptiles, birds, and mammals share a common ancestry and presumably a common neural pattern. Because aspects of neural pattern can be recognized in the fossil record, the timing of neural innovations that differentiate major *bauplans* can hypothetically be recognized. Ulinski (1986) correlated changes in patterns of locomotion with gross changes in the forebrain during the transition from reptiles to mammals. Neural canal anatomy may also offer evidence of neural changes during this transition. The consistent enlargement of the cervical neural canals of mammals relative to those of reptiles and birds has been mentioned above. It appears to be correlated with expansion of cerebral control of locomotion and an increase in numbers of long ascending and descending fibers to and from the anterior cord. Jenkins and Parrington (1976) noted the presence of enlarged cervical neural canals in early mammals, and Rowe (1988, p. 250) identified "neural canal diameter in cervical greater than in thoracic vertebrae" as an unequivocal synapomorphy of his Mammaliamorpha. Examination of therapsid and early mammal cervical vertebrae is presently underway and should supplement our understanding of character transitions in synapsids.

Reexamination of function in recent taxa

Examination of neural canal anatomy in fossil vertebrates has had the unexpected result of raising questions about locomotor patterns of living taxa. Extensive sampling of two groups of marine mammals to establish comparative series for fossil taxa revealed that a wider range of neural anatomy, and presumably of locomotor pattern, exists than had been previously reported. Phocine and monachine phocids (seals) display variations on the phocid pattern of neuroanatomy, although consistent differences in their swimming patterns have not been documented (King 1983). A wide range in postcranial neuroanatomy also exists in mysticete and odontocete whales, suggesting either a greater diversity in cetacean swimming pattern than is presently recognized, or a looser link between neuroanatomy and locomotor mode in cetaceans than in pinnipeds.

THE LIMITS OF INTERPRETATION: WHAT THE NEURAL CANAL CANNOT TELL US

Variations in neural canal design involve a relatively small number of parameters. This condition can be

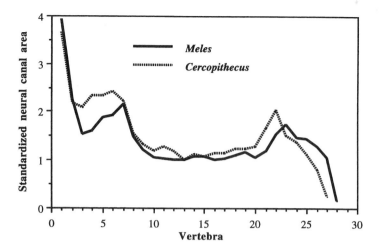

Figure 14.10. Neural canal area, by segment, of two mammals of different limb use but similar gross neuroanatomy. The badger, *Meles meles*, is specialized for digging; the diana monkey, *Cercopithecus diana*, is an arboreal quadruped with a power grip.

viewed as a constraint imposed by the conservative nature of spinal anatomy and by our ability to observe only the gross anatomy of a system with intricate cellular structure. Characters of gross anatomy that change with variations in phylogenetic pattern and/or with locomotor pattern include size of neural canal by segment, position of enlargements relative to vertebrae and to each other, extent of frontal accumulation, and pattern of tail innervation. From these we may infer not only vertebrate class but also the extent of a given somatic segment's innervation. However, such variations cannot tell us whether we are viewing changes in limb territory or in pattern (density of innervation). Neither can they tell us the particular usage of the musculature. Such interpretations must be based on the osteological record of the limbs themselves, insofar as that is possible (see Lauder, this volume).

Within a given neural *bauplan*, animals with markedly different ways of life may possess many similarities of gross neuroanatomy despite differences in cellular neuroanatomy. Among mammals, several taxa display forelimb innervation considerably more extensive than that of torso or hindlimb. The size-standardized traces in Figure 14.10 belong to a badger (*Meles*) and a diana monkey (*Cercopithecus*). Although minor differences exist between the curves, especially in extent of cervical enlargement attributable to long ascending and descending fibers, there is little basis for predicting digging or an arboreal life-style and power grip from the curves alone. Each curve could be more usefully compared with others generated from members of the same mammalian order or family, so that range of variation of a given pattern could be documented.

Despite such restrictions, postcranial neural anatomy offers a relatively powerful tool with which to supplement existing paleoneurological methods of predicting life-style in extinct vertebrates.

ACKNOWLEDGMENTS

NSF grants EAR 8819581 and EAR 9105546 funded parts of the original work described in this chapter.

REFERENCES

Ariens Kappers, C.U., Huber, G.C., & Crosby, E.C. 1936. *The Comparative Anatomy of the Nervous System of Vertebrates, Including Man*. New York: Macmillan.

Barnes, L.G. 1972. Miocene Desmatophocinae (Mammalia: Carnivora) from California. *University of California Publications in Geological Science* 89, 168.

Cruce, W.L.R. 1979. Spinal cord in lizards. In *Biology of the Reptilia*, ed. C. Gans, C.R.G. Northcutt, & P. Ulinski, vol. 10, pp. 111–131. London: Academic Press.

De Gennaro, L.D. 1982. The glycogen body. *Avian Biology* 6, 341–371.

Edinger, T. 1926. The brain of *Archaeopteryx*. *Annals and Magazine of Natural History* 18, 151–156.

Edinger, T. 1942. The pituitary body in giant animals fossil and living: a survey and a suggestion. *Quarterly Review of Biology* 17, 31–45.

Edinger, T. 1948. Evolution of the horse brain. *Memoirs of the Geological Society of America* 25, 1–177.

Edinger, T. 1962. Anthropocentric misconceptions in Paleoneurology. *Proceedings of the Rudolf Virchow Medical Society, City of New York* 19, 56–107.

Edinger, T. 1966. Brains from 40 million years of camelid history. In *Evolution of the Forebrain*, ed. R. Hassler & H. Stephan, pp. 153–161. Stuttgart: Georg Thieme.

English, A.W. 1976. Limb movements and locomotor function in the California sea lion (*Zalophus californianus*). *Journal of Zoology, London* 178, 341–364.

Galton, P. 1990. Stegosauria. In *The Dinosauria*, ed. D.B. Weishampel, P. Dodson, & H. Osmólska, pp. 435–455. Berkeley: University of California Press.

Gauthier, J.A. 1984. A cladistic analysis of the higher systematic categories of the Diapsida. Unpublished Ph.D. Thesis, University of California at Berkeley.

Giffin, E.B. 1990. Gross spinal anatomy and limb use in living and fossil reptiles. *Paleobiology* 16, 448–458.

Giffin, E.B. 1991. Endosacral enlargements in dinosaurs. *Modern Geology* 16, 101–112.

Giffin, E.B. 1992. Functional implications of neural canal anatomy in Recent and fossil marine carnivores. *Journal of Morphology* 214, 357–374.

Gould, S.J., & Lewontin, R.C. 1979. The spandrels of San Marco and the Panglossian paradigm: a critique of the adaptationist programme. *Proceedings of the Royal Society of London* B205, 581–598.

Hopson, J.A. 1977. Relative brain size and behavior in archosaurian reptiles. *Annual Review of Ecology and Systematics* 8, 429–448.

Hopson, J.A. 1979. Paleoneurology. In *Biology of the Reptilia*, ed. C. Gans, R.G. Northcutt, & P. Ulinski, vol. 9, pp. 39–146. London: Academic Press.

Howell, A.B. 1928. Contribution to the comparative anatomy of the eared and earless seals (Genera *Zalophus* and *Phoca*). *Proceedings of the United States National Museum* 73, 1–142.

Imhof, G. 1905. Anatomie und Entwicklungsgeschichte des Lumbalmarkes bei den Voegeln. *Archiv für Mikroskopische Anatomie und Entwicklungsgeschichte* 65, 498–610.

Janensch, W. 1939. Der sakrale Neuralkanal einiger Sauropoden und anderer Dinosaurier. *Palaeontologische Zeitschrift* 21, 171–194.

Jenkins, F.A., Jr., & Goslow, G.E. 1983. The functional anatomy of the shoulder of the Savannah monitor lizard (*Varanus exanathematicus*). *Journal of Morphology* 175, 195–216.

Jenkins, F.A., Jr. & Parrington, F.R. 1976. The postcranial skeletons of the Triassic mammals *Eozostrodon, Megazostrodon* and *Erythrotherium*. *Philosophical Transactions of the Royal Society of London B. Biological Sciences* 273, 387–431.

Jerison, H.J. 1971. More on why birds and mammals have big brains. *American Naturalist* 105, 185–189.

Jerison, H.J. 1973. *Evolution of the Brain and Intelligence.* New York: Academic Press.

King, J. W. 1983. *Seals of the World*, 2nd ed. Ithaca: Cornell University Press.

Kirsch, J.A. W., & Johnson, J.I. 1983. Phylogeny through brain traits: trees generated by neural characters. *Brain, Behavior and Evolution* 22, 60–69.

Krause, W. 1881. Zum Sacralhirn der Stegosaurier. *Biologisches Centralblatt* 1, 461.

Kusuma, A., ten Donkelaar, H.J., & Nieuwenhuys, R. 1979. Intrinsic organization of the spinal cord. In *Biology of the Reptilia*, ed. C. Gans, R.G. Northcutt, and P. Ulinski, vol. 10, pp. 102–109. London: Academic Press.

Leonard, R.B., & Cohen, D.H. 1975. A cytoarchitectonic analysis of the spinal cord of the pigeon (*Columba livia*). *Journal of Comparative Neurology* 163, 159–180.

Marsh, O.C. 1874. Small size of the brain in Tertiary mammals. *American Journal of Science* 8, 66–67.

Mitchell, E. 1966. The Miocene pinniped *Allodesmus*. *University of California Publications in Geological Sciences* 61, 1–105.

Nieuwenhuys, R. 1964. Comparative anatomy of the spinal cord. *Progress in Brain Research* 11, 1–57.

Ostrom, J.H. 1969. Osteology of *Deinonychus antirrhopus*, an unusual theropod from the Lower Cretaceous of Montana. *Bulletin of the Peabody Museum of Natural History* 30, 1–165.

Peterson, J.A. 1984. The locomotion of *Chamaeleo* (Reptilia: Sauria) with particular reference to the forelimb. *Journal of Zoology, London* 202, 1–42.

Petras, J.M. 1976. Comparative anatomy of the tetrapod spinal cord: dorsal root connections. In *Evolution of Brain and Behavior in Vertebrates*, ed. R.B. Masterton, C.B.G. Campbell, M.E. Bitterman, & N. Hotton III, pp. 345–381. New Jersey: Lawrence Erlbaum Associates.

Radinsky, L. 1968a. A new approach to mammalian cranial analysis, illustrated by examples of prosimian primates. *Journal of Morphology* 124, 167–180.

Radinsky, L. 1968b. Evolution of somatic sensory specialization in otter brains. *Journal of Comparative Neurology* 134, 495–506.

Radinsky, L. 1969. Outlines of canid and felid brain evolution. *Annals of the New York Academy of Science* 167, 277–288.

Radinsky, L. 1971. An example of parallelism in carnivore brain evolution. *Evolution* 25, 518–522.

Radinsky, L. 1973. Evolution of the canid brain. *Brain, Behavior and Evolution* 7, 169–202.

Radinsky, L. 1975. Viverrid neuroanatomy: phylogenetic and behavioral implications. *Journal of Mammalogy* 56, 130–150.

Rexed, B. 1954. A cytoarchitectonic atlas of the spinal cord in the cat. *Journal of Comparative Neurology* 100, 297–379.

Roth, J.J., Roth, E.C., & Hotton, N., III. 1986. The parietal foramen and eye: their function and fate in therapsids. In *The Ecology and Biology of Mammal-like Reptiles*, ed. N. Hotton III, P.D. MacLean, J.J. Roth, & E.C. Roth, pp. 173–184. Washington & London: Smithsonian Institution Press.

Rowe, T. 1988. Definition, diagnosis, and origin of Mammalia. *Journal of Vertebrate Paleontology* 8, 241–264.

Seymour, R.S. 1982. Physiological adaptations to aquatic life. In *Biology of the Reptilia*, ed. C. Gans & F.H. Pough, vol. 13, pp. 1–51. London: Academic Press.

Snyder, R.C. 1962. Adaptations for bipedal locomotion of lizards. *American Zoologist* 2, 191–203.

Steel, R. 1970. Saurischia. In *Handbuch der Paläoherpetologie*, ed. O. Kuhn, part 13, 87 pp. Stuttgart: Gustav Fischer Verlag.

Stensiö, E. 1963. The brain and the cranial nerves in fossil, lower craniate vertebrates. *Skrifter utgitt av Det Norske Videnskaps-Akademi i Oslo* 13, 1120.

Switzer, R.C., Johnson, J.I., & Kirsch, J.A.W. 1980. Phylogeny through brain traits: relation of lateral olfactory tract fibers to the accessory olfactory formation as a palimpsest of mammalian descent. *Brain, Behavior and Evolution* 17, 339–363.

Tarasoff, F.J., Bisaillon, A., Pierard, J., & Whitt, A. 1972. Locomotory patterns and external morphology of the river otter, sea otter and harp seal (Mammalia). *Canadian Journal of Zoology* 50, 915–927.

Thomson, K.S. 1988. *Morphogenesis and Evolution*. New York: Oxford University Press.

Ulinski, P.S. 1983. *Dorsal Ventricular Ridge; a Treatise on Forebrain Organization in Reptiles and Birds*. New York: John Wiley and Sons.

Ulinski, P.S. 1986. Neurobiology of the therapsid–mammal transition. In *The Ecology and Biology of Mammal-like Reptiles*, ed. N. Hotton III, P.D. MacLean, J.J. Roth, &

E.C. Roth, pp. 149–171. Washington and London: Smithsonian Institution Press.

Voris, H.C. 1928. The morphology of the spinal cord of the Virginia opossum. *Journal of Comparative Neurology* 46, 407–459.

Wake, D.B. 1991. Homoplasy: the result of natural selection, or evidence of design limitations? *American Naturalist* 138, 543–567.

15

To what extent can the mechanical environment of a bone be inferred from its internal architecture?

JEFFREY J. THOMASON

ABSTRACT

Living bones modify their structure at gross, tissue, and molecular levels in response to the force patterns acting on them. On this basis, the preserved structure of fossil bones may contain direct information on the forces they may have experienced during life. I review the difficulties in functionally interpreting bone structure, current areas of success, and the future potential for expanding such analyses. Despite the intense scrutiny to which bones are subjected during functional reconstruction, we may be missing important information encoded in their internal structure.

INTRODUCTION

How fast could mastodons run? Could the largest sauropods have stood upright on their hindlimbs? In answering biomechanical questions for extinct vertebrates we combine observation and inference in several different ways, each combination providing part of the answer. Consider as an example Alexander's (1989) attempt to answer the question of bipedal stance in sauropods. He collected a number of estimates of the body weight of a large sauropod (based on dimensions of the whole skeleton or measurements of the volume of models), and then calculated the approximate bending and compressive strengths of the femora from external dimensions of the bones. The forces required to break the femora appeared greater than the animal's weight, giving a tentative indication that the quadrupedal dinosaur could have stood on its hindlimbs alone. The observations were the dimensions of the skeleton and of the femora, and the inferences were body weight and bone strength.

We are largely limited to information provided by the bones of extinct vertebrates, but the variety of data they provide allow us to construct several combinations of observation and biomechanical inference. In the sauropod example, we could add observations on the muscle scars on the bones and dissect extant reptiles to infer muscle positions and the possible magnitude and orientation of maximal forces they might generate. We could measure the lever arms of muscles about joints, and describe the shape and orientation of the joints themselves to infer possible positions of the legs. Each of these combinations of observation and inference could be assembled using the rigorous methods of Bryant and Russell (1992) and Witmer (this volume). However, no combination alone answers the question, and any given combination may provide inaccurate results. But viewed in concert, they may help us first to limit the range of possibilities, and then perhaps to define a plausible set of probabilities. Given that we can never test function in fossils, we should be thorough in our attempts to limit the set of probable functional answers.

In addition to the external features of bones, the internal architecture, or distribution of osseous material within bones, provides another source of primary data on which to base inferences in biomechanical reconstruction. One architectural feature, the cross-sectional shape of long bones, has been used extensively by physical anthropologists in calculations of absolute or relative bone strength (Ruff 1989; Demes, Jungers & Selpien 1991), and to a lesser extent in paleontological studies (Thomason 1985a; Alexander 1989), and I will review the technique below. But bone architecture may carry considerably more information than estimates of the maximum forces a bone can withstand. It has the potential to tell us something of the mechanical environment of the bone: the forces that may have acted on it during life. Calculations from cross sections help define the possible forces on a bone; other aspects of the architecture may help limit the probable forces. If this is the case, then bone architecture has been an underexploited tool in paleontological biomechanics.

A considerable amount of current orthopedic research is directed at elucidating the way in which bone material is formed and remodeled in response to stress patterns in vivo. The results of this research indicate the degree to which the internal architecture of bones encodes the stress patterns. If the code were simple, we could "read" mechanical function directly from the structure of fossil bones. Unfortunately, the code is complex – an example of loose coupling between form and function – but we can extract more information from bone architecture than we do now. I have previously demonstrated that the internal structure of equid metacarpals contributes to our understanding of the functional changes during the evolution of equid limb anatomy (Thomason 1985b). In this chapter I review the mechanical information currently inferred from internal architecture, and discuss the potential for extracting further information.

First, I start by setting the limits of what we might expect to gain from interpretations of bone structure. This is a discussion of the difficulties, assumptions and *caveats* involved in such interpretation. The coding of mechanical function into bone form is complex and has many sources of noise. Second, I review the information that can be retrieved by analyzing the cross-sectional shape of long bones, giving examples of areas where such methods have been used successfully, and warnings about when not to use them. Third, I review two of the many areas of current research in orthopedic biomechanics that hold great promise for the future, both in their own context and for interpreting the structure of fossil bone. One area demonstrates a way to reconstruct the physical properties of cancellous bone from its architecture. The other uses finite-element methods to simulate the processes of bone remodeling on a computer, and indicates how the reverse process of reconstructing loading from structure may be achieved. Finally, I discuss briefly my views on the methodology of biomechanical analyses of fossil vertebrates, and show how inferences from internal architecture can be incorporated.

THE CODING OF MECHANICAL INFORMATION IN BONE ARCHITECTURE

Every bone has to bear external forces from some combination of weight-bearing and muscular activity (Figure 15.1). These forces induce in the material of the bone patterns of stress and strain that are dependent upon the magnitudes and orientations of the forces, and on the structural organization (internal architecture) and material properties (strength, stiffness, etc.) of the bone. The material properties

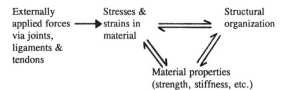

Figure 15.1. Flow chart illustrating the influence of external forces on bone. Our aim is to work from structure on the right to force patterns on the left.

themselves are partly dependent on the structural organization, hence the connecting arrows in Figure 15.1. If bone were inert, a given external force would produce the same stress patterns every time, but bone remodels in response to the stress pattern (or, more probably, to the strains induced by the stresses). Remodeling alters the structural organization; the material properties change in consequence; and the stress pattern itself is modified. The right hand part of Figure 15.1 is a feedback loop. Our aim is to work from structural organization at the right, through material properties to stress patterns and finally out of the loop to the external forces. The loading pattern of forces is the signal we are trying to rebuild; the structure is a noisy form of the signal from which we are working.

The feedback loop is where information on stress patterns is coded into the bone's structure, so I will consider this first and then discuss how to infer forces from stress patterns. The difficulties we encounter in the loop are derived from the number of levels of structural organization in bone, the way in which bone responds to stress (i.e., how the signal is coded), and the several sources of noise.

Levels of structure

Bone material is structurally organized at many different levels of size (Currey 1984). At the smallest level relevant to this discussion are the sizes and shapes of the molecules that comprise bone matrix, predominantly collagen and calcium phosphate, but with contributions from other organic and inorganic compounds. The absolute and relative amounts of the two main molecules vary considerably among bones and in different regions within a bone. For convenience, I include the degree of mineralization under the broad category of structural organization. In addition to quantity, the spatial organization of molecules of each type within the matrix is variable and functionally important (Boyde & Riggs 1990; Martin & Ishida 1989).

Reducing the magnification, we move up to the cell and tissue level at which several histological configurations have been described (Martin & Burr 1989): woven, primary lamellar, laminar (plexiform), primary osteonal, and secondary osteonal. At a still lower magnification we can make broad distinctions between compact and cancellous bone, and finally we can consider the geometry and external form of the whole unit, the bone itself.

What difficulties do these levels impose on structural analysis? They certainly add a degree of complexity; the effects of each level have to be taken into account, or evidence must be produced to support the assumption that any given level can be ignored. For example, the effects of molecular arrangement and degree of mineralization are often ignored in calculations of long bone strength on the grounds that strength is influenced more by cross-sectional shape than by the mineral composition of the bone in the section (Ruff 1989). Ironically, many difficulties in mechanical analysis arise at the largest levels. It has only recently been possible to analyze stress distribution in trabecular bone and in some whole bones such as vertebrae and cranial bones, because of their intricate shapes.

For the paleontologist, the presence of these levels raises questions of a familiar kind: How well is each structural level preserved, and can we access them? The first question needs to be asked of each specimen of interest in turn, because of the extreme variation in quality of preservation among fossils. In most cases the degree of mineralization cannot be assessed because of the intrusion of petrifying minerals, or replacement of bone mineral by them. The orientations of collagen fibers and calcium phosphate crystals are not usually preserved. There are, however, many published examples of excellent structural preservation of bone at the tissue level and above in different vertebrates (e.g., Enlow & Brown 1956, 1957, 1958; Reid 1985; Thomason 1985b; Varricchio 1993). Most workers have used geological saws and polishing wheels to physically section and prepare histological specimens from fossils. Details ranging from osteonal morphology to the gross distribution of cancellous and compact bone are evident in such specimens. As an example at an intermediate magnification, Figure 15.2 shows the excellent trabecular detail in a thin transverse section through a vertebral centrum from an ichthyosaur. The sectioning method is appropriate if small, redundant fragments are available, or if a curator decides to sacrifice a complete specimen.

Standard radiographic techniques have also proved effective with the advantage that the specimen remains whole (Snure 1924; Tobien 1964), and computed tomography (CT or CAT) on medical-grade machines is being increasingly used on fossils (Conroy & Vannier 1984; Ruff & Leo 1986; Joeckel 1992). High-resolution CT is now available, which represents a combination of the best of sectioning and radiographic techniques. Rowe & Carlson (1992)

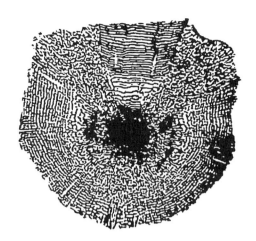

Figure 15.2. Trabecular organization in a thin, transverse section of an ichthyosaur vertebra. The trabecular pattern is mainly tangential, but is interrupted under articulations for neural arches by diverging "wings" of more randomly oriented trabeculae. (Specimen courtesy of C. McGowan.)

used this method to develop a digital-image atlas on compact disc (CD-ROM) of the skull of *Thrinaxodon*, a mammal-like reptile. The resolution of the industrial CT machine they used is of the order of 100 μm, and a newer version will improve that to 20 μm – considerably less than the diameter of individual osteones – in the near future.

I conclude that, while problems of preservation and access may hinder the analysis of individual bones, they are not ubiquitous obstacles at most of the structural levels.

Structure, material properties, and stress patterns

The interaction among the three parameters in the feedback loop (Figure 15.1) holds the key for unlocking the mechanical secrets of bone structure. Research on this interaction has at least a century of history. It is currently very extensive because of its relevance to human orthopedics and bone diseases, and it has been widely and well reviewed (Murray 1936; Currey 1984; Martin & Burr 1989; and contributors to the supplement of volume 11 of the *Journal of Biomechanics* 1991). In summarizing this work I will focus on the problems hampering attempts to "crack the code" of bone's response to mechanical loading.

The current paradigm of how bone responds to mechanical stress is the trajectorial theory, also known as Wolff's law of bone transformation. It originated in the observations of physicians and engineers in the last century (Meyer 1867; Wolff 1870, 1892) and has been modified by sometimes intense debate in the intervening years. Trajectories

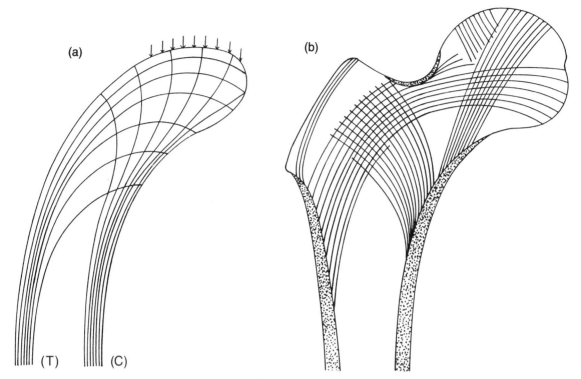

Figure 15.3. Illustration of the trajectorial theory. (a) Compressive (C) and tensile (T) stress trajectories calculated for a curved beam under a distributed vertical load (arrows). (b) Diagram of main trabecular systems in a frontal section through the proximal end of the human femur. (Redrawn after Wolff 1870 and Meyer 1867.)

are lines (isobars) of compressive and tensile stress that can be calculated from beam theory to act in the material of a beam under load (Figure 15.3a). The theory has two parts, one about product, the other about process. The first part states that the trabeculae of cancellous bone lie along trajectories, and that bone density reflects stress intensity (hence the compact bone corresponding to closely spaced trajectories in Figure 15.3). The second part states that, if the stress patterns changes (reflecting changes in external loads), the bone remodels toward a structural optimum: Trabeculae are realigned with the new trajectories, and material is added or removed where stresses are high or low, respectively. This theory was largely developed from qualitative comparisons such as that in Figure 15.3, combined with observations on the remodeling of misaligned fractures. Current researchers recognize that the theory is an oversimplification of how bone is constructed and how it responds to loading (Martin & Burr 1989). However, the trajectorial theory remains the paradigm under which most biomechanical research on bone is conducted (Bertram & Swartz 1991). In trying to elucidate the response of bone to loading, there are two main categories of difficulty: (1) the pragmatic problems of quantifying bone structure, and the forces, stresses, and strains acting on a given

bone in vivo, and (2) the number of different ways in which bone structure initially develops and is subsequently modified.

PRACTICAL DIFFICULTIES. When the trajectorial theory was established, it was impossible to evaluate bone loadings in vivo and, therefore, impossible to test the "theory." This difficulty has now been partly remedied by the development of measurement techniques using force plates, tendon force transducers, pressure transducers, strain gauges, and electromyography (Lanyon & Smith 1969; Cochran 1972; Barnes & Pinder 1974; Cavagna 1985; Loeb & Gans 1986). In combination, these methods can provide information on overall forces acting on the limbs, or in tendons or muscles, and local strains at selected sites on bones. From this information, forces on individual bones can be calculated and, most importantly, data on how these forces vary in magnitude and direction during different activities can be obtained.

A second problem, which was solved independently, was how to calculate detailed stress patterns throughout a bone, given all the levels of structural composition described above. The methods of beam theory used to generate Figure 15.3a were based on the assumption of material homogeneity. As we shall see in a section on the cross-sectional geometry of bones, violation of this assumption is not critical when calculating stresses in the midshaft cortex of long bones, but beam theory cannot be used for trabecular bone. Continuum theory, which was developed fairly early in this century but did not become practical until the advent of digital computers, provided much of the solution to the problem. The rest of the solution came from numerous in vitro experiments on the mechanical properties of bone, in particular those relating the properties of cancellous bone to its trabecular architecture. (Describing this architecture has its own problems which I will discuss in the next section.) The parts of the solution come together in finite element modeling, a computer-based technique derived from continuum theory (Zienkiewicz 1971). It is the method of choice in much engineering research and development, and is now widely used in human biomechanics (Simon 1980). It requires data on the shape of a structure, on the material properties of its constituents (in this case, cancellous and compact bone), and on the loads on the structure. These data are now available for several types of bone from a few mammals.

THE DEVELOPMENT AND MODIFICATION OF BONE STRUCTURE. Bone responds to mechanical stress during formation (osteogenesis) and growth, and as its structure is continually modified throughout life.

Our understanding of stress's influence on osteogenesis is clouded by the fact that bone is formed in two ways: Dermal bones ossify directly in mesenchymal membranes, whereas endochondral bones are first formed in cartilage that ossifies as it grows. Stress is influential in both cases but in different ways. The mechanical interaction among dermal bones as they ossify may in fact help determine their final form (see Thomson, this volume, for a novel interpretation). The general shape of endochondral bones is set by that of the cartilage precursor under both genetic and developmental control (osteogenesis is largely epigenetic; Smith & Hall 1991). Stress patterns appear to have little effect on the form but a strong effect on the internal architecture as it develops in the ossifying cartilage. Glucksmann (1941) was probably the first to demonstrate that mechanical stress is an important osteogenic influence. Although it is probably not the primary stimulus for osteogenesis, stress is important in organizing the structure of the bone as it ossifies (Burger, Klein-Nulend, & Veldhuijzen 1991). In a recent elegant application of previous experimental work, Wong and Carter (1988) used finite element modeling to generate stress patterns that emulated the commonly observed patterns of ossification in the human sternum. The different stress patterns were produced simply by altering the external loads on the cartilaginous sternum. The models incorporated assumptions of the ways ossifying cartilage integrates a response to cyclically varying stress levels, and of different types of loads on the cartilage cells (shear and hydrostatic pressure).

Postosteogenic modification of bone structure is not a single process (definitions to follow are from Martin & Burr 1989). The primary actions are deposition and resorption of bone matrix. Osteoclasts resorb matrix, osteoblasts deposit it, and both types of cell may be embedded in a lacuna in the matrix, or on an endosteal (inner) or periosteal (outer) surface. *Modeling* is defined as either deposition or resorption occurring alone at a given site. For example, osteoclasts and osteoblasts on endosteal, and periosteal surfaces have the ability to alter the gross form of the bone by making the cortices thicker or thinner or of a different size or shape. On the endosteum of trabeculae, these cells can vary the density of the cancellous bone by depositing or resorbing bone matrix, and have some ability to reorient trabeculae. *Remodeling* is resorption followed by deposition, and can occur on surfaces or within the material. Secondary osteonal bone, for example, is always the product of remodeling by embedded cells, not primary formation. Tissue-type remodeling occurs normally during growth, and also in response to stress which is termed *adaptive remodeling*. Osteocytes continually exchange calcium with the matrix, as part of the mechanism of calcium homeostasis (not strictly included in modeling or remodeling) and in response to mechanical stress.

Bone is modeled and remodeled in a number of ways, and in response to a variety of factors, only one of which is loading. Adaptive remodeling, on its own, is a complex process and it is not yet clear how this response is mediated (Martin & Burr 1989). Given the variety of ways in which the response occurs, it may not be a single process. Certainly the stress thresholds at which remodeling is initiated vary widely among bones: They are considerably lower in the bones of the braincase than in long bones of the limbs. Added to this is the problem of signal integration over time. At any given time, the aspects of a bone's structure that reflect its response to stress do not just indicate the most recent stresses but a summation of its modeling and remodeling responses to stress, growth, calcium balance, etc., since the onset of osteogenesis.

The number of different ways in which bone form can be modified, therefore, makes it difficult to determine general responses to changes in stress patterns. Furthermore, the sum total of all these activities does not appear to achieve or maintain the optimal structural configuration suggested by the trajectorial theory. For example, Woo et al. (1981) found significant increases in cortical thickness in the limb bones of exercised pigs compared with non-exercised controls. There are no surprises here, except that the bone was laid down primarily on the endosteal surface rather than periosteally which would have been mechanically more optimal for resisting the loads known to act on these bones. Bertram and Swartz (1991) review this and other cases that illustrate the difficulties in interpreting the functional adaptation of bone, and suggest that functional adaptation may be only an indirect response to stress. They argue that many of the experiments seen as providing important evidence for the trajectorial theory do not irrefutably demonstrate that the response was to loading rather than to, say, increased blood flow. Their work may stimulate a long overdue examination of the paradigmatic status of the trajectorial theory. From our perspective, examples such as the results of Woo et al. (1981) emphasize the problems involved with functionally interpreting bone structure. The coding of functional information in bone structure is not a single or simple process.

Sources of noise

In addition to the complexity of the relationship between stress patterns and bone structure, two categories of noise add to the problems of structural interpretation: (1) nonmechanical influences on structure, and (2) mechanical influences only indirectly related to the loading regimes we are trying to reconstruct.

NONMECHANICAL INFLUENCES ON STRUCTURE. Included here are aging, nutrition, hormonal state, gender, and disease. Apart from disease, all may be regarded as consequences of our evolutionary history, as phylogenetic constraints. For example, the use of the skeleton as a store of calcium salts is an important physiological function that probably originated with the earliest vertebrates. From the perspective of the mechanical functions of the skeleton, this physiological response is a liability. Mineral matrix may be removed in times of low dietary intake of calcium or of hormonal imbalance, regardless of structural consequences. Conditions such as osteoporosis, in which bone loss is strongly linked to age and gender, certainly fall into this category. In humans, at least, the situation is compounded by differences in remodeling capacity between the sexes (Martin 1991). In males, endosteal and periosteal modeling can alter the cross-sectional shape of long bones to compensate partially for the strength reduction concomitant with osteoporotic mineral loss. Females have almost negligible capacity for such compensatory remodeling. Putting these important medical considerations aside, the physiological

response of bone to nonmechanical factors represent an important source of noise in structural interpretation.

NOISE OF MECHANICAL ORIGIN. Bones may experience forces other than those transmitted via joints, tendons, ligaments, and fleshy muscular attachments. An example is the pressure exerted by lateral expansion during contraction of muscle bellies adjacent to the diaphysis of a long bone. The cross-sectional shape of the diaphysis can be quite strongly influenced by such pressure during growth. This is one reason that limits the information carried in cross sections, which is unfortunate because they have the potential to tell us a lot about bone loading, as we shall in the section below on cross-sectional geometry.

To summarize so far, the feedback loop itself presents many difficulties to hinder us in reconstructing bone function from its structure. But the difficulties do not end there; the step from stress patterns to the forces that induced them poses problems of its own.

Difficulties in reconstructing forces from stress patterns

The primary difficulty here is that a single stress pattern in a material may be produced by more than one pattern of externally applied forces. The problem is exacerbated for stress patterns at a distance from where the forces are applied, and may be reduced closer to the site of loading (St. Venant's principle). A second aspect of this problem is that forces acting on bone are rarely static. They change in magnitude, direction, and duration with time. We met time averaging above when considering stress patterns. Reconstructing forces from stress patterns only compounds the difficulties. Not only does the structure encode stress patterns integrated over time, but each pattern may have been produced by a number of different applied loads.

A FINAL CAVEAT. Most of the research into the structure and mechanical behavior of bone has been on human or other mammalian bone. The most active areas are in physical anthropology and orthopedic biomechanics. Although the histological structure of nonmammalian bone has been well described (Enlow & Brown 1956, 1957), relatively few workers have investigated the mechanical properties of bone from taxa other than mammals (Hamilton et al. 1981; Biewener 1982; Currey 1988, 1990). Furthermore, experiments on bone remodeling in birds (Rubin & Lanyon 1984; Biewener, Swartz, & Bertram 1986) are included in the literature on the trajectorial theory without confirmation that the

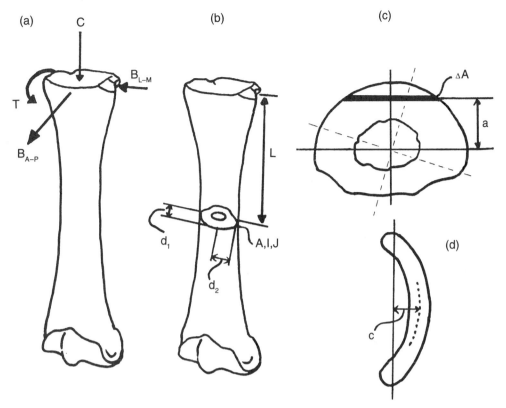

Figure 15.4. (a) Possible components of force and moment that can act on the end of a bone (in this case, the third metacarpal of a horse). Balancing components on distal end have been omitted for clarity. (b) Measurements necessary to calculate maximum values for each component in (a). (c) One procedure for calculating second moments of area. (d) How the moment arm of curvature is measured for curved bones. See text for descriptions of all parts of the figure.

mechanisms of functional adaptation are comparable across divergent vertebrate taxa.

THE CROSS-SECTIONAL GEOMETRY OF LONG BONES

With only a few assumptions, the strength and stiffness of long bones can be estimated using the long-established engineering principles of beam theory. This is currently the most successful area of functional analysis of bone architecture. The primary assumptions in this application of beam theory are that (1) the length of the beam must be at least twice its maximum diameter; (2) forces must be applied away from region of interest, and (3) the material must be homogeneous (isotropic).

Any force acting on a long bone can be reduced to components oriented along the bones anatomical axes (Figure 15.4a). These components include axial compression (C), lateromedial and anteroposterior bending (B_{L-M}, B_{A-P}), and a moment (T) exerting torsion about the longitudinal axis. (Each of the arrows in Figure 15.4 could be reversed, converting

compression to tension and reversing the direction of bending and torsion.) A few geometrical measurements allow us to calculate the possible maximal values of each component if it were acting alone (Figure 15.4b): distance L is half the length of the bone; A is the cross-sectional area at the midshaft; d_1 is the maximal anteroposterior distance from the centroid of the area to the bone's surface, and d_2 is the comparable lateromedial measurement. If the bone is curved, we need to measure the moment arm of curvature (Figure 15.4d; c). Parameter I is the second moment of area which quantifies whether the material of the cross section is near to or far from the centroid. It has a different value for each axis through the cross section and is a measure of bending strength about a given axis. Second moments of area can be evaluated by dividing the section into narrow strips parallel to an axis through the centroid (Figure 15.4c; ΔA). The area of each strip is multiplied by the *square* of its distance to the centroidal axis (Figure 15.4c; a), and all such products are summed. The procedure is repeated for other axes of interest. For most long bones, the maximum and minimum values of I occur about axes (Figure

15.4c; dashed lines) that do not correspond to anatomical axes. The final parameter, J, is the polar moment of inertia and is used to calculate torsional strength. For circular or elliptical sections, it is the sum of I_{A-P} and I_{L-M} or of I_{min} and I_{max}. For irregular sections, it is evaluated in a manner similar to that for I. Several computer programs and graphical techniques exist for evaluating cross-sectional areas and second and polar moments of inertia from physical sections, CT images, and density curves produced by photon absorptiometry (Lovejoy & Barton 1980; Nagurka & Hayes 1980; Martin & Burr 1984).

If cross sections cannot be obtained for a given bone, approximate values for A, I, and J can be calculated from the diameters of the midshaft by considering the cross section to be elliptical (formulae in Alexander 1989). Such values must be used with caution because asymmetry in cross-sectional shape and in the position of the medullary cavity may introduce large errors (Piziali, Hight, & Nagel 1980; Biknevicius & Ruff 1992).

Maximal values for each force component or moment in Figure 15.4a, assuming each acted alone, can be calculated from:

Compressive force
for a straight bone $C_s = \sigma_{max}.A$ (1a)
Compressive force
for a curved bone $C_c = (\sigma_{max}.I_{A-P})/(c.d_1)$ (1b)
Anteroposterior
bending force $B_{A-P} = (\sigma_{max}.I_{A-P})/(L.d_1)$ (2a)
Lateromedial
bending force $B_{L-M} = (\sigma_{max}.I_{A-P})/(L.d_1)$ (2b)
Torsional moment $T = \tau_{max}.J/d_{max}$ (3)

The newly introduced parameters in these equations are the maximum distance from the centroid to the surface (d_{max}), and the maximal stresses that the material can withstand in compression and bending (σ_{max}) and in torsion (τ_{max}). The maximal stresses are physical properties of the material and would have single values if bone were homogeneous. But σ_{max} and τ_{max} vary with tissue type and mineral density in the compact bone of the diaphysis. Furthermore, σ_{max} has different values for compressive and tensile loading. During bending, one side of a bone is stressed in tension and the other in compression, so Equations (1b), (2a), and (2b) have to be evaluated with tensile and compressive stress maxima.

We cannot estimate values of maximal stresses for individual fossil bones with any degree of certainty, but might be able to if it were possible to evaluate their mineral density. Fortunately, in vitro tests show limited variability in the material properties of compact bone from a variety of mammals (Biewener 1983). In applying Equations (1a)–(3) we can either use a single mean or modal value for the maximal stresses, or use the upper and lower values

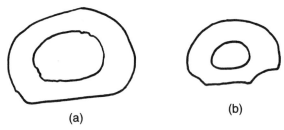

Figure 15.5. Midshaft cross sections through the (a) radius and (b) third metacarpal of a horse. Ratio of lateromedial second moment of area to anteroposterior value is approximately 2 in both cases.

of the known range. Most workers have used single values. The possible set of maximal values for each component might be better expressed by using the extreme maximal stresses. We can then begin refining this possible range to a more probable one, and to consider how the components might have combined.

Biewener (1983) observed that the maximal stresses for mammalian compact bone are 2 to 5 times the peak operating stresses during locomotion. This excess of strength is known as a *safety factor*. Dividing the range of force components by 2 and 5 gives new extremes that approximate the peak operating values, still considering each component to act alone. Can we infer the relative proportion of components in combination?

Bone curvature helps predict the predominant direction of bending in mammalian long bones, largely because axial compressive loads tend to further bend the bone (Equ. [1b]). Bertram and Biewener (1988) suggested the curvature may be a mechanism to ensure bending in a predetermined plane so that each cortex usually had only one mode of loading, either compression or tension. The advantage, they surmised, was that the remodeling mechanisms in each cortex would need to be sensitive to only one mode, not both. The potential disadvantage is a reduction in safety factor, because long bones are considerably more vulnerable to bending than axial compression.

In principle, the second moments of area are also of use in predicting primary bending directions. In practice, they must be used with caution. The midshaft cross sections of the radius and third metacarpal (MCIII) of a horse have a greater lateromedial than anteroposterior diameter (Figure 15.5). This is reflected in the greater lateromedial second moments of area that are approximately twice the anteroposterior values for both bones in Figure 15.5. Piotrowski, Sullivan, and Colahan (1983) suggested that the MCIII was constructed primarily to resist lateromedial bending, and we might infer the same for the radius, but this is not the case. Strain-gauge studies in vivo demonstrate that the radius is bent posteriorly (as its curvature indicates) and the metacarpus is loaded (at slow-to-medium gaits) primarily in axial compression with a little posterior bending superimposed (Rybicki et al. 1977; Biewener et al. 1983).

It is difficult to rationalize why a bone is weaker in the direction it is primarily bent, and different factors may be involved for each of the bones in Figure 15.5. Many curved bones are similar to the equine radius: narrower in the direction of curvature than at right angles to it. Bertram and Biewener (1988) suggested this might be another part of the mechanism to regulate the primary direction of bending; if compressive forces are not exactly aligned with the bone's axis they will bend it preferentially in the direction of the smallest second moment of area. In the case of the equine metacarpus, which is quite straight, I suspect another factor comes into play. The bone is considerably flared towards its ends (Figure 15.4a) because the joint surfaces are expanded lateromedially (which restricts the degrees of freedom of joint motion and increases joint stability). It may be that the bone is not long enough to change proportions along its length, from being lateromedially widened at its ends to being anteroposteriorly widened at its midshaft. If this is the case, the midshaft shape may be partly a function of providing stability at the joints. This argument may be only part of the story for the metacarpus (and certainly does not hold for the metatarsus), but is supported by the shape of the phalanges. They are considerably shorter than MCIII and have similar lateromedial flaring of the joint surfaces and cross-sectional dimensions. They are, however, unlikely to be loaded in lateromedial bending, given the distribution of forces in the equine digit.

As I suggested in the section on sources of noise, cross-sectional shape may reflect loads other than those of interest. For example, the approximately triangular cross section of the tibia in many mammals may be due to pressure from the adjacent muscle bellies (Lanyon 1980).

Most students of cross-sectional properties in physical anthropology and vertebrate biology wisely avoid making absolute statements about forces acting on bones unless they can be measured or estimated. For example, the scaling of cross-sectional properties of limb bones with overall size has been examined without explicit reference to force or stress (Ruff 1987b; Bertram & Biewener 1990). (Biewener 1983, 1989 has also included experimentally determined stresses in his scaling analyses.) Many studies are based on comparisons: for example, between fore- and hindlimb bones to determine their relative use in locomotion (Schaffler et al. 1985; Demes et al. 1991; Biknevicius 1993), among bones at different stages of development (Carrier & Leon 1990), or

among homologous bones in different populations to compare robusticity (Ruff 1987a). The changes in robusticity in an age series of dinosaurs, presented by Weishampel in this volume, is an elegant example of the comparative use of cross sections.

Information in histological sections

The distribution of tissue types in cross sections and osteonal orientation in longitudinal sections appear to contain information that may assist in reconstructing the probable proportions of each loading component, but once again the interpretation is equivocal. Secondary osteones are produced during remodeling, and Schaffler and Burr (1985) demonstrated correlations between the percentage of osteonal bone in the femur and the locomotory behavior of various primates. They inferred that force patterns on the femur vary with behavior differences among species and induce different degrees of remodeling. Martin and Burr (1989), however, reviewed a number of studies showing how secondary osteones accumulate with age in distributions unrelated to that of stress or strain.

Lanyon and Bourn (1979) studied secondary osteones in longitudinal sections of sheep tibiae for which they also had in vivo strain data. The mean orientation of the osteones lay between that of the largest principal strain and the long axis of the bone, which indicates that the osteonal orientation may give a conservative estimate of the direction of principal strains. The significance to the present discussion is that the orientation of the principal strains indicates the relative amounts of torsional and axial strain (due to compression and bending). Thus, we potentially have a way of estimating the proportion of torsional moments to bending and compressive forces. The drawback is that this is the only study in which structural and strain orientations have been compared, and data on other species and bones are needed before possible nonmechanical causes of the correlation can be ruled out. Given that osteones are often extremely well preserved in fossils, a more complete data base on in vivo strain and osteonal orientation in extant vertebrates might be of great value to paleontologists.

TWO AREAS OF RECENT ADVANCES IN THE ANALYSIS OF BONE ARCHITECTURE

The literature on adaptive remodeling in bone is likely to be of interest to future historians and philosophers of science interested in the mechanisms of scientific progress. The last decade or so of research on adaptive remodeling has been particularly

(a)

Figure 15.6. (a) Thin section through a sample of trabecular bone with a superimposed stereological grid of parallel lines (only drawn over bone, not pores). MIL is mean length of all lines shown. Angle α is increased incrementally and MIL is recorded at each increment. (b) A nonlinear regression of MIL on α gives the principal orientation (θ) of trabeculae to the anatomical axis and degree to which trabeculae are oriented along axis (expressed as ratio of axes of ellipse). (Redrawn after Turner 1992.)

(b)

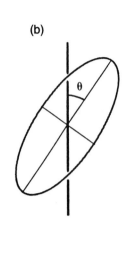

exciting, with alternating advances in theory and experimental results providing mutual stimulus. I have selected just two examples because of their relevance to my aims here.

The structure and mechanical properties of cancellous bone

Turner (1992) reviewed an elegant methodology relating the mechanical properties of cancellous bone to its architecture. The relationship of properties to structure should be the simplest arm of the feedback loop to quantify (Figure 15.1), but has proven elusive for cancellous bone because of its intricate three-dimensional form. Turner traced the development of his work back to that of Whitehouse (1974) who described a technique for quantifying cancellous bone architecture. This technique is itself derived from the work of metallurgists describing granular structure in metal alloys (Underwood 1970; Weibel 1979), and all such techniques comprise the field of *stereology.*

Whitehouse (1974) superimposed a grid of parallel lines over sections of cancellous bone, varying the orientation of the lines with respect to a predefined anatomical axis (Figure 15.6a). At each angle of orientation α he measured the length of line segments (intercepts) lying over trabecular bone and calculated the mean intercept length (MIL). A nonlinear regression of MIL on α fitted the data to an ellipse whose orientation (Figure 15.6b; θ) indicated the primary direction of trabeculae in the section. The ratio of the axes of the ellipse described the degree of nonuniformity of the bone or its anisotropy. A ratio of 1 implied random orientation of trabeculae, or isotropy; a ratio of 0 indicated the

trabeculae are parallel columns, that is, highly oriented or anisotropic. Most human cancellous bone falls between 0.2 and 0.59 (Turner 1992).

In a similar stereological procedure (Weibel 1979), a grid of points is superimposed over the section, and the ratio is calculated of points lying over bone to the total number of points. This gives a measure known as the *solid volume fraction* which is inversely related to the porosity of the bone. The amount of bone mineral present is related to the solid volume fraction and degree of mineralization of the matrix, and is called the *apparent density, ρ.*

Stereological methods allow cancellous bone architecture to be described in terms of orientation, degree of organization, and solid volume fraction. Cowin (1985) went back to the literature of metallurgical stereology and extended this process to three dimensions by way of what he termed the *fabric parameter.* To generate this parameter, the ellipses from orthogonal surfaces of a cube of cancellous bone are combined to form an ellipsoid. The lengths of the ellipsoid's axes define the fabric parameter, *H.* These lengths are influenced by the solid volume fraction and may be normalized to remove its effect. They may be then projected onto the anatomical axes of the specimen using the inverse of a calculation of eigenvectors and values. (Eigenvectors determine the axes of an ellipsoid from data measured with respect to arbitrary external axes. In this case, the axes of the ellipsoid are being projected onto external, anatomical axes; hence the calculation is reversed.) The normalized values of *H* projected onto each anatomical axis are referred to as H_1, H_2, and H_3.

Turner, Cowin, and co-workers applied the theoretical construct of the fabric parameter to compare measured fabric and apparent density values with experimentally determined values of compressive strength and stiffness for cubes for cancellous bone (Turner et al. 1990). They found that better than 90 percent of the variance in yield strength in compression, and 70 percent for the Young's modulus (stiffness), was explained by the term $\rho^2.H_i^3$ (where $i = 1$ to 3, indicating the plane of orientation of the bony cube). In other words, the square of the apparent density and the cube of the fabric value for a given plane can be used to predict the mechanical properties of cancellous bone samples with a high degree of accuracy. Turner (1992) extended these results to suggest that the anisotropy in cancellous bone structure is equal to that in the stress patterns normally acting in this material, and that the primary adaptive principle for the construction of cancellous bone is to achieve a constant yield strain.

This work is important in the context of the trajectorial theory, and from our perspective in that it potentially gives a means of estimating the mechanical properties of fossil cancellous bone from an analysis of its structure. One slight drawback in applying this technique to fossils is that the apparent density ρ is partly dependent on the degree of mineralization, which is unobtainable for petrified bone. However, ρ has a linear relationship with the solid volume fraction (Currey 1988) which can be measured stereologically from fossil specimens.

The line of progress from Whitehouse (1974) to Turner (1992) is one of several attempts at solving the same problem. Alternative stereological techniques exist (Hahn et al. 1992), and the ellipse approach has been criticized because it describes trabecular bone in terms of orthogonal axes when trabeculae are rarely orthogonal (Kuo & Carter 1991). However, the work of Turner and colleagues is of interest here because it demonstrates the strong dependence of the mechanical properties of cancellous bone on its structural density and architectural organization. We now have the potential to infer the mechanical properties of cancellous bone in fossil mammals, at least, and refinements to the stereological procedures should only serve to improve the probability of accurate inference.

Computer simulations of bone's response to stress

When Wolff and others devised the trajectorial theory in the last century, contemporary engineering theory allowed them to compute only static stress patterns in shapes that were not particularly accurate representations of bones (Figure 15.3). In the last few years the sophistication of finite-element techniques has allowed the development of computer models that simulate bone's response to stress.

Carter and his colleagues at Stanford have devised a series of finite-element models that progressively incorporate more complex external forces, experimentally supported assumptions on how stresses are integrated over time, cyclic changes in stress levels (as happen in normal daily activity), and relationships of mechanical properties to bone architecture similar to those proposed by Turner (1992). These models simulate the effects of stress on bone and joint formation and adaptive remodeling at all ontogenetic stages (reviewed in Carter, Fyhrie, & Whalen [1987] and Carter, Wong & Orr [1991]). The simulations of adaptive remodeling are of particular interest here.

Recent models of the proximal third of the human femur (Figure 15.7a) are loaded by three pairs of forces on the head and greater trochanter to represent estimates of forces during difference common activities (Carter, Fyhrie & Whalen 1987; Fyhrie & Carter 1990). Each pair of forces is applied repeatedly, according to observations of the number of cycles usually occurring daily in each activity. The stress patterns generated in the bone are summed for each loading and integrated over the total number of cycles to give a factor termed the effective stress. Initially stresses are calculated by assuming the structure has uniform density which is then increased in areas of high stress, and reduced where stress is low, according to the power law relationship (Carter & Hayes 1977) between Young's modulus and apparent density. The "remodeled" model is iteratively loaded a number of times.

After 30 cycles (Figure 15.7c), the variation in "apparent density" of the model corresponds very closely to that in the femur. The model has the highest density in areas corresponding to the compact bone of the shaft and neck. It has a high density band running obliquely through the head, corresponding to a dense band of trabeculae in the real bone. Areas of lower density occur in the greater trochanter, and in the neck, the latter corresponding to a region known as Ward's triangle in which trabecular density is reduced.

Other models developed by Carter and his students predict predominant orientations of trabecular density in the femur, and the distribution of cartilage and cancellous and compact bone under joint surfaces (Carter, Orr, & Fyhrie 1989). Together, these models demonstrate the power of finite-element analyses in simulating adaptive remodeling in bone. Other workers have used finite-element methods to simulate adaptive remodeling (Huiskes et al. 1987), and there is some debate as to how the response of bone to stress should be incorporated in each model. This debate emphasizes the difficulties in interpreting the complex remodeling process, but does not invalidate the finite-element approach.

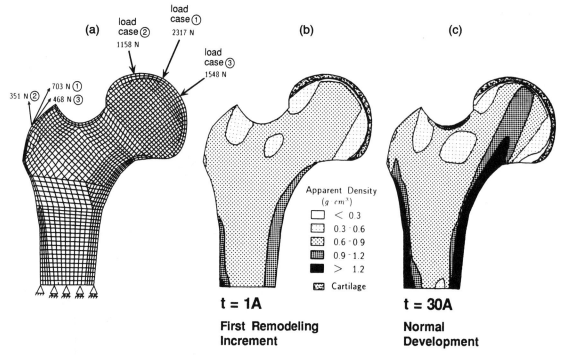

Figure 15.7. (a) Finite-element model of the proximal end of the human femur showing the three load cases that are repeated cyclically. Structural density of model is uniform. (b) After one round of loading, structural density of the model is reduced in areas of low stress and increased where stress in high. (c) After 29 more iterations, the density distribution of the model quite closely resembles that in the real femur. (From Carter et al. 1991. Used with kind permission from Pergamon Press, Ltd.)

The most exciting model from my perspective here is one where the calculations were done both ways: first, to "remodel" the bone in response to a pattern of applied forces, and second, to integrate the derived structure in an attempt to reconstruct the force pattern (Fischer, Jacobs, & Carter 1993). This work is aimed at providing a way to reconstruct in vivo loads on bones from cadaveric specimens, exactly the goal we are trying to reach for fossil bones. The models used to test this two-way process of loading-to-structure-to-loading are rudimentary. But I suspect it will not be long before they are sufficiently sophisticated to allow the prediction of force patterns from real specimens. At that point, the value of bone architecture as an information source in the functional reconstruction of fossils will be exponentially enhanced. Given the level of engineering sophistication in the finite-element models, I do not expect to see immediate application of the methods to paleontological problems. But the recent collaboration of Carter's group with Kevin Padian (Carter, van der Meulen, & Padian 1992) is encouraging for future interactions. Forces exerted on the bone during life are complex, and such detailed information is not supplied by any other method of biomechanical reconstruction.

METHODOLOGY OF BIOMECHANICAL ANALYSES OF FOSSIL VERTEBRATES

I view biomechanical analysis of fossil vertebrates in terms of first assessing the possible range of function, and then refining that estimate, using all available combinations of observation and inference, to arrive at as narrow a range of probable functions as is possible. This is akin to the principles of achieving consilience discussed by Padian (this volume): Inferences are combined to arrive at a consensus. In biomechanical studies, however, a unique consensus is unlikely: Bone function cannot be reduced to "the resistance of a single force." I picture the results of a comprehensive biomechanical study as a kind of Venn diagram of intersecting sets of probable ranges in force within a superset of the possible range. The most inclusive intersection of sets defines the narrowest probable range, or a region where consilience is most closely approached. The methods of arriving at each probable set were outlined in the introduction and in the section on cross-sectional geometry. They involve procedures such as weight estimates, muscle reconstruction, and bone-strength calculations. Patterns of force inferred from internal architecture would constitute another such set, but one that might be more heavily weighted as it would reflect actual loadings.

A truly comprehensive biomechanical study of a given fossil bone, limb, or skeleton would become quite a monumental task. Some of the combinations of observation and inference are fairly major projects in themselves, particularly those based on muscle reconstruction.

CONCLUSIONS

My main conclusion is that bone structure does encode information of potential use in reconstructing patterns of force that probably acted on fossil bones in vivo. The nature of the code is sufficiently complex and noisy to have precluded all but a few attempts to extract loading information (beam theory analyses of long bones being the exception). Recent advances and continuing work in the area of the trajectorial theory indicate that methods will be available in the next few years to permit the inference of the major force pattern acting on a bone in vivo from its internal architecture. At that time functional morphology of vertebrate fossils will acquire a powerful tool for biomechanical reconstruction.

ACKNOWLEDGMENTS

Discussions with J.E.A. Bertram and A. Biknevicius helped formulate the ideas in this chapter. Thanks are due to C. McGowan, A. Biknevicius, R. MacDougall, D.R. Carter, and C.H. Turner for reading and commenting upon the manuscript.

REFERENCES

Alexander, R. McN. 1989. *Dynamics of Dinosaurs & Other Extinct Giants.* New York: Columbia University Press.

Barnes, G.R.G., & Pinder, D.N. 1974. In vivo tendon tension and bone strain measurement and correlation. *Journal of Biomechanics* 7, 35–42.

Bertram, J.E.A., & Biewener, A.A. 1988. Bone curvature: sacrificing strength for load predictability? *Journal of Theoretical Biology* 131, 75–92.

Bertram, J E.A., & Biewener, A.A. 1990. Allometry and curvature in the long bones of quadrupedal mammals. *Journal of Zoology, London* 226, 455–467.

Bertram, J.E.A., & Swartz, S.M. 1991. The "Law of bone transformation": a case of crying Wolff? *Biological Reviews* 66, 245–273.

Biewener, A.A. 1982. Bone strength in small mammals and bipedal birds: do safety factors change with body size. *Journal of Experimental Biology* 98, 289–301.

Biewener, A.A. 1983. Locomotory stresses in the limb bones of two small mammals: the ground squirrel and chipmunk. *Journal of Experimental Biology* 103, 131–154.

Biewener, A.A., Swartz, S.M., & Bertram, J.E.A. 1986. Bone modeling during growth: dynamic strain equilibrium in the chick tibiotarsus. *Calcified Tissue International* 39, 390–395.

Biewener, A.A., Thomason, J., Goodship, A., & Lanyon, L.E. 1983. Bone stress in the horse forelimb during locomotion at different gaits: a comparison of two experimental methods. *Journal of Biomechanics* 16, 565–572.

Biknevicius, A.R. 1993. Biomechanical scaling of limb bones and differential limb use in caviomorph rodents. *Journal of Mammalogy* 74, 95–107.

Biknevicius, A.R., & Ruff, C.B. 1992. Use of biplanar radiographs for estimating cross-sectional geometric properties of mandible. *Anatomical Record* 232, 157–163.

Boyde, A., & Riggs, C.M. 1990. The quantitative study of the orientation of collagen in compact bone slices. *Bone* 11, 35–39.

Bryant, H.N., & Russell, A.P. 1992. The role of phylogenetic analysis in the inference of unpreserved attributes of extinct taxa. *Philosophical Transactions of the Royal Society of London* B337, 405–418.

Burger, E.H., Klein-Nulend, J., & Veldhuijzen, J.P. 1991. Modulation of osteogenesis in fetal bone rudiments by mechanical stress in vitro. *Journal of Biomechanics* 24, 101–109.

Carrier, D.R., & Leon, L.R. 1990. Skeletal growth and function in the California gull *(Larus californicus). Journal of Zoology, London* 222, 375–389.

Carter, D.R., Fyhrie, D.P., & Whalen, R.T. 1987. Trabecular bone density and loading history: regulation of connective tissue biology by mechanical energy. *Journal of Biomechanics* 20, 785–794.

Carter, D.R., & Hayes, W.C. 1977. The compressive behavior of bone as a two phase porous structure. *Journal of Bone and Joint Surgery* 49A, 954–962.

Carter, D.R., Orr, T.E., & Fyhrie, D.P. 1989. Relationships between loading history and femoral cancellous bone architecture. *Journal of Biomechanics* 22, 231–244.

Carter, D.R., van der Meulen, M.C.H., & Padian, K. 1992. Historical and functional factors in the construction of pterosaur bones: a preliminary analysis. *Journal of Vertebrate Paleontology* 12 (supplement), 21A.

Carter, D.R., Wong, M., & Orr, T.E. 1991. Musculoskeletal ontogeny, phylogeny, and functional adaptation. *Journal of Biomechanics* 24, 3–16.

Cavagna, G.A. 1985. Force platforms as ergometers. *Journal of Applied Physiology* 39, 174–179.

Cochran, G.B.V. 1972. Implantation of strain gauges on bone in vivo. *Journal of Biomechanics* 5, 119–123.

Conroy, G.C., & Vannier, M.W. 1984. Noninvasive three-dimensional computer imaging of matrix-filled fossil skulls by high-resolution computed tomography. *Science* 226, 456–458.

Cowin, S.C. 1985. The relationship between the elasticity tensor and the fabric tensor. *Mechanics of Materials* 4, 137–147.

Currey, J.D. 1984. *Mechanical Adaptations of Bones.* Princeton: Princeton University Press.

Currey, J.D. 1988. The effect of porosity and mineral content on the Young's modulus of elasticity of compact bone. *Journal of Biomechanics* 21, 131–139.

Currey, J.D. 1990. Physical characteristics affecting the tensile failure properties of compact bone. *Journal of Biomechanics* 23, 837–844.

Demes, B., Jungers, W.L., & Selpien, K. 1991. Body size, locomotion, and long bone cross-sectional geometry in indriid primates. *American Journal of Physical Anthropology* 86, 537–547.

Enlow, D.H., & Brown, S.O. 1956. A comparative study of fossil and recent bone tissues. Part I. Introduction, methods, fish and amphibian bone tissues. *Texas Journal of Science* 8, 405–443.

Enlow, D.H., & Brown, S.O. 1957. A comparative study of fossil and recent bone tissues. Part II. Reptilian and bird bone tissues. *Texas Journal of Science* 9, 186–214.

Enlow, D.H., & Brown, S.O. 1958. A comparative study of fossil and recent bone tissues. Part III. Mammalian bone tissues. *Texas Journal of Science* 10, 187–230.

Fischer, K.J., Jacobs, C.R., & Carter, D.R. 1993. Determination of bone and joint loads from bone density distributions. *Transactions, Orthopaedic Research Society* 18, 529.

Fyhrie, D.P., & Carter, D.R. 1990. Femoral head apparent density distribution predicted from bone stresses. *Journal of Biomechanics* 23, 1–10.

Glucksmann, A. 1941. The role of mechanical stresses on bone formation in vitro. *Journal of Anatomy* 76, 231–249.

Hahn, M., Vogel, M., Pompesius-Kempa, M., & Delling, G. 1992. Trabecular bone pattern factor – a new parameter for simple quantification of bone microarchitecture. *Bone* 13, 327–330.

Hamilton, S.J., Mehrle, P.M., Mayer, F.L., & Jones, J.R. 1981. Method to evaluate mechanical properties of bone in fish. *Transactions, American Fisheries Society* 110, 708–717.

Huiskes, R., Weinans, H. Grootenboer, H.J., Dalstra, M., Fudala, B., & Sloof, T.J. 1987. Adaptive bone-remodeling theory applied to prosthetic design analysis. *Journal of Biomechanics* 201, 1135–1150.

Joeckel, R.M. 1992. Comparative anatomy and function of the leptaucheniine oreodont middle ear. *Journal of Vertebrate Paleontology* 12, 505–523.

Kuo, A.D., & Carter, D.R. 1991. Computational methods for analyzing the structure of cancellous bone in planar sections. *Journal of Orthopaedic Research* 9, 918–931.

Lanyon, L.E. 1980. The influence of function on the development of bone curvature. An experimental study on the rat tibia. *Journal of Zoology, London* 192, 457–466.

Lanyon, L.E., & Bourn, S. 1979. The influence of mechanical function on the development and remodeling of the tibia. *Journal of Bone and Joint Surgery* 61A, 263–273.

Lanyon, L.E., & Smith, R.N. 1969. Measurements of bone strain in the walking animal. *Research in Veterinary Science* 10, 93–94.

Loeb, G.E., & Gans, C. 1986. *Electromyography for Experimentalists*. Chicago: University of Chicago Press.

Lovejoy, C.O., & Barton, T.J. 1980. A simple, rapid method of obtaining geometrical properties from sections or laminograms of long bones. *Journal of Biomechanics* 13, 65–67.

Martin, R.B. 1991. Determinants of the mechanical properties of bones. *Journal of Biomechanics* 24, 79–88.

Martin, R.B., & Burr, D.B. 1984. Non-invasive measurement of long bone cross-sectional moment of inertia by photon absorptiometry. *Journal of Biomechanics* 17, 195–201.

Martin, R.B., & Burr, D.B. 1989. *Structure, Function, and Adaptation of Compact Bone*. New York: Raven Press.

Martin, R.B., & Ishida, J. 1989. The relative effects of collagen fiber orientation, porosity, density and mineralization on bone strength. *Journal of Biomechanics* 22, 419–426.

Meyer, H. 1867. Die Architektur der Spongiosa. *Archiv für Anatomie und Physiologie* 1867, 615–628.

Murray, P.D.F. 1936. *Bones*. Cambridge University Press.

Nagurka, M.L., & Hayes, W.C. 1980. An interactive graphics package for calculating cross-sectional properties of complex shapes. *Journal of Biomechanics* 13, 59–64.

Piotrowski, G., Sullivan, M., & Colahan, P.T. 1983. Geometric properties of equine metacarpi. *Journal of Biomechanics* 16, 129–139.

Piziali, R.P., Hight, T.K., & Nagel, A. 1980. Geometric properties of human leg bones. *Journal of Biomechanics* 13, 881–885.

Reid, R.E.H. 1985. On the supposed Haversian bone from the hadrosaur *Anatosaurus*, and the nature of compact bone in dinosaurs. *Journal of Paleontology* 59, 140–148.

Rowe, T., & Carlson, W. 1992. Digital map of the skull of *Thrinaxodon liorhinus*: high resolution CT imagery published on CD-ROM. *Journal of Vertebrate Paleontology* 3 (supplement), 49A.

Rubin, C.T., & Lanyon, L.E. 1984. Regulation of bone formation by applied dynamic loads. *Journal of Bone and Joint Surgery* 66A, 397–402.

Ruff, C.B. 1987a. Sexual dimorphism in human lower limb bone structure: relationship to subsistence strategy and sexual division of labor. *Journal of Human Evolution* 16, 391–416.

Ruff, C.B. 1987b. Structural allometry of the femur and tibia in Hominoidea and *Macaca*. *Folia Primatologica* 48, 9–49.

Ruff, C.B. 1989. New approaches to structural evolution of limb bones in primates. *Folia Primatologica* 53, 142–159.

Ruff, C.B., & Leo, F.P. 1986. Use of computed tomography in skeletal structure research. *Yearbook of Physical Anthropology* 29, 181–196.

Rybicki, E.F., Mills, A. S., Turner, A.S., & Simonen, F.A. 1977. In vivo and analytical studies of forces and moments in equine long bones. *Journal of Biomechanics* 10, 701–705.

Schaffler, M.B., & Burr, D.B. 1985. Primate cortical bone microstructure: relationship to locomotion. *American Journal of Physical Anthropology* 65, 191–197.

Schaffler, M.B., Burr, D. B., Jungers, W.L., & Ruff, C.B. 1985. Structural and mechanical indicators of limb specialization in primates. *Folia Primatologica* 45, 61–75.

Simon, B.R., ed. 1980. *Finite Elements in Biomechanics*. Tucson: University of Arizona.

Smith, M.M., & Hall, B.K. 1991. Development and evolutionary origins of vertebrate skeletogenic and odontogenic tissues. *Biological Review* 65, 277–373.

Snure, H. 1924. A roentgen-ray study of the La Brea (California) fossils. *American Journal of Roentgenology* 11, 351–354.

Thomason, J.J. 1985a. Estimation of locomotory forces and stresses in the limb bones of Recent and extinct equids. *Paleobiology* 11, 209–220.

Thomason, J.J. 1985b. The relationship of trabecular architecture to inferred loading patterns in the third metacarpals of the extinct equids *Merychippus* and *Mesohippus*. *Paleobiology* 11, 323–335.

Tobien, H. 1964. Paläontologische Forschungen mit radiologischen Methoden. In *Deutscher Röntgencongress 1963*. Stuttgart: Georg Thieme Verlag.

Turner, C.H. 1992. On Wolff's Law of trabecular architecture. *Journal of Biomechanics* 25, 1–9.

Turner, C.H., Cowin, S.C. Rho, J.Y., Ashman, R.B., & Rice, J.C. 1990. The fabric dependence of the orthotropic elastic constants of cancellous bone. *Journal of Biomechanics* 23, 549–561.

Underwood, E.E. 1970. *Quantitative Stereology*. Reading, MA: Addison-Wesley.

Varricchio, D.J. 1993. Bone microstructure of the Upper Cretaceous theropod dinosaur *Troodon formosus*. *Journal of Vertebrate Paleontology* 13, 99–104.

Weibel, E.R. 1979. *Stereological Methods*. London: Academic Press.

Whitehouse, W.J. 1974. The quantitative morphology of anisotropic trabecular bone. *Journal of Microscopy, Oxford* 101, 153–168.

Wolff, J. 1870. Ueber die innere Architektur der Knochen und ihre Bedeutung für die Frage vom Knochenwachstum. *Virchows Archiv für pathologische Anatomie und Physiologie* 50, 389–450.

Wolff, J. 1892. *Das Gesetz der Transformation der Knochen*. Translated as *The Law of Bone Remodelling*, 1986, P. Maquet & R. Furlong. Berlin: Springer-Verlag.

Wong, M., & Carter, D.R. 1988. Mechanical stress and morphogenetic endochondral ossification of the sternum. *Journal of Bone and Joint Surgery* 70A, 992–1000.

Woo, S.L.-Y., Kuei, S.C., Gomez, M.A., Hayes, W.C., White, F.C., Akeson, W.H., & Amiel, D. 1981. The effect of prolonged physical training on the properties of long bone: a study of Wolff's Law. *Journal of Bone and Joint Surgery* 63A, 780–787.

Zienkiewicz, O.C. 1971. *The Finite Element Method in Engineering*. New York: McGraw-Hill.

16

Form versus function: the evolution of a dialectic

KEVIN PADIAN

ABSTRACT

The relationship of form to function in the interpretation of animal design is a question that can be traced back to Aristotle in the Western tradition. The dialectic, if it can be called that, between how form constrains function and how function influences form, appears in the writings of major biologists from the fifth century BC to the Victorian Era. It is often couched as a consequence of how strongly one interprets the determination of a Prime Mover to create a perfect (immanent) world, as opposed to the plasticity of indeterminate biological material to yield to ecological pressures and exigencies.

Darwin and Wallace shifted the theater of this debate to a materialistic world in which Prime Movers were less important or less interesting than the secondary causes of functional needs or historical legacies. In the following years, particularly through most of the twentieth century, the study of function was centered on adaptationism and lacked a strong historical context. The advent of cladistics provided this context, but the search for pattern has often obscured the processes of evolution or found them uninteresting. Recent works in comparative biology stress the use of all potentially informative lines of evidence, working closely within a phylogenetic framework, to test hypotheses, and such works can be regarded a priori as having decisive influence over others. Otherwise phylogenetic hypotheses are not tested by any means but internal phylogenetic evidence, and functional hypotheses have no independent content if they must stand or fall based on a given phylogeny.

INTRODUCTION

The question of how form relates to function concerns not only vertebrate paleontology, but biological science through history. Aristotle devoted several books to it (*De Partibus Animalium, De Motu Animalium, De Incessu Animalium*). Through the ages, perspectives on form and function have varied widely, according to philosophical and cultural considerations in Western thought. We are no less prisoner of these considerations today than other scientists have been in the past, and it is of increasing interest to historians of science to understand not just how our present knowledge evolved, but what scientists in the past have thought and what influenced their thinking. How has the question of the relation of form to function changed through time, and what have been the principal historical and philosophical influences on it? This is a complex question because form and function have not always been seen in the same terms that they are today. Without reading too much into the preoccupations of past scientific communities, an acknowledgment of some of their influences may help to shed light on why we frame the terms of this argument as we do today.

In this chapter I want to look at some historical patterns in the analysis of form and function. Some treatments of this analysis have very long histories; some historical periods have movements that build on what has come before; and some seem to break tradition completely. I will contend that the Judaeo-Christian and Classical Greek traditions, despite many similarities and integrations, maintained some fundamental distinctions that informed functional morphology into the Victorian Era. The principal break with these traditions came with the advent of Darwin's theory, which brought with it a materialism that marginalized their essentialist and transcendental approaches. The neo-Darwinian Synthesis of the 1930s and later decades constructed an extrapolationist view of macroevolutionary, and hence functional, change based on processes seen at populational levels. Phylogenetic systematics shifted the emphasis from process to pattern, and current approaches to the evolution of form and function have advocated rooting hypotheses of functional evolution in phylogenetic terms. I argue that functional hypotheses should be proposed and tested independently of phylogenetic ones, but that the robusticity of a complex historical scenario depends on consilience of independent lines of evidence. For vertebrate paleontologists in particular, neontological studies of form and function can be complemented by both transformational studies of fossil lineages and morphogenetic mechanisms known in living organisms.

Central to the present argument is the fact that modern evolutionary biology has its roots in the late eighteenth and early nineteenth centuries in France and England. For Anglo-American science in particular, the early Victorian Era serves as a nexus of ideas and traditions that both culminated biological thought and germinated new growth (Russell 1916; Hull 1973; Desmond 1982, 1989; Desmond & Moore 1992; Sloan 1992). The issues of Victorian science were of course different from those of today; so were the issues and standards of philosophy of science, and even the meanings of many scientific words, including "evolution." But Victorian biology spawned many other contemporary concepts besides Darwinism.

Today we think of the diversity of form and function largely as a dynamic between the demands of ecology and the potential of phylogenetic history. We recognize that form and function have evolved through time via the transmutation of species. But in the absence of this working understanding, as in pre-Darwinian biology, how was the relationship of form to function considered?

Several recent reviews have focused on the question of form and function, particularly as it relates to current approaches, including those of functional morphology, biomechanics, theoretical morphology, and phylogenetically entrained studies of evolutionary morphology (Lauder 1981, 1982, 1991; Hickman 1988, 1991; Rieppel 1988, 1990; D. Wake 1982; Liem & D. Wake 1985; Lauder & Liem 1991; D. Wake & Roth 1991; M. Wake 1991, 1992; see especially what M. Wake 1992 calls "integrative evolutionary morphology"). For the most part I will avoid duplication of the issues discussed in these works, but hope that they will be seen as necessary complements to this chapter. For example, holistic morphology, constructional morphology, bioengineering, and the question of whether form and function have any predictable correspondence are all fascinating and crucial to functional morphology, but in this chapter I want to concentrate on the historical and historicistic approaches to form and function.

THE JUDAEO-CHRISTIAN AND CLASSICAL GREEK TRADITIONS

Until a generation or two ago a classical education was central to the training of any person who was expected to make a mark on Western society. It is not only important that the history, culture, literature, language, and philosophy of the Ancients were inculcated into centuries of students: These traditions had the import of morality and authority that shaped the thinking of the intellectual oligarchy. Some of these influences are obvious, but others are more subtle. These traditions influenced how some biological questions were framed, and why some questions were never asked, or satisfactorily answered, at all.

There are many similarities between the Judaeo-Christian and Classical Greek intellectual traditions, many of which were forged for Western culture during the period of Medieval Scholasticism. As Will Provine (1982, p. 501) remarks, "The Greek and Judaeo-Christian heritages share a fundamental assumption: that the physical world and human ethics could not possibly have arisen through the sheer mechanical workings of things," and that in both traditions nature and ethics are inseparable. As fundamental as Reason and Necessity are to Design in both cultures, there is also a considerable difference between the two traditions of dealing with the natural world, and hence with biological questioning. This difference has significantly informed the basic issue of morphology, the dynamic between form and function. The difference, in short, is that the Aristotelian tradition presumes that there are laws governing the expression of biological diversity, as there are governing the phenomena of physics, chemistry, astronomy, and so on. The Judaeo-Christian tradition (as we see it particularly in the Old Testament), in contrast, presumes that an active Divinity shapes not only our ends, but also the individual forms of organisms and many of the individual events of life. The difference is partly one of emphasis on secondary or primary causes.

The Judaeo-Christian epic called the Bible begins with divine action. God in the first verses of Genesis sets in motion the parts of the universe and its life.[1] The laws that structure the natural world are not explained, nor is there any attempt in the Hebrew tradition to understand them rationally. With hindsight, we can cast Adam as the first taxonomist, Noah as the first conservationist of endangered species, and Jonah as an accomplished amateur of cetacean studies, but this is caricature. There is little in the Bible of interest in nature for its own sake (except in some of the Psalms), nor even as a way of understanding God. Knowledge, in the Bible, comes through revelation. The generalizations are moral and spiritual, not philosophical or empirical. The Bible gives ethical *rules* but not natural *laws*. Scientists and natural philosophers are conspicuously absent from the Bible; the notable protagonists are rulers of men and men of God. The Judaeo-

Christian tradition, rooted in the Old Testament, stresses revelation, divine guidance, and individual explanations for natural phenomena. In the Old Testament there is little sense of natural system or predictability apart from the cycle of seasons. Floods, pestilences, and droughts are commonly seen as acts of God that are often difficult to understand. The Bible functions as a fundamental moral, historical, and ethical vessel of Hebrew culture, and these, after all, were its intended purposes.

The classical Greek tradition, in the elements that survived to become a standard part of European education, is quite different from that of the Judaeo-Christian tradition. Plato's Demiurgus can be seen as a model for the Judaeo-Christian God, but the latter is far more personified and active. The Greek philosophical tradition, following the lead of Plato, Aristotle, Theophrastus, and others, assumed that the operation of the universe is lawful, and that therefore generalizations can be made about such matters as the creative tension between function and form. In Aristotle's sense, of course, both have a role, as he explores (in *De Partibus Animalium* and elsewhere) the relationship of function to structure, correlation of parts, comparative morphology, and relative imperfection of organs in a series. These ideas invested the philosophy of later scholars, ranging from Cuvier's biological theory to the arrangement of specimens and organs in John Hunter's anatomical demonstration collection, which later became the centerpiece of the Royal College of Surgeons. There is clearly a sense of system, of order, and of lawfulness in this Classical approach to understanding nature.

The theology of ancient Greece, by contrast, functions in the ethical, moral, and historical senses more as metaphor or parable, by exploring the lives, actions, and fortunes of half-legendary people and gods, assembled from the local traditions of many cultures that came under the aegis of Greek civilization (Hamilton 1940). In this way it holds much in common with Judaeo-Christian tradition. Members of the Greco-Roman pantheon had foibles and failures that ordinary humans could relate to even as they feared them. In the Bible, Adam and Eve were commanded not to eat from the Tree of Knowledge, and were punished for their transgression. Likewise, Prometheus was punished for giving humans fire (a most important tool of knowledge), and on another occasion for tricking Zeus into agreeing that the worst, not the best, parts of animals should be sacrificed to the gods (Zeus's retaliation was to create Pandora and arrange the assembly of the contents of her infamous box). So in both cultures the authoritarian theistic tradition can be seen (I hope not too elliptically) to conspire to withhold knowledge from humans and to teach them to accept divine

acts and commands unquestioningly (the Book of Job provides further examples). Nevertheless, the Greek culture also boasts an exceptional rationalist tradition in natural philosophy, the counterpart of which (if it ever existed) in ancient Hebraic culture has long since been lost. There seems to be no a priori reason why it should have been in one tradition and not the other, apart from historical accident, the aleatory element of human culture.

Aristotle's writings on form and function, the apotheosis of Greek rationality in biological philosophy, emphasize the plasticity of shape in the service of function (Russell 1916; Rieppel 1990). While recognizing the form of the animal as a final cause or purpose, he understands what we would today call adaptations, and has the first sense of what we would call homologous and analogous parts. He recognizes the functions of different organs and how they vary among animals according to use. He orders the structures of organisms as relatively perfect or imperfect according to whether their organs have a full, excessive, or defective development. And he shows that excessive development of one organ system (e.g., horns in some bovids) is accompanied by less perfect development in another system (e.g., reduction in the number of teeth). He clearly sees patterns in these structures and functions in animals, but resists a Platonic idealization of these patterns. His concepts of the *genos* and *eidos* are (more so than for Plato) practical guides to discovering the essence of a taxon by listing its characteristics; but they are not taxonomic criteria of definition or diagnosis in the sense that we would use them today (see Rowe 1987). Overall, Aristotle is both a formalist and a functionalist: He knows ideals and essences, but sees as interesting the ways in which form is harnessed to material exigencies, to be shaped by function. The organism is a product of both formal and final causes, and so its "purpose" is teleological as well as material.

INTELLECTUAL TRADITIONS OF THE VICTORIAN ERA

It may seem a large jump from pre-Christian cultures to the nineteenth century of England, but in fact between the fourth century B.C.E., when Aristotle wrote, and the 1800s C.E. we have not skipped much of crucial import to the question of form and function. Certainly the medieval Christian scholastics were the first to fuse the rationality of Greek thought with the veneration of Judaeo-Christian tradition, and so they should not be ignored. In particular, the Scholastics asked rational (Platonic) questions about the relationship between divinity (particularly in the Christian sense) and the natural world. They sought explicit means of interpreting Nature in terms of a Divine rationality, a point of view that extended to Owen as much as to Linnaeus. But two points are relevant here. First, scholastic rationality required the rediscovery of a classicism that would have been almost completely lost from Western tradition if not for its importation and extensive critical commentary by scholars from the Near East. And second, the rationality of seeing the world as a *lawful* extension of God's will extended through physics and chemistry, but not (in most senses) to biology. This made the philosophy of biology anomalous because, as Alfred North Whitehead (1926, p. 21) has said, the Middle Ages formed one long training of the intellect of Western Europe in the sense of order. But in a deeper sense, Linnaeus's pigeonholing of natural diversity brought little in the way of theoretical advancement or even reference, though it has often been hailed as a milestone in ordering nature according to the rational plan of God. This most enduring of pre-Darwinian natural systems – in contrast to some contemporary functionalist or utilitarian classifications – had nothing to say about the relationship of form to function; scholars still turned to Aristotle for illumination on those points. Moreover, it is difficult to demonstrate the direct influence of Scholasticism on the terms of morphological debates in the Victorian Era, even though some of the arguments are very similar.

More relevant to our central question is the development of German *Naturphilosophie* and transcendentalism. German philosophers such as Goethe, Kant, and Schelling revived the concept of transcendental forms that could be sought rationally by studying the material world. Goethe appears to have been the first to articulate the vertebral theory of the skull (Russell 1916; Rieppel 1988), along with Carus, Spix, Kielmeyer, and Oken (among many others); and the idea that shared ontogenetic characters implied formal relationships among different animals was developed by Oken, Meckel, Carus, Haeckel, and later Von Baer, whose work was of some influence on Darwin (R. Richards 1992). Kant's rationalism was very much in vogue among the Scottish school of anatomists and morphologists in early Victorian England, notably Robert Edmond Grant (Desmond 1982, 1989) and Joseph Henry Green (Sloan 1992), who studied extensively in Germany.

Transcendentalism brought a revival of Platonic concepts, and thus a resurgence of Greek philosophical tradition, to the study of morphology. In the Platonic world, material things are merely manifestations of the ideal, which represents the true reality. What is real can be known only through pure logic; empiricism can only generate hypothetical knowledge. The ideal can be analyzed by studying its empirical manifestations and their patterns. The rational mind can then organize and sift through these patterns to come to an understanding of the ideal, which is not evident from empiricisms alone. In this sense the organisms of the material world are variations on an ideal theme, an ideal form altered by functional and ecological habits. Sloan (1992, p. 29) quotes a revealing passage from an early paper of Kant's in which the philosopher compares invidiously the taxonomy of Linnaeus with the evolutionary phylogenetic scheme of Buffon: "One produces an arbitrary system for the memory, the other a natural system for the understanding [*Verstand*]. The first has only the intention of bringing creation under titles; the second intends to bring it under laws." The rationalist viewpoint stresses that it is not enough to see patterns; one must look beyond them to the ideal. In morphology, then, functional exigencies disguise ideal form, and part of the business of morphology is to discover archetypal form.

Before finally turning to Victorian biology, it is necessary to acknowledge the crucial development of ideas about form and function in France, contributed by Buffon, Lamarck, Cuvier, and Geoffroy St.-Hilaire. The complexity of these debates in both scientific and political terms has been explored by Burkhardt (1977), Appel (1987), Rieppel (1990), and others; these works provide a much fuller perspective than can be given here. From a Darwinian standpoint, the French debates of this era are often seen as a chronicle of nascent transmutationism, with Buffon and Lamarck frankly discussing transitions among species, and Cuvier staunchly resisting interspecific change, emphasizing the hegemony of function over form even as Geoffroy advocated the opposite. On closer inspection the views of each scholar were much less simple than this picture would imply, and fraught with what could be seen as internal inconsistencies from a later, Whiggish viewpoint. But a clear line of difference was drawn between the functionalist ideas of Cuvier – who followed Aristotle in his concepts of correlation of parts, excess and defect, and the subjugation of form to function – and Geoffroy, who emphasized the transcendental properties of form over function.

Debates about morphological form that took place in France and Germany were reflected in the complex world of Victorian biology in England and Scotland (Desmond 1982, 1989; Sloan 1992; Desmond & Moore 1992). In Victorian England all these currents of thought found their apotheosis in Richard Owen who, as E.S. Russell so clearly perceived, embodied so many of these traditions and put his own stamp on so many of these questions.

Russell's (1916) elegant opening sentence in his masterful chapter on Owen notes that "Owen is the epigonos of transcendental morphology; in him its guiding ideas find clear expression, and in his writings are no half-truths struggling for utterance."[2] But Owen was also pragmatic, as Russell pointed out, and in an eclectic way typical of great synthesizers. Owen took what was best or most workable from debates in France and Germany – the vertebral theory from Goethe and Carus, the sense of homology and analogy from Cuvier, the sense of archetypal form from Buffon, Aristotle, and the German transcendentalists, and so on. But he always had a new twist on past formulations, and always the ideas were made to fit his very complex (often Procrustean) and still largely obscure ideas about morphology and evolution. Owen dominated the Victorian biological scene, as Desmond (1982, 1985, 1989) has argued, partly because his ideas so well fitted the status quo of society, politics, and philosophy embodied in the ruling Tory class in England.

I would like to turn the argument to slightly different terms: that Owen's biology (and he is not unique in this, only predominant in his field and time) represents a new kind of synthesis of the two great traditions that dominated Western thought – the Judaeo-Christian and the Classical. The synergy and tension between its two major philosophical elements (revelation and rationalism) in large part structured the terms of the study of form and function in the nineteenth century, particularly in England.[3]

The Classical tradition is based on the search for lawful order in nature, with the presumption that there are absolutes, ideals of morphological understanding elucidated by natural laws. This tradition was congenial to early nineteenth-century naturalists, as it had been to medieval Scholastics, because they saw the need to bring biology into line with other sciences such as physics and chemistry (Hull 1973; Sloan 1992). The lawful operation of nature, that is the understanding of the mechanics that proceeded from primary causes, was the principal philosophical focus of all the sciences at that time. And it is easy to see that the question of form and function would be investigated in terms of overarching laws. Owen's discussion of the vertebral theory, which he revised and championed, was just such an investigation. Though coy about the mechanics by which form was altered among species, Owen saw morphology as an elaboration of ideal principles and forms regulated by the laws of structure. The vertebral theory can be seen as an example of serial homology argued from the principle of the Archetype, an ideal and frankly neo-Platonic construct (Padian 1985a; Rieppel 1988; Sloan 1992).

Judaeo-Christian tradition also contributed certain terms to Victorian treatments of form and function. Principally, I would argue, it contributed the resurgence of an active and omnipotent Deity, who was to be understood by recourse to his works: the tradition in early nineteenth-century England of "argument from design," reading the mind of God through what He has wrought. Primary examples of this literary genre include William Paley's *Natural Theology* (1802), standard issue for Cambridge undergraduates from the early 1800s until at least the 1920s, and a great influence on Darwin's thought and rhetoric (Desmond 1989: pp. 403–406; Gould 1993); the *Bridgewater Treatises* (or "Bilgewater Treatises") of the 1830s; and Gideon Mantell's (1844) *Medals of Creation.*[4] Many other examples could be listed.

In the literature of natural theology, which survives today among American creationists as the "argument from design" (e.g., Bird 1989; Davis, Kenyon, & Thaxton 1989), it is not material laws alone, but the individuation of natural phenomena, the sense of a divine Primary Cause putting all the laws into motion, creating all the matter in the universe, and individually structuring the forms of life, that the Judaeo-Christian tradition contributed. In this world-view there is a natural sense of uncertainty, and of unpredictability, based on the difficulty of reading the mind of God.[5] Accordingly, the understanding of natural laws (i.e., secondary causes) is all the more difficult. It is given that God created and runs everything; but it is equally given that the secondary laws, the mechanics, cannot be reduced to random processes. This explains why John Herschel, the foremost Victorian philosopher of science, referred to natural selection as "the law of higgledy-piggledy," a slight that crushed Darwin (Hull 1973).

It is within this tradition that Richard Owen can search for laws that work according to God's (or his) plan. Yet these laws appear to lose their Unity of Plan in some cases, only to turn up in the guise of another natural law in another place. For example, one of the things that frustrated Seeley so much about Owen's (1870) description of the pterosaur *Dimorphodon* (Padian 1980, 1987b) is that Owen constantly compared the features of this so highly derived animal to those of lizards, amphibians, and even fishes. Seeley (1870) was shocked that Owen did not see the obvious relation to birds, which pterosaurs resemble so uncannily in so many aspects of their anatomy (down to the ultrathin bones and pneumatic foramina). But Owen was guided by different principles. Having established that pterosaurs were reptiles, he was able to postulate all the physiological correlates of reptilian anatomy, including cold-bloodedness (Desmond 1975, 1982; Padian 1980, 1987b). The other features, Owen proclaimed, were "purely adaptive." In other words, the taxon gives the characters; the applications of the laws of chemistry and physics, at least in their biological manifestations, are subservient to Archetypal design. The alteration of form to serve a particular function is merely expedience, because the essential characters of the form persist.

To us, Owen's principles often seem to be invocable on demand, but in fact, as the above example shows, he was highly consistent, though his priorities differ from today's. To be fair, Darwin was seen in his time to argue in equally slippery terms about the interplay between what we would now call adaptation and historical constraint. No one gets upset about that now, but they did in 1859, as the reception of Darwin's magnum opus shows (Hull 1973; Glick 1988). The point here is that Owen was constantly wrestling with a morphological tradition that had to serve two masters: the search for natural laws and the search for a divine plan. To the extent that he was able to balance these two (often tortuously disparate) traditions, he did it better than nearly anyone, as his scientific and political success shows. But Owen himself had too many masters to serve, both politically and scientifically, as Desmond (1982, 1989) has shown. I would further suggest that this is one reason why Owen never fully explicated his theories of morphological evolution. There was a broad spectrum of religious and philosophical thought among the Tories who sponsored Owen's scientific and personal advancement (Desmond 1989; Sloan 1992), some of it quite conservative. Owen had seen what happened to ideas that were too venturesome for some clerics (e.g., Chambers' 1844 anonymous *Vestiges of Creation*, with which Owen broadly sympathized; E. Richards 1987), and he wanted to take no chances. Darwin was left having to answer only criticisms of his own work, instead of having to respond to an articulated model of organic design that would have been at least equally plausible to the Victorian (especially Tory) sensibility. Owen never rose to this challenge, which his supporters repeatedly urged him to meet (Desmond 1989).

The failure of transcendental anatomists to articulate the form–function relationship according to any but the most abstract principles (without serious empirical errors, as Cuvier and the young Owen showed) allowed Darwin and his allies to step in and shatter the traditional terms of the dialectic. It is likely that this was entirely unintentional, merely a by-product of the paradigm shift of issues that were Darwin's home turf. Darwin and most of his correspondents were not primarily interested in traditional debates about morphology: They were investigating the patterns of differentiation within and among species according to geography and artificial selection.[6]

What shattered both the classical, transcendental, Archetype/Bauplan-oriented tradition, and the individuating, idiographic, divinely directed tradition of Judaeo-Christian "natural theology" was a theory that only metaphorically saw a guiding hand in nature. Natural selection was empirical, mechanistic, and historically based. It is easy to see why it was anathema to Owen and his allies.

First and foremost, Darwin's theory was materialist and nonessentialist. It presumed no abstract ideals. For this reason beyond all others, Owen hated it. It denied the Unity of Plan that Owen saw in the Archetype; his transcendental entities bore no relevance to Darwin's mechanics of nature. For Darwin, the "Unity of Type" was explained by the unity of descent, modified by the conditions of life.

Second, the role of the Deity was uncertain or even irrelevant in Darwin's world. At most, in Darwin's (1859) self-styled "truckling" passages, God is restricted to the role of a Creator who set the laws of the universe in motion. This was not new to physicists, who had long since made peace with the clerics on that score (Hull 1973; Sloan 1992). The biological phenomena of the universe were now no less subject to the operation of laws than the physical phenomena. This implied that God did not individuate entities in biology; that is, there were not separate laws and plans for each group of organic beings. So, for example, it is telling that in the years before the *Origin of Species* was published, Owen (e.g., 1846, 1849) was permitted to expound on the *vertebrate* Archetype without having to answer why Archetypes did not equally apply to the invertebrate phyla, as well as to other organic beings. God could have created a vertebrate Archetype in Owen's world, but not in Darwin's.

Third, Darwin's theory is historically based: It implies a central role for genealogical history, leaving its mark on the descendants of a common ancestor. It was enough to draw Owen's rage simply to postulate that species were not fixed (which Darwin, 1859, clearly thought they were not), and could transmutate through time. But Darwin's entire theory implies a further heresy: It depends on an acceptance of vast quantities of time in order to accomplish its purpose. We may think, and Eldredge and Gould (1972) have said as much, that Darwin's theory was congenial to the Victorian mind because it implied slow and steady change through untold ages. True as this pace may be to Victorian predispositions, the depth of this time, and the renunciation of the fixity of species, posed a threat that far outweighed any comfort to the Tory audience, and even to the Whigs that numbered Darwin's family among them.

These three considerations formed a great part of the biological materialism that shocked traditional Victorian thought. They are so intrinsic to science today that it is difficult for most scientists to conceive that they were ever controversial. Yet the traditions that materialism supplanted are anything but subliminal in the early nineteenth-century British scientific literature. Darwin's answer to the question of form and function is that Natural Selection shapes the modifications of inherited features of organisms to perform new functions. These functions, in turn, are limited by historical, genealogical compromises on the possibilities of form. And yet Darwin is continually amazed in his books by the power of Natural Selection to use available biological materials to craft new adaptive solutions.

The difference is that all other explanations and primary causes have been utterly discarded: Darwin replaced the Great Traditions with a single new one, Materialism. From this point on, biologists would give almost no further attention to divine design or neo-Platonic ideals. What mattered was the mechanics of Nature. And so, at last, biology was brought onto the same terms as physics and chemistry, though contemporary philosophers did not generally see it that way (Hull 1973).

FORM AND FUNCTION IN THE MODERN SYNTHESIS

In Darwin's time, and for many decades after, limitations in the understanding of genetics and of patterns of change in the fossil record slowed progress in understanding the mechanics by which organisms change and the rate and timing of morphological change through geologic history. Interest in these issues proceeded as logical consequences of Darwin's theory, and morphological change among taxa, which could not be addressed in Owen's terms, took on important macroevolutionary overtones.[7] In the Modern Synthesis of genetics, paleontology, and population biology that began in the 1930s (Jepson, Mayr, & Simpson 1949; Mayr & Provine 1980), functional studies that involved major evolutionary changes took their general form from Simpson's pioneering works such as *Tempo and Mode in Evolution* (1944). In these cases, "major groups" were identified by their taxonomic diversity and adaptational persistence through time. (These are, of course, two different qualities, as Lauder 1981 has noted.) Simpson (1944) observed that in the early phases of radiation of a major group, form and function initially appear to be highly diverse; but soon, differential extinction of lineages follows from more successful adaptation and competition among lineages (Simpson 1944, figs. 29, 30). As a result, the transition from an original adaptive zone to a new one is relatively rapid, and the corpses of unsuccessful intermediate stages litter the fairways (Simpson 1944, fig. 35).

A closer look at Simpson's argument reveals just how much the terms of argument about morphology have shifted since Darwin's time. Darwin established the importance of small, incremental morphologic changes selected through time. The Modern Synthesis affirmed the genetic basis of that change and described how these processes were structured in natural populations. Simpson talked about macroevolutionary change through time in terms of population genetic models and statistics. In particular, he used Sewall Wright's "shifting balance" model of the "genetic landscape" (Provine 1971; Gould 1980a; Wright 1986) developed to describe population genetic changes in living animals. Simpson (1944) moved the analogy to a grander scale of the "adaptive landscape" experienced by clades evolving in macroevolutionary time, for example as Miocene horses gained deeper grazing teeth and hooves for running on plains. These features were adaptive modifications of those of their browsing, parkland ancestors. In this way Simpson's concept of the "grand analogy" follows Darwin, who drew "natural selection" from the well-known "artificial selection" of contemporary breeders and farmers. Darwin, in turn, had taken his cue from Paley (1802), who in *Natural Theology* developed an argument from design based on the analogy that, just as the complexity of a watch implies a watchmaker, so the complexity and perfection of organic design implies a Creator. Simpson was carrying on a rhetorical tradition even as he broke new ground in explaining the patterns of morphological radiation and change through time, basing morphology in population genetics.

The process by which the Neo-Darwinian patterns described by Simpson and others were typically explained took the form of a "key adaptation" evolving early in a lineage, elaborated and improved, and often coupled with other adaptive features that may or may not have been functionally related (Simpson 1944, 1953; see also some enormously influential papers by Bock 1965; Schaeffer 1965; Bock & Von Wahlert 1965), which Cuvier would have called "correlation of parts." The evolution of hooves and teeth in Simpson's horses is a good example. Toes became reduced in number and unguals enlarged; teeth became deeper and more convoluted in cusp pattern. This set of adaptations, by way of Sewall Wright's theory, was seamlessly grafted onto the evolution of grasslands during the Miocene, allowing the radiation and success of these new horses. In these scenarios there were typically no experiments or data on the genetic basis of these changes: the genetic mechanics were simply assumed to underlie large-scale morphologic change because some variation in natural populations had been demonstrated to be genetically related. Macroevolutionary trends could be reduced to the workings of infinite events of natural selection at the genetic level.

In the literature of the Modern Synthesis there is therefore a natural genetic relationship between form and function. There is no discussion of Unity of Plan or of Archetypes, and the mechanics of evolution do not discuss Divine intervention. This triumph of materialism, following Darwin's establishment of natural selection, contrasts sharply with the situation during the Victorian Era. There is a strong sense that, instead of the functional diversity of animals being ephemeral modifications of an ideal plan, form follows function. Diversification of shape according to progressive adaptation is not only the general pattern seen in the fossil record, but is also necessary to keep up with an ever-changing environment. A nearly limitless store of mutation-induced genetic variation is assumed to provide the basis for all this change (Mayr 1963). Hence shape is plastic, function is supreme. Function can be seen as the manifestation of the possibilities of form. Paleontological studies of functional evolution followed the directives of progressive improvement and selection extrapolated from populational studies of the Modern Synthesis.

PHYLOGENETIC ANALYSIS AND THE EVOLUTION OF FORM

The advent of cladistics forced a radical change from the approach of the Modern Synthesis by emphasizing the discovery of pattern above everything – certainly pattern over process, and form over function. Evolution in the phylogenetic system is documented by nested sets of shared derived characters. Functional morphology was at least initially of little use to cladistic practitioners because it concentrated so much on process and so little on pattern: The evolutionary pattern was assumed but not explicitly demonstrated in most analogical models that compared function in different animals without demonstrating phylogenetic intermediacy. In recent years, many comparative biologists (e.g., Lauder 1981, 1982, 1991, this volume; Lauder & Liem 1991; Padian 1982, 1987a, 1991; Fisher 1982, 1985; Greene 1986; Coddington 1988; O'Hara 1988) have emphasized that a precise understanding of phylogeny has to go hand-in-hand with explanations of functional evolution. This has set up an entirely new dynamic of the old question of form and function.

In some of this newer literature (e.g., Coddington 1988; O'Hara 1988), there is a strong sense that explanations of functional evolution depend on well-corroborated hypotheses of phylogeny, which must be based on many characters not involved with the functions under consideration. For the most part this division has been strictly maintained (though see Hickman 1991 for an interesting counterpoint). Bock

(1986) has recently taken a quite different point of view, arguing that fossils are too incomplete and phylogenies are not critical to studying the evolution of adaptations, and that therefore an investigator should assemble a "pseudo-phylogeny" of functionally intermediate (though not at all necessarily related) types among living animals in order to study the evolution of a particular adaptation. There appear to be certain limitations to this kind of research program unless the phylogeny is already known, because otherwise it risks the circularity of demonstrating an adaptive pathway that has been pre-ordained by the investigator. It also appears to dismiss some potentially important lines of evidence.

An important issue to be raised is the relationship between testable hypotheses of phylogeny and testable hypotheses of functional evolution. Some current workers propose that functional hypotheses are entirely the handmaidens of phylogenies. That is, the best procedure is to derive the best phylogeny possible from the character evidence, and then graft to the phylogeny a functional explanation that is congruent with it. This has the effect of making functional explanations almost completely subservient to those of form, at least in the analysis of patterns.

This is not entirely unreasonable, given that any explanation of historical process must follow what is best understood of the historical pattern. In much of the literature of the Modern Synthesis (e.g., Bock & Von Wahlert 1965), selective values or adaptive pathways are assumed a priori as matters of process. Statements to the effect that the ancestors of birds would have needed stereoscopic vision to clamber about in trees, and feathers to keep them warm at night among the cold branches, make process assumptions without analyzing the phylogenetic distribution of features (e.g., the terrestrial theropod sister-groups to birds already had stereoscopic vision, and the display function of integumentary structures in reptiles is more generally distributed than thermoregulatory or aerodynamic ones). However, a strongly supported phylogeny can test conflicting hypotheses of functional evolution (and, Hickman 1991 argues, the converse can also apply). Without suggesting in any way a return to ad hoc hypotheses of biological process and a priori statements about the relative ease or likelihood of certain pathways in evolution, I would contend that functional explanations can be developed without recourse to specific phylogenies. In fact they *should*, because they depend on more than phylogenetic plausibility. Functional hypotheses, if they have any intellectual content, have to be able to stand or fall on their own merits, and must be testable on their own terms. Then they can be compared to plausible phylogenies in order to arrive at the most robust explanation of all the available evidence. If the functional hypothesis fails when the cladogram fails, then there is not much of a functional hypothesis (Padian 1982, 1987a).

The implication of this line of thought is much closer to at least one reading of Whewell's (1840, 1847) philosophical concept of *consilience*, and perhaps analogically in some respects to Kluge's (1989) phylogenetic concept of *total evidence*, than it is to conventional cladistic notions of *parsimony* (a concept that has taken on a wholly new meaning in cladistic usage: see Johnson 1982), or to the statistically based treatment of history known as *maximum likelihood*. These concepts are all heuristically useful, and each has its place in the analysis of some kinds of evidence and methodology. What needs emphasis here, perhaps, is Whewell's argument that separate lines of evidence converging on a single conclusion produce a more robust argument than a single line of evidence alone:

the cases in which inductions from classes of facts altogether different have thus jumped together, belong only to the best established theories which the history of science contains. . . . I call this the consilience of inductions. (Whewell 1840; II, p. 230).

The consilience of inductions takes place when an induction, obtained from one class of facts, coincides with an induction, obtained from another different class. This consilience is a test of the truth of the theory in which it occurs. (Whewell 1847; II, p. 469).[8]

Elsewhere (Padian 1982, 1987a, 1991) I have tried to detail some ideas about the analysis of the evolution of functional adaptations, a primary focus of paleobiologists concerned with functional morphology (similar methods are laid out by Greene 1986 and Lauder & Liem 1991). First, an adaptation should be diagnosed and defined, much as monophyletic taxa are, in strict functional terms. The taxa that share these adaptations should be identified. Second, these groups are analyzed phylogenetically to determine two things: how the individual functional characters in the complex evolved from outgroups that lack the adaptation under study, and how the functional complex evolved further within the group. Third, from these analyses, the steps in the phylogenetic sequence can be interpreted functionally. Fourth, the adaptational sequences can be compared among the groups that share the adaptation. This is frequently helped by bringing in a fifth step, additional lines of evidence that may be important to consider in the case of particular adaptations.

An extended example

A useful example in this context is the origin of flight in tetrapods, about which much has been written in recent decades (e.g., for birds see Hecht et al. 1985; Padian 1986, 1987a). How can functional morphologists study the transition between locomotory modes of animals now completely extinct, lacking most intermediate forms and all direct physiological evidence?

As the analytical method just explained indicates, we start with definitions and diagnoses, both phylogenetic and functional. Traveling through the air takes several forms (parachuting, gliding, flapping, and soaring), and each of these adaptations has been recognized and differentiated by specific morphological and aerodynamic features (Padian 1985b, 1987a; Norberg 1990; Rayner 1991). As far as the evidence seems to allow, there is no necessary continuum among any of these features, with two possible exceptions. All soaring animals seem to have evolved from flapping ancestors, and probably all gliders evolved from parachuters, depending on whether gliding is defined aerodynamically or behaviorally (Padian 1985b).

Several functional features are characteristic of gliders (Padian 1985b, 1987a). The gliding membrane generally has a weaker internal structure than a flapping wing does, and is usually stretched between limbs or ribs, and sometimes the tail. There are few skeletal or physiological changes for gliding, apart from slight elongation of propodials and zeugopodials compared to nongliding relatives. But fliers are quite different. Their wings have strong internal structure and power of manipulation. The wing is principally controlled by the forelimb, which provides the characteristic flight stroke absent in gliders. The forelimb is usually elongated, and its outer segments are especially so; there are expanded sites of origin and insertion for the enlarged flight muscles. The shoulder girdle is buttressed, and the joints are modified and usually restricted in individual motions to those that accomplish the flight stroke. In birds and pterosaurs many bones are pneumaticized. The characteristic flight stroke produces a vortex wake, shed in rings, that propels the animal forward (Rayner 1979). These features comprise a functional diagnosis of flight; the phylogenetic definitions of the three groups of tetrapods that bear this adaptation are birds, bats, and (by inference from structure) pterosaurs.

Next, phylogenetic analysis of flying groups is used to examine both the nonflying relatives and the details of evolution within each group. Bats, birds, and pterosaurs are the only tetrapods known or inferred to have evolved flapping flight. When we reconstruct their phylogenetic histories, we see that birds and pterosaurs differ from bats in some important respects.

The closest relatives of bats are disputed. Recently there has been a strong challenge to the monophyly of bats (Pettigrew et al. 1989; Pettigrew 1991a, 1991b), concluding that megabats are related to primates and dermopterans, and that they evolved independently of microbats. Other evidence (e.g., Baker, Novacek, & Simmons 1991; Simmons, Novacek, & Baker 1991; Novacek 1992) holds that bats are monophyletic after all, and that their closest relatives are dermopterans (there is other dissent; see Thewissen & Babcock [1992]). If dermopterans turn out to be the sister-group of one or both bat groups, it would hardly be surprising from a functional point of view, because bats share so many of the features of gliders such as dermopterans (though bats themselves rarely glide.) Both groups are arboreal, nocturnal, and given to hanging upside-down, and share many other morphological and behavioral features related to arboreal life.

Considering the origin of flight in bats, there are three possibilities. Bats could have evolved flight without a gliding stage; or bats and dermopterans might have evolved gliding independently, and bats went on to evolve powered flight; or their common ancestor evolved gliding, and bats went on to evolve powered flight. On the basis of morphological evidence (to which could be added the energetic improbability of evolving flight from an arboreal habitat without being able to glide), the first choice has never received wide support. The second and third choices are both plausible functionally, and could only be discriminated depending on the outcome of the phylogenetic analysis. In either case, the functional analysis determines that both groups share the features of arboreal gliders, and so bats presumably passed through an arboreal gliding stage, regardless of their precise ancestry. This illustrates why it is so important to develop a functional hypothesis independent of the precise phylogenetic hypothesis, even though the two must be tested against each other to satisfy the criterion of consilience. Consilience gives both hypotheses strength.

So much has been written about the origin of bird flight (see Hecht et al. [1985] and Padian [1986] for some recent reviews) that no brief summary can do the problem justice. It is now established, though not unanimously, that birds evolved from small carnivorous dinosaurs (see Gauthier & Padian 1985; Gauthier 1986). The steps in the evolution of this phylogenetic sequence also show us many important adaptational features of flight changing through time (Gauthier & Padian 1985; Padian 1985b, 1987a; Rayner 1991) – such as the reduction, elongation, and fusion of fingers and wrist elements. The theropod relatives of birds were terrestrial, active, agile predators. They show no gliding or arboreal features. And, although intuitively it seems easier to evolve flight through a gliding stage, the question must be whether there is any direct evidence for such features. An alternative model (Padian 1987a; Rayner 1991) would suggest that flight could evolve incrementally from the ground up, involving flight little by little. The most often cited problem (Rayner 1991) is building up enough speed to make flight viable from the ground up; but this is only a restriction if the object is to launch immediately into full-scale flight, as opposed to incremental flapping leaps. More compelling is the phylogenetic evidence of the sister-groups of birds, and their morphological features that suggest an ecology unrelated to arboreal habits and gliding.

Pterosaurs, like birds, seem to have been strong fliers and have many features of bipedal, terrestrial runners (Padian 1983, 1985b). Their hips and legs share with birds many features of joint articulations, bone orientations, and morphologies that reflect these features, and are entirely different in bats, which cannot walk with the normal parasagittal locomotion of other mammals. All the immediate relatives of pterosaurs, such as *Scleromochlus* (Padian 1984), appear to have been small, agile bipeds, as can be seen in Sereno's (1991) cladogram including the locomotory tendencies of the animals in question. The burden of evidence is thus on those who maintain that pterosaurs could not walk erect and bipedally, because both the functional evidence and the phylogenetic evidence suggest otherwise.

It would be less than straightforward to imply that these conclusions have been universally accepted by other workers. However, alternative scenarios have yet to analyze the same questions and evidence systematically (using "systematically" both in the phylogenetic and methodological denotations). These questions have often been treated as matters of opinion or taste, or even as insoluble because of lack of hard evidence. But I think that they are soluble, and that if we do not bring specific methods to bear onto these questions, we avoid analysis of the basic and ultimate question of our field, which is how major adaptations evolve. We also abrogate any responsibility to develop a methodology capable of answering these questions, which amounts to a denial that a philosophy of biology is relevant to our field. Consilience should be a major feature of this philosophy.

In vertebrate history, the construction of defensible scenarios about the evolution of form and function depends on well-corroborated hypotheses not only of phylogeny – the basic pattern of history – but of function, the basic process of change. Vertebrate paleontologists have recently reset the debate about the evolution of early archosaur locomotion through further phylogenetic and mechanical analyses of the taxa involved. It had generally been thought that the mesotarsal (hinge-ankled) archosaurs such as pterosaurs and dinosaurs had to evolve from the crurotarsal (swing-ankled) archosaurs such as phytosaurs, aetosaurs, crocodiles, and others (see review in Sereno & Arcucci 1990). Difficulties with constructing a scenario in which the crocodiloid heel was reduced and the tarsal movement straightened led some workers to propose a much earlier and independent origin of the dinosaur clade from other archosaurs. As it turned out, the latter hypothesis now seems better corroborated, as new phylogenetic analyses split the archosaurs into two major clades (Gauthier 1986), united by different ankle patterns. Further analyses (e.g., Sereno & Arcucci 1990; Sereno 1991) have refined the phylogeny and made the functional pattern even clearer. This kind of work is exemplary to solving the major functional problems of vertebrate evolution. Other examples ripe for further analysis can be found in Schultze and Trueb (1991), but none are likely to see progress without adherence to more explicit functional methodology.

PROSPECTUS FOR FORM AND FUNCTION?

This returns us to our original question. Stephen Jay Gould suggested in a retrospective essay some years ago (1980b, pp. 101, 111–112) that functional morphology was not likely to provide paleobiology with any striking new insights, except that organisms work pretty well, which we already knew. This viewpoint is difficult to argue with if the prospectus of functional morphology does not change, especially regarding extinct animals. But on the other hand, the research program of functional morphology has already started to change (M. Wake 1992; Hickman 1988, 1991), as its fusion with cladistic analysis shows. There is now a superb opportunity to initiate a new research program that returns to the root of evolutionary biology, which is how things evolve – or how form changes.

It is important to see the considerable achievements of population genetics, developmental biology, and speciation theory as vital to a great many pieces of the evolutionary puzzle that have been put in place since Darwin's theory was first proposed. However, as far as understanding the paths by which organisms actually change – in response to external conditions of the environment, in modification of their morphogenetic programs, and in coordinating and fine-tuning genetically and developmentally independent changes without a hitch – we are scarcely farther along than we were in 1859. We regard as anomalous or trivial the problems that obsess observers outside current evolutionary biology. How, for example, do the cusp patterns of mammalian

teeth evolve in coordination between upper and lower jaws, if they belong to different developmental and probably genetic fields? What is the mechanism by which callosities appear precociously on the feet of human embryos and the ischia of unhatched ratite birds? The process of developmental "capture" of these originally adult features needs a great deal of further study in both morphogenetic and evolutionary terms.[9] Because so many of these salient features are functional, they can be analyzed as an assembly of morphological characters that are collected one by one into an evolutionary pattern. There is ample evidence that developmental patterns and anatomical tissues are far more plastic than the original formulation of Mendelian genetics might have anticipated. Why not accept that flexibility in behavior and morphology permits the behavioral changes that precede much morphologic "hardwiring," and use that pattern of hard-wiring, as revealed from phylogenetic analysis, to develop and test hypotheses of the evolution of functional and morphogenetic processes?

It is time for another fusion of development and evolution, as many have said, and functional morphology has an important role to play. The sequences of assembly of functional complexes that are seen in the fossil record of both living and extinct groups enable the mapping of morphological change through lineages more explicitly than ever before, because these changes can be interpreted against an explicitly testable phylogenetic framework.

The major morphological changes that we recognize in evolution are almost tautologically functional. But without understanding the mechanisms of morphogenesis and its evolutionary change, it is impossible to conceive of functional change. How can we get at this in fossils? We can analyze the sequence of morphological change in the assembly and modification of adaptations. We can interpret the behavior that might have led to the "hardwiring" of these features (Waddington 1975; Riedl 1985; Rachootin & Thomson 1981); the fusion of the avian hand and wrist during the refinement of flapping is an obvious example (Gauthier & Padian 1985). Understanding these sequences and these modifications of behavior provides questions to ask of morphogenetic developmental biologists, who have studied myriad molecular pathways and developmental sequences.

Analyses of functional evolution provide an evolutionary pattern on which to hang knowledge of morphogenetic processes and test their validity and distribution. Without needing to deny the formulations of the workers of the Modern Synthesis regarding the importance of major adaptations, we can go beyond the terms of the 1960s to ask how we can work with developmental and behavioral biologists to forge a newly synthetic approach to how the major features of evolution are assembled.

The question of form and function is not a dichotomy of determinant influences. New insights are bound to emerge from considering how function influences historically inherited form; how it is mediated through development and behavior; and how it crafts and shapes new roles for continually evolving organisms. The most exciting prospect, to me, is the potential reciprocal illumination of the processes of developmental morphogenetic change played against the patterns of functional evolution seen in the fossil record.

ACKNOWLEDGMENTS

It is a pleasure to thank Wendy Olson, Dave Polly, Marvalee Wake, David Goines, and the students in my graduate seminar on pre-Darwinian biology, who contributed many stimulating discussions that helped me to refine the ideas in this paper. Particular thanks go to Will Provine, Olivier Rieppel, and two anonymous reviewers for extensive comments and helpful criticisms. As usual, these generous colleagues must be held blameless for the shortcomings of my analysis of these complex subjects, and I only hope they can read it without wincing too often. This is contribution No. 1598 from the University of California Museum of Paleontology, which supported much of this work.

NOTES

1. Bloom (1990) analyzes the character and formulation of this Deity by "J", the author of the first three divisions of Torah.
2. I am indebted to David Lance Goines for the observation that Russell, by the context of his chapter, could not have meant to use the term *epigonos* in its usual connotation, which is "the inferior descendant of illustrious ancestors." Russell clearly meant to say that Owen is the intellectual descendant of the transcendentalists (true in a post-1837 context), but he regarded him more as an apotheosis than an *epigonos*.
3. M.J.S. Rudwick remarks in his recent *Scenes from Deep Time* (1992, p. 229) that fossil animal and landscape restorations – which he argues have a strong historical tie to Biblical panels – were first promulgated "in the one major European country where – as contemporary observers often noted – the practice of religion was not regarded as antithetical to the practice of science."
4. The last two authors, coincidentally, described the first two known dinosaurs, which Owen first recognized as a group in 1842 (Desmond 1979; Torrens 1992).
5. We can argue in hindsight, of course, that the absence of understanding of deep time and phylogenetic legacy makes the patterns of the biological world seem much more chaotic and arbitrary than they otherwise seem.
6. For a perspective on developmental morphology that interested Darwin, see R. Richards 1992.
7. See Waisbren (1988) for a review of the question of

Darwinian and post-Darwinian morphology, notably in the contributions of the "Oxford Group" – a fascinating lesson in how morphology became caught in the *fin de siècle* paradigm shift to experimentation.

8. A corollary is that functional hypotheses must be consistent with phylogenies drawn from a variety of character analyses, because our current understanding is that even the most elegant functional hypotheses do not have a sufficiently strong explanatory power to override well-corroborated phylogenies. However, this does not obviate the fact that multiple functional hypotheses can and should be generated and tested independently of phylogenetic hypotheses.
9. Waisbren (1988, pp. 294–298) describes the Haeckelian mechanisms of "terminal addition" and "condensation" that persisted into the early twentieth century. Even translated into von Baerian terms, we still need answers to how the timing shifts, and how this change is "hardwired" genetically.

REFERENCES

Appel, T.A. 1987. *The Cuvier-Geoffroy Debate: French biology in the decades before Darwin.* New York: Oxford University Press.

Baker, R.J., Novacek, M.J., & Simmons, N.B. 1991. On the monophyly of bats. *Systematic Zoology* 40, 216–231.

Bird, W.R. 1989. *The Origin of Species Revisited: The theories of evolution and of abrupt appearance.* New York: Philosophical Library.

Bloom, H. 1990. *The Book of J.* New York: Grove Weidenfeld.

Bock, W. 1965. The role of adaptive mechanisms in the origin of higher levels of organization. *Systematic Zoology* 14, 272–287.

Bock, W.J. 1986. The arboreal origin of avian flight. In *The Origin of Birds and the Evolution of Flight*, ed. K. Padian, pp. 57–82. Memoirs of the California Academy of Sciences Number 8.

Bock, W.J., & Von Wahlert, G. 1965. Adaptation and the form–function complex. *Evolution* 19, 269–299.

Burkhardt, R.W. 1977. *The Spirit of System: Lamarck and evolutionary biology.* Cambridge, MA: Harvard University Press.

[Chambers, R.] 1844. *Vestiges of the Natural History of Creation*, 2nd ed. London: Churchill.

Coddington, J.A. 1988. Cladistic tests of adaptational hypotheses. *Cladistics* 4, 3–22.

Darwin, C. 1859. *On the Origin of Species by Means of Natural Selection; or, the Preservation of Favoured Races in the Struggle for Existence.* London: John Murray.

Davis, P., Kenyon, D.H., & Thaxton, C.B. 1989. *Of Pandas and People: The central question of biological origins.* Dallas, TX: Haughton.

Desmond, A.J. 1975. *The Hot-Blooded Dinosaurs: A revolution in palaeontology.* London: Blond & Briggs.

Desmond, A.J. 1979. Designing the Dinosaur: Richard Owen's response to Robert Edmond Grant. *Isis* 70, 224–234.

Desmond, A.J. 1982. *Archetypes and Ancestors: Palaeontology in Victorian London, 1850–1875.* London: Blond & Briggs.

Desmond, A.J. 1985. Richard Owen's reaction to transmutation in the 1830's. *British Journal of the History of Science* 18, 25–50.

Desmond, A.J. 1989. *The Politics of Evolution: Morphology, medicine, and reform in Radical London.* Chicago: University of Chicago Press.

Desmond, A.J., & Moore, J. 1992. *Darwin.* New York: Time Warner.

Eldredge, N., and S.J. Gould. 1972. Punctuated equilibrium: an alternative to phyletic gradualism. In *Models in Paleobiology*, ed. T.J.M. Schopf, pp. 82–115. San Francisco: Freeman.

Fisher, D.C. 1982. Phylogenetic and macroevolutionary patterns within the Xiphosurida. *Proceedings of the Third North American Paleontological Convention* 1, 175–180.

Fisher, D.C. 1985. Evolutionary morphology: beyond the analogous, the anecdotal, and the ad hoc. *Paleobiology* 11, 120–138.

Gauthier, J.A. 1986. Saurischian monophyly and the origin of birds. In *The Origin of Birds and the Evolution of Flight*, ed. K. Padian, pp. 3–55. Memoirs of the California Academy of Sciences 8.

Gauthier, J.A., & Padian, K. 1985. Phylogenetic, functional, and aerodynamic hypotheses of the origin of birds and their flight. In *The Beginnings of Birds*, ed. M.K. Hecht, J.H. Ostrom, G. Viohl, & P. Wellnhofer, pp. 185–197. Eichstätt, Germany: Freunde des JuraMuseums.

Glick, T.F., ed. 1988. *The Comparative Reception of Darwinism.* Chicago: University of Chicago Press.

Gould, S.J. 1980a. G.G. Simpson, paleontology, and the Modern Synthesis. In *The Evolutionary Synthesis: Perspectives on the unification of biology*, ed. E. Mayr & W. Provine. Cambridge, MA: Harvard University Press.

Gould, S.J. 1980b. The promise of paleobiology as a nomothetic, evolutionary discipline. *Paleobiology* 6, 96–118.

Gould, S.J. 1993. Modified grandeur. *Natural History* 102 (3), 14–20.

Greene, H.W. 1986. Diet and arboreality in the emerald monitor, *Varanus prasinus*, with comments on the study of adaptation. *Fieldiana* 1370, 1–12.

Hamilton, E. 1940. *Mythology.* Boston: Little, Brown & Co.

Hecht, M.K., Ostrom, J.H., Viohl, G., & Wellnhofer, P., ed. 1985. *The Beginnings of Birds.* Eichstätt, Germany: Freunde des JuraMuseums.

Hickman, C.S. 1988. Analysis of form and function in fossils. *American Zoologist* 28, 775–793.

Hickman, C.S. 1991. Functional analysis and the power of the fourth dimension in comparative evolutionary studies. In *The Unity of Evolutionary Biology*, ed. E.C. Dudley, pp. 548–554. Portland, OR: Dioscorides Press.

Hull, D.L. 1973. *Darwin and His Critics: The reception of Darwin's theory of evolution by the scientific community.* Chicago: University of Chicago Press.

Jepsen, G.L., Mayr, E., & Simpson, G.G., ed. 1949. *Genetics, Paleontology, and Evolution.* Princeton: Princeton University Press.

Johnson, R. 1982. Parsimony principles in phylogenetic systematics: a critical re-appraisal. *Evolutionary Theory* 6, 79–90.

Kluge, A.J. 1989. A concern for evidence and a phylogenetic hypothesis of relationships among *Epicrates* (Boidae, Serpentes). *Systematic Zoology* 38, 7–25.

Lauder, G.V. 1981. Form and function: structural analysis in evolutionary morphology. *Paleobiology* 7, 430–442.

Lauder, G.V. 1982. Historical biology and the problem of design. *Journal of Theoretical Biology* 97, 57–67.

Lauder, G.V. 1991. Biomechanics and evolution: integrating physical and historical biology in the study of complex systems. In *Biomechanics and Evolution*, ed. J.M.V. Rayner & R. J. Wootton, pp. 1–19. Cambridge University Press.

Lauder, G.V., & Liem, K.F. 1991. The role of historical factors in the evolution of complex organic functions. In *Complex Organismal Functions: Integration and evolution in vertebrates*, ed. D.B. Wake & G. Roth, pp. 63–78. Chichester: John Wiley.

Liem, K.F., & Wake, D.B. 1985. Morphology: current concepts and approaches. In *Functional Vertebrate Morphology*, ed. M. Hildebrand, D.M. Bramble, K.F. Liem, & D.B. Wake, pp. 366–377. Cambridge, MA: Belknap Press (Harvard University Press).

Mantell, G. 1844. *Medals of Creation.* London: H.G. Bohn.

Mayr, E. 1963. *Animal Species and Evolution.* Cambridge, MA: Belknap Press (Harvard University Press).

Mayr, E., & Provine, W., ed. 1980. *The Evolutionary Synthesis: Perspectives in the Unification of Biology.* Cambridge, MA: Harvard University Press.

Norberg, U.M. 1990. *Vertebrate Flight.* Berlin: Springer-Verlag.

Novacek, M. 1992. Mammalian phylogeny: shaking the tree. *Nature* 356, 121–125.

O'Hara, R.J. 1988. Homage to Clio, or Toward an historical philosophy for evolutionary biology. *Systematic Zoology* 37, 142–155.

Owen, R. 1846. Report on the Archetype and homologies of the vertebrate skeleton. *Reports of the British Association for the Advancement of Science* 1845, 169–340. (Reprinted 1847 by Richard & John E. Taylor, London.)

Owen, R. 1849. *On the Nature of Limbs.* London: John Van Voorst.

Owen, R. 1870. *A Monograph of the Fossil Reptilia of the Liassic Formations. III.* London: Monographs of the Palaeontographical Society.

Padian, K. 1980. Studies of the structure, evolution, and flight of pterosaurs (Reptilia: Pterosauria). Unpublished Ph.D. Thesis, Department of Biology, Yale University.

Padian, K. 1982. Macroevolution and the origin of major adaptations: vertebrate flight as a paradigm for the analysis of patterns. *Proceedings, Third North American Paleontological Convention*, 387–392.

Padian, K. 1983. A functional analysis of flying and walking in pterosaurs. *Paleobiology* 9, 218–239.

Padian, K. 1984. The origin of pterosaurs. In *Third Symposium on Mesozoic Terrestrial Ecosystems: Short Papers*, ed. W.-E. Reif & F. Westphal, pp. 163–168. Tubingen, Germany: ATTEMPTO.

Padian, K. 1985a. On Richard Owen's Archetype, Homology, and the Vertebral Theory: Interrelationships and implications. *Abstracts, XVII International Congress of the History of Science* 1, Bj 5.P.

Padian, K. 1985b. The origins and aerodynamics of flight in extinct vertebrates. *Palaeontology* 28, 423–433.

Padian, K., ed. 1986. *The Origin of Birds and the Evolution of Flight. Memoirs of the California Academy of Sciences* 8.

Padian, K. 1987a. A comparative phylogenetic and functional approach to the origin of vertebrate flight. In *Recent Advances in the Study of Bats*, ed. B. Fenton, P.A. Pacey, & J.M.V. Rayner, pp. 3–22. Cambridge University Press.

Padian, K. 1987b. The case of the bat-winged pterosaur: typological taxonomy and the influence of pictorial representation on scientific perception. In *Dinosaurs Past and Present*, ed. E.C. Olson & S.M. Czerkas, vol. 1, pp. 65–81. Los Angeles: Los Angeles County Museum of Natural History.

Padian, K. 1991. Pterosaurs: were they functional birds or functional bats? In *Biomechanics and Evolution*, ed. J.M.V. Rayner & R.J. Wootton, pp. 145–160. Cambridge University Press.

Paley, W. 1802. *Natural Theology*. London: C. Knight.

Pettigrew, J.D. 1991a. Wings or brains? Convergent evolution in the origin of bats. *Systematic Zoology* 40, 199–215.

Pettigrew, J.D. 1991b. A fruitful, wrong hypothesis? Response to Baker, Novacek, and Simmons. *Systematic Zoology* 40, 231–238.

Pettigrew, J.D., Jamieson, B.G.M., Robson, S.K., Hall, L.S., McNally, K.I., & Cooper, H.M. 1989. Phylogenetic relations between microbats, megabats and primates (Mammalia: Chiroptera and Primates). *Philosophical Transactions of the Royal Society of London* B325, 489–559.

Provine, W. 1971. *Origins of Theoretical Population Genetics*. Chicago: University of Chicago Press.

Provine, W. 1982. Influence of Darwin's ideas on the study of evolution. *BioScience* 32, 501–506.

Rachootin, S.P., & Thomson, K.S. 1981. Epigenetics, paleontology, and evolution. In *Evolution Today*, ed. G.G.E. Scudder & J.L. Reveal, pp. 181–191. Pittsburgh: Carnegie-Mellon University.

Rayner, J.M.V. 1979. A new approach to animal flight mechanics. *Journal of Experimental Biology* 80, 17–54.

Rayner, J.M.V. 1991. Avian flight evolution and the problem of *Archaeopteryx*. In *Biomechanics and Evolution*, ed. J.M.V. Rayner & R.J. Wootton, pp. 183–212. Cambridge University Press.

Richards, E. 1987. A question of property rights: Richard Owen's evolutionism revisited. *British Journal of the History of Science* 20, 129–171.

Richards, R. 1992. *The Meaning of Evolution: The morphological construction and ideological reconstruction of Darwin's theory*. Chicago: University of Chicago Press.

Riedl, R. 1985. A systems-analytical approach to macroevolution. *Quarterly Review of Biology* 52, 351–370.

Rieppel, O. 1988. *Fundamentals of Comparative Biology*. Basel: Birkhauser.

Rieppel, O. 1990. Structuralism, Functionalism, and the four Aristotelian causes. *Journal of the History of Biology* 23, 291–320.

Rowe, T. 1987. Definition and diagnosis in the phylogenetic system. *Systematic Zoology* 36, 208–211.

Rudwick, M.J.S. 1992. *Scenes from Deep Time*. Chicago: University of Chicago Press.

Russell, E.S. 1916. *Form and Function: A contribution to the history of animal morphology*. London: John Murray.

Schaeffer, B. 1965. The role of experimentation in the origin of higher levels of organization. *Systematic Zoology* 14, 318–336.

Schultze, H.-P, & Trueb, L., ed. 1991. *Origins of the Higher Groups of Tetrapods*. Ithaca: Cornell University Press.

Seeley, H.G. 1870. Remarks on Prof. Owen's monograph on *Dimorphodon. Annals and Magazine of Natural History*, series 4: 6, 129–152.

Sereno, P.C. 1991. Basal archosaurs: phylogenetic relationships and functional implications. *Society of Vertebrate Paleontology Memoir* 2, 1–53.

Sereno, P.C., & Arcucci, A.B. 1990. The monophyly of crurotarsal archosaurs and the origin of bird and crocodile ankle joints. *Neues Jahrbuch für Geologie und Paläeontologie, Abhandlungen* 180, 21–52.

Simmons, N.B., Novacek, M.J., & Baker, R.J. 1991. Approaches, methods, and the future of the chiropteran monophyly controversy: a reply to J.D. Pettigrew. *Systematic Zoology* 40, 239–243.

Simpson, G.G. 1944. *Tempo and Mode in Evolution*. New York: Columbia University Press.

Simpson, G.G. 1953. *The Major Features of Evolution*. New York: Columbia University Press.

Sloan, P.R., ed. 1992. *Richard Owen: The Hunterian Lectures in Comparative Anatomy, May and June 1837*. Chicago: University of Chicago Press.

Thewissen, J.G.M., & Babcock, S.K. 1992. The origin of flight in bats. *BioScience* 42, 340–345.

Torrens, H. 1992. When did the dinosaur get its name? *New Scientist*, 4 April 1992, pp. 40–44.

Waddington, C.H. 1975. *Evolution of an Evolutionist*. Ithaca: Cornell University Press.

Waisbren, S.J. 1988. The importance of morphology in the evolutionary synthesis as demonstrated by the contributions of the Oxford Group: Goodrich, Huxley, and de Beer. *Journal of the History of Biology* 21, 291–330.

Wake, D.B. 1982. Functional and evolutionary morphology. *Perspectives in Biology and Medicine* 25, 603–620.

Wake, D.B., & Roth, G. 1991. The linkage between ontogeny and phylogeny in the evolution of complex systems. In *Complex Organismal Functions: Integration and evolution in vertebrates*, ed. D.B. Wake & G. Roth, pp. 361–377. Chichester: John Wiley.

Wake, M.H. 1991. The impact of research in functional morphology and biomechanics on studies of evolution. In *The Unity of Evolutionary Biology*, ed. E.C. Dudley, pp. 555–557. Portland, OR: Dioscorides Press.

Wake, M.H. 1992. Morphology, the study of form and function, in modern evolutionary biology. *Oxford Surveys in Evolutionary Biology* 8, 289–346.

Whewell, W. 1840. *The Philosophy of the Inductive Sciences, Founded upon Their History*. London: John W. Parkes.

Whewell, W. 1847. *History of the Inductive Sciences*. London: John W. Parkes.

Whitehead, A.N. 1926. *Science and the Modern World*. Cambridge University Press.

Wright, S. 1986 *Evolution: Selected Papers*, ed. W. Provine. Chicago: University of Chicago Press.